Metal-Air Batteries

Metal-Air Batteries: Principles, Progress, and Perspectives covers the entire spectrum of metal-air batteries, their working principles, recent advancement, and future perspectives. Leading international researchers address materials design, electrochemistry, and architectural aspects. The fundamentals of metal-air materials for cathode and anode, their synthetic approaches, chemistries to modify their properties to provide high energy and power densities, along with long life and stable electrochemical characteristics are detailed.

Key Features:

- Covers materials, chemistry, and technologies for metal-air batteries.
- Reviews state-of-the-art progress and challenges in metal-air batteries.
- Provides fundamentals of the electrochemical behavior of various metal-air batteries.
- Offers insight into tuning the properties of materials to make them suitable for metal-air batteries.
- Provides new direction and a better understanding to scientists, researchers, and students working in diverse fields.

This is a unique offering and a valuable resource for a wide range of readers including those in academia and industries worldwide.

Metal-Air Batteries

Principles, Progress, and Perspectives

Edited by Ram K. Gupta

CRC Press is an imprint of the
Taylor & Francis Group, an **informa** business

First edition published 2023
by CRC Press
6000 Broken Sound Parkway NW, Suite 300, Boca Raton, FL 33487–2742

and by CRC Press
4 Park Square, Milton Park, Abingdon, Oxon, OX14 4RN

CRC Press is an imprint of Taylor & Francis Group, LLC

© 2023 selection and editorial matter, Ram K. Gupta; individual chapters, the contributors

Reasonable efforts have been made to publish reliable data and information, but the author and publisher cannot assume responsibility for the validity of all materials or the consequences of their use. The authors and publishers have attempted to trace the copyright holders of all material reproduced in this publication and apologize to copyright holders if permission to publish in this form has not been obtained. If any copyright material has not been acknowledged please write and let us know so we may rectify in any future reprint.

Except as permitted under U.S. Copyright Law, no part of this book may be reprinted, reproduced, transmitted, or utilized in any form by any electronic, mechanical, or other means, now known or hereafter invented, including photocopying, microfilming, and recording, or in any information storage or retrieval system, without written permission from the publishers.

For permission to photocopy or use material electronically from this work, access www.copyright.com or contact the Copyright Clearance Center, Inc. (CCC), 222 Rosewood Drive, Danvers, MA 01923, 978–750–8400. For works that are not available on CCC please contact mpkbookspermissions@tandf.co.uk

Trademark notice: Product or corporate names may be trademarks or registered trademarks and are used only for identification and explanation without intent to infringe.

Library of Congress Cataloging-in-Publication Data
Names: Gupta, Ram K., editor.
Title: Metal-air batteries : principles, progress and perspectives / edited by Ram K. Gupta.
Description: First edition. | Boca Raton : CRC Press, 2023. | Includes bibliographical references.
Identifiers: LCCN 2022046437 (print) | LCCN 2022046438 (ebook)
Subjects: LCSH: Storage batteries—Materials. | Metals. | Oxygen.
Classification: LCC TK2941 .M485 2023 (print) | LCC TK2941 (ebook) |
    DDC 621.31/2424—dc23/eng/20230111
LC record available at https://lccn.loc.gov/2022046437
LC ebook record available at https://lccn.loc.gov/2022046438

ISBN: 978-1-032-28208-4 (hbk)
ISBN: 978-1-032-28212-1 (pbk)
ISBN: 978-1-003-29576-1 (ebk)

DOI: 10.1201/9781003295761

Typeset in Times
by Apex CoVantage, LLC

Contents

Preface ix
Biography xi
List of Contributors xiii

Chapter 1 Metal-Air Batteries: An Introduction 1

Felipe M. de Souza, Anuj Kumar, and Ram K. Gupta

Chapter 2 Materials and Electrochemistry of Metal-Air Battery 15

Anubha Tomar, Sakshee Chandel, and Alok Kumar Rai

Chapter 3 Electrochemical Fundamentals and Issues of Metal-Air Batteries 29

Shasha Li, Enze Li, Xiaowei An, and Guoqing Guan

Chapter 4 Mathematical Modeling for Enhanced Electrochemical Properties 45

Yuhui Tian, Shanqing Zhang, and Yun Wang

Chapter 5 Materials and Technologies of Mg-Air Primary Batteries 65

Xingrui Chen, Qichi Le, Jeffrey Venezuela, and Andrej Atrens

Chapter 6 Materials and Technologies of Al-Air Batteries 77

Weng Cheong Tan, Lip Huat Saw, Ming Chian Yew, and Ming Kun Yew

Chapter 7 Materials and Technologies of Other Metal-Air Batteries 89

Dhavalkumar N. Joshi and Vinod Kumar

Chapter 8 Novel Architectural Designs for Improved Performance 105

José Béjar, Minerva Guerra-Balcázar, Lorena Álvarez-Contreras, and Noé Arjona

Chapter 9 Phase-Engineered Materials as Efficient Electrocatalysts for Metal-Air Batteries 121

Mengyang Dong, Huajie Yin, Porun Liu, and Huijun Zhao

Chapter 10 Noble Metal-Based Electrocatalysts for Metal-Air Batteries 135

Jun Mei

Chapter 11 1D Materials as Electrocatalysts for Metal-Air Batteries 151

Merve Gençtürk and Emre Biçer

Chapter 12 2D Materials as Electrocatalysts for Metal-Air Batteries 165

Eren Kursun, Abdullah Uysal, and Solen Kinayyigit

Chapter 13 3D Materials as Electrocatalysts for Metal-Air Batteries 179

Demet OZER

Chapter 14 Carbon-Based Electrocatalysts for Metal-Air Batteries 195

Zongge Li and Guoxin Zhang

Chapter 15 Metal Oxide-Based Electrocatalysts for Metal-Air Batteries 209

Bhugendra Chutia, Chiranjita Goswami, and Pankaj Bharali

Chapter 16 Enhanced Performance of Lithium-Air Batteries by Improved Cathode Materials 227

B. Jeevanantham and M. K. Shobana

Chapter 17 Aqueous Electrolytes 249

Rijith S, Sarika S, Akhila M, and Sumi V S

Chapter 18 Non-Aqueous Electrolytes in Metal-Air Batteries 265

Pravin N. Didwal, An-Giang Nguyen, Satyanarayana Maddukuri, and Rakesh Verma

Chapter 19 Ionic Electrolytes 281

Vandana, Fabeena Jahan, Anjali Paravannoor, and Baiju Kizhakkekilikoodayil Vijayan

Chapter 20 Hybrid-Electrolyte Metal-Air Batteries 291

Yifei Wang, Xinhai Xu, Mingming Zhang, Meng Ni, and Dennis Y.C. Leung

Chapter 21 Polymer Electrolytes 305

Changlin Liu, Shasha Li, Abuliti Abudula, and Guoqing Guan

Chapter 22 Hydrogel Electrolytes 317

Siyuan Zhao, Tong Liu, and Meng Ni

Chapter 23 Wearable Metal-Air Batteries 335

Arpana Agrawal and Chaudhery Mustansar Hussain

Chapter 24 Flexible Metal-Air Batteries... 347

Runwei Mo

Chapter 25 Challenges in Metal-Air Batteries... 361

Alexander Kube and Dennis Kopljar

Index.. 375

Preface

There is a need to develop high-performance and sustainable energy devices which could meet the future demand for energy on the consumer as well as industrial scale. Most of the current energy needs are fulfilled by fossil fuels which are damaging our environment. Global warming and air pollution are some of the major issues caused by the use of excessive fossil fuels, and therefore researchers need to find some alternatives. Approaches such as the use of hybrid electric vehicles and electric vehicles are a few solutions to these problems. Adaptation of electric vehicles has some challenges such as low performance of the batteries, their life, cost, and range per charge. Metal-air batteries offer higher energy density compared to many other battery technologies. However, such technology has many challenges such as the use of advanced electrochemical materials, porous cathode, electrolytes, device architecture, etc.

The main purpose of this book is to discuss current, state-of-the-art knowledge and future challenges in metal-air batteries. This book covers the fundamentals of metal-air materials for cathode and anode, their synthetic approaches, and chemistries to modify their properties to provide high energy and power densities, along with long life and stable electrochemical characteristics. All the chapters are covered by experts in these areas around the world making this a suitable textbook for students and providing new guidelines to researchers and industries working in energy, materials, and nanotechnology.

Ram K. Gupta
Pittsburg State University
Pittsburg, Kansas, United States

Biography

Dr. Ram K. Gupta is an associate professor at Pittsburg State University. Before joining Pittsburg State University, he worked as an assistant research professor at Missouri State University, Springfield, MO, then as a senior research scientist at North Carolina A&T State University, Greensboro, NC. Dr. Gupta's research focuses on green energy production and storage using conducting polymers and composites, electrocatalysts, fuel cells, supercapacitors, batteries, nanomaterials, optoelectronics and photovoltaics devices, organic-inorganic hetero-junctions for sensors, nanomagnetism, bio-based polymers, bio-compatible nanofibers for tissue regeneration, scaffold and antibacterial applications, and bio-degradable metallic implants. Dr. Gupta has published over 250 peer-reviewed articles, made over 350 national/international/regional presentations, and chaired many sessions at national/international meetings, as well as writing and editing many books/chapters for the American Chemical Society, CRC, and Elsevier publishers. He has received several million dollars for research and educational activities from external agencies, and is serving as associate editor, guest editor, and editorial board member for various journals.

Contributors

Abuliti Abudula
Hirosaki University
Japan

Arpana Agrawal
Shri Neelkantheshwar Government
  Post-Graduate College
India

Lorena Álvarez-Contreras
Complejo Industrial Chihuahua
México

Xiaowei An
Taiyuan University of Technology
China

Noé Arjona
Parque Tecnológico Querétaro Sanfandila
México

Andrej Atrens
The University of Queensland
Australia

José Béjar
Complejo Industrial Chihuahua
México

Pankaj Bharali
Tezpur University
India

Emre Biçer
Sivas University of Science and Technology
Türkiye

Sakshee Chandel
University of Delhi
India

Xingrui Chen
The University of Queensland
Australia

Bhugendra Chutia
Tezpur University
India

Felipe M. de Souza
Pittsburg State University
United States

Pravin N. Didwal
University of Oxford
United Kingdom

Mengyang Dong
Griffith University
Australia

Merve Gençtürk
Sivas University of Science
  and Technology
Türkiye

Chiranjita Goswami
Tezpur University
India

Guoqing Guan
Hirosaki University
Japan

Minerva Guerra-Balcázar
Universidad Autónoma de Querétaro
Querétaro, México

Ram K. Gupta
Pittsburg State University
United States

Chaudhery Mustansar Hussain
New Jersey Institute of Technology
United States

Fabeena Jahan
Kannur University
India

B. Jeevanantham
Vellore Institute of Technology
India

Dhavalkumar N. Joshi
Indian Institute of Technology Delhi
India

Solen Kinayyigit
Gebze Technical University
Turkey

Dennis Kopljar
Institute TT-ECE
Germany

Alexander Kube
Institute TT-ECE
Germany

Anuj Kumar
GLA University
India

Vinod Kumar
Dambi Dollo University
Ethiopia

Eren Kursun
Gebze Technical University
Turkey

Qichi Le
Northeastern University
China

Dennis Y.C. Leung
The University of Hong Kong
China

Enze Li
Shanxi University
China

Shasha Li
Taiyuan University of Science and Technology
China

Zongge Li
Shandong University of Science
 and Technology
China

Changlin Liu
Hirosaki University
Japan

Porun Liu
Griffith University
Australia

Tong Liu
The Hong Kong Polytechnic University
P.R. China

Akhila M
Sree Narayana College
India

Satyanarayana Maddukuri
Bar-Ilan University
Israel

Jun Mei
Queensland University of Technology
Australia

Runwei Mo
East China University of Science and Technology
China

An-Giang Nguyen
Chonnam National University
Republic of Korea

Meng Ni
The Hong Kong Polytechnic University
China

Demet OZER
Hacettepe University
Turkey

Anjali Paravannoor
Kannur University
India

Alok Kumar Rai
University of Delhi
India

Rijith S
Sree Narayana College
India

Sarika S
Sree Narayana College
India

Lip Huat Saw
Lee Kong Chian Faculty of Engineering
 and Science
Malaysia

Contributors

M. K. Shobana
Vellore Institute of Technology
India

Sumi V S
Government College
India

Weng Cheong Tan
Lee Kong Chian Faculty of Engineering
 and Science
Malaysia

Yuhui Tian
Griffith University
Australia

Anubha Tomar
University of Delhi
India

Abdullah Uysal
Gebze Technical University
Turkey

Vandana
Kannur University, India

Jeffrey Venezuela
The University of Queensland
Australia

Rakesh Verma
University of Allahabad
India

Baiju Kizhakkekilikoodayil Vijayan
Kannur University
India

Yifei Wang
Harbin Institute of Technology
China

Yun Wang
Griffith University
Australia

Xinhai Xu
Harbin Institute of Technology
China

Ming Chian Yew
Lee Kong Chian Faculty of Engineering
 and Science
Malaysia

Ming Kun Yew
Lee Kong Chian Faculty of Engineering
 and Science
Malaysia

Huajie Yin
Chinese Academy of Sciences
P. R. China

Guoxin Zhang
Shandong University of Science
 and Technology
China

Mingming Zhang
Harbin Institute of Technology
China

Shanqing Zhang
Griffith University
Australia

Huijun Zhao
Griffith University
Australia

Siyuan Zhao
The Hong Kong Polytechnic University
P.R. China

1 Metal-Air Batteries
An Introduction

Felipe M. de Souza, Anuj Kumar, and Ram K. Gupta

CONTENTS

1.1 Introduction 1
1.2 Design and Working Principles of Metal-Air Batteries 3
1.3 Recent Advancement in Metal-Air Batteries 6
1.4 Future Perspective 11
References 11

1.1 INTRODUCTION

The proper distribution and use of the various forms of energy, particularly electricity, is one of the core aspects of the proper functioning of society. Because of that, it seems necessary to develop a stable and sustainable way to harvest, store, and deliver energy to decrease the dependence on non-renewable sources. One of the reasonable technologies to address this challenge lies in power sources provided through electrochemical processes and energy storage devices. One of the scientific advancements that landed one of the strongest impacts on modern society was the advent of lithium-ion batteries (LIBs) which is currently present in most portable electronic devices. This technology enabled the push toward electric vehicles which can greatly reduce the number of fossil fuels that are being consumed. However, the energy requirements for a vehicle engine are more demanding when compared to smaller portable devices. Hence, the performance of LIBs had to be considerably improved to allow their application in this sector. Figure 1.1 shows the theoretical and practical energy values at which the current batteries present compared to petrochemical fuel [1].

It can be noted that reaching practical specific energy that is comparable to gasoline can be a considerable challenge since there is still a considerable gap between these technologies. There is a need to increase the overall practical energy density of batteries to make them more competitive against petrochemical-based fuels. Alongside this factor, it is also worth noting that Li is not an abundant source as it presents around 0.0017 wt.% over the planet's crust. Hence, exploring other metals that are more readily available such as Na, K, and Mg can be another viable approach. Introducing larger amounts of other metals aside from Li is one of the answers to properly manage its use and further improve the material's electrochemical properties. An example of this case is the layered materials incorporated with Ni, Mn, and Co (NMC) such as $Li_{1+x}Ni_zCo_wMn_yO_2$ where $x + z + w + y = 1$ as their specific capacitance can range from 240 to 260 mAh/g at a maximal potential charge of 4.3 V. However, they usually suffer from low stability which hinders their long-term use along with requiring relatively high operating voltages which makes it unsuitable for most commonly known electrolytes [2–4]. Some of these issues can be addressed through doping with hetero atoms as well as coatings over the layered structure that can prevent their degradation [5–7]. Following that, graphite is commercially used as the anode for LIBs which presents a theoretical capacity of around 372 mAh/g. That value can be further increased with the use of Si-based anodes which present a specific theoretical capacitance of around 3000 mAh/g [8–10]. However, these materials are unstable due to the formation of a solid electrolyte interface (SEI). Another approach proposed by Hwang *et al*. [11] consisted of using Li metal as the anode which functioned as an alternative to graphite along with NMC as the cathode. The battery assembled through these electrodes

DOI: 10.1201/9781003295761-1

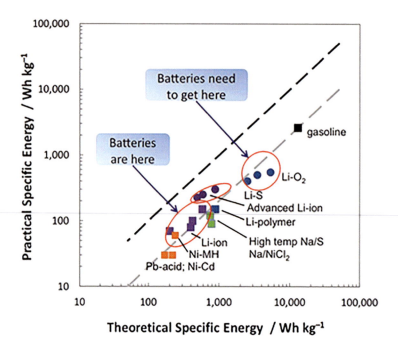

FIGURE 1.1 Graph displaying the relation between theoretical and practical energy densities of different types of batteries in comparison to petrochemical-based fuel. (Adapted with permission [1]. Copyright 2012, Royal Society of Chemistry.)

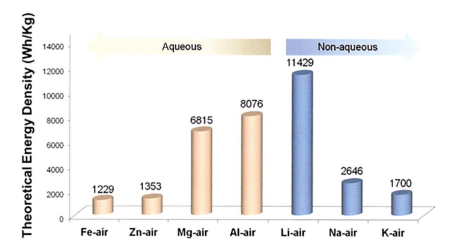

FIGURE 1.2 Various metal-air batteries with their theoretical energy density values at aqueous and non-aqueous media. (Adapted with permission [12]. Copyright 2017, American Chemical Society.)

presented high capacitance and operational voltage, as well as faster kinetics. However, it raises safety concerns due to the dendritic growth of Li within the system which can cause it to short-circuit. Based on these aspects it is highly required to further optimize the current technologies to attend to the required demands. Metal-air batteries appear as an interesting alternative as they presented a design that consists of a combination of aspects of a battery and a fuel cell. In this sense, an air electrode works as a cathode and O_2 as an anode. Other metallic sources aside from Li-based can be used for their fabrication which includes Na, K, Mg, Fe, Zn, or Al. Figure 1.2 presents the values of theoretical energy densities (Wh/kg) for several metal-air batteries as some are presented in aqueous and non-aqueous media.

Based on the theoretical values presented it is expected that values around 500 to 1,000 Wh/kg could enable an electric vehicle to travel a distance of around 500 km [13]. Some of the metals such as Fe, Zn, Mg, and Al are unstable in aqueous media. However, they undergo a passivation process that converts their surface into metal oxides or hydroxides, hence offering some degree of protection and enabling their use in an aqueous-based environment. Based on that, during the discharge process, there is the oxidation of the metal at the anode whereas the neighboring O_2 in the air is reduced on the cathode's surface. The reactions to this process are presented in equations 1 and 2.

$$M \leftrightarrow M^{n+} + ne^- \text{ (Anode)} \tag{1}$$

$$O_2 + ne^- + 2H_2O \leftrightarrow 4OH^- \text{ (Cathode)} \tag{2}$$

Under this scope, optimizing the properties of metal-air batteries such as improving energy efficiency, cyclability, bifunctional stability, and proper adsorption of O_2 are some of the challenges that the scientific community, as well as industry, are trying to overcome to remove some of the obstacles that are hindering the use of this technology into a larger scale.

1.2 DESIGN AND WORKING PRINCIPLES OF METAL-AIR BATTERIES

The components of a Li-O_2 battery consist of metallic Li that works as the anode, the negative electrode whereas O_2 works as the cathode, the positive electrode, and an electrolyte that contains Li$^+$ ions for transportation. One of the main focuses of the design of Li-O_2 batteries is based on the development of a proper cathode, which is usually a composite that is a conducting porous structure that can adsorb O_2 to allow its contact with Li$^+$ from the electrolyte. In this sense, the processes that take place are based on a deposition/dissolution of Li metal at the anode's surface with simultaneous oxygen evolution reaction (OER)/oxygen reduction reaction (ORR) at the cathode. The schematic for a traditional Li-O_2 is presented in Figure 1.3.

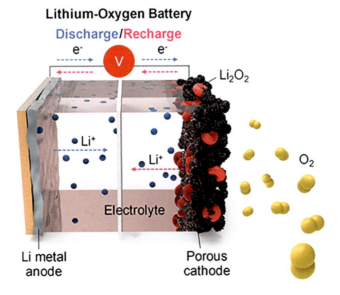

FIGURE 1.3 Scheme depicting the main components of a traditional Li-O_2 battery. (Adapted with permission [14]. Copyright 2020, American Chemical Society.)

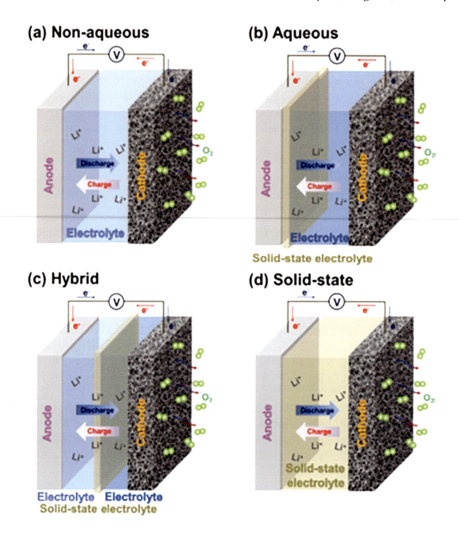

FIGURE 1.4 Different configurations for the Li-O$_2$ batteries are based on the employment of varying electrolytes based on (a) non-aqueous or aprotic, (b) aqueous, (c) hybrid, and (d) solid-state. (Reproduced with permission [22]. Copyright 2017, Elsevier.)

Furthermore, Li-O$_2$ can be divided into four types which are aqueous, aprotic, solid-state, and hybrid, in a sense that they differ based on the type of electrolyte employed. The schematics for each of these configurations of Li-O$_2$ batteries are presented in Figure 1.4. Such changes lead to variations in the electrochemical reactions related to both charging and discharging processes. The aprotic Li-O$_2$ batteries (Figure 1.4a) are based on a metallic Li anode along with a porous air cathode which is connected through a Li-based salt that is dissolved in an organic solvent. One of the reactions that take place in the system consists of the reduction of O$_2$ into lithium superoxide (LiO$_2$) which is presented in equation 3. Following that process, the lithium superoxide can be further reduced to lithium peroxide (Li$_2$O$_2$) which is the main discharging product. Such a process can occur through either disproportionation or receiving a second electron as presented in equations 4 and 5, respectively.

$$O_2 + Li^+ + e^- \rightarrow LiO_2 \tag{3}$$

$$\text{Disproportionation: } 2LiO_2 \rightarrow Li_2O_2 + O_2 \tag{4}$$

$$\text{Second electron acceptance: } LiO_2 + Li^+ + e^- \rightarrow Li_2O_2 \tag{5}$$

Based on this process the formed Li_2O_2 as a discharging product is not soluble in organic solvents which makes it deposit over the cathode's surface in the form of an insulating film or as toroidal-shaped crystallites. Because of that, one of the issues that emerges is the overall battery's capacitance may decrease based on the thickness of the Li_2O_2 film. During the charging, the reverse process takes place in which Li_2O_2 goes through an oxidation process that occurs at the surface of the positive electrode, leading to the regeneration of O_2 [15]. Based on this situation, another design to avoid the formation of an insulating Li_2O_2 was proposed by researchers from the company PolyPlus Battery which used an aqueous-based electrolyte that contained Li-based salts such as Li acetates along with a conducting membrane of Ohara glass, which prevented the corrosion of the Li anode (Figure 1.4b) [14, 16, 17]. The employment of an aqueous electrolyte led to the formation of lithium hydroxide (LiOH) as the discharge product instead of solid Li_2O_2. The reaction that takes place in an aqueous media for the formation of LiOH is presented in equation 6. However, this system is still in the works to reach the market as one of the issues lies in the saturation of LiOH in water which can reach a maximum of 5.25 mol/L at room temperature. Further addition of LiOH leads to the formation of $LiOH·H_2O$ which can precipitate and therefore, cause the overall specific capacitance to decrease to around 130 mAh/g [18].

$$4Li + O_2 + 2H_2O \rightarrow 4Li^+ + 4OH^- \qquad (6)$$

The use of liquid electrolytes carries the inherent risk of leaking of toxic solutions. To address that, another design for $Li-O_2$ batteries was proposed which consists of a hybrid LOB as depicted in Figure 1.4c. This system consists of a relatively more complex arrangement of components in a way that the anode side is filled with an organic solvent, whereas on the cathode side, the air electrode, is filled with an aqueous electrolyte. Following that, these two electrolyte solutions are separated through a solid electrolyte membrane. Through that, direct contact between the aprotic and aqueous electrolyte can be avoided which can result in an enhancement in the anionic conductivity of Li^+.

The last design of a $Li-O_2$ battery consists of a solid-state-based electrolyte (Figure 1.4d) which even though has gained some attention from the scientific community still has some challenges mostly related to the proper storage of Li_2O_2 in the cathode [19]. Despite that, some progress has been made since there are a considerable number of solid-state electrolytes available such as glass-ceramics, polymers, and single-crystalline silicon, among others [20]. Some of the examples in that class include $Li_{1+x}Al_xTi_{2-x}(PO_4)_3$ (LATP), $Li_{1+x}Al_xGe_{2-x}(PO_4)_3$ (LAGP), and NASICON-type glass ceramics. One of the interests in these materials is attributed to their inherent property of adsorbing O_2 alongside promoting its reduction of lithium peroxides as presented in equations 7–9 [21].

$$2LAGP\text{-}Li^+ + O_2 \rightarrow 2LAGP\text{-}Li^+ : O \qquad (7)$$

$$2LAGP\text{-}Li^+ : O + 2e^- \rightarrow 2LAGP\text{-}Li^+ + 2O^- \qquad (8)$$

$$2Li^+ + 2O^- \rightarrow Li_2O_2 \qquad (9)$$

Among those different designs of $Li-O_2$ batteries, the aprotic configuration was the one that presented an appreciable degree of rechargeability. Therefore, attracting some attention from the scientific community. Furthermore, there are also functional additives that can be added to the electrolyte system to improve the cyclability, performance, and stability of the $Li-O_2$ battery. Some examples can include the introduction of polydopamine which acts as a superoxide radical scavenger, quinone-based species that induce the formation of Li_2O_2, redox mediators, addition of TPFPB and H_2O to aid in the oxidation of Li_2O_2 into O_2 during charging process [22].

1.3 RECENT ADVANCEMENT IN METAL-AIR BATTERIES

There are several factors related to metal-air batteries that scientists are working on improving. One of them lies in optimizing the performance of electrolytes in terms of their stability, particularly related to their behavior when in contact with superoxide and peroxide species which are obtained during the reactions with oxygen. In this sense, there are some polar aprotic solvents such as N, N-dimethylformamide (DMF), dimethylacetamide (DMA), or propylene carbonate, for instance, which are susceptible to reacting with highly reactive reduced oxygen species (RROS) which include $HO^{\bullet}$, $HOO^{\bullet}$, $O_2^{\bullet-}$, and O_2^{2-} under the presence of Li ions [23]. Some of the reactions that can occur between the RROS, and the polar aprotic solvents are presented in equations 10–13.

$$RX + O_2^{\bullet-} \rightarrow X^- + ROO^{\bullet} \tag{10}$$

$$RCOX + O_2^{\bullet-} \rightarrow RCOO^- + XO^- \tag{11}$$

$$RCH_2CH_2X + O_2^{\bullet-} \rightarrow RCH=CH_2 + HO_2^- \tag{12}$$

$$A + O_2^{\bullet-} \rightarrow A^{\bullet-} \rightarrow A^{\bullet-} + O_2 \tag{13}$$

It has been observed that alkyl carbonates such as polyether tend to react with the RROS species as presented in equations 14 and 15 [23].

$$(CH_3O(CH_2)_2)_2O + Li^+ {}^-OO^- Li^+ \rightarrow CH_3OO^- Li^+ + CH_3O(CH_2)_2O(CH_2)_2O^- Li^+ \tag{14}$$

$$(CH_3OCH_2CH_2)_2O + Li^+ {}^-OO^- Li^+ \rightarrow CH_3O^- Li^+ + CH_3O(CH_2)_2O(CH_2)_2OO^- Li^+ \tag{15}$$

Such reactivity of these species can be attributed to the high nucleophile character of both peroxide and superoxide due to their neighboring electron pairs of non-bonding electrons [24]. Following that, the Li^+ can be strongly bonded to the lone pairs of electrons of O from the ether moieties as they are hard Lewis bases. Through that, the alkoxy groups at the polyether can be converted to better-leaving groups of Li-alkoxide such as CH_3OLi which can be followed by an abstraction of a proton by the peroxide or superoxide groups. Based on these aspects it can be noted that the lack of stability of polyether-based solvents during the ORR process in the presence of Li^+ could be attributed due to the combination of a nucleophilic substitution along with elimination reactions which are considered the major side reactions in this system [23].

As an attempt to improve the inherent instability of ethereal systems, Aurbach et al. [23] performed methylation at an ethylene oxide backbone of dimethoxyethane (DME). This approach led to an improvement in cyclability of $Li-O_2$ cells along with a smaller population of side products when compared to the non-methylated DME. Along that, DMSO-based electrolytes at an Au or TiC substrate can present a satisfactory cycling behavior. DMSO is also reactive toward both peroxide and superoxide radicals which leads to the generation of reactive species such as dimsyl anion hydroperoxyl radical nucleophiles [25]. In this sense, the latter can react with the S atom from the sulfoxide group which leads to the formation of dimethyl sulfone $((CH_3)_2SO_2)$ along with LiOH. Hence, it is necessary to employ electrolytes that can withstand the highly reactive nature of RROSs. Since these species are prompt to react with organic compounds there is another trend of research that points to the use of inorganic electrolytes. Based on that, solid inorganic ceramics have been used since they can act as proper protection for Li-metal against water and O, likely diminishing side reactions [26, 27]. The use of molten salt-based electrolytes composed of $LiNO_3$-KNO_3 at an intermediate temperature of around 150°C has presented appreciable stability when in contact with $Li-O_2$ cathode [28]. Lastly, another common approach adopted to address the reactivity of peroxides and superoxides consists of using fluorinated solvents. However, it can considerably reduce the solubility of a Li-based salt leading to hindrances in performance.

The cathode of $Li-O_2$ in a battery is another major component and a strong topic of study among scientists. In this sense, during the discharging process Li_2O_2 is formed which is insulating, insoluble in organic solvents, and a strong oxidant agent. Based on that, it is required that a cathode to present high electronic conductivity, allow the proper transport of Li+ and O_2, along with presenting a chemical inertness towards the strong oxidizing nature of Li_2O_2 and avoid side reactions. In this sense, three main factors should be considered in the design of a cathode: proper electron transport, electrolytes, as well as cathode stability.

The cathodes used in Li-O_2 batteries can be composed of carbon-based materials which can provide a high surface area along with high conductivity and relatively lower cost. The theoretical specific capacitance of discharge related to carbon-based cathodes in Li-O_2 may reach values of around 2000 mAh/g [29]. Ding *et al.* [30] proposed the synthesis of a hierarchically organized carbon structure with pore sizes with different dimensions. Through that, a correlation between the volume of the pore and discharge capacity could be observed. It was also observed that mesoporous carbon with an ultrahigh surface area was not accessible for the storage of Li_2O_2. Based on that, large macropores that maintain their size after several cycles along with enabling the flow of oxygen through the cathode are some of the desired properties [31]. Graphene-based cathodes are a promising material that can accommodate Li_2O_2. On top of that, considerably high specific capacitances of around 10000 mAh/g_{carbon} have been reported [32]. It is worth noting that a more accurate measurement of the specific capacitance has been obtained when the mass of Li_2O_2 is considered, which has been demonstrated by Mitchell *et al.* [33] who obtained a specific capacitance of 4720 mAh/g_{carbon} when considering carbon against 944 mAh/$g_{carbon+Li_2O_2}$. Despite the desirable properties of carbon-based materials, they may suffer from degradation at the operational stage of Li-O_2 batteries. In this sense, the release of CO_2 was observed when a carbon black-based cathode was employed suggesting the occurrence of side reactions [34, 35]. It has been observed that carbon's surface which presents a larger number of defects as well as hydrophilic groups leads to a faster degradation process when in contact with Li_2O_2. Also, it has been observed through the work of Gallant *et al.* [36] who reported the presence of Li_2CO_3 located at the interface between the carbon-based electrode and Li_2O_2, the schematics for this process are presented in Figure 1.5. In this sense, X-ray absorption near edge structure (XANES) was performed on the discharged cathodes which confirmed the presence of Li_2CO_3, which as the discharge capacity increases gave way to Li_2O_2. In this sense, the larger concentration of Li_2CO_3 at higher cycles made it more difficult for its large agglomerates to oxidize which led to a diminishment in capacitance in the following cycles.

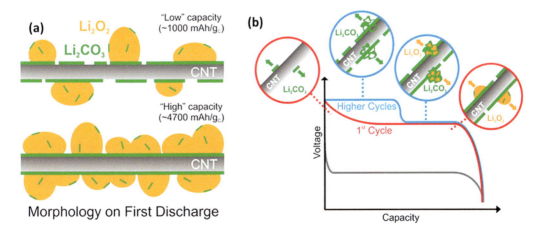

FIGURE 1.5 (a) Scheme for the formation of Li_2CO_3 in between the vertically aligned carbon nanotube-based cathode and Li_2O_2 formed as discharge product at both "low" and "high" capacitances. (b) Plot demonstrating the effect on the charging behavior due to the change in morphology caused by the discharge product presented at the 1st and higher cycles. (Adapted with permission [36]. Copyright 2012, American Chemical Society.)

One of the strategies adopted by the researchers to diminish the degradation of carbon-based cathodes consists of introducing a redox mediator. Gao *et al.* [37] proposed the use of a dual mediator system for the discharge and charge based on 2,5-di-tert-butyl-1,4-benzoquinone (DBBQ) and 2,2,6,6-tetramethyl-1-piperidinyloxy (TEMPO), respectively. After the incorporation of mediators, the carbon instability during cycles went from 0.12% to around 0.008%. Through that, the Li-O_2 became more influenced by the O_2 transport related to the pore-clogging at the interface between the cathode and gas rather than the insolubility and insulating properties of Li_2O_2. Based on that, it seems important to understand the mechanism at which DBBQ aids in the O_2 reduction whereas TEMPO oxidizes Li_2O_2. During discharge, DBBQ is reduced to LiDBBQ at the positive electrode's surface. Then, it can react with O_2 in the solution which leads to $LiDBBQO_2$ intermediate. Following that, the intermediate can go through further reduction leading to the regeneration of DBBQ and the formation of Li_2O_2. On the other hand, during charging, TEMPO undergoes oxidation at the positive electrode's surface forming the TEMPO$^+$ intermediate. Then, the TEMPO$^+$ can oxidize Li_2O_2 and therefore regenerate to TEMPO. Hence, these redox cycling processes prevent the degradation of the carbon-based electrode instead.

Other options for cathode materials aside from carbon-based have also been proposed such as Au and TiC which demonstrated relatively better cycling stability [38]. It was observed that stable metal carbides such as TiC can form a chemically stable and conducting thin layer of TiO_2 [39]. Through that, Li_2O_2 can be oxidized at a lower thermodynamical barrier which leads to a decrease in overpotential. Yet, there is still a need to further investigate other materials that can act in the same way as TiC since metallic nitrides tend to suffer from the complete oxidation of Li_2O_2 under an anodic current which hinders the electron transfer steps [40]. In addition, Mo_2C, for instance, can be converted to MoO_2 during discharge which leads to a low charge overpotential [41]. However, it further converts to Li_xMoO_3 which can degrade the electrode. Other approaches which included depositing metal oxides such as Al_2O_3 over a carbon substrate have been proposed as an alternative to protect it from degradation. Another important issue that scientists are trying to address is the passivation that occurs at the cathode which limits the cell overall capacitance. In this sense, the clogging of the cathode's pores due to the deposition of Li_2O_2 prevents the proper adsorption of O_2 as well as charge transport leading to an increase in overall resistance, hence jeopardizing the performance [42, 43].

Through this discussion, it has been noted that heterogeneous electrocatalysis may present some challenges mostly related to the formation of Li_2O_2 which hinders the electron transfer steps as well as Li$^+$ due to its inherent resistance and low solubility, respectively. Homogenous catalysis that employs redox mediators (RM) has been proposed. In this sense, the RM functions as a mobile specie in the electrolyte that can perform the electron transfer at the electrode/electrolyte interface as demonstrated in Figure 1.6. During discharge the process may consist of the formation of an RM$^-$ specie which interacts with O_2 forming an O_2^- intermediate that further interacts with Li$^+$ in solution forming a Li$^+O_2^-$ intermediate that later converts to Li_2O_2. During charge the opposite takes place as the RM can oxidize Li_2O_2 converting it into RM$^-$ and Li$^+$ and O_2, then the RM$^-$ performs the electron transfer process at the electrode's surface, regenerating the RM.

Even though the incorporation of an RM may improve the instability issue it may also create other obstacles such as, in some cases, being not stable against Li metal. Also, since RM is organic-based material it can be likely to decompose likewise the carbon-based electrode or the electrolyte. Another phenomenon that has been observed is the "redox shuttle" which can diminish the Coulombic efficiency. Redox intermediate RM species formed during the charging step may diffuse towards the anode and go through a reduction reaction when in contact with Li metal, leading to a self-discharge process. Ha *et al.* [45] proposed the use of *o*-methyl-phenothiazine (MPT). It was observed that MPT could react with Li_2O/Li_2O_2 as well as with metallic Li which led to a permanent decomposition of the metal (Figure 1.7). It can be proposed that the Li metal could be protected by a solid electrolyte interface that must be added into the system or a solid electrolyte. Lim *et al.* [46] showed that the ionization energy is a proper parameter that can predict the redox potential of an RM, which can suggest at which level it can interact with the reactive oxygen species

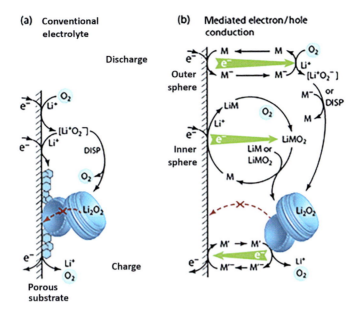

FIGURE 1.6 Schematics for the electrocatalysis at the Li-O_2 cathode (a) without RM and (b) with RM. (Adapted with permission [44]. Copyright 2017, Royal Society of Chemistry. This article is licensed under a Creative Commons Attribution 3.0 Unported License.)

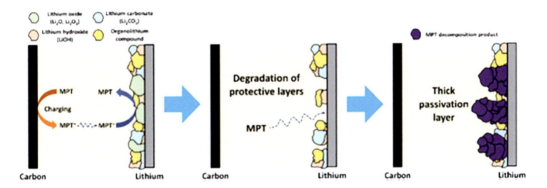

FIGURE 1.7 Scheme displaying the degradative effect of RM into a Li-O_2 which can lead to the formation of a thick passivation layer that diminishes the overall cell's efficiency. (Adapted with permission [45]. Copyright 2017, Royal Society of Chemistry.)

without degrading the electrolyte. It has been observed that RM presenting singly occupied molecular orbitals (SOMO) state of an oxidized form that is similar in energy values to the highest occupied molecular orbitals (HOMO) of the solvent may tend to extract the electrons from the solvent, leading to its decomposition. The authors concluded that dimethylphenazine (DMPZ) can function as a low-voltage RM that is inert to the solvent.

There has been some progress in the scientific community in regards to optimizing the performance of Li-O_2. Li *et al.* [47] proposed the design of a Li-O_2 that could be safely operated in an open environment with a continuous flow of O_2. The authors employed a nonflammable and nonvolatile sulfolane (TMS) along with $LiNO_3$. Through that, a stable SEI could be obtained which allowed a stable cyclability of around 1000h and reversible deposition of Li_2O_2 as well as a stripping process accompanied by a low charge overpotential of 80 mV. This work demonstrated a relatively simple approach to obtaining a stable environment that can carry on a continuous redox process between metallic Li and O_2 into Li_2O_2. Another approach performed by Li et al. [48] synthesized a composite cathode for Li-O_2 battery based on a structure of $MoS_2 x$ and carbon nanotubes (CNTs), $MoS_2 x$@CNT which was synthesized through a hydrothermal approach followed by an annealing process with $NaBH_4$ for the post-reducing process. The procedure led to the formation of MoS_2 nanoflakes that were evenly dispersed. One of the major aspects of their work was attributed to the inherent properties of CNT which promoted an increase in surface area which presented an optimal space for the proper cyclability of Li_2O_2 over the $MoS_2 x$@CNT cathode's surface as it could accommodate the volume changes. Also, the conductivity compensated for the insulation properties of Li_2O_2 which facilitated the electron transfer processes along with the decomposition of the discharge products. It was observed that the defective structure of MoS_2 served as a protective layer as it could prevent the contact between CNTs with electrolytes as well as act as a catalytic site.

Another work performed by Wu et al. [49] proposed an approach to tackle the instability issues related to Li-O_2 batteries. For that, a fluorinated graphene-based structure (CF_x) where x = 1 then F-Gr was incorporated into an ether-based electrolyte. It was observed that F-Gr could catalyze the ORR process which led to an improved discharge capacity along with a restrain on O_2^- derived side reaction. It could be likely that the polarization induced by the F atoms at the graphene's structure led to polarization at the neighboring C atoms which allowed the O_2^- to be held into it, hence preventing it from reacting with the electrolyte. The schematics for this process are presented in Figure 1.8.

Despite the considerable research on Li-O_2 batteries, there is also the need to use alternative metallic sources that are more abundant and likely more sustainable than Li to allow a more controlled consumption. In this sense, Na-O_2 is another point of interest for research, since even though it presents a relatively lower formal specific energy when compared to Li-O_2 it also presents a lower charge overpotential which is usually below 0.2 V, and appreciable reversibility. Following that, Na^+ may be more effective to stabilize O_2 based on the Hard-Soft Acid-Base (HSAB) theory in which Na^+ has a larger radius accompanied by a lower charge density in comparison to Li^+. This feature of Na^+ can suppress the spontaneous disproportionation of sodium superoxide (NaO_2) to sodium peroxide (Na_2O_2). It has been proposed that NaO_2 is formed at the cathode's surface during discharge and is then transferred to the electrolyte. After saturation is reached, NaO_2 precipitates into cubical crystals. During charge, NaO_2 redissolves into the electrolyte and flows to the cathode to undergo oxidation. The reversibility of this process presents a lower thermodynamic barrier which allows it to present a large discharge capacity along with a low charge overpotential.

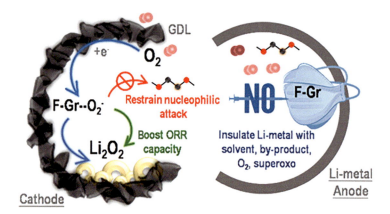

FIGURE 1.8 Schematics for the F-Gr-based cathode that enable longer adsorption of O_2^- to enhance the catalytic process of ORR in a Li-O_2 battery. (Adapted with permission [49]. Copyright 2022, American Chemical Society.)

Aside from Na-O_2, the K-O_2 batteries are also being reached as it presents the formal specific capacitance of 935 Wh/kg. One of the aspects of using K instead of Li or Na is that the formation of potassium superoxide (KO_2) is thermodynamically favorable in comparison to LiO_2 and NaO_2. The K-O_2 can be more selective towards the O_2/O_2^- redox couple which may prevent some of the side reactions from occurring. On top of that, KO_2 has been demonstrated to be more stable after one month of aging when compared to NaO_2 which can suffer from irreversible decomposition when in contact with the electrolyte. There is still some research required since KO_2 tends to be reactive towards ether-based electrolytes as it can promote some loss in performance over time. Even though there is a considerable number of challenges needed to be overcome, metal-O_2 batteries hold great potential as an alternative to energy production and storage.

1.4 FUTURE PERSPECTIVE

Throughout this chapter, it has been discussed that there is a need to develop technologies that can serve as an alternative to the excessive use of non-renewable petrochemical-based sources due to an environmental as well as economical matter since the dependence on only one source can lead to complications in the energy distribution and further issues. At the same time, LIBs have been a widely explored technology that led to considerable consumption of Li resources. Because of that, there is the need to further optimize its use to meet the required demands as well as find other alternative metallic sources to diminish the strain on the use of Li. Hence, the need to obtain devices that can potentially surpass the performance of combustion engines led to the development of Li-O_2 which can potentially offer higher values of energy density as well as a relatively eco-friendlier approach to the generation as well as storage of energy. Following that, other metals have also been researched to address the issue of large consumption of Li which led to the development of Fe-, Na-, K-, Mg-, Zn-, and Al-O_2 batteries. Despite the exciting promises and achievements of these technologies, there is still a considerable amount of work to further elucidate the complex mechanism between the oxygenated species and metallic ions which leads to the formation of superoxides and peroxides that can hinder the stability, cyclability, and overall performance of the metal-air cell due to their inherent low conductivity and solubility. For that, one of the approaches based on the use of solid-state electrolytes has been proposed as it can hinder the formation of a thick film of super or peroxides over the electrode's surface. Also, the relatively lower stability of the porous carbon-based cathodes remains another target for improvement in this field. In this sense, the strategies consist of using RMs which can improve the operational time of a metal-air battery. Another challenge lies in improving the overall electrochemical stability of aprotic electrolytes to prevent their reaction with both the M^+ ions as well as O_2 which has scientists looking for materials suitable as electrolytes that can be nonflammable, nonvolatile, and chemically inert under considerably highly reactive environments due to the presence of oxygen-based radical intermediates. The reward for overcoming such obstacles can propose a novel, efficient, and eco-friendlier way to generate energy without the consumption of petrochemicals which can diminish the strain on their use.

REFERENCES

[1] M.M. Thackeray, C. Wolverton, E.D. Isaacs, Electrical energy storage for transportation: Approaching the limits of, and going beyond, lithium-ion batteries, Energy Environ. Sci. 5 (2012) 7854–7863.
[2] E. Markevich, G. Salitra, P. Hartmann, J. Kulisch, D. Aurbach, K.-J. Park, C.S. Yoon, Y.-K. Sun, New insights related to rechargeable lithium batteries: Li metal anodes, Ni Rich $LiNi_xCo_yMn_zO_2$ cathodes and beyond them, J. Electrochem. Soc. 166 (2019) A5265–A5274.
[3] J. Kim, H. Lee, H. Cha, M. Yoon, M. Park, J. Cho, Prospect and reality of Ni-Rich cathode for commercialization, Adv. Energy Mater. 8 (2018) 1702028.
[4] W. Liu, P. Oh, X. Liu, M.-J. Lee, W. Cho, S. Chae, Y. Kim, J. Cho, Nickel-Rich layered lithium transition-metal oxide for high-energy lithium-ion batteries, Angew. Chemie Int. Ed. 54 (2015) 4440–4457.

[5] H.-H. Ryu, K.-J. Park, C.S. Yoon, Y.-K. Sun, Capacity fading of Ni-Rich Li[NixCoyMn1—x—y]O2 (0.6 ≤ x ≤ 0.95) cathodes for high-energy-density lithium-ion batteries: Bulk or surface degradation?, Chem. Mater. 30 (2018) 1155–1163.

[6] B. Han, B. Key, S.H. Lapidus, J.C. Garcia, H. Iddir, J.T. Vaughey, F. Dogan, From coating to dopant: How the transition metal composition affects alumina coatings on Ni-Rich cathodes, ACS Appl. Mater. Interfaces. 9 (2017) 41291–41302.

[7] M. Dixit, B. Markovsky, F. Schipper, D. Aurbach, D.T. Major, Origin of structural degradation during cycling and low thermal stability of Ni-Rich layered transition metal-based electrode materials, J. Phys. Chem. C. 121 (2017) 22628–22636.

[8] X. Zuo, J. Zhu, P. Müller-Buschbaum, Y.-J. Cheng, Silicon based lithium-ion battery anodes: A chronicle perspective review, Nano Energy. 31 (2017) 113–143.

[9] K. Feng, M. Li, W. Liu, A.G. Kashkooli, X. Xiao, M. Cai, Z. Chen, Silicon-based anodes for lithium-ion batteries: From fundamentals to practical applications, Small. 14 (2018) 1702737.

[10] J.W. Choi, D. Aurbach, Promise and reality of post-lithium-ion batteries with high energy densities, Nat. Rev. Mater. 1 (2016) 16013.

[11] J.-Y. Hwang, S.-J. Park, C.S. Yoon, Y.-K. Sun, Customizing a Li-metal battery that survives practical operating conditions for electric vehicle applications, Energy Environ. Sci. 12 (2019) 2174–2184.

[12] Y. Li, J. Lu, Metal-air batteries: Will they be the future electrochemical energy storage device of choice?, ACS Energy Lett. 2 (2017) 1370–1377.

[13] J. Christensen, P. Albertus, R.S. Sanchez-Carrera, T. Lohmann, B. Kozinsky, R. Liedtke, J. Ahmed, A. Kojic, A critical review of Li/Air batteries, J. Electrochem. Soc. 159 (2011) R1–R30.

[14] W.-J. Kwak, Rosy, D. Sharon, C. Xia, H. Kim, L.R. Johnson, P.G. Bruce, L.F. Nazar, Y.-K. Sun, A.A. Frimer, M. Noked, S.A. Freunberger, D. Aurbach, Lithium-oxygen batteries and related systems: Potential, status, and future, Chem. Rev. 120 (2020) 6626–6683.

[15] B.D. McCloskey, R. Scheffler, A. Speidel, G. Girishkumar, A.C. Luntz, On the mechanism of nonaqueous Li-O2 electrochemistry on C and its kinetic overpotentials: Some implications for Li-air batteries, J. Phys. Chem. C. 116 (2012) 23897–23905.

[16] Y. Shimonishi, T. Zhang, P. Johnson, N. Imanishi, A. Hirano, Y. Takeda, O. Yamamoto, N. Sammes, A study on lithium/air secondary batteries: Stability of NASICON-type glass ceramics in acid solutions, J. Power Sources. 195 (2010) 6187–6191.

[17] T. Zhang, N. Imanishi, Y. Shimonishi, A. Hirano, Y. Takeda, O. Yamamoto, N. Sammes, A novel high energy density rechargeable lithium/air battery, Chem. Commun. 46 (2010) 1661–1663.

[18] J.P. Zheng, P. Andrei, M. Hendrickson, E.J. Plichta, The theoretical energy densities of dual-electrolytes rechargeable Li-Air and Li-Air flow batteries, J. Electrochem. Soc. 158 (2011) A43.

[19] S.H. Oh, L.F. Nazar, Oxide catalysts for rechargeable high-capacity Li-O2 batteries, Adv. Energy Mater. 2 (2012) 903–910.

[20] Y. Sun, Lithium ion conducting membranes for lithium-air batteries, Nano Energy. 2 (2013) 801–816.

[21] B. Kumar, J. Kumar, Cathodes for solid-state Lithium-oxygen cells: Roles of nasicon glass-ceramics, J. Electrochem. Soc. 157 (2010) A611.

[22] P. Tan, H.R. Jiang, X.B. Zhu, L. An, C.Y. Jung, M.C. Wu, L. Shi, W. Shyy, T.S. Zhao, Advances and challenges in lithium-air batteries, Appl. Energy. 204 (2017) 780–806.

[23] D. Aurbach, B.D. McCloskey, L.F. Nazar, P.G. Bruce, Advances in understanding mechanisms underpinning Lithium-air batteries, Nat. Energy. 1 (2016) 16128.

[24] M. Nava, A.E. Thorarinsdottir, N. Lopez, C.C. Cummins, D.G. Nocera, Chemical challenges that the peroxide dianion presents to rechargeable Lithium-air batteries, Chem. Mater. 34 (2022) 3883–3892.

[25] L. Wang, H. Noguchi, Oxygen reduction reaction mechanism in highly concentrated lithium nitrate-dimethyl sulfoxide: Effect of lithium nitrate concentration, J. Phys. Chem. C. 126 (2022) 11457–11467.

[26] A. Paolella, W. Zhu, D. Campanella, S. Kaboli, Z. Feng, A. Vijh, NASICON lithium ions conductors: materials, composites and batteries, Curr. Opin. Electrochem. (2022) 101108.

[27] T. Zhang, M. Yu, J. Li, Q. Li, X. Zhang, H. Sun, Effect of porosity gradient on mass transfer and discharge of hybrid electrolyte lithium-air batteries, J. Energy Storage. 46 (2022) 103808.

[28] Y.G. Zhu, G. Leverick, L. Giordano, S. Feng, Y. Zhang, Y. Yu, R. Tatara, J.R. Lunger, Y. Shao-Horn, Nitrate-mediated four-electron oxygen reduction on metal oxides for lithium-oxygen batteries, Joule. 6(8) (17 August 2022) 1887–1903.

[29] H. Woo, J. Kang, J. Kim, C. Kim, S. Nam, B. Park, Development of carbon-based cathodes for Li-air batteries: Present and future, Electron. Mater. Lett. 12 (2016) 551–567.

[30] N. Ding, S.W. Chien, T.S.A. Hor, R. Lum, Y. Zong, Z. Liu, Influence of carbon pore size on the discharge capacity of Li-O2 batteries, J. Mater. Chem. A. 2 (2014) 12433–12441.
[31] H. Nie, Y. Zhang, J. Li, W. Zhou, Q. Lai, T. Liu, H. Zhang, Synthesis of a meso-macro hierarchical porous carbon material for improvement of O2 diffusivity in Li-O2 batteries, RSC Adv. 4 (2014) 17141–17145.
[32] J. Xiao, D. Mei, X. Li, W. Xu, D. Wang, G.L. Graff, W.D. Bennett, Z. Nie, L. V Saraf, I.A. Aksay, J. Liu, J.-G. Zhang, Hierarchically porous graphene as a Lithium-air battery electrode, Nano Lett. 11 (2011) 5071–5078.
[33] R.R. Mitchell, B.M. Gallant, C.V Thompson, Y. Shao-Horn, All-carbon-nanofiber electrodes for high-energy rechargeable Li-O2 batteries, Energy Environ. Sci. 4 (2011) 2952–2958.
[34] C.-L. Li, G. Huang, Y. Yu, Q. Xiong, J.-M. Yan, X. Zhang, A low-volatile and durable deep eutectic electrolyte for high-performance Lithium-oxygen battery, J. Am. Chem. Soc. 144 (2022) 5827–5833.
[35] P. Zhang, B. Han, X. Yang, Y. Zou, X. Lu, X. Liu, Y. Zhu, D. Wu, S. Shen, L. Li, Y. Zhao, J.S. Francisco, M. Gu, Revealing the intrinsic atomic structure and chemistry of amorphous LiO2-containing products in Li-O2 batteries using cryogenic electron microscopy, J. Am. Chem. Soc. 144 (2022) 2129–2136.
[36] B.M. Gallant, R.R. Mitchell, D.G. Kwabi, J. Zhou, L. Zuin, C.V Thompson, Y. Shao-Horn, Chemical and morphological changes of Li-O2 battery electrodes upon cycling, J. Phys. Chem. C. 116 (2012) 20800–20805.
[37] X. Gao, Y. Chen, L.R. Johnson, Z.P. Jovanov, P.G. Bruce, A rechargeable Lithium-oxygen battery with dual mediators stabilizing the carbon cathode, Nat. Energy. 2 (2017) 17118.
[38] Y. Wang, S. Pan, Y. Guo, S. Wu, Q.-H. Yang, A bidirectional phase-transfer catalyst for Li-O2 batteries with high discharge capacity and low charge potential, Energy Storage Mater. 50 (2022) 564–571.
[39] Z. Wang, J. Sun, Y. Cheng, C. Niu, Adsorption and deposition of Li2O2 on TiC{111} surface, J. Phys. Chem. Lett. 5 (2014) 3919–3923.
[40] M.M. Ottakam Thotiyl, S.A. Freunberger, Z. Peng, Y. Chen, Z. Liu, P.G. Bruce, A stable cathode for the aprotic Li-O2 battery, Nat. Mater. 12 (2013) 1050–1056.
[41] W. Devina, H. Setiadi Cahyadi, I. Albertina, C. Chandra, J.-H. Park, K. Yoon Chung, W. Chang, S. Kyu Kwak, J. Kim, High-energy: Density carbon-coated bismuth nanodots on hierarchically porous molybdenum carbide for superior lithium storage, Chem. Eng. J. 432 (2022) 134276.
[42] S. Dong, S. Yang, Y. Chen, C. Kuss, G. Cui, L.R. Johnson, X. Gao, P.G. Bruce, Singlet oxygen and dioxygen bond cleavage in the aprotic lithium-oxygen battery, Joule. 6 (2022) 185–192.
[43] Z. Zhang, X. Xiao, W. Yu, Z. Zhao, P. Tan, Modeling of a non-aqueous Li-O2 battery incorporating synergistic reaction mechanisms, microstructure, and species transport in the porous electrode, Electrochim. Acta. 421 (2022) 140510.
[44] N. Mahne, O. Fontaine, M.O. Thotiyl, M. Wilkening, S.A. Freunberger, Mechanism and performance of Lithium-oxygen batteries: A perspective, Chem. Sci. 8 (2017) 6716–6729.
[45] S. Ha, Y. Kim, D. Koo, K.-H. Ha, Y. Park, D.-M. Kim, S. Son, T. Yim, K.T. Lee, Investigation into the stability of Li metal anodes in Li-O2 batteries with a redox mediator, J. Mater. Chem. A. 5 (2017) 10609–10621.
[46] H.-D. Lim, B. Lee, Y. Zheng, J. Hong, J. Kim, H. Gwon, Y. Ko, M. Lee, K. Cho, K. Kang, Rational design of redox mediators for advanced Li-O2 batteries, Nat. Energy. 1 (2016) 16066.
[47] Z. Li, C. Song, P. Dai, X. Wu, S. Zhou, Y. Qiao, L. Huang, S.-G. Sun, Nonvolatile and nonflammable sulfolane-based electrolyte achieving effective and safe operation of the Li-O2 battery in open O2 environment, Nano Lett. 22 (2022) 815–821.
[48] D. Li, L. Zhao, Q. Xia, J. Wang, X. Liu, H. Xu, S. Chou, Activating MoS2 nanoflakes via sulfur defect engineering wrapped on CNTs for stable and efficient Li-O2 batteries, Adv. Funct. Mater. 32 (2022) 2108153.
[49] X. Wu, X. Wang, Z. Li, L. Chen, S. Zhou, H. Zhang, Y. Qiao, H. Yue, L. Huang, S.-G. Sun, Stabilizing Li-O2 batteries with multifunctional fluorinated graphene, Nano Lett. 22 (2022) 4985–4992.

2 Materials and Electrochemistry of Metal-Air Battery

Anubha Tomar, Sakshee Chandel, and Alok Kumar Rai

CONTENTS

2.1 Introduction 15
2.2 Materials and Electrochemistry..16
    2.2.1 Air as a Cathode Electrode..17
        2.2.1.1 Electrocatalyst Layer..19
        2.2.1.2 Gas Diffusion Layer (GDL) 21
    2.2.2 Metal Electrode (Anode)..21
        2.2.2.1 Zn Electrode 21
        2.2.2.2 Li Electrode 22
    2.2.3 Electrolytes...23
        2.2.3.1 Aqueous Electrolytes...24
        2.2.3.2 Organic Electrolytes 24
        2.2.3.3 Other Electrolytes 25
2.3 Conclusions 26
References...26

2.1 INTRODUCTION

There is certainly a high demand for energy supply in the near future, which has petrified the globe due to the gradual depletion of fossil fuels. To ameliorate this picture, the inevitable transition from fossil fuels to clean, efficient, and safe renewable energy is of vital importance. However, renewable energy sources such as solar, wind, etc. are geographically intermittent, which eventually directs the investigation to search for an alternative reliable electrochemical energy storage (EES) system for stable and efficient power transmission [1–2]. EES systems are highly crucial to reserve renewable energy by storing and then discharging it during off-peak times and peak times, respectively to minimize the overall power plant generation, consumption of fuel, etc., resulting in acceleration of electric transportation [3]. Among various EES systems, rechargeable batteries are one of the promising energy storage devices, which is heavily used nowadays for portable electronic devices and are soon to be used for electric vehicle application. Currently, lithium-ion batteries (LIBs) are extensively used in the market. However, it is still a challenge to satisfy the growing demands in the areas of electric vehicles, electric grids, and portable energy devices due to the limited energy density and scarcity of Li metal [1, 4]. Metal-air batteries (MABs) among several EES systems have also gained significant interest as next-generation energy storage devices with the ability to replace LIBs in near future because of their high theoretical capacity, easy fabrication, eco-friendliness, etc. More importantly, it also does not contain any hazardous metals such as cobalt, manganese, and nickel on which LIBs heavily rely.

MABs were first developed in the late nineteenth century by Maiche, who created the first primary non-rechargeable Zn-air battery in 1878, and then commercialized in 1932 [4]. However, though the conventional batteries stored the reagents within their cells, metal-air batteries use atmospheric oxygen at the cathode and make them a hybrid of batteries and fuel cells. MABs exhibit

DOI: 10.1201/9781003295761-2

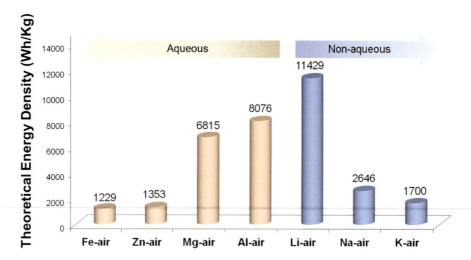

FIGURE 2.1 Theoretical energy density of various metal electrodes in metal-air batteries. (Adapted with permission [1]. Copyright (2017), American Chemical Society.)

high theoretical energy densities of around ~1353–11429 Wh/kg (such as Zn, Al, Mg, and Li-air batteries), which are about 3 to 30 times higher than that of LIBs, as displayed in Figure 2.1 [1]. In this chapter, the fundamentals of materials and electrochemistry of MABs are summarized based on aqueous and non-aqueous (organic) electrolytes.

2.2 MATERIALS AND ELECTROCHEMISTRY

MABs are comprised of the metal anode and air as cathode, which generates electricity via a redox reaction between the metal and atmospheric oxygen. They have an open cell construction that constantly allows endless delivery of the oxygen on the cathode from the air rather than encapsulated within the cell [5]. The atmospheric oxygen as a cathode material ensures a huge specific capacity (~3.35 Ah/g) and high redox potential (~1.23 V vs RHE). Since the air cathode is lightweight and acts as a strong oxidizing agent, it significantly reduced the weight of MABs with more capacity than LIBs [6].

Generally, metal is oxidized into ions on the anode and oxygen transforms into the hydroxide ions at the cathode under MABs. The rechargeable MABs electrode reactions altered with the change of the metal electrodes and the type of the electrolytes, as displayed in Figure 2.2. The following equations represent the electrode reactions of MABs under aqueous electrolyte [2]:

$$\text{Metal electrode: } M \leftrightarrow M^{n+} + ne^- \quad (1)$$

$$\text{Air electrode: } O_2 + 2H_2O + 4e^- \leftrightarrow 4OH^- \quad (2)$$

(Where M denotes the metals such as Al, Zn, Fe, etc. and n shows the oxidation number of the metal ions). Equation 1 is reversible during the discharge and charge process on the metal electrode. In contrast, the oxygen reduction reaction (ORR) and oxygen evolution reactions (OER) generally occurred during the conversion between the oxygen and water on the cathode, as shown in equation 2. The general electrode reactions for non-aqueous MABs are given below [2]:

$$\text{Metal electrode: } M \leftrightarrow M^{n+} + ne^- \quad (3)$$

$$\text{Air electrode: } xM^+ + O_2 + xe^- \leftrightarrow M_xO_2 \ (x = 1 \text{ or } 2) \quad (4)$$

Materials and Electrochemistry of Metal-Air Battery

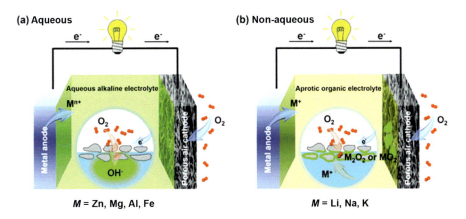

FIGURE 2.2 Schematic illustration of metal-air batteries with operational characteristics: (a) aqueous and (b) non-aqueous electrolytes. (Adapted with permission [7], Copyright (2020), Royal Society of Chemistry.)

Here, the metals generally used are Li, Na, and K due to their sensitivity to the water and CO_2 in the air atmosphere. However, they can directly react with pure O_2 and form metal superoxides (MO_2) or peroxides (M_2O_2) on the cathode during the discharge process as shown in the equations (3) and (4) [2]. More importantly, it also should be noted that the involved electrochemical reactions followed by the obtained products may differ from each other for the various types of metal-air batteries, which depend on the opted-specific metals, electrolytes, and catalytic materials. The following sections describe the electrochemistry and materials used for different components of metal-air batteries:

2.2.1 AIR AS A CATHODE ELECTRODE

The air cathode with a porous structure allows a diffusion path for atmospheric oxygen, which significantly helps to determine the battery performance. Since oxygen is readily available in the ambient air, it does not require to be stored within the battery, which significantly reduces the weight and cost of MABs. The air cathode mainly offers reaction sites for both ORR during discharge and OER during charge. More importantly, these reactions occur on the gas (O_2)-liquid (electrolytes)-solid (electrocatalysts) interface and thus named a three-phase zone, as demonstrated in Figure 2.3 [8].

Heise and Schumacher of the National Carbon Company constructed a classical sandwich design of the Zn-air battery in 1932, which is still the same [3]. The design of an air cathode contained three layers such as a gas diffusion layer (GDL), a catalyst layer with an oxygen electrocatalyst, and a current collector, as shown in Figure 2.4.

Each component in an air electrode has a specific function. The first layer "GDL" is exposed to the outside environment. Since GDL is hydrophobic with various pores to prevent the leakage of internal liquid electrolytes, it also allows O_2 to reach the catalyst layer. The second layer of the current collector is embedded between the GDL and catalyst layer to transfer the electrons from or to the catalyst layer. Moreover, the catalyst layer is hydrophilic, which generally faces the internal liquid electrolytes and offers sites for oxygen reduction or evolution processes at the interface of gas-solid electrocatalyst and the liquid electrolytes. An ideal cathode should have high electronic conductivity, a porous structure for the accommodation of discharge product, high permeability for electrolytes, and excellent catalytic activity with high stability [3].

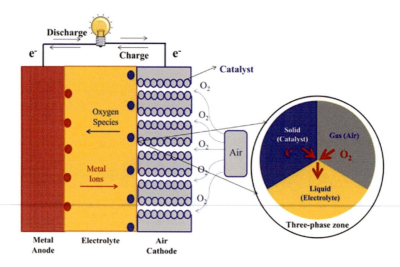

FIGURE 2.3 Schematic diagram of metal-air batteries with a simple solid-liquid-gas three-phase zone. (Adapted with permission [8]. Copyright (2018), Springer Nature.)

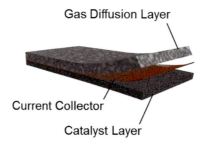

FIGURE 2.4 Design of an air cathode. (Adapted with permission [3]. Copyright (2020), Elsevier.)

In general, the ORR during discharge and OER during charge are the main reactions of oxygen at the air cathode. The ORR is a sequence of sophisticated electrochemical reactions, which involve multistep electron-transfer mechanisms with oxygen-containing intermediates such as O*, HO*, and HOO* [9]. Generally, the ORR process can be divided into two pathways: (a) a direct four-electron pathway (O_2 to OH^-) and (b) a two-electron pathway (O_2 to H_2O_2 or HO_2^- as a reaction intermediate). During the four-electron pathway, one oxygen molecule receives four electrons and is reduced to four OH^- (Eq. 2). In contrast, in the two-electron pathway, which is also known as the indirect four-electron pathway, the oxygen molecule is initially reduced to an intermediate of H_2O_2 or HO_2^- (Eq. 5) and transferred into an electrolyte [3, 10]. Either H_2O_2 or HO_2^- undergoes disproportionation reaction (Eq. 6) or accepts $2e^-$ to form OH^- (Eq. 7) [10].

$$O_2 + H_2O + 2e^- \rightarrow HO_2^- \qquad (5)$$

$$HO_2^- \rightarrow 2HO^- + O_2 \qquad (6)$$

$$HO_2^- + H_2O + 2e^- \rightarrow 3OH^- \qquad (7)$$

On the other hand, the OER during charge showed high reversibility of the same reactions involved in ORR. And the same intermediates (such as O*, HO*, and HOO*) are also observed at

Materials and Electrochemistry of Metal-Air Battery

high overpotential. The detailed equations for the aforementioned reactions under different electrolytes are given next [10, 11]:

Reaction Pathways in Aqueous MABs

Alkaline Solution

Two electron pathways:

$$O_2(g) + H_2O(l) + 2e^- \rightarrow O^* + 2OH^-$$

$$O^* + 2OH^- + H_2O(l) + 2e^- \rightarrow 4OH^-$$

Four electron pathways:

$$O_2(g) + * \rightarrow O_2^*$$

$$O_2(g) + H_2O(l) + e^- \rightarrow OOH^* + OH^-$$

$$OOH^* + e^- \rightarrow O^* + 2OH^-$$

$$O^* + H_2O(l) + e^- \rightarrow OH^* + OH^-$$

$$OH^* + e^- \rightarrow * + OH^-$$

Acidic Solution

Four electron pathways:

$$O_2(g) + * \rightarrow O_2^*$$

$$O_2(g) + H^+ + e^- \rightarrow OOH^*$$

$$OOH^* + H^+ + e^- \rightarrow O^* + H_2O(l)$$

$$O^* + H^+ + e^- \rightarrow OH^*$$

$$OH^* + H^+ + e^- \rightarrow H_2O(l) + *$$

Reaction Pathways in Non-Aqueous MABs

Two electron pathways:

ORR

$$O_2(g) + e^- \rightarrow O_2^-$$

$$O_2^- + Li^+ \rightarrow LiO_2$$

OER

$$Li_2O_2 \rightarrow 2Li^+ + O_2(g) + e^-$$

$$2LiO_2 \rightarrow Li_2O_2 + O_2(g)$$

Where * is the active sites on the electrocatalyst surface
(*l*) and (*g*) denote the liquid and gas phases, respectively and
O*, HO*, and HOO* are adsorbed intermediates

The ORR process occurring at the air electrode of a MABs comprises by various steps:

i. Diffusion of oxygen from the environment followed by absorption on the catalyst surface
ii. Electron transfer from anode to the oxygen molecules through an external circuit
iii. Weakening and breaking of the bond between the oxygen molecule
iv. Detachment of hydroxyl ion from the catalyst surface followed by diffusion into the electrolyte

It was also found that the kinetics of the oxygen electrochemistry is sluggish during the discharge and charge processes due to the strong bond between the oxygen molecule. Therefore, it is necessary to insert the oxygen electrocatalyst to break the bond and lower the overpotential, which can finally accelerate both ORR and OER reactions [12].

2.2.1.1 Electrocatalyst Layer

To consider mass and charge transport, the catalyst layer has a hierarchical pores structure with many macropores and mesopores that serve as the gas transport channels as well as offer sites for oxygen reduction and evolution under the three-phase zone. In general, the catalyst layer is hydrophilic and hydrophobic on the electrolyte and gas sides, respectively [13]. There are two types of a catalyst such as oxygen reducing catalysts and oxygen bifunctional catalysts, which are generally used for primary and rechargeable batteries, respectively. Since the kinetics of oxygen electrochemistry is slow due to the high activation barriers during the charge and discharge processes, it is thus necessary to insert an oxygen electrocatalyst at the air cathode to promote ORR and OER mechanisms by decreasing the activation energy, which eventually helps to increase the conversion rate [14]. A variety of materials have been utilized till now as the cathode catalysts to actively stimulate oxygen reduction or evolution, such as (a) noble metals and alloys (b) transition metal oxides, (c) carbonaceous materials, and (d) conductive polymers. The most commonly used electrocatalysts are discussed next:

2.2.1.1.1 Noble Metals

Platinum (Pt) is a noble metal that acts as a catalyst to facilitate ORR reaction and is often chosen as a benchmark material for the studies of alternative catalysts due to its superior electrocatalytic activity and high stability. However, it shows high performance towards the ORR but not in OER because of its dissolution under OER reactions, resulting in a limited operational life. Thereby, it is necessary to maximize the activity of Pt-based catalysts by tuning the size and morphology or by forming alloys to reduce the cost and improve the performance using other appropriate noble metals or transition metals. The PtM catalysts (M = Ni, Co, Fe, Pd, and Au) showed excellent improvement in the activity and stability of the ORR due to the loose binding of oxygen than that of other PtM alloys or pure Pt materials [15]. However, inexpensive noble metals such as palladium (Pd), gold (Au), silver (Ag), and their alloys have received considerable attention due to their modest activity and relatively higher abundance.

2.2.1.1.2 Transition Metal Oxides

Recently, noble metals are being substituted by the transition metal oxides in MABs as they possess multiple valences, high catalytic activities, and the capability to form different crystal structures. For example, manganese (Mn) exhibits different oxidation states of (II), (III), and (IV) by forming MnO, Mn_2O_3, and MnO_2, respectively [16]. The other metal oxides such as Co_3O_4, NiO, CuO, and Fe_xO_y are also extensively investigated along with Mn_xO_y due to their incomplete filled d-orbitals along with numerous possible oxidation states, leading to high intrinsic activity towards bifunctional ORR and OER electrocatalysts. Moreover, the transition metal oxides with spinel structure such as $A_xB_{3-x}O_4$ where divalent A^{2+} ions (i.e., Co, Ni, Fe, and Mn) and trivalent B^{3+} ions (i.e., Al, Fe, and Co) may partially or completely occupy all the tetrahedral and octahedral sites, respectively, resulting acts as a bi-functional oxygen electrocatalyst for both the ORR and OER activity. The next electrocatalyst is perovskite with a general formula of ABO_3, where A and B represent rare or alkaline earth metals (e.g., La, Ca, Sr, etc.) and transition metals (e.g., Co, Mn, Fe, etc.), respectively that usually serve as the catalyst active centers [17]. Perovskites have been generally recognized as an efficient bifunctional catalyst for both ORR/OER activity. A typical perovskite (such as $LaMnO_3$ $LaNiO_3$ $LaCoO_3$ etc.) showed a better intrinsic ORR activity compared to the Pt/C catalyst. More importantly, the catalytic activity also can be tuned by partially replacing the cation A and/or B with other metals to form the $AA'BB'O_3$ structure [15].

2.2.1.1.3 Carbonaceous Materials

Carbonaceous materials also have been studied extensively as air electrode electrocatalysts in MABs because of their high electrical conductivity, large specific surface area, low cost, and porous structure compared to the noble metal catalyst. In general, the pristine carbon materials such as carbon nanotube, carbon fiber, activated carbon, carbon black, and graphene are electrocatalytically inactive towards the ORR. Since the hetero-atoms doping (such as N, P, S, and B) into carbon materials make them metal-free catalysts, the catalytic activity is also improved due to the size and electronegativity difference between the dopants and carbon materials. N-doped carbons are one of the earliest materials to be investigated, and act as a bifunctional oxygen catalyst due to the presence of both electron-donating quaternary N sites (n-type doping) for ORR and electron-withdrawing Pyridine N moieties (p-type doping) for OER [9].

2.2.1.1.4 Conductive Polymers

The conductive organic polymers such as polyaniline, polypyrrole, polythiophene, poly(3-methyl) thiophene, and poly(3,4-ethylene dioxythiophene) exhibit high potential to substitute noble metals as cathode electrocatalysts in MABs due to their low cost, high electrical conductivity, distinct redox properties, and mix-metal with a polymer-like behavior [5, 18].

2.2.1.2 Gas Diffusion Layer (GDL)

The GDL in the air electrode of MABs ensures an even supply of oxygen to the electrode from the environment and also avoids the leakage of electrolytes from the air cathode. It should be also thin, porous, and hydrophobic to enhance the oxygen movement in the air electrode. It is majorly composed of porous active carbon and hydrophobic polymers such as PTFE (polytetrafluoroethylene), PVDF (polyvinylidene fluoride), etc. [17]. The hydrophobic agents are not only working as water-proofing agents and air diffusing medium but also resist the paths of CO_2 gas into the electrolyte. To enhance the waterproofness and gas diffusion rate, the polymer content must be optimized. By adding 30 to 70 wt% of hydrophobic polymers to the carbon black, acetylene black, active carbon, etc., the GDL becomes wet-proofed and prevents electrolyte flooding, resulting in improvement in the electrical and thermal conductivities along with surface wettability [3]. Toray Industries, Inc. has successfully created a commercial GDL for polymer electrolyte fuel cells using Toray carbon paper, which also can be used for MABs [19]. The most commonly used materials for GDL are woven carbon cloths, which are similar to carbon paper with numerous micropores. Furthermore, since carbon paper and cloth exhibit excellent electronic conductivities, it allows them to act as the current collectors [11].

However, the carbon materials of the GDL of rechargeable aqueous MABs are suffered due to the side reactions such as carbon corrosion at high oxidation potential, resulting in degradation of the electrode materials [20]. Hence, to fix this issue, the high potential corrosion-resistant materials such as high-graphitized carbons (graphene and carbon nanotubes) and porous materials are highly recommended. Metal-based GDL (such as Ti meshes, stainless steel, Ni foam, etc.) also has been investigated for both aqueous and non-aqueous MABs as they possess high electrical conductivity as well as high resistivity to the corrosive environment due to the formation of passivation layers on the surface. Sun et al. have fabricated a binder-free carbon cathode with hierarchical $MnCo_2O_4$ nanowire bundles on 3D nickel foam with a porous structure, which shows better rate capability and large discharge capacity up to 12919 mAh/g at 0.1 mA/cm^2 [21].

2.2.2 Metal Electrode (Anode)

Metal plates are used as an anode in most of the MABs, which primarily determine the configuration of these batteries. The type of metal used as an anode also decides the electrolyte nature and design of the air electrode. The metallic anode should possess strong reducing behavior with light weight and compactness. It also should be inexpensive as the cost of the battery significantly depends on the metal anode in MABs though oxygen is free on the cathode side. Generally, the metals such as lithium, zinc, aluminum, sodium, magnesium, etc. are used as anodes. Each metal has its own set of advantages and disadvantages when it comes to being used as anodes. The common problems with metallic anodes are corrosion, passivation, and dendrite formation.

2.2.2.1 Zn Electrode

Zn metal is extensively known as active anode material for Zn-air batteries. The following chemical reactions are generally involved during discharge at the Zn electrode in an alkaline medium:

$$Zn \rightarrow Zn^{2+} + 2e^- \tag{8}$$

$$Zn + 4OH^- \rightarrow Zn(OH)_4^{2-} \tag{9}$$

$$Zn(OH)_4^{2-} \rightarrow ZnO + H_2O + 2OH^- \tag{10}$$

$$Zn + H_2O \rightarrow Zn(OH)_2 + H_2 \uparrow \quad (11)$$

The discharge process showed oxidation of metallic Zn to Zn^{2+}, solvation of ions in the solution, ions diffusion in the electrolyte, and then precipitation of ZnO and $Zn(OH)_2$ due to the supersaturation. The deposition of passive layers of ZnO and $Zn(OH)_2$, dendrites formation, and corrosion of Zn in aqueous electrolytes are the main issues associated with Zn metal as an anode. In addition, the corrosion of Zn in an aqueous electrolyte also leads to self-discharge due to the coupling of the Zn oxidation and hydrogen evolution reaction. To fix these issues, the nanostructuring of the active materials, formation of protective coatings, and use of electrolyte additives were found to be primary strategies. Various metal oxides such as Al_2O_3, Bi_2O_3, CuO, SiO_2, etc. have been reported as coating agents on Zn anode [22–25]. Lee et al. have reported that the Al_2O_3 coated Zn electrode prevented the electrode exposure to the KOH electrolyte, resulting in improved active material utilization with a low self-discharge rate [22]. On the other hand, silica-based coatings were also employed to improve the discharge capacity of battery-grade Zn electrodes by reducing the early precipitation of passive ZnO layers. It is also observed that the hydrogen evolution reaction (HER) could be reduced by 40% using SiO_2 coating on the Zn electrode due to the inhibition of the exposure of Zn metal to the alkaline electrolyte. Simultaneously, the silica layers also get swelled in aq. KOH electrolyte and formed gel of $Si(OH)_4$ with the availability of OH- ions, as shown in equations (12) and (13) [23].

$$Zn_2SiO_4 + 2H_2O \rightarrow Si(OH)_4 + 2ZnO \quad (12)$$

$$SiO_2 + 2H_2O \rightarrow Si(OH)_4 \quad (13)$$

Furthermore, it is also found that lithium boron oxide coatings ($Li_2O-2B_2O_3$) could also increase the discharge capacity, and suppress the HER and self-discharge rate [24]. In contrast, the CuO coating on the Zn electrode prevents dendrite formation, minimizes the corrosion rate, and improves the reversible cycling stability of the Zn anode [25]. Different kinds of surfactants (such as anionic, cationic, and non-ionic) and organic additives (such as derivatives of benzene) are also used to improve the electrochemical performances of the Zn electrode [26]. Despite the advantages of organic additives, they also act as insulators and impurities in electrolytes and electrodes. In contrast, inorganic additives (such as bismuth oxide, PbO, In_2O_3, and SnO) have also been investigated to mitigate Zn dendrite growth [27]. In addition, the alloying metals such as Pb, In, Bi, and Ni formed alloy with Zn metal and thus improved the electrochemical performances of Zn anode under Zn air batteries by suppressing the HER [28]. The carbon-containing conducting polymer (such as polyaniline) could also effectively shield the corrosion reaction of the Zn anode [2]. The electrochemical behavior of Zn deposited on Ag, Bi, Cu, Fe, Ni, and Sn substrates was also studied [29]. Bi and Cu substrates were found to be more suitable as current collectors for Zn anode due to their anti-corrosion properties and compact Zn deposition.

2.2.2.2 Li Electrode

Li metal anode is generally used in Li-air batteries. Li-air batteries with non-aqueous electrolytes are more common using the following reactions [5]:

$$2Li(s) + O_2 \leftrightarrow Li_2O_2 \qquad E° = 2.96 \text{ V vs. } Li^+/Li \quad (14)$$

$$4Li(s) + O_2 \leftrightarrow 2Li_2O \qquad E° = 2.91 \text{ V vs. } Li^+/Li \quad (15)$$

The preceding reactions can be divided into anodic and cathodic reactions as given next:

$$\text{Li}(s) \leftrightarrow \text{Li}^+ + e^- \quad \text{(anodic reaction)} \tag{16}$$

$$\text{Li}^+ + 12\text{O}_2 + e^- \leftrightarrow 12\text{Li}_2\text{O}_2 \quad \text{(cathodic reaction)} \tag{17}$$

$$\text{Li}^+ + 14\text{O}_2 + e^- \leftrightarrow 14\text{Li}_2\text{O} \quad \text{(cathodic reaction)} \tag{18}$$

The anodic reaction is reversible; however, the reversibility of cathodic reactions depends on the type of battery. If Li-air battery is rechargeable, the cathodic reactions should be reversible. It can be noticed that the oxidation occurs at the anode during discharge with the release of electrons, which flow through an external circuit and the obtained Li$^+$ ions reduced the oxygen to form Li$_2$O$_2$ on the cathode. Since Li is much more reactive than Zn, Li anodes react with O$_2$, H$_2$O, CO$_2$, and electrolytes, which can cause serious safety issues. In addition, the Li anode also suffers from dendrite formation, which leads to reducing the specific capacity, cycling stability, coulombic efficiency, etc. Apart from this, the diffusion of O$_2$ through the cathode caused the decomposition of organic electrolytes, resulting the formation of LiOH and carbonates on the Li electrode.

To inhibit the Li dendrite growth and improve the electrochemical performances of Li metal anode, several approaches such as surface coating, applying a 3D Li host, tuning the electrode composition, and design of the alloy anode to enhance the corrosion resistance ability are being considered. It is also reported that the solid electrolyte interface (SEI) layer enhanced the surface stability of the Li metal anode, while the use of additives in the electrolyte stabilized the SEI layer formed on the surface of Li metal. For example, Lithium nitrate (LiNO$_3$) is used as an additive to form a passivating layer of Li$_2$O on the Li-metal anode, as shown in equation (19), which served as a stable SEI and substantially inhibits the reaction between the Li-metal anode and the electrolyte [30].

$$2\text{Li}(s) + \text{LiNO}_3 \rightarrow \text{Li}_2\text{O} + \text{LiNO}_2 \tag{19}$$

An artificial protecting layer composed of LiF, Li$_2$CO$_3$, polyene, and C-F bond-based compounds derived from fluoroethylene decomposition is also used, which effectively protects the Li-metal anode from corrosion by dissolved oxygen. A Ge-based protective layer is also constructed by immersing Li foil in GeCl$_4$-tetrahydrofuran steam. The obtained layer is composed of Ge, GeO$_x$, Li$_2$CO$_3$, LiOH, LiCl, and Li$_2$O that allow stable cycling of Li electrodes, especially in "moist" electrolytes [31]. Furthermore, 2,2,6,6-tetramethylpiperidinyl-1-oxyl (TEMPO) as a redox mediator along with Al$_2$O$_3$ and poly(vinylidenefluoride-co-hexafluoropropylene) (PVDF-HFP) acts as a composite protective layer reported for Li-metal anode, which enhanced the cycling performance of non-aqueous Li-air batteries [32]. The reduced graphene oxide (rGO) with molten Li is not only guaranteed the uniform plating of Li$^+$ on the rGO/Li anode, which significantly mitigates Li dendrite formation but also effectively enhanced the flexibility and mechanical strength of the anode [33].

2.2.3 Electrolytes

The electrolyte generally provides a medium for the movement of ions between cathode and anode throughout the redox reactions. To choose suitable electrolytes for MABs, the following points should be considered:

i. Stability against reduced O$_2$ species and metal electrodes
ii. High Boiling point, low volatility, and viscosity
iii. High solubility in O$_2$ with excellent diffusivity
iv. Low moisture adsorption
v. Non-toxic

2.2.3.1 Aqueous Electrolytes

The aqueous electrolytes are categorized into three solutions based on the pH scale i.e. alkaline solution (7< pH ≤ 13), acidic solution (2 ≤ pH < 7), and neutral salt solution (pH = 7). The aqueous alkaline electrolytes are generally used by the water-stable metal electrodes such as Zn, Al, etc. MABs with alkaline electrolytes have shown higher potential to resist metal corrosion with better oxygen electrocatalysis than acidic and neutral electrolytes. The common alkaline electrolytes used in Zn-air batteries are KOH, NaOH, and LiOH. However, KOH is extensively used among all due to its high ionic conductivity (e.g. K$^+$: ~73.50 Ω^{-1} cm^2/equiv. than Na$^+$: ~50.11 Ω^{-1}cm^2/equiv.) and low viscosity [34]. In addition, alkaline electrolytes are also very sensitive toward CO_2, indicating the formation of carbonates. Then the carbonate ions react with hydroxyl ions to form the insoluble precipitate as shown in equations (20) and (21), which eventually blocks the air diffusion pathway of cathode, resulting in the degradation of MABs performance [35].

$$CO_2 + OH^- \rightarrow CO_3^{2-} + H_2O \tag{20}$$

$$CO_2 + KOH \rightarrow K_2CO_3 + H_2O \tag{21}$$

On the contrary, the acidic electrolytes (such as sulfuric acid, nitric acid, hydrochloric acids, methane sulfonic acid, etc.) have been barely used in rechargeable Zn-air batteries due to the excessive release of H$^+$ in the solution, which directly reacts with metal anode and reduces the battery efficiencies.

Furthermore, neutral electrolytes are also widely used in Zn-air batteries. The neutral electrolytes generally bring down the Zn solubility with less CO_2 absorption. A neutral electrolyte is composed of one or more inorganic salts such as HCl, NaCl, NH_4Cl, $ZnCl_2$, BF_4^-, $CH_2SO_3^-$ etc. The following equations elucidate the Zn dissolution mechanism in neutral electrolytes [36].

$$Zn + H_2O \rightarrow ZnOH + H^+ + e^- \tag{22}$$

$$ZnOH \rightarrow ZnO + H^+ + e^- \tag{23}$$

$$ZnO + H_2O \rightarrow Zn^{2+} + 2OH^- \tag{24}$$

As can be noticed that there are no supersaturated species of $Zn(OH)_4^{2-}$ formed in neutral electrolytes though it is formed in the alkaline electrolyte, as shown in equations (25) and (26).

$$Zn + OH^- \rightarrow Zn(OH)_4^{2-} + 2e^- \tag{25}$$

$$Zn(OH)_4^{2-} \rightarrow ZnO + OH^- + H_2O \tag{26}$$

Therefore, the absence of $Zn(OH)_4^{2-}$ in a neutral electrolyte reduced the dendrite formation as well as avoided the carbonization of the electrolyte [37].

2.2.3.2 Organic Electrolytes

In general, the aprotic organic electrolytes are significantly used with the metal electrodes, which are unstable in water (Li and Na) for MABs application. Continuous efforts have been made for designing the electrolytes with the aim of high stability and decrease of the oxygen crossover to improve the air/metal electrode performances [30]. The electrolytes have been modified by optimizing the organic solvents, metal salts, and electrolyte additives. Firstly, carbonate-based solvents such as ethylene carbonate and propylene carbonate were studied to use in the electrolytes. However, the generation of superoxide species on air electrodes makes them unstable and limits their further use as solvents [38]. Then, the ether-based solvents (such as tetrahydrofuran, triglyme, and tetraglyme, etc.) were tested, which exhibited the decomposition of electrolytes due to the oxygen crossover, resulting in low performance of the battery [2, 30]. Then, other types of organic solvents such as amide-based (N, N-Dimethylacetamide) and sulfoxide-based (dimethyl sulfoxide) have also been investigated for MABs, which have offered sufficient stability toward oxygen-reduction species [30]. However, these electrolytes are not always found to be compatible with lithium-based anodes due to the production of high vapor pressure. On the other hand, metal salts are another essential component of organic electrolytes. Highly soluble metal salts with stable anions are recommended for MABs. For example, lithium-containing salts ($LiClO_4$, $LiPF_6$, $LiBr$, $LiNO_3$, $LiTFSI$, and $LiCF_3SO_3$) have been studied for Li-air batteries [39]. The lithium salts especially with fluoride anions tend to decompose due to the discharge product of Li_2O_2 along with other oxygen species. The stability of metal salts can be optimized with different types of electrolyte solvents and their concentrations. A protective layer is usually required on the surface of metal electrodes for better electrolyte stability. Wu et al. have reported an electrolyte of FEC/G4-LiTFSI, which produced a unique stable SEI film on the lithiated Si anode surface and acts as a barrier to block the oxygen crossover as well as inhibit the parasitic reactions at the interface of the lithiated Si anode and the electrolyte [40].

2.2.3.3 Other Electrolytes

The shortcomings of the aqueous and organic electrolytes have led the researchers to investigate a suitable electrolyte for MABs. Therefore, a room temperature ionic liquid with promising salts, whose melting point is below 100°C, are explored as electrolytes for MABs owing to their high ionic conductivity and negligible vapor pressure [2, 30]. Ionic liquids turned out to be promising alternatives as electrolyte solvents for both aqueous and organic electrolytes due to their wide potential window, high ionic conductivity, nonflammability, and low volatility. Thomas et al. have demonstrated a Li-air battery with improved stability towards ambient moisture using [Li(G4)][TFSA] electrolyte (TFSA = bis(trifluoromethanesulfonyl)amide) [32]. Ionic liquids have also been studied as electrolyte additives. A TEMPO-grafted ionic liquid is employed as an electrolyte additive in Li-air batteries [41]. The TEMPO group acts as a redox mediator and oxygen shuttle, in which the imidazolium moiety enables the formation of a stable SEI layer. The IL-TEMPO can also serve as the electrolyte solvent to enable the operation under atmospheric air. In addition, a Li-air battery with a mixed electrolyte composed of 25% of 1-ethyl-3-methylimidazolium tetrafluoroborate (EMIM-BF_4) and 75% of dimethyl sulfoxide is reported [42].

Recently, solid-state electrolytes have also been considered to overcome the drawbacks of liquid electrolytes such as leakage, flammability, and volatility. Apart from this, the solid-state electrolytes are also mechanically strong to suppress the growth of lithium dendrites. Thus, there are mainly three types of solid-state electrolytes which are employed now in metal-air batteries: (i) solid polymer; (ii) gel polymer, and (iii) ceramic electrolytes. The solid polymer electrolytes consist of ion-conducting polymer matrix such as polyacrylonitrile, poly(carbonate-ether), and poly(ethylene oxide) [5]. However, the low ionic conductivity with high interfacial resistance limits their practical applications in MABs. Furthermore, the gel polymer electrolyte also consists of a polymer matrix encapsulated with liquid electrolyte, which seems to be a smart choice. In addition, the gel polymer electrolytes provide an additional benefit of processability and scalability compared to solid polymer and ceramic electrolytes for flexible metal-air batteries. Hence, the gel polymer electrolytes have been widely explored for both aqueous and organic flexible metal-air batteries such as poly(vinylidene fluoride-co-hexafluoropropylene) (PVDF-HFP) and natural rubber polymer for Li-air battery and sodium polyacrylate hydrogel (PANa) for flexible Zinc-air batteries [2, 43]. The ceramic fillers (TiO_2, SnO_2, MgO, Al_2O_3, and SiO_2) also have been incorporated into the host polymers for improving the ionic conductivity and interfacial properties at the metal electrode [43].

Inorganic or ceramic electrolytes have also demonstrated their application in metal-air batteries owing to their satisfactory ionic conductivity and high mechanical properties. In general, Li$^+$ superionic conductor (LISICON), Na$^+$ superionic conductor (NASICON), perovskite, garnet, and sulfides have been used as electrolytes in MABs [2, 44]. Liu et al. have reported NASICON type solid electrolyte named $Li_{1.5}Al_{0.5}Ge_{0.5}P_3O_{12}$ (LAGP), which exhibits high ionic conductivity and excellent mechanical strength [44]. Hence, it can be noticed that the electrolyte stability and the interfacial contact between the electrode and the solid-state electrolyte have great importance to design solid-state metal-air batteries. However, the electrochemical stability window of ceramic electrolyte is still narrow thereby further investigations of solid-state metal-air batteries with satisfactory performances are highly recommended for practical applications.

2.3 CONCLUSIONS

In summary, MABs exhibit a significant role in the research of next-generation energy storage devices due to their promising electrochemical performances, low cost, and ultrahigh energy density. However, the practical applications of MABs are still limited due to the numerous issues of metal electrode reversibility, ORR, OER, HER, and electrode/electrolyte stability. In this chapter, we have summarized nearly all the information like the advantages/disadvantages of materials used for electrodes and electrolytes followed by their electrochemistry involved in both aqueous and non-aqueous MABs. The research on MABs has been carried out as per the following aspects: the oxygen electrocatalyst, gas diffusion layer, and pore structures of the air electrode; the compositions, additives, and protective layers of the metal electrode and the type of electrolytes and their compositions. It can be observed that these modifications/strategies have improved the power/energy density, specific capacity, coulombic efficiency, and cycle life of metal-air batteries.

REFERENCES

[1] L. Yanguang, L. Jun, Metal-air batteries: Will they be the future electrochemical energy storage device of choice? ACS Energy Lett., 2 (2017) 1370–1377.
[2] H.F. Wang, X. Qiang, Materials design for rechargeable metal-air batteries, Matter, 1 (2019) 565–595.
[3] L. Qianfeng, P. Zhefei, W. Erdong, A. Liang, S. Gongquan, Aqueous metal-air batteries: Fundamentals and applications, Energy Storage Mater. 27 (2020) 478–5054.
[4] D. Ahuja, V. Kalpna, P.K. Varshney, Metal air battery: A sustainable and low cost material for energy storage. J. Phys.: Conf. Ser. 1913 (2021) 012065.
[5] L. Li, Z.W. Chang, X.B. Zhang, Recent progress on the development of metal-air batteries, Adv. Sustainable Syst. 1 (2017) 1700036.
[6] C. Wang, Y. Yu, J. Niu, Y. Liu, D. Bridges, X. Liu, J. Pooran, Y. Zhang, A. Hu, Recent progress of metal-air batteries-a mini review, Appl. Sci. 9 (2019) 2787.
[7] J. Zhou, J. Cheng, B. Wang, H. Peng, J. Lu, Flexible metal-gas batteries: A potential option for next-generation power accessories for wearable electronics, Energy Environ. Sci. 13 (2020) 1933–1970.
[8] Y.J. Wang, B. Fang, D. Zhang, A. Li, D.P. Wilkinson, A. Ignaszak, L. Zhang, J. Zhang, A review of carbon-composited materials as air-electrode bifunctional electrocatalysts for metal—air batteries, Electrochem. Energy Rev., 1 (2018) 1–34.
[9] D. Liu, Y. Tong, X. Yan, J. Liang, S.X. Dou, Recent advances in carbon-based bifunctional oxygen catalysts for zinc-air batteries, Batteries & Supercaps, 2 (2019) 743–765.
[10] R. Ma, G. Lin, Y. Zhou, Q. Liu, T. Zhang, G. Shan, M. Yang, J. Wang, A review of oxygen reduction mechanisms for metal-free carbon-based electrocatalysts, Npj Comput. Mater., 5 (2019) 1–15.
[11] A.G. Olabi, E.T. Sayed, T. Wilberforce, A. Jamal, A.H. Alami, K. Elsaid, S.M.A. Rahman, S.K. Shah, M.A. Abdelkareem, Metal-air batteries: A review, Energies. 14 (2021) 7373.
[12] H. Li, L. Ma, C. Han, Z. Wang, Z. Liu, Z. Tang, C. Zhi, Advanced rechargeable zinc-based batteries: Recent progress and future perspectives, Nano Energy. 62 (2019) 550–587.
[13] J. Pan, Y.Y. Xu, H. Yang, Z. Dong, H. Liu, B.Y. Xia, Advanced architectures and relatives of air electrodes in Zn-air batteries. Adv. Sci. 5 (2018) 1700691.

[14] Z.L. Wang, D. Xu, J.J. Xu, X.B. Zhang, Oxygen electrocatalysts in metal-air batteries: From aqueous to nonaqueous electrolytes. Chem. Soc. Rev. 43 (2014) 7746–7786.

[15] X. Cai, L. Lai, J. Lin, Z. Shen, Recent advances in air electrodes for Zn-air batteries: Electrocatalysis and structural design, Mater. Horiz. 4 (2017) 945–976.

[16] D.U. Lee, P. Xu, Z.P. Cano, A.G. Kashkooli, M.G. Park, Z. Chen, Recent progress and perspectives on bi-functional oxygen electrocatalysts for advanced rechargeable metal—air batteries, J. Mater. Chem. A, 4 (2016) 7107–7134.

[17] X. Chen, Z. Zhou, H.E. Karahan, Q. Shao, L. Wei, Y. Chen, Recent advances in materials and design of electrochemically rechargeable zinc-air batteries, Small, 14 (2018) 1801929.

[18] M. Yuasa, A. Yamaguchi, H. Itsuki, K. Tanaka, M. Yamamoto, K. Oyaizu, Modifying carbon particles with polypyrrole for adsorption of cobalt ions as electrocatatytic site for oxygen reduction, Chem. Mater. 17 (2005) 4278–4281.

[19] Y. Li, H. Dai, Recent advances in zinc-air batteries, Chem. Soc. Rev. 43 (2014) 5257–5275.

[20] D.U. Lee, J.Y. Choi, K. Feng, H.W. Park, Chen, Z. Advanced extremely durable 3D bifunctional air electrodes for rechargeable zinc-air batteries, Adv. Energy Mater. 4 (2014) 1301389.

[21] H.T. Wu, W. Sun, Y. Wang, F. Wang, J.F. Liu, X.Y. Yue, Z.H. Wang, J.S. Qiao, D.W. Rooney, K.N. Sun, Facile synthesis of hierarchical porous three-dimensional free-standing $MnCo_2O_4$ cathodes for long-life $Li-O_2$ batteries, ACS Appl. Mater. Interfaces. 9 (2017) 12355–12365.

[22] S.M. Lee, Y.J. Kim, S.W. Eom, N.S. Choi, K.W. Kim, S.B. Cho, Improvement in self-discharge of Zn anode by applying surface modification for Zn-air batteries with high energy density, J. Power Sources. 227 (2013) 177–184.

[23] M. Schmid, M. Willert-Porada, Electrochemical behavior of zinc particles with silica based coatings as anode material for zinc air batteries with improved discharge capacity, J. Power Sources. 351 (2017) 115–122.

[24] Y.D. Cho, G.T.K. Fey, Surface treatment of zinc anodes to improve discharge capacity and suppress hydrogen gas evolution, J. Power Sources. 184 (2008) 610–616.

[25] Y.J. Kim, K.S. Ryu, The surface-modified effects of Zn anode with CuO in Zn-air batteries, Appl. Surf. Sci. 480 (2019) 912–922.

[26] R.K. Ghavami, Z. Rafiei, Performance improvements of alkaline batteries by studying the effects of different kinds of surfactant and different derivatives of benzene on the electrochemical properties of electrolytic zinc, J. Power Sources. 162 (2006) 893–899.

[27] D.J. Park, E.O. Aremu, K.S. Ryu, Bismuth oxide as an excellent anode additive for inhibiting dendrite formation in zinc-air secondary batteries, Appl. Surf. Sci. 456 (2018) 507–514.

[28] A.R. Kannan, S. Muralidharan, K.B. Sarangapani, V. Balaramachandran, V. Kapali, Corrosion and anodic behaviour of zinc and its ternary alloys in alkaline battery electrolytes, J. Power Sources. 57 (1995) 93–98.

[29] X. Wei, D. Desai, G.G. Yadav, D.E. Turney, A. Couzis, S. Banerjee, Impact of anode substrates on electrodeposited zinc over cycling in zinc-anode rechargeable alkaline batteries, Electrochim Acta. 212 (2016) 603–613.

[30] H. Song, H. Deng, C. Li, N. Feng, P. He, H. Zhou, Advances in lithium-containing anodes of aprotic $Li-O_2$ batteries: Challenges and strategies for improvements, Small Methods. 1 (2017) 1700135.

[31] K. Liao, S. Wu, X. Mu, Q. Lu, M. Han, P. He, Z. Shao, H. Zhou, Developing a "water-defendable" and "dendrite-free" lithium-metal anode using a simple and promising $GeCl_4$ pretreatment method, Adv. Mater. 30 (2018) 1705711.

[32] J. Zhang, B. Sun, Y. Zhao, A. Tkacheva, Z. Liu, K. Yan, X. Guo, A.M. McDonagh, D. Shanmukaraj, C. Wang, T. Rojo, M. Armand, Z. Peng, G. Wang, A versatile functionalized ionic liquid to boost the solution-mediated performances of lithium-oxygen batteries, Nat. Commun. 10 (2019) 602.

[33] Z. Guo, J. Li, Y. Xia, C. Chen, F. Wang, A.G. Tamirat, Y. Wang, Y. Xia, L. Wang, S. Feng, A flexible polymer-based Li—air battery using a reduced graphene Oxide/Li composite anode, J. Mater. Chem. A. 6 (2018) 6022–6032.

[34] J.S. Lee, S.T. Kim, R. Cao, N.S. Choi, M. Liu, K.T. Lee, J. Cho, Metal—air batteries with high energy density: Li-air versus Zn-air, Adv. Energy Mater. 1 (2011) 34–50.

[35] S. Hosseini, S.M. Soltani, Y.Y. Li, Current status and technical challenges of electrolytes in zinc—air batteries: An in-depth review, Chem. Eng. J. 408 (2021) 127241.

[36] J.W. Johnson, Y.C. Sun, W.J. James, Anodic dissolution of Zn in aqueous salt solutions, Corros. Sci. 11 (1971) 153–159.

[37] C. Wang, J. Li, Z. Zhou, Y. Pan, Z. Yu, Z. Pei, S. Zhao, L. Wei, Y. Chen, Rechargeable zinc-air batteries with neutral electrolytes: Recent advances, challenges, and prospects, EnergyChem. 3 (2021) 100055.

[38] Z. Zhao, J. Huang, Z. Peng, Achilles' heel of lithium-air batteries: Lithium carbonate, Angew. Chem. Int. Ed. 57 (2018) 3874–3886.

[39] R. Younesi, G.M. Veith, P. Johansson, K. Edstrom, T. Vegge, Lithium salts for advanced lithium batteries: Li-metal, Li-O2, and Li-S, Energy Environ. Sci. 8 (2015) 1905–1922.

[40] S. Wu, K. Zhu, J. Tang, K. Liao, S. Bai, J. Yi, Y. Yamauchi, M. Ishida, H. Zhou, A long-life lithium ion oxygen battery based on commercial silicon particles as the anode, Energy Environ. Sci. 9 (2016) 3262–3271.

[41] M.L. Thomas, Y. Oda, R. Tatara, H.M. Kwon, K. Ueno, K. Dokko, M. Watanabe, Suppression of water absorption by molecular design of ionic liquid electrolyte for Li-air battery, Adv. Energy Mater. 7 (2017) 1601753.

[42] M. Asadi, B. Sayahpour, P. Abbasi, A.T. Ngo, K. Karis, J.R. Jokisaari, C. Liu, B. Narayanan, M. Gerard, P. Yasaei, X. Hu, A. Mukherjee, K.C. Lau, R.S. Assary, F. Khalili-Araghi, R.F. Klie, L.A. Curtiss, A. Salehi-Khojin, A lithium-oxygen battery with a long cycle life in an air-like atmosphere, Nature. 555 (2018) 502–507.

[43] J. Yi, X. Liu, S. Guo, K. Zhu, H. Xue, H. Zhou, Novel stable gel polymer electrolyte: Toward a high safety and long life Li-air battery, ACS Appl. Mater. Interfaces, 7 (2015) 23798–23804.

[44] Y. Liu, C. Li, B. Li, H. Song, Z. Cheng, M. Chen, P. He, H. Zhou, Germanium thin film protected lithium aluminum germanium phosphate for solid-state Li batteries, Adv. Energy Mater. 8 (2018) 1702374.

3 Electrochemical Fundamentals and Issues of Metal-Air Batteries

Shasha Li, Enze Li, Xiaowei An, and Guoqing Guan

CONTENTS

3.1 Introduction 29
3.2 Electrochemical Fundamentals of Metal-Air Batteries 31
    3.2.1 Battery Structure and Working Mechanism 31
    3.2.2 Aqueous Systems 32
    3.2.3 Nonaqueous Aprotic Systems 36
    3.2.4 Hybrid and Solid-State Systems..39
3.3 Summary and Outlook 40
Acknowledgments..41
References...41

3.1 INTRODUCTION

Because the over-exploitation and consumption of fossil fuels have caused severe global energy crisis and environmental damages, it is urgent to develop more renewable and sustainable energy conversion and storage devices such as metal-air batteries, supercapacitors, water splitting, fuel cells, and so on to solve these issues [1–4]. Among them, metal-air batteries (MABs) as a family of electrochemical cells (e.g., zinc-air, aluminum-air, lithium-air, potassium-air, sodium-air batteries, and so on) have gained great interest owing to their high theoretical energy density, based on unlimited oxygen supply from the atmosphere and that they do not have to store reactants at air cathode [5–7]. Especially, the lithium-air (Li-air) batteries exhibit the largest theoretical energy density ($\approx$ 5928 Wh/kg), which is extremely higher than the commercial Li-ion batteries ($\approx$ 400 Wh/kg) (Figure 2.1a). Meanwhile, a series of metals such as Li, sodium (Na), magnesium (Mg), aluminum (Al), zinc (Zn), and iron (Fe) are appropriate to serve as anode materials in MABs. Particularly, more and more research on Zn-air batteries and Li-air batteries has been reported [6] (as shown in Figure 2.1b). Notably, according to the different metal anode species and rechargeability, MABs can be classified into three categories including primary, secondary, and mechanically rechargeable batteries. Furthermore, based on the different types of electrolytes and cell configurations, they also can be divided into four types, which are aqueous, nonaqueous aprotic, hybrid, and solid-state MABs [8, 9].

However, Li is not an abundant element on the earth and is so active that the Li-air system inevitably faces security risks. Additionally, side reactions are always caused by Li electrode and electrolyte decomposition during the charging process [10, 11]. Na is one of the most abundant elements on the earth in comparison with Li, and Na is the second-smallest and -lightest alkali metal next to lithium. Unfortunately, the theoretical energy density of the Na-air battery is far lower than the Li-air battery [4, 12]. Furthermore, Li, Na, and K react violently with H_2O and therefore are usually operated in nonaqueous aprotic electrolytes. Other metals, such as Zn, Mg, Al, and Fe, which are more abundant and less reactive, are more feasible at present for application in MABs.

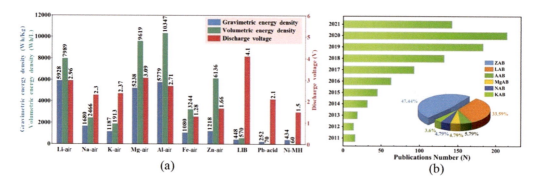

FIGURE 3.1 (a) Theoretical gravimetric energy density, volumetric energy density, and discharge voltage of various MABs; (b) number of publications that include flexible MABs according to the Web of Science (as of September 2021), and the keywords of publication retrieval are flexible M-air/O_2 batteries (M = metal, Li, Na, Mg, Al, K, Fe, or Zn). (Adapted with permission [6]. Copyright (2021), Wiley-VCH.)

The Mg-air and Al-air batteries display higher theoretical energy densities and are safer in aqueous electrolytes, but the lower reduction potentials of Mg and Al easily cause rapid self-discharge of the batteries and eventually reduce the practical energy densities [6, 13, 14]. The metals of Zn and Fe are highly abundant with low cost; thus the Zn-air and Fe-air batteries have received increasing attention. Although Fe-air batteries have long cycle lifetimes, Zn-air batteries have a higher theoretical energy density (1218 Wh/kg, 6136 Wh/L) and can be charged more efficiently in comparison with Fe-air batteries in alkaline electrolytes [15, 16]. In addition, currently, as a relatively mature technology, the primary Zn-air batteries have been used in hearing-aid devices. Therefore, with the rapid development of material science and energy technology, more types of MABs will serve as next-generation high-performance and environmental-benign power sources in the energy storage systems in the near future, such as wearable devices, mobile energy fields, aerospace industry, and so on.

For the aqueous electrolytes for MABs (commonly utilized alkaline electrolytes), the anode metal is easily oxidized to oxides, hydroxides, or other species at the anode/electrolyte interface to retard electrolyte access, thereby preventing the discharging of remaining active material [4, 5, 17]. Furthermore, some metal anodes in MABs suffer from passivation and corrosion, and as such, the discharge product or self-discharge product could increase the internal resistance, decrease the coulombic efficiency of the anode and shorten cycle life. In comparison, as alternatives to traditional aqueous electrolytes irreversible with the negative standard potential, nonaqueous electrolytes can be recharged and corrosion and passivation of metal electrodes can be largely reduced [8, 18]. Unfortunately, such systems are usually suffering from low ionic conductivity, high viscosity, and interfacial properties, thereby limiting the energy or power density. Therefore, the aqueous electrolytes are still the first choice in most cases. For the cathode (i.e., air electrode), the major issues are the intrinsic slow reactions, high overpotentials, and poor reversibility of oxygen chemistry [19, 20]. Although using oxygen from the surrounding air in the cathode is regarded as an intriguing aspect, the cathode reactions involve complex processes and the mechanism is also complicated. It should be noted that the electrochemical performance of MABs largely depends on the electrochemical characteristics of cathode materials.

In this chapter, the progress in the development of those aqueous and/or nonaqueous, hybrid, or solid-state battery systems as shown in Figure 3.2 is briefly reviewed. Also, the major issues mainly including the metal anode, the air cathode, and electrolyte will be discussed. In addition, the current challenges and outlook of MABs will be presented. It is expected to offer a better understanding of the technical issues of MABs and provide some guidance for the development of next-generation MABs, thus stimulating more novel research in the field of batteries.

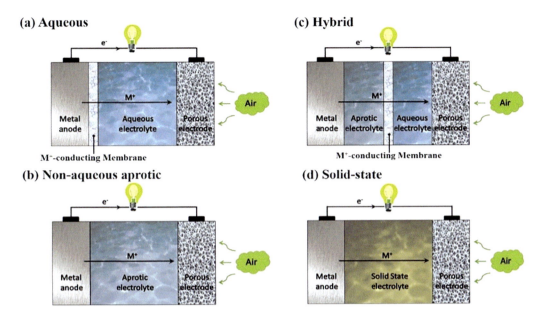

FIGURE 3.2 Schematic cell configurations for the four different types of metal-air batteries (M^+ stands for metal cations). (Adapted with permission [4]. Copyright (2017), Wiley-VCH.)

3.2 ELECTROCHEMICAL FUNDAMENTALS OF METAL-AIR BATTERIES

3.2.1 Battery Structure and Working Mechanism

Generally, the basic structure of MABs is composed of a metal anode, electrolyte, and air cathode. A separator sometimes is used in the middle to avoid short circuits between the electrodes. The basic working mechanism of MABs is the dissolution and deposition of metal at the metal anode, and oxygen reduction reaction (ORR) or/and oxygen evolution reaction (OER) occur at the air cathode. It is noteworthy that the metal anode functions as an energy storage device, which determines the output capacity while the air cathode serves as an energy converter, which determines the output power density [6, 21]. Meanwhile, the working principle of MABs is different from the traditional ionic batteries. The traditional ionic batteries involve the transformation of metallic ions from the anode to the cathode. However, for MABs, metals are oxidized at the anode and release electrons to the external circuit to transform metallic ions during the discharging, while, at the cathode oxygen diffused accepts the electrons from the anode and is reduced (referring to ORR) to oxygen-containing species. Then, the metallic ions and reduced-oxygen species migrate across the electrolyte and couple to form the discharge products of metal oxides. In the charging process, the process is reversed, where the metal cations are reduced at the anode and oxygen evolves at the cathode (referring to OER) under applied voltage [13, 14, 18]. Therefore, compared to the metal anode, the air cathode plays an important role in the performance of MABs, including permission oxygen passing, proceeding ORR/OER, and storing discharge products. Moreover, the rechargeability and energy efficiency of MABs are mainly limited by the ORR/OER electrocatalysts at the air cathode. Herein, it is significantly important to have an in-depth understanding of the mechanism of oxygen electrochemical (ORR and OER) and further promote ORR or/and OER processes and design of electrocatalysts. Furthermore, it should be mentioned that the electrochemical reactions and the overall product may differ from each other among various types of MABs based on the specific metal, the electrolyte, and the electrocatalytic materials. In the following, the basic electrochemical cathode reactions of MABs are discussed based on different electrolytes.

3.2.2 Aqueous Systems

An aqueous MABs system is commonly composed of a metal anode, a porous carbon cathode with an alkaline electrolyte (Figure 3.2a). It is well known that the advantage of aqueous MABs is that the capacity is not restricted by the capability of the cathode, since the discharge products are soluble hydroxides rather than stored in the pore structures of the cathode. This can ensure continuous gas diffusion through the porous air cathode. Therefore, the performance of an aqueous MABs system is heavily dependent on the mass transport properties of the electrolyte [4, 22, 23]. Moreover, it should be pointed out that if there is no effective air selective membrane, using an alkaline electrolyte could result in potential cell failure because of the clogging of electrode pores caused by the accumulation of poor solubility carbonates in the alkaline electrolyte.

A metal anode is also a major factor affecting the performance of MABs. Notably, metals such as Zn, Fe, Al, and Mg are thermodynamically unstable in acidic electrolytes, and they can react violently to produce H_2 with severe corrosion, which can cause short shelf-life and remarkably decline the performance of MABs. Therefore, alkaline electrolytes are usually used because metal anodes are relatively stable. Meanwhile, these metals may be passivated to form the corresponding anodic oxides or hydroxides (e.g., ZnO, $Fe(OH)_2$, $Mg(OH)_2$, or $Al(OH)_3$) in the alkaline media, leading to a serious self-discharge problem and further premature degradation of MABs. Herein, to alleviate such issues and reduce the anodic polarization to improve the Faradaic efficiency of the anode reaction, various approaches such as increasing the purities of metal anodes, preparing alloy with other metals, coating the anode surface, fabricating the porous and micro/nanostructures of an anode, or optimizing the electrolyte compositions have been considered [4, 8, 20]. For the aqueous MABs, these metals are oxidized at the anode during the discharging; meanwhile, O_2 from the surrounding air is reduced on the active sites of electrocatalysts supported on the gas-diffusion cathode as the following [17, 18]:

$$\text{Anode: } M \leftrightarrow M^{n+} + ne^- \tag{1}$$

$$\text{Cathode: } O_2 + ne^- + 2H_2O \leftrightarrow 4OH^- \tag{2}$$

where M stands for the different types of metals and n represents the oxidation number of the metal ions. These electrochemical reactions could be reversed upon recharge with metals plated on the anode and O_2 evolving at the cathode. Consequently, besides the rational design of anode materials, an in-depth investigation of oxygen electrochemistry at the air cathode should be important for the improvement of battery performance. The air cathode reaction mainly takes place at the triple-phase boundary where the solid electrode is simultaneously interfaced with liquid electrolyte and gaseous O_2.

It is noteworthy that ORR at the air cathode is the performance-limiting electrode for MABs since it is a rather sluggish reaction. Thus, an insight understanding of the mechanisms of ORR is needed for further rationally guiding the design of electrocatalysts to promote and accelerate the ORR process. The ORR includes a series of complex electrochemical reactions involving multistep electron-transfer processes with the generation of complicated oxygen-containing species such as OH^-, O_2^-, and HO_2^-. In addition, there are either four-electron or two-electron pathways for O_2 electron reduction at different pH values of electrolytes [3, 22, 24, 25].

In acidic electrolytes, the related cathode reactions can be briefly given as:

(1) A four-electron reduction to generate H_2O:

$$O_2 + 4H^+ + 4e^- \rightarrow 2H_2O \; (E^0 = 1.229 \text{ V}) \tag{3}$$

(2) A two-electron pathway reduction to form of H_2O_2 intermediates, the reactions are as follows:

$$O_2 + 2H^+ + 2e^- \rightarrow H_2O_2 \ (E^0 = 0.7 \text{ V}) \quad (4)$$

$$H_2O_2 + 2H^+ + 2e^- \rightarrow 2H_2O \ (E^0 = 1.76 \text{ V}) \quad (5)$$

In an alkaline environment, the reactions can be presented as follows:

(1) A four-electron pathway directly to produce OH^-:

$$O_2 + 2H_2O + 4e^- \rightarrow 4OH^- \ (E^0 = 0.41 \text{ V}) \quad (6)$$

(2) A two-electron pathway involving reduction to from HO_2^- intermediates:

$$O_2 + H_2O + 2e^- \rightarrow HO_2^- \ (E^0 = -0.07 \text{ V}) \quad (7)$$

$$HO_2^- + H_2O + 2e^- \rightarrow 3OH^- \ (E^0 = 0.87 \text{ V}) \quad (8)$$

From the preceding reactions, one can see that the ORR in the alkaline electrolyte is more favorable than that in acidic media owing to better kinetics and lower overpotentials. Importantly, it should be mentioned that the 2e⁻ and 4e⁻ reaction schemes may occur simultaneously and compete with each other in alkaline electrolytes. Recently literature indicates that two mechanisms of ORR comprising both direct 4e⁻ and series 2 × 2e⁻ pathways exist on the carbon-supported Pt nanoparticles [19, 25]. The direct path of 4e⁻ reduction reaction at a high potential is dominating whereas the reduction reaction at a low potential is the peroxide-mediated pathway. Herein, the direct 4e⁻ ORR is desirable owing to its high energy efficiency, and is commonly recognized that this reduction process predominates on most noble metals. The 2e⁻ ORR is undesirable since the corrosive peroxide species could be generated which can cause premature degradation of the electrochemical cell. Generally, such a 2e⁻ ORR pathway participates in those carbonaceous materials [16, 26].

In the case of oxygen electrochemistry at the surface of metal oxides, it follows the same principle, however, the surface charge distribution is different from that of metal electrocatalysts. Anion coordination is completed by the oxygen in the H_2O molecule in an aqueous solution since the surface cations of stoichiometric oxides are not fully coordinated with oxygen. Thus, the reduction of a surface cation should be charge-compensated by the protonation of a surface oxygen ligand. Accordingly, a four-step ORR pathway mechanism is proposed on the surface of those transition metal oxides, as indicated in the following equations [4, 15, 20, 27]. It includes steps of surface hydroxide displacement, surface peroxide formation, surface oxide formation, and surface hydroxide regeneration. Herein, the competition between the O_2^{2-}/OH^- displacement and OH^- regeneration is the rate-limiting for the ORR kinetics in alkaline media.

$$M^{m+}\text{—}O^{2-} + H_2O + e^- \rightarrow M^{(m-1)+}\text{—}OH^- + OH^- \quad (9)$$

$$O_2 + e^- \rightarrow O^-_{2,\text{ads}} \quad (10)$$

$$M^{(m-1)+}\text{—}OH^- + O^-_{2,\text{ads}} \rightarrow M^{m+}\text{—}O\text{—}O^{2-} + OH^- \quad (11)$$

$$M^{m+}\text{—}O\text{—}O^{2-} + H_2O + e^- \rightarrow M^{(m-1)+}\text{—}O\text{—}OH^- + OH^- \quad (12)$$

$$M^{(m-1)+}\text{—}O\text{—}OH^- + e^- \rightarrow M^{m+}\text{—}O^{2-} + OH^- \quad (13)$$

For different electrocatalysts, various ORR pathways exist, relying on special electronic structure, composition, transporting effects such as desorption and readsorption of intermediates, and the interactions between the hydrated alkali-metal cations and adsorbed species on the active sites of electrocatalysts, and so on. As has been well documented, the ORR pathways and mechanisms may vary with the electrocatalytic materials and behave as a structure-sensitive reaction, even for the same electrocatalyst [3, 12, 19]. Hence, the determination of an explicit ORR mechanism remains challenging. It has been known that a series of reduction reactions or chemical decomposition of intermediate species can proceed extremely quickly, the produced peroxide would be simultaneously reduced to hydroxide, and the overall process can be considered as an apparent 4e⁻ reduction pathway. However, the real surface reactions are still unclear owing to the limitation of analysis techniques.

As for rechargeable MABs, the air cathode should be able to efficiently electro-catalyze both ORR and its reverse reaction, i.e., OER, which occur at discharge and charge processes for the MABs, respectively. It is well known that Pt-based materials demonstrate the highest activity for ORR. However, when applied as OER electrodes, they undergo oxidation at a high positive potential, thus leading to shorter cycle life. Likewise, the most efficient electrocatalyst for the OER does not have the best electrocatalytic activity for ORR [11, 28]. Therefore, it is urgent to develop active and durable bifunctional oxygen electrocatalysts for the facilitating of both ORR and OER. Furthermore, many researchers have proposed that the bifunctional electrocatalysts for ORR and OER generally involve four elementary reaction steps, as displayed in the following equations [3, 29]:

$$* + O_2(g) + H_2O(l) + e^- \leftrightarrow HOO^* + OH^- \tag{14}$$

$$HOO^* + e^- \leftrightarrow O^* + OH^- \tag{15}$$

$$O^* + H_2O(l) + e^- \leftrightarrow HO^* + OH^- \tag{16}$$

$$HO^* + e^- \leftrightarrow OH^- + * \tag{17}$$

The overall reaction in an alkaline solution is:

$$O_2 + 2H_2O(l) + 4e^- \leftrightarrow 4OH^- \tag{18}$$

One can see that the ORR proceeds *via* diffusion followed by the adsorption of O_2 onto the electrocatalyst surface. Electrons drawn from the anode are subsequently transferred to the adsorbed oxygen with the breaking of the oxygen double bond at the three-phase boundaries among oxygen (gas), electrolyte (liquid), and active material (solid). Then, hydroxide ions are removed from the electrocatalyst surface to the alkaline electrolyte. Unlike ORR, the OER proceeds much more difficultly due to a series of complex electrochemical reactions with multi-step electron-transfer processes at the electrolyte-electrode interface. Furthermore, it should be pointed out that the electrocatalytic ORR process is mainly restricted by the O_2 reduction step (blue line) and OH^* reduction step (black line) whereas the OER is usually limited by the OOH^* (green line) and O^* formation steps (red line), as depicted in Figure 3.3. Therefore, it is difficult to develop efficient and stable bifunctional electrocatalysts for ORR/OER. The ORR and OER processes will produce O^*, OH^*, and OOH^* species, which are always bound to the electrocatalyst surface as intermediates. The relative stability of these intermediates and the activation barrier among them affect the reaction rate of the corresponding step, which finally determines the rate of ORR/OER. Additionally, the properties of these generated intermediates depend on the types of different electrocatalysts and various reaction conditions.

Among the types of aqueous Zn-, Fe-, Al-, or Mg-air batteries, Zn-air batteries have been widely investigated. The basic structure of a Zn-air battery is composed of a zinc electrode, a porous air electrode with active materials, and an alkaline electrolyte. During the discharging process, oxygen diffuses into the cathode and is reduced *via* ORR to form hydroxyl ions with the electrons released from zinc metal. Subsequently, the hydroxide ions migrate to the anode to combine with the zinc metal-generated Zn ions to form zincate ions ($Zn(OH)_4^{2-}$), which further decompose to insoluble zinc oxide (ZnO) (Figure 3.4). The electrochemical reactions of Zn-air battery with alkaline electrolytes during discharge can be described as follows [30, 31]:

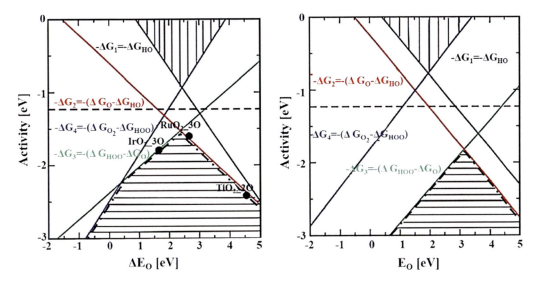

FIGURE 3.3 Early activity volcano trends of the ORR and OER. The theoretical activity on the metal oxide surface (a) and the metal surface (b) is described as a function of the oxygen binding energy. (Adapted with permission [3]. Copyright (2019), The Royal Society of Chemistry.)

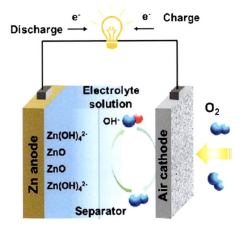

FIGURE 3.4 Operation principles of rechargeable Zn-air batteries. (Adapted with permission [14]. Copyright (2021), American Chemical Society.)

$$\text{Air electrode: } O_2 + 4e^- + 2H_2O \rightarrow 4OH^- \tag{19}$$

$$\text{Zinc electrode: } Zn \rightarrow Zn^{2+} + 2e^- \tag{20}$$

$$Zn + 4OH^- \rightarrow Zn(OH)_4^{2-} \tag{21}$$

$$Zn(OH)_4^{2-} \rightarrow ZnO + H_2O + 2OH^- \tag{22}$$

$$\text{Overall reaction: } 2Zn + O_2 \rightarrow 2ZnO \; (E_{OCV} = 1.65 \text{ V}) \tag{23}$$

During the charging process, the preceding electrochemical reactions are reversed, in which zinc is deposited at the zinc electrode and oxygen is released by the OER at the electrolyte-electrode interface as shown in the following equations [4, 32]:

$$\text{Zinc electrode: } ZnO + H_2O + 2OH^- \rightarrow Zn(OH)_4^{2-} \quad (24)$$

$$Zn(OH)_4^{2-} + 2e^- \rightarrow Zn + 4OH^- \quad (25)$$

$$\text{Air electrode: } 4OH^- \rightarrow O_2 + 2H_2O + 4e^- \quad (26)$$

$$\text{Overall reaction: } 2ZnO \rightarrow 2Zn + O_2 \quad (27)$$

The Zn-air battery has a theoretical energy density of 1353 Wh/kg with an equilibrium potential of 1.65 V. Unfortunately, to date, the practically estimated attainable energy densities of Zn-air batteries are 350–700 Wh/kg with a voltage less than 1.65 V, and thus most of them are applied for low power needed devices such as hearing aids batteries because the inadequate rate capability is restricted by the inefficiency of air electrocatalysts. Meanwhile, the hydrogen evolution reaction (HER) may occur in the case of zinc anode corrosion ($Zn + 2H_2O \rightarrow Zn(OH)_2 + H_2$), which leads to the degradation of Zn-air batteries. During the charging process, the zinc metal anode may suffer from irreversible shape change such as the formation of zinc dendrites, which could lead to short circuits of Zn-air batteries [8, 17]. Meanwhile, the generated white solid ZnO powder on the surface of the anode acts as an insulator, further resulting in the weakening of rechargeability. As a result, the rechargeable Zn-air batteries suffer from a low energy efficiency of < 60% and poor cycle life, especially owing to the lack of efficient and stable ORR/OER bifunctional electrocatalysts and cyclable metal anode. Therefore, to develop a fully rechargeable Zn-air battery, it is of vital importance to investigate in-depth the electrochemical behaviors of Zn anode and oxygen electrochemistry in alkaline electrolytes.

Besides Zn-air batteries, other aqueous MABs are confronted with similar challenges. Although both Al-air batteries and Mg-air batteries have significantly higher theoretical specific energy density (8100 and 6800 Wh/kg, respectively) than the Zn-air batteries, they can only be mechanically recharged rather than electrically recharged due to the severe corrosion of the metallic anodes in contact with the aqueous electrolyte. Nevertheless, Fe-air batteries can be electrically rechargeable, but the practical energy density (60–80 Wh/kg) is much lower, which is not comparable to the current lithium-ion batteries.

3.2.3 Nonaqueous Aprotic Systems

Compared with the aqueous systems, the nonaqueous aprotic systems have been triggering intensive interest in the last decade. Generally, a nonaqueous aprotic MABs system is composed of an anode, a porous cathode with an open structure, and an aprotic solvent dissolved with metal salt as an electrolyte (Figure 3.2b). Despite the MABs using aqueous electrolytes being featured with advantages including low cost, wide availability, and high ionic conductivity, the range of practical voltages are much lower than those theoretical voltages of metal anodes during the electrochemical processes because H_2O is stable against HER or OER. Particularly, for the highly reactive metal anode, such as Li, Na, and K metals, it is dangerous in an aqueous system, making them often use a nonaqueous aprotic electrolyte. Moreover, for the aqueous Li/Na/K-system, a special membrane or water-stable layer is required to protect the reactive metal anode, which could significantly increase the cost and complicate the fabrication of a cell [4, 5, 33, 34]. Therefore, the Li/Na/K-air batteries have been widely investigated in a nonaqueous aprotic system rather than in aqueous systems. Notably, the mechanism of ORR in the aprotic electrolyte is dramatically different from that in aqueous electrolytes as indicated in the following equations [17, 33].

$$\text{Anode: } M \leftrightarrow M^+ + e^- \quad (28)$$

$$\text{Cathode: } xM^+ + O_2 + xe^- \leftrightarrow M_xO_2 \; (x = 1 \text{ or } 2) \quad (29)$$

Unlike that in the aqueous electrolytes, in which the ORR process undergoes a four- or two-electron reduction pathway corresponding to the formation of H_2O or H_2O_2, in the nonaqueous electrolyte, the ORR involves several oxygen-containing species, which could generate insoluble discharge product deposits on the cathode surface. As such, it is reported that the ORR involves an initial one-electron reduction of O_2 on the electrocatalyst surface to produce a superoxide anion O^{2-}, and then further reacts with alkali metal cation to form peroxide MO_2 [11, 25, 30]. As for Li^+, its small size cannot stabilize the superoxide anion (O^{2-}), thus subsequent disproportionate reaction of LiO_2 occurs to form peroxide Li_2O_2 as the discharge product. In contrast, the larger cations (Na^+ and K^+), can stabilize the O^{2-} efficiently based on the hard-soft acid theory. Generally, the discharge products of Na-air batteries are usually a mixture of NaO_2 and Na_2O_2 whereas that of K-air batteries is predominantly KO_2. It should be noted that the generated superoxides or peroxides have drastically less solubility in the electrolyte which can accumulate on the air cathode, leading to the gradual blockage of the efficient cathode surface area and eventually shutting off the battery. Hence, the discharge capacity of non-aqueous metal-air batteries is strongly dependent on the storage capacity of the air cathode toward the discharge product, and as a result, the practical capacity is far lower than the theoretical value. Since the aprotic Li-air batteries demonstrate the highest theoretical energy density (5928 Wh/kg) with a high cell voltage (2.96 V), in the following, the issues of the nonaqueous aprotic Li-air batteries will be mainly discussed.

For the aprotic Li-air batteries system, during the discharging, the Li metal is oxidized at the anode electrode to form Li ions and electrons whereas O_2 dissolves into the aprotic electrolyte and then is reduced *via* the ORR process at the two-phase boundaries between electrolyte (liquid) and active material (solid) to produce superoxide ions (O_2^-). Subsequently, the superoxide ions couple with Li ions to generate lithium superoxide (LiO_2). However, the LiO_2 is unstable, which further undergoes a one electron-transfer electrochemical process and/or disproportionated reaction to form more stable Li_2O_2 (Figure 3.5) as shown in the following equations [8, 35].

$$\text{Lithium electrode: } Li \rightarrow Li^+ + e^- \tag{30}$$

$$\text{Air electrode: } O_2 + e^- \rightarrow O_2^- \tag{31}$$

$$LiO_2 + Li^+ + e^- \rightarrow Li_2O_2 \tag{32}$$

$$2LiO_2 \rightarrow Li_2O_2 + O_2 \tag{33}$$

$$\text{Total reaction: } 2Li + O_2 \rightarrow Li_2O_2 \ (E_{OCV} = 2.96 \text{ V}) \tag{34}$$

In the charging process, the electrochemical process is reversed, and the discharge products are always Li ions and released O_2 (referring as the OER process).

Firstly, although the metal anode for Li-batteries is rarely reported, it could also be a limiting factor for battery performance. The major issue for the application of Li anode is the dendrite formation, which would induce irreversible cycle performance and short-circuits for the Li-air batteries [36].

Secondly, to date, the suitable electrolytes for non-aqueous MABs are yet to be identified since the formed discharge product of superoxide or peroxide is oxidative, causing the decomposition of most common electrolytes used in conventional Li-ion batteries. Abraham *et al.* [37] first reported the rechargeable Li-air batteries in 1996; they used carbonate-based gel electrolyte to replace the aqueous electrolyte. Meanwhile, they found that the discharge reaction is $2Li + O_2 \rightarrow Li_2O_2$, which is different from the aqueous metal-air batteries in which the discharge product of Li_2O_2 is always deposited on the cathode rather than the anode. Furthermore, they found that the rechargeability of Li-air batteries failed within just three cycles since the organic carbonates easily suffer from nucleophilic attack by those superoxide or peroxide species. Some evidence has proven that the reaction between superoxide/peroxide and propylene carbonate electrolyte could result in the formation of the ring opening of the heterocycle and subsequent generation of lithium alkyl carbonates, Li_2CO_3, CO_2, and H_2O rather than Li_2O_2 as the discharge products on the cathode, which have been confirmed with the help of X-ray diffraction analysis, *in situ* surface-enhanced Raman spectroscopy (SERS), *in situ* differential electrochemical mass spectroscopy (DEMS), NMR and FTIR techniques [4, 8, 9,

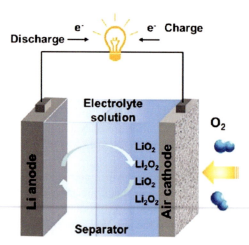

FIGURE 3.5 Operation principle of rechargeable Li-air batteries. (Adapted with permission [14]. Copyright (2021), American Chemical Society.)

20]. Herein, during the recharging process, the oxidation of these carbonate species occurs with the evolution of H_2O and CO_2. The charge/discharge pathway without the reversible Li_2O_2 formation is inconsistent with most of the previously proposed mechanisms, which could also explain the degrading of the electrode performance and the causing of poor rechargeability of Li-air batteries. In this regard, alternative electrolytes such as ether-based solvents as the electrolyte-tetraethylene glycol dimethyl ether (TEGDME) or dimethyoxyethane (DME) have been further deeply investigated due to their poor electrophilic characteristics. Meanwhile, dimethyl sulfoxide (DMSO) and acetonitrile electrolytes have also drawn many researchers' interests [17, 38]. Unfortunately, these electrolytes are insufficiently stable and are often decomposed during the battery cycling testing. Therefore, it is desirable to explore better electrolytes to realize the long-term cycling stability (>1000 cycles) for non-aqueous MABs.

Thirdly, to boost the discharge performance of Li-air batteries, the optimization of both electrolyte and cathode is necessary. It is well known that the solvent donor number (DN) greatly influences the solubility toward the reaction intermediates, thus impacting the discharge characteristic of Li-air batteries [17, 20]. Generally, those high-DN solvents could provide the increased stability of the $Li^+(solvent)_n$-O_2^-, leading to reversible O_2/O_2^- with the further formation of the Li_2O_2 on the cathode. In the solvents with low DN, O_2^- may be fully electrochemically reduced to O^{2-} or quickly decompose to O_2^{2-}. Herein, the Li-air batteries using high-DN solvents usually demonstrate higher specific capacities than those using low-DN solvents. Furthermore, the optimized cathode structure should have a suitable pore volume to accommodate the Li-air batteries for the generation of intermediates and products. Additionally, after the growth of intermediates and products on the cathode surface, it should also have a high surface area with sufficient pores size to maintain mass transfer [4]. If the pore size is too small, the pores on the air electrode could be easily blocked by the accumulation of insoluble solids, making the electrode's inner surface area inaccessible, and further limiting the discharge performance of Li-air batteries.

Forth, the reversible Li_2O_2 decomposition requires a lower overpotential (< 3.5 V) to improve the energy efficiency and cycle life of Li-air batteries. A large overpotential is generally needed to reverse the reaction and electrochemically decompose Li_2O_2 at the cathode, which can not only decrease the energy efficiency < 75% of Li-air batteries but also cause electrolyte or electrode degradation, thus leading to poor cycle life. Currently, most Li-air battery cathodes use carbon materials as the electrocatalysts or electrocatalyst supports, but their electrochemical stability insufficiency arouses concern

[26, 30]. Carbon could be oxidized at the positive potential, leading to the accumulation of Li_2CO_3 on the cathode surface, which could passivate the electrode so that it is extremely difficult to decompose. Bruce et al. [39] developed a Li-O_2 battery comprised of carbon-free nano-porous gold as the electrocatalyst and dimethyl sulfoxide as electrolyte. The capacity of this Li-O_2 battery remained 95% after 100 cycles and it is found that the battery was dominated by reversible formation/decomposition of Li_2O_2 at the cathode upon the cycling capacity. Meanwhile, the reaction relating to Li_2O_2 was reversible in the discharge and charge processes, and the kinetics of Li_2O_2 oxidation on charge is approximately 10 times faster than that on carbon electrodes. This work provided important guidance for the design of O_2 electrodes for rechargeable Li-O_2 batteries. However, the gold is not suitable for practical application. Thus, numerous efforts have been devoted to developing other low-cost electrocatalysts to facilitate the reversible Li_2O_2 formation/decomposition, which includes Co_3O_4, MnO_2, RuO_2, and recently developed single-atom-catalysts, and some of them have been proven to very effective for the Li-air batteries [16, 40–42]. Recently, it is reported that the rechargeability of Li-air batteries might also be dramatically improved if their reaction pathway could be altered. Generally, the discharge product of Li_2O_2 is thermodynamically favored for Li-air batteries, however, the pure solid LiO_2 is difficult to be synthesized due to its unstable to disproportionation with the generation of Li_2O_2. Hence, Lu and coworker [43] found that the crystalline LiO_2 can be selectively formed as the discharge product by using an iridium-graphene (Ir-rGO) composite electrocatalyst. In this case, the LiO_2 is more facilely decomposed during the recharging process owing to its half-metallic characteristics and the effect of Ir nanoparticles on the cathode surface. As a result, the Li-air batteries can be repeatedly charged and discharged with a relatively low charge potential (about 3.2 volts) and great cycle life. Therefore, the development of low-cost, highly efficient, and stable bifunctional electrocatalysts is required to enhance kinetics, lower the overpotential during discharge and charge processes, and improve rate capability and cycling performances.

Besides, there are other challenges for Li-air batteries. For example, they are greatly sensitive to moisture and CO_2, and usually have to be tested in pure O_2 rather than ambient air, and thus literature calling them the "Li-O_2" batteries. The aforementioned problems are not specific to Li-air batteries. They are the common issues for other non-aqueous metal-air batteries including Na-air and K-air batteries.

3.2.4 Hybrid and Solid-State Systems

Hybrid and solid-state MABs are rarely reported. To date, appropriate electrolyte and membrane materials are still lacking for them. They also display inferior battery performances in comparison with the aqueous and nonaqueous aprotic MABs. In the hybrid MABs, two distinct electrolytes are used, where the nonaqueous electrolyte is used at the anolyte, which could protect the metal anode while an aqueous electrolyte is applied at the cathodic area, and separated by an ionic conductive membrane to prevent water and gas diffusions during the transferring of cations (Figure 3.2c) [4, 44]. The hybrid MABs also have some advantages, including the sharing of the same oxygen electrochemistry with an aqueous system, no need to concern about the corrosion of metal anode with the aqueous electrolyte, and the clogging of air cathode with nonaqueous aprotic systems. To date, the excellent conductivity and long-term stability of the membrane materials remain a great challenge for the application of hybrid MABs.

Recently, the solid-state MABs using solid electrolytes have also attracted rapid attention. The reaction mechanism of solid-state MABs (Figure 3.2d) is similar to that of the aprotic ones. In such MABs, the ionically conductive solid electrolyte can also serve as a separator. The applicable electrolytes include organic solid polymers, inorganic ceramics, and organic-inorganic hybrid/composite materials [45, 46]. Generally, thermal properties, chemical and electrochemical stability, and mechanical strength of solid-state electrolytes are better than those of liquid ones. Besides, compared to the liquid electrolyte, the solid-state electrolyte allows much stable ion transport, thus resulting in high performance and stable cycling of the high-energy MABs.

For example, Zhou et al. [47] proposed Li-air batteries consisting of a Li anode, a solid Li-ion conductive electrolyte, and a gel cathode, which sustained repeated cycling in ambient air for 100 cycles (~ 78 days) with a discharge capacity of 2000 mAh/g. Guo et al. [48] developed the solid-state Li-air batteries using garnet (i.e., $Li_{6.4}La_3Zr_{1.4}Ta_{0.6}O_{12}$, LLZTO) ceramic disks with high density and high ionic conductivity as the electrolytes, which showed high discharge capacity in real air at medium temperatures. However, the poor ionic conductivity, bad interfacial properties, and low mass transportation kinetics of the electrolyte materials are the major obstacles for the application of solid-state MABs.

3.3 SUMMARY AND OUTLOOK

Theoretically, MABs possess some advantages over other batteries such as Li-ion, lead-acid, and Ni-HM batteries. Especially, Li-air batteries could provide higher specific energy. Notably, the semi-open structure of MABs can effectively reduce the battery weight, and therefore further lead to a high gravimetric energy density. With the full exploitation of the electrochemical performance of various MABs, it is believed that they could serve as next-generation power sources for large-scale energy storage, aerospace industry, and mobile energy. In this chapter, the recent advances in aqueous and/or nonaqueous MABs (especially, Zn-air and/or Li-air batteries) and hybrid and solid-state MABs are reviewed and discussed. Several technical issues for MABs should be solved before their large-scale applications.

The aqueous MABs, such as Zn-air, Al-air, and Mg-air batteries, mainly suffer from metal anode corrosion or passivation and sluggish electrochemical reactions at air cathodes, leading to inferior rate performance and poor charge-discharge cycle stability. Herein, it is urgent to improve the purity of anode materials, use the techniques such as alloying for the enhancement of corrosion resistance, and explore active and stable bifunctional ORR/OER electrocatalysts for the improvement of rechargeability. For nonaqueous MABs, especially for the Li-air batteries, (i) dendrite formation could induce poor cycle performance or even short-circuit risk, which may be addressed by using a robust metallic composite electrode or a stable protective layer on the electrode surface; (ii) the fundamental understanding of the reaction mechanism of the ORR and OER electrochemical processes in nonaqueous aprotic electrolytes should be fully understood, in which it is necessary to clearly study the oxygen reaction mechanism (e.g., electron pathway in the presence or absence of oxygen-containing species), which is prerequisite to guide the design of highly active and stable air electrodes to increase kinetics and lower the overpotentials during the recharging process and improve rate capability and cycling life; (iii) the instability of electrolytes (including organic solvents and salts) and carbon cathodes electrode can lead to a series of parasitic reactions rather than the formation/decomposition of expected discharge products, which could ultimately reduce the cycling and capacity performances of metal-air batteries, and thus it is necessary to rationally design the structure of air cathode for the realization of the full utilizing porous nanostructure, and thus increasing specific capacity and the electrochemical reaction kinetics of nonaqueous MABs. Additionally, for the hybrid and solid-state MABs, the poor conductivity and sluggish mass transportation kinetics of electrolyte materials should be solved.

Hence, to make further breakthroughs in the future research of MABs, some challenges as indicated in the following need to be addressed:

(i) Research on oxygen electrochemistry and electrocatalytic mechanism at the air cathode is the major route toward the high-performance MABs. Especially, the combination of theoretical calculation and *in situ* experimental technologies should be applied to deepen the mechanistic understanding of the oxygen electrochemistry in different types of MABs, and simultaneously, rational design of efficient and stable oxygen electrocatalysts is also important.

(ii) The low rechargeability of present MABs always leads to low round-trip efficiency as well as coulombic efficiency with the highly energy-consuming and quick capacity loss, which could be addressed by improving metal electrode reversibility and suppressing side reactions. Moreover, for the practical applications of MABs, the issues for the whole cell system including metal dendrite growth, electrolyte instability, impure gas CO_2 influence, the flexible current collector and encapsulating materials, and so on should be systematically considered.

(iii) It is well known that the MABs theoretically have significant advantages in energy density, but also show great disadvantages relating to rechargeability. As such, besides developing advanced materials to improve battery performance, it is also crucial for finding suitable techniques to realize high-capacity energy storage.

(iv) Currently, many researchers concentrate on the material design for the MABs. In fact, it is also important to focus on the technical designs of the cell configuration; for instance, the flow batteries with circulating electrolytes have been considered. Thus, a proper cell configuration is expected to take full advantage of the rationally designed materials for the MABs.

ACKNOWLEDGMENTS

This work is supported by the National Nature Science Foundation of China (No. 22278251), the Ph.D. Scientific Research Foundation of Taiyuan University of Science and Technology P. R. China (No. 20192052), a research project supported by the Shanxi Scholarship Council of China (No. HGKY2019086 and HGKY2019005) and the National Nature Science Foundation of China (21908137).

REFERENCES

[1] T.W. Chen, G. Anushya, S.M. Chen, P. Kalimuthu, V. Mariyappan, P. Gajendran, R. Ramachandran, Recent advances in nanoscale based electrocatalysts for metal-air battery, fuel cell and water-splitting applications: An overview, Materials. 15 (2022) 458.

[2] L. Du, L. Xing, G. Zhang, S. Sun, Metal-organic framework derived carbon materials for electrocatalytic oxygen reactions: recent progress and future perspectives, Carbon. 156 (2020) 77–92.

[3] S. Li, X. Hao, A. Abudula, G. Guan, Nanostructured co-based bifunctional electrocatalysts for energy conversion and storage: current status and perspectives, J. Mater. Chem. A. 7 (2019) 18674–18707.

[4] L. Li, Z.W. Chang, X.B. Zhang, Recent progress on the development of metal-air batteries, Adv. Sustain. Syst. 1 (2017) 1700036.

[5] M. Li, X. Bi, R. Wang, Y. Li, G. Jiang, L. Li, C. Zhong, Z. Chen, J. Lu, Relating catalysis between fuel cell and metal-air batteries, Matter. 2 (2020) 32–49.

[6] T. Li, X. Peng, P. Cui, G. Shi, W. Yang, Z. Chen, Y. Huang, Y. Chen, J. Peng, R. Zou, X. Zeng, J. Yu, J. Gan, Z. Mu, Y. Chen, J. Zeng, J. Liu, Y. Yang, Y. Wei, J. Lu, Recent progress and future perspectives of flexible metal-air batteries, SmartMat. 2 (2021) 519–553.

[7] D.U. Lee, P. Xu, Z.P. Cano, A.G. Kashkooli, M.G. Park, Z. Chen, Recent progress and perspectives on bifunctional oxygen electrocatalysts for advanced rechargeable metal-air batteries, J. Mater. Chem A. 4 (2016) 7107–7134.

[8] P. Tan, B. Chen, H. Xu, H. Zhang, W. Cai, M. Ni, M. Liu, Z. Shao, Flexible Zn- and Li-air batteries: Recent advances, challenges, and future perspectives, Energy Environ. Sci. 10 (2017) 2056–2080.

[9] Y. Sun, X. Liu, Y. Jiang, J. Li, J. Ding, W. Hu, C. Zhong, Recent advances and challenges in divalent and multivalent metal electrodes for metal-air batteries, J. Mater. Chem. A. 7 (2019) 18183–18208.

[10] C. Wang, Y. Yu, J. Niu, Y. Liu, D. Bridges, X. Liu, J. Pooran, Y. Zhang, A. Hu, Recent progress of metal-air batteries-a mini review, Appl. Sci. 9 (2019) 2787.

[11] H.F. Wang, Q. Xu, Materials design for rechargeable metal-air batteries, Matter. 1 (2019) 565–595.

[12] X. Chen, I. Ali, L. Song, P. Song, Y. Zhang, S. Maria, S. Nazmus, W. Yang, H.N. Dhakal, H. Li, M. Sain, S. Ramakrishna. A review on recent advancement of nano-structured-fiber-based metal-air batteries and future perspective, Renew. Sust. Energy Rev. 134 (2020) 110085.

[13] X. Peng, T. Li, L. Zhong, J. Lu, Flexible metal-air batteries: An overview, SmartMat. 2 (2021) 123–126.
[14] Y. Wang, F. Chu, J. Zeng, Q. Wang, T. Naren, Y. Li, Y. Cheng, Y. Lei, F. Wu, Single atom catalysts for fuel cells and rechargeable batteries: Principles, advances, and opportunities, ACS Nano. 15 (2021) 210–239.
[15] A.K. Worku, D.W. Ayele, N.G. Habtu, Recent advances and future perspectives in engineering of bifunctional electrocatalysts for rechargeable zinc-air batteries, Mater. Today Adv. 9 (2021) 100116.
[16] Q. Xia, Y. Zhai, L. Zhao, J. Wang, D. Li, L. Zhang, J. Zhang, Carbon-supported single-atom catalysts for advanced rechargeable metal-air batteries, Energy Mater. 2 (2022) 200015.
[17] Y. Li, J. Lu, Metal-air batteries: Will they be future electrochemical energy storage of choice?, ACS Energy Lett. 2 (2017) 1370–1377.
[18] D. Zhang, H. Zhao, F. Liang, W. Ma, Y. Lei, Nanostructured arrays for metal-ion battery and metal-air battery applications, J. Power Sources. 493 (2021) 229722.
[19] X. Cai, L. Lai, J. Lin, Z. Shen, Recent advances in air electrodes for Zn-air batteries: Electrocatalysis and structural design, Mater. Horiz. 4 (2017) 945–976.
[20] F. Cheng, J. Chen, Metal-air batteries: From oxygen reduction electrochemistry to cathode catalysts, Chem. Soc. Rev. 41 (2012) 2172–2192.
[21] Q. Liu, Z. Chang, Z. Li, X. Zhang, Flexible metal-air batteries: Progress, challenges, and perspectives, Small Methods. 2 (2018) 1700231.
[22] Q. Liu, Z. Pan, E. Wang, L. An, G. Sun, Aqueous metal-air batteries: Fundamentals and applications, Energy Storage Mater. 27 (2020) 478–505.
[23] S. Chen, M. Zhang, P. Zou, B. Sun, S. Tao, Historical development and novel concepts on electrolytes for aqueous rechargeable batteries, Energy Environ. Sci. 15 (2022) 1805–1839.
[24] Y.L. Zhang, K. Goh, L. Zhao, X.L. Sui, X.F. Gong, J.J. Cai, Q.Y. Zhou, H.D. Zhang, L. Li, F.R. Kong, D., M. Gu, Z.B. Wang, Advanced non-noble materials in bifunctional catalysts for ORR and OER toward aqueous metal-air batteries, Nanoscale. 12 (2020) 21534–21559.
[25] F. Wang, X. Li, Y. Xie, Q. Lai, J. Tan, Effects of porous structure on oxygen mass transfer in air cathodes of nonaqueous metal-air batteries: A mini-review, ACS Appl. Energy Mater. 5 (2022) 5473–5483.
[26] Z. Zhao, M. Li, L. Zhang, L. Dai, Z. Xia, Design principles for heteroatom-doped carbon nanomaterials as highly efficient catalysts for fuel cells and metal-air batteries, Adv. Mater. 27 (2015) 6834–6840.
[27] H. Huang, D. Yu, F. Hu, S.C. Huang, J. Song, H.Y. Chen, L.L. Li, S. Peng, Clusters induced electron redistribution to tune oxygen reduction activity of transition metal single-atom for metal-air batteries, Angew. Chem. Int. Ed. Engl. 61 (2022) e202116068.
[28] S. Ghosh, R.N. Basu, Multifunctional nanostructured electrocatalysts for energy conversion and storage: Current status and perspectives, Nanoscale. 10 (2018) 11241–11280.
[29] Z.F. Huang, J. Wang, Y. Peng, C.Y. Jung, A. Fisher, X. Wang, Design of efficient bifunctional oxygen reduction/evolution electrocatalyst: Recent advances and perspectives, Adv. Energy Mater. 7 (2017) 1700544.
[30] Q. Zhang, C. Wang, Z. Xie, Z. Zhou, Defective/doped graphene-based materials as cathodes for metal-air batteries, Energy Environ. Mater. 0 (2022) 1–14.
[31] Y. Zhang, Y.P. Deng, J. Wang, Y. Jiang, G. Cui, L. Shui, A. Yu, X. Wang, Z. Chen, Recent progress on flexible Zn-air batteries, Energy Storage Mater. 35 (2021) 538–549.
[32] K.W. Leong, Y. Wang, M. Ni, W. Pan, S. Luo, D.Y.C. Leung, Rechargeable Zn-air batteries: Recent trends and future perspectives, Renew. Sust. Energy Rev. 154 (2022) 111771.
[33] W. Li, C. Han, K. Zhang, S. Chou, S. Dou, Strategies for boosting carbon electrocatalysts for the oxygen reduction reaction in non-aqueous metal-air battery systems, J. Mater. Chem. A. 9 (2021) 6671–6693.
[34] X. Zou, Q. Lu, K. Liao, Z. Shao, Towards practically accessible aprotic Li-air batteries: Progress and challenges related to oxygen-permeable membranes and cathodes, Energy Storage Mater. 45 (2022) 869–902.
[35] P. Zhang, Y. Zhao, X. Zhang, Functional and stability orientation synthesis of materials and structures in aprotic Li-O_2 batteries, Chem. Soc. Rev. 47 (2018) 2921–3004.
[36] C. Zhou, K. Lu, S. Zhou, Y. Liu, W. Fang, Y. Hou, J. Ye, L. Fu, Y. Chen, L. Liu, Y. Wu, Strategies toward anode stabilization in nonaqueous alkali metal-oxygen batteries, Chem Commun (Camb). 58 (2022) 8014–8024.
[37] K. Abraham, Z. Jiang, A polymer electrolyte-based rechargeable lithium/oxygen battery, J. Electrochem. Soc. 143 (1996) 1–5.

[38] M. Salado, E. Lizundia, Advances, challenges, and environmental impacts in metal-air battery electrolytes, Mater. Today Energy. 28 (2022) 101064.

[39] Z. Peng, S.A. Freunberger, Y. Chen, P.G. Bruce, A reversible and higher-rate Li-O_2 battery, Science. 337 (2012) 563–566.

[40] L. Grande, E. Paillard, J. Hassoun, J.B. Park, Y.J. Lee, Y.K. Sun, S. Passerini, B. Scrosati, The lithium/air battery: Still an emerging system or a practical reality? Adv. Mater. 27 (2015) 784–800.

[41] L. Liu, H. Guo, L. Fu, S. Chou, S. Thiele, Y. Wu, J. Wang, Critical advances in ambient air operation of nonaqueous rechargeable Li-air batteries, Small. 17 (2021) e1903854.

[42] B. Liu, Y. Sun, L. Liu, S. Xu, X. Yan, Advances in manganese-based oxides cathodic electrocatalysts for Li-air batteries, Adv. Funct. Mater. 28 (2018) 1704973.

[43] J. Lu, Y.J. Lee, X. Luo, K.C. Lau, M. Asadi, H.H. Wang, S. Brombosz, J. Wen, D. Zhai, Z. Chen, D.J. Miller, Y.S. Jeong, J.B. Park, Z.Z. Fang, B. Kumar, A. Salehi-Khojin, Y.K. Sun, L.A. Curtiss, K. Amine, A lithium-oxygen battery based on lithium superoxide, Nature. 529 (2016) 377–382.

[44] X. Xu, K.S. Hui, D.A. Dinh, K.N. Hui, H. Wang, Recent advances in hybrid sodium-air batteries, Mater. Horiz. 6 (2019) 1306–1335.

[45] L. Fan, S. Wei, S. Li, Q. Li, Y. Lu, Recent progress of the solid-state electrolytes for high-energy metal-based batteries, Adv. Energy Mater. 8 (2018) 1702657.

[46] J. Yi, S. Guo, P. He, H. Zhou, Status and prospects of polymer electrolytes for solid-state Li-O_2 (air) batteries, Energy Environ. Sci. 10 (2017) 860–884.

[47] T. Zhang, H. Zhou, A reversible long-life lithium-air battery in ambient air, Nat. Commun. 4 (2013) 1817.

[48] J. Sun, N. Zhao, Y. Li, X. Guo, X. Feng, X. Liu, Z. Liu, G. Cui, H. Zheng, L. Gu, H. Li, A rechargeable Li-air fuel cell battery based on garnet solid electrolytes, Sci. Rep. 7 (2017) 41217.

4 Mathematical Modeling for Enhanced Electrochemical Properties

Yuhui Tian, Shanqing Zhang, and Yun Wang

CONTENTS

4.1 Introduction 45
4.2 Methods and Models 46
    4.2.1 Adsorption Model and Sabatier Principle 46
    4.2.2 Computational Hydrogen Electrode Model 47
    4.2.3 *d*-band Center Theory 50
    4.2.4 Aprotic MABs 51
4.3 Applications of the DFT-Based Mathematical Modeling 53
    4.3.1 Atomistic Understanding 53
        4.3.1.1 Identification of Active Sites 53
        4.3.1.2 Reaction Mechanisms 54
        4.3.1.3 Solvent Effect 56
    4.3.2 Design of Electrocatalysts 56
        4.3.2.1 Alloying 57
        4.3.2.2 Multisite Functionalization 57
        4.3.2.3 Heteroatom-Doping 58
    4.3.3 Machine Learning-Based High-Throughput Screening 59
4.4 Conclusions and Outlook 60
Acknowledgments 60
References 61

4.1 INTRODUCTION

With a growing consensus for reducing carbon emissions to establish low-carbon economies, the modern energy structure needs to transition to a cleaner and greener one. Tremendous efforts and investments have been devoted to clean energy technologies to enable reliable and sustainable energy supplies, including electrochemical energy storage and conversion. Due to their high theoretical energy densities and zero carbon emission, metal-air batteries (MABs) have been considered a promising option for electric vehicles and grid storage [1]. As an electrochemical power source, MABs can generate electricity via the redox reaction between O_2 from ambient air and metals, such as Li, Na, Zn, Mg, Al, etc. [1]. However, the practical performances of MABs are significantly influenced by the electrocatalytic oxygen reduction reaction (ORR) during discharging and oxygen evolution reaction (OER) during charging occurring on the cathode. The sluggish kinetics of both ORR and OER leads to large charge/discharge polarization, low energy efficiency, and poor cycling stability [2]. To address these crucial issues, high-performance electrocatalysts for reducing reaction energy barriers and expediting cathode redox kinetics become a prerequisite for realizing the large-scale application of MABs.

DOI: 10.1201/9781003295761-4

In the past decade, a wide range of functional materials has been investigated as ORR/OER electrocatalysts, including noble metals and their alloys, transition metal oxides/chalcogenides/carbides, organometallic complexes, carbon-based materials, and single-atoms catalysts (SACs) [3]. The catalyst design has been predominantly conducted by trial-and-error approaches because it is challenging to gain a comprehensive insight into the reaction mechanism of the complicated electrochemical process at the atomic level. Thus, scientists tried to use mathematical modeling as an alternative strategy to reveal the structure-property-performance relationship of catalyst materials, which can further be applied to guide the rational design of catalysts with the desired electrocatalytic properties.

Mathematical modeling approaches for computational electrochemistry can be multi-scale, including density functional theory (DFT) calculations, classical molecular dynamics (MD) simulations, and microkinetic modeling. However, the classical MD and microkinetic modeling are dependent on the availability and quality of force fields to determine the interaction between atoms. Unfortunately, only a very few force fields are available to simulate the ORR/OER systems. As a result, the most widely used modeling method is the DFT, which is based on the Hohenberg–Kohn theorem to compute the structural, electronic, and mechanical properties of materials in terms of the ground-state electron density [4]. Attributed to the advancement in computer science and well-developed approximation of exchange-correlation (XC) functionals, the energetic information, i.e., adsorption energies and reaction barriers of heterogeneous catalysis, can be estimated quantitatively with good chemical accuracy (2–3 kcal/mol). Consequently, the surface reactions on catalysts can be reliably investigated to reveal their mechanisms, identify the active sites, and determine the key factors of reaction efficiency. Such insights at the atomic level can be validated via comparison with the available experimental data. The confirmed mechanisms and principles of electrocatalytic reactions, in turn, help to guide the catalyst design and optimization in experiments. In this regard, DFT simulations bridge the catalyst structures and underlying fundamentals in electrocatalysis and have become one of the most important mathematical simulation tools to accelerate the commercialization of MABs.

This chapter introduces DFT-based mathematical modeling strategies for the rational design of MAB electrocatalysts, including fundamentals, calculation methods, theories, and principles. The role of theoretical investigations in identifying active sites and understanding the reaction mechanisms and activity origin is highlighted. Some recent cases are utilized to show a physical picture of advanced theoretical modeling for enhanced electrochemical properties of ORR/OER electrocatalysts in MABs. Finally, some future perspectives for this field are outlined.

4.2 METHODS AND MODELS

4.2.1 Adsorption Model and Sabatier Principle

The adsorption energy of reactants/intermediates/products on electrocatalyst surfaces is an important property for reflecting the electrochemical reactivity of the catalyst [5]. In general, weak adsorption will limit the electron transfer from the surface to the adsorbed species. The activation of reactants will become the rate/potential determination step. However, too strong adsorption will lead to sluggish desorption of products, which blocks the active site for further electrochemical processes. The ideal electrocatalysts should exhibit appropriate adsorption energies (neither too strong nor too weak) to achieve the optimal catalytic activity [6]. The function of the adsorption strength between catalysts and adsorbates to their catalytic activity is visualized in a volcano-shaped plot, as shown in Figure 4.1a. This volcano-shaped relationship is defined as the Sabatier principle, which is widely used to explain the performance of electrocatalysts [7].

The adsorption energy (ΔE_{ads}) can be calculated as follows:

$$\Delta E_{ads} = \Delta E_{total} - \Delta E_{adsorbate} - \Delta E_{substrate} \tag{1}$$

where $E_{adsorbate}$, $E_{substrate}$, and E_{total} represent the total energy of an isolated adsorbate molecule, bare substrate surface, and substrate surface with an adsorbate, respectively. Based on this definition,

Mathematical Modeling for Enhanced Electrochemical Properties

the more negative ΔE_{ads} represents stronger adsorption of the adsorbate on the catalyst surface. DFT methods can describe adsorption energies on the catalyst surface and the adsorbate-adsorbate interactions with the desired chemical accurate and relatively low cost [8]. The calculated ΔE_{ads} provides quantitative descriptions for surface reactions and reactivity at the fundamental level.

4.2.2 Computational Hydrogen Electrode Model

The computational hydrogen electrode (CHE) approach developed by Nørskov and co-workers represents an affordable way to model the electrochemical process involving multi-electron transfer processes [9]. In this model, the overall electrocatalysis reaction process is divided into several proton-coupled electron transfer (PCET) steps with the formation/breakage of key reaction intermediates on the catalyst surface. The chemical potential (μ) of the H$^+$/e$^-$ pair is referenced to half of gaseous H$_2$ at the standard condition (T = 298.15 K, H$^+$ activity = 1, and H$_2$ pressure = 1 bar). The electrode potential, U relative to the standard hydrogen electrode (SHE), is taken into account by shifting the electron by $-eU$ when an electron is transferred:

$$\mu(H^+) + \mu(e^-) = 1/2\, \mu(H_2) - eU \qquad (2)$$

Using the CHE method, the Gibbs free energy change of the adsorption can be computed using the adsorption energy via thermal correction with zero-point energy and entropy as follows:

$$\Delta G_{ads} = \Delta E_{ads} + \Delta ZPE + T\Delta S + \Delta G_U + \Delta G_{pH} + \Delta G_{field} \qquad (3)$$

$$\Delta G_U = -neU \qquad (4)$$

where ΔZPE is the zero-point energy of adsorbate or free molecules calculated by the vibrational frequency of adsorbed species; T is the temperature, and ΔS is the entropy change. U in equation (4) is the electrode applied potential *vs.* SHE; e is the transferred charge, and n is the number of proton-electron transferred pairs. ΔG_{pH} is the correction of the H$^+$ free energy based on the pH value of the electrolyte and can be expressed as:

$$\Delta G_{pH} = -k_B T ln\left[H^+\right] = pH * k_B T ln10 \qquad (5)$$

where k_B is the Boltzmann constant. ΔG_{field} originates from the electrical double layer effect and is normally ignored due to its small value. For surface adsorbates, only vibration modes of the adsorbates are considered for thermodynamic correction, assuming that changes in the vibrations of the catalyst surface caused by the presence of the adsorbate are minimal [10]. For the electrochemical ORR/OER process in the aqueous solution, there are mainly two possible reaction pathways, namely the associative pathway and dissociative pathway, as shown next:

Pathways	Acidic solution	Alkaline solution
Associative Pathway [4e$^-$]	O_2 (g) + * + H$^+$ + e$^-$ ↔ OOH*	O_2(g)+*+H$_2$O+e$^-$↔OOH*+OH$^-$
	OOH* + H$^+$ + e$^-$ ↔ O* + H$_2$O	OOH* + e$^-$ ↔ O* + OH$^-$
	O* + H$^+$ + e$^-$ ↔ OH*	O* + H$_2$O + e$^-$ ↔ OH* + OH$^-$
	OH* + H$^+$ + e$^-$ ↔ * + H$_2$O	OH* + e$^-$ ↔ * + OH$^-$
Dissociative Pathway [2e$^-$ + 2e$^-$]	O_2 (g) + 2* ↔ O* + O*	O_2 (g) + 2* ↔ O* + O*
	O* + H$^+$ + e$^-$ ↔ OH*	O* + H$_2$O + e$^-$ ↔ OH* + OH$^-$
	OH* + H$^+$ + e$^-$ ↔ * + H$_2$O	OH* + e$^-$ ↔ * + OH$^-$

Here, * represents the active site on the catalyst surface for adsorption. OOH*, O*, and OH* are reaction intermediates during electrocatalytic ORR/OER. In alkaline solutions, H_2O is the proton source for forming the H^+/e^- pair. Due to the limitation of the DFT, the accurate calculation of the total energy of O_2 is still difficult. Thus, the free energy of the O_2 molecule is usually calculated from the total energies of H_2 and H_2O via the reaction $O_2 + 2H_2 = 2H_2O$, given the free energy change of the reaction is 4.92 eV. The adsorption free energies of reaction intermediates (ΔG_{OOH^*}, ΔG_{O^*}, and ΔG_{OH^*}) can be calculated from the free energy of stoichiometrically appropriate amounts of H_2O (g) and H_2 (g) as follows:

$$\Delta G_{OOH^*} = \Delta G(2H_2O\ (g) + * \rightarrow OOH^* + \frac{3}{2} H_2\ (g))$$
$$= \mu_{OOH^*} + 1.5 \times \mu_{H_2} - 2 \times \mu_{H_2O} - \mu_*$$
$$= (E_{OOH^*} + 1.5 \times E_{H_2} - 2 \times E_{H_2O} - E^*) + (E_{ZPE(OOH^*)} + 1.5 \times E_{ZPE(H_2)} - 2 \times E_{ZPE(H_2O)} - E_{ZPE(*)})$$
$$- T \times (S_{OOH^*} + 1.5 \times S_{H_2} - 2 \times S_{H_2O} - S_*) \quad (6)$$

$$\Delta G_{O^*} = \Delta G(H_2O\ (g) + * \rightarrow O^* + H_2\ (g))$$
$$= \mu_{O^*} + \mu_{H_2} - \mu_{H_2O} - \mu_*$$
$$= (E_{O^*} + E_{H_2} - E_{H_2O} - E^*) + (E_{ZPE(O^*)} + E_{ZPE(H_2)} - E_{ZPE(H_2O)} - E_{ZPE(*)})$$
$$- T \times (S_{O^*} + S_{H_2} - S_{H_2O} - S_*) \quad (7)$$

$$\Delta G_{OH^*} = \Delta G(H_2O\ (g) + * \rightarrow OH^* + \frac{1}{2} H_2\ (g))$$
$$= \mu_{OH^*} + 0.5 \times \mu_{H_2} - \mu_{H_2O} - \mu_*$$
$$= (E_{OH^*} + 0.5 \times E_{H_2} - E_{H_2O} - E^*) + (E_{ZPE(OH^*)} + 0.5 \times E_{ZPE(H_2)} - E_{ZPE(H_2O)} - E_{ZPE(*)})$$
$$- T \times (S_{OH^*} + 0.5 \times S_{H_2} - S_{H_2O} - S_*) \quad (8)$$

Thus, the Gibbs free energy changes of each elementary step for ORR/OER (equations 9–12) can be calculated based on the adsorption free energy changes of OH*, O*, and OOH* intermediates as follows:

$$\Delta G_1^{ORR/OER} = 4 \times 1.23 - \Delta G_{OOH^*} \quad (9)$$

$$\Delta G_2^{ORR/OER} = \Delta G_{OOH^*} - \Delta G_{O^*} \quad (10)$$

$$\Delta G_3^{ORR/OER} = \Delta G_{O^*} - \Delta G_{OH^*} \quad (11)$$

$$\Delta G_4^{ORR/OER} = \Delta G_{OH^*} \quad (12)$$

Figure 4.1b shows a typical example of the ORR free energy diagram on the Pt(111) surface with the associative pathway. At $U = 0$ V, all the reaction steps are downhill. At the electrode potential of 0.75 V, the adsorption free energy of each intermediate is shifted by $-n \times 0.75$ based on the corresponding electron transfer number (n) [11]. At the equilibrium potential (U_{eq}) of 1.23 V vs. SHE, the formation of OOH* and OH*, and reduction of OH* to H_2O become uphill, thus potentially determining the ORR rate. The thermodynamic limiting potential (U_L) is calculated as follows:

$$U_L^{ORR} = \frac{\min[|\Delta G_1|, |\Delta G_2|, |\Delta G_3|, |\Delta G_4|]}{e} \qquad (13)$$

and estimated to be 0.8 V for Pt(111). Based on the definition of theoretical overpotential,

$$\eta^{ORR} = 1.23 \text{ V} - U_L \qquad (14)$$

the η^{ORR} for Pt(111) is calculated to be 1.23 − 0.8 V = 0.43 V.

Likewise, the OER with the evolution of the O_2 molecule from H_2O can be regarded as the inverse process of ORR, involving the same electron transfer process and intermediates. The maximum value of U_L^{OER}, will define the limiting potential for the full reaction:

$$U_L^{OER} = \frac{\max[|\Delta G_1|, |\Delta G_2|, |\Delta G_3|, |\Delta G_4|]}{e} \qquad (15)$$

$$\eta^{OER} = U_L - 1.23 \text{ V} \qquad (16)$$

For a thermodynamically ideal ORR/OER catalyst, ΔG for each elementary step should be the same (4.92 eV/4 = 1.23 eV) at the zero potential. Thus, the overall free energy diagram is completely flat at $U = 1.23$ V with zero overpotential [12]. Therefore, the computational design of efficient catalysts aims to adjust the free energy of each step close to 1.23 eV with the optimal catalytic activity.

However, it has been found that the adsorption free energies of structurally similar intermediates, such as OH* and OOH*, linearly correlate with each other over a variety of catalyst materials. For ORR/OER, ΔG_{OOH^*} and ΔG_{OH^*} were found to approximately obey the scaling relation as $\Delta G_{OOH^*} = \Delta G_{OH^*} + 3.2 \pm 0.2$ eV (Figure 4.1c) [11, 13]. It should be noted that 3.2 refers to the most often reported scaling intercept, and deviations from this value have been reported in the literature [14, 15]. Due to such scaling relations, adsorption energies of most adsorbed species can be correlated to one or two key intermediates. As a result, the complex multistep catalytic reaction can be simplified and understood by the defined key descriptors.

Take OER as an example, ΔG_2^{OER} ($\Delta G_{O^*} - \Delta G_{OH^*}$) and ΔG_3^{OER} ($\Delta G_{OOH^*} - \Delta G_{O^*}$) can be expressed as $\Delta G_2^{OER} + \Delta G_3^{OER} = 3.2 \pm 0.2$. Thus, the adsorption energy of O* determines the variation of OER overpotential (η^{OER}). When the step OH* → O* + H+ + e− or H_2O + O* → OOH* + H+ + e− is the potential rate-determining step (RDS), η^{OER} can be calculated as follows:

$$\eta^{OER} = \frac{\max[(\Delta G_{O^*} - \Delta G_{OH^*}), 3.2 \pm 0.2 - (\Delta G_{O^*} - \Delta G_{OH^*})] - 1.23 \text{ V}}{e} \qquad (17)$$

Consequently, a universal volcano plot can be established as a function of the $\Delta G_{O^*} - \Delta G_{OH^*}$ to understand and screen the OER activity trend of various catalysts [16]. As shown in Figure 4.1d, on the left side of the volcano, the reaction kinetics is limited by the formation of OOH* due to the strong O* binding. Oppositely, the formation of OH* is the RDS on the right resulting from the weak binding of O*.

Such scaling relation imposes a severe restriction on obtaining the optimal activity by forbidding the freedom for tuning the adsorption energy of each intermediate. Suppose a certain electrocatalyst exhibits ideal thermodynamics for ΔG_2^{OER} ($\Delta G_{O^*} - \Delta G_{OH^*} = 1.23$ eV at $U = 0$ V), the third OER elementary step (O* + H_2O (l) → OOH* + H+ + e−) cannot be thermodynamically ideal due to $\Delta G_3^{OER} = 1.97 \pm 0.2$ eV. As a result of such nonideal scaling relation between OOH* and OH*, no catalyst materials that fulfill the requirements of a thermodynamically ideal catalyst have been

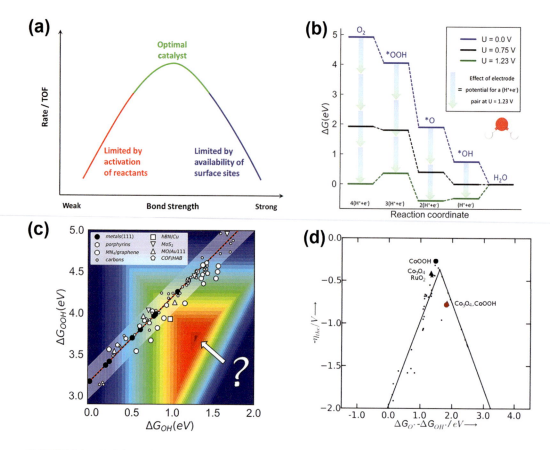

FIGURE 4.1 (a) Schematic representation of the Sabatier principle. (Adapted with permission [7], Copyright (2020), American Chemical Society.) (b) Typical ORR free energy diagram on Pt(111). (c) Scaling relationship ($\Delta G_{OOH^*} = \Delta G_{OH^*} + 3.2 \pm 0.2$ eV) for OOH* and OH* on different catalyst materials. (Adapted with permission [11], Copyright (2018), American Chemical Society.) (d) The established activity volcano plot of metal oxides as the function of $\Delta G_{O^*} - \Delta G_{OH^*}$. (Adapted with permission [16], Copyright (2012), American Chemical Society.)

identified [13]. A compromise is distributing the free-energy changes into $\Delta G_2^{ORR/OER} = \Delta G_3^{ORR/OER} = 3.2/2 = 1.6$ eV. In this case, the theoretically lowest OER overpotential is $1.6 - 1.23$ eV/$e = 0.37$ V. Therefore, breaking scaling relations is of vital importance to improve the energy conversion efficiency for electrocatalytic OER/ORR.

4.2.3 D-Band Center Theory

The computational screening and design of suitable catalyst materials rely on the qualitative descriptions of the structure-property relationship. Based on the CHE model, the different adsorption strength of the reaction intermediates on various catalyst surfaces is closely associated with their electronic structures. The d-band model developed by Hammer and Nørskov can approximately describe adsorbate interaction strength and activity trend on transition metal surfaces [17]. The electronic structures of the transition metal surfaces can be illustrated by the density of states (DOS), as shown in Figure 4.2. The s-band from metal valence s-orbitals results in a broad overlapping band, while the more localized valence d-orbitals form narrow bands. The interaction of the adsorbate states with the narrow d-band will split into bonding and antibonding states.

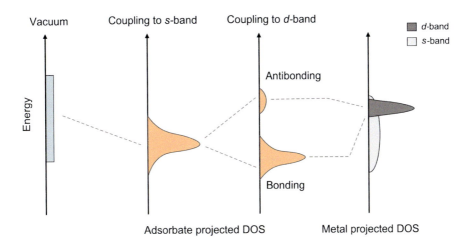

FIGURE 4.2 Schematic diagram of the d-band model showing the electronic state at the transition metal surface.

The bond strength depends on the relative occupancy of these states. If only the bonding states are filled, there will be a strong bond. In contrast, the filling of antibonding states gives rise to a weakened adsorbate-metal interaction [18]. The energy of the d-states relative to the Fermi level varies substantially on different metals. The d-band center (ε_d), defined as the energy-weighted average of the density of d-states, can be a good indicator of the bond strength and can be written as follows:

$$\varepsilon_d = \frac{\int_{-\infty}^{+\infty} \rho E dE}{\int_{-\infty}^{+\infty} \rho dE} \tag{18}$$

where ρ is the DOS of the d states; E is the energy of the d electrons, and ρdE is the number of states. In a series of metals, the higher the ε_d position relative to the Fermi energy suggests a stronger interaction of the electrode surface to the intermediates and vice versa [19]. Consequently, the ε_d can be referenced for understanding the activity trends of transition metal-based catalysts.

4.2.4 Aprotic MABs

In contrast to ORR and OER in the aqueous solution, the discharge products in aprotic electrolytes, such as Li_2O_2 and Na_2O_2, are not soluble and accumulate on the cathode. The sluggish reaction kinetics for the formation and decomposition of these insulating discharge products render high overpotentials and poor cycle stability of aprotic MABs. The performance of cathodic electrocatalysts is important for decreasing ORR/OER overpotentials and achieving better energy conversion efficiencies. DFT computations can also be used to investigate the electrochemistry of oxygen reactions in MABs with aprotic media, i.e., Li-O_2 and Na-O_2 batteries.

Similar to the CHE model, the reduction and evolution of O_2 in aprotic media can be divided into several elementary steps based on the reaction mechanism, too. Taking the Li-O_2 battery as an example, O_2 is reduced to O_2^- in aprotic media and then combined with Li^+ to form insoluble Li_2O_2 during discharge, while O_2 is evolved with the decomposition of Li_2O_2 during charge. The Gibbs free energy change (ΔG) for all elemental steps (i) can be calculated with the following equations [20]:

$$\Delta G = G_{Li_xO_y} - (x-u)G_{Li}^+ - \frac{(y-v)}{2}G_{O_2} - G_{Li_uO_v} + neU \qquad (19)$$

where n is the number of electrons involved in the electrochemical reaction, e is the elementary charge, and U is the applied electrode potential.

The total Gibbs free energy change of the reaction can be calculated by subtracting the free energy of the isolated Li and O_2 from the total free energy of Li_xO_y as described as follows:

$$\Delta G_f = G_{Li_xO_y} - xG_{Li} - \frac{y}{2}G_{O_2} + neU \qquad (20)$$

The equilibrium potential (U_{eq}) is employed to drive the total ORR/OER to occur spontaneously, which can be calculated from the Nernst equation:

$$\Delta U_{eq} = -\frac{\Delta G_f}{ne} \qquad (21)$$

The analyses of the electrocatalytic activity for the discharge and charge commonly rely on the assessment of thermodynamic discharge/charge potentials and thermodynamic overpotentials. The discharge potential ($U_{discharge}$), the maximum voltage to make all intermediate steps downhill, can be obtained by equation (22), as determined by the step with the smallest free energy drop ($\min[|\Delta G_i|]$) during the ORR when U is 0 V.

$$U_{discharge} = \frac{\min[|\Delta G_i|]}{e} \qquad (22)$$

Similarly, the charge potential (U_{charge}), the minimum potential to make the free energies of all intermediates downhill, can be calculated using equation (23). This is determined by the step with the largest energy increase ($\max[|\Delta G_i|]$) during the OER when U is equal to 0 V.

$$U_{charge} = \frac{\max[|\Delta G_i|]}{e} \qquad (23)$$

Theoretical overpotential for the ORR and OER is the difference between the thermodynamic discharge/charge potentials and the U_{eq}, and could be expressed as:

$$\eta_{discharge}^{ORR} = U_{eq} - U_{discharge} \qquad (24)$$

$$\eta_{charge}^{ORR} = U_{discharge} - U_{eq} \qquad (25)$$

Hummelshøj et al. first investigated the reaction mechanism in an aprotic Li-O_2 battery via DFT calculations [20]. They assumed the thermodynamic equilibrium potential (U_0) of 2.47 V based on the formation of Li_2O_2. From the calculated free energy diagram in Figure 4.3, the discharge potential ($U_{discharge}$) is identified as 2.03 V, which can make all intermediate steps downhill. Thus, the theoretical discharge overpotential ($\eta_{discharge}$) is 0.44 V via $\eta_{discharge} = U_0 - U_{discharge}$. Similarly, the charge steps (from right to left) are all downhill at a potential of 3.07 V. Thus, the charge overpotential (η_{charge}) is 3.07 − 2.47 = 0.60 V. The calculated results are approximate to the experimental observations. Via such a computational approach, more candidate materials have been theoretically screened. By analyzing the origin of overpotentials during charge and discharge, theoretical results can provide a satisfactory explanation for the experimentally observed phenomenon.

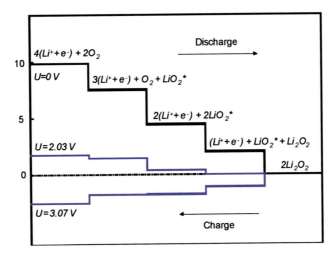

FIGURE 4.3 Free energy diagram of the reactions on the cathode in an aprotic Li-O_2 battery. (Adapted with permission [20], Copyright (2010), AIP Publishing.)

4.3 APPLICATIONS OF THE DFT-BASED MATHEMATICAL MODELING

4.3.1 Atomistic Understanding

Theoretical calculations with predictive quality have emerged as a key contributor in electrocatalysis studies. The dedicated computation thermodynamics, energetics, and activity descriptors have been increasingly employed for mechanistic investigations at the atomic scale. The remarkable knowledge and mechanistic insights gained from computational studies in turn guide the design and screening of high-performance catalyst materials in the real world. In this section, some examples of recent advancements will be given to show how theoretical models accelerate catalyst design and discovery.

4.3.1.1 Identification of Active Sites

The catalytic properties of catalysts are strongly related to their compositions and structures. In practice, multiple active structures may exist in the catalyst materials. It is difficult to investigate the electrocatalytic activity of a specific catalyst structure due to the challenge of preciously synthesizing the catalyst with a fixed structural configuration. Attributed to the computational simulations and catalyst modeling, the catalytic performance of a specific catalyst site can be solely investigated via DFT simulations. For example, intrinsic defects, including edges (armchair and zigzag), vacancies, and a series of topological defects (i.e., pentagons, heptagons, and octagons), are inseparable in carbon materials. They act as potential active sites for electrocatalytic reactions. To understand the influence of different defective configurations on ORR activity, Hu's group constructed four types of defect models for theoretical calculations (Figure 4.4a) [21]. The results revealed that zigzag edge defects and pentagonal defects possessed lower energy barriers for ORR than these of pristine sp^2 carbon, armchair edge, and hole defects (Figure 4.4b), thus acting as potential active sites during the ORR process. In this case, the established structure-property relation can facilitate the development of high-performance electrocatalytic materials.

Our recent studies also revealed that the N-doped sp^3/sp^2 nanocarbon exhibited outstanding ORR performance for Zn-air batteries [22]. The DFT computations showed that the active site was the sp^2-hybridized C atom next to the N dopant. The electronic property analysis revealed that the spin densities of the catalytic sites at the N-doped sp^3/sp^2 carbon interface were greatly enhanced due to the existence of the sp^3 C as electron-repelling groups (Figure 4.4c).

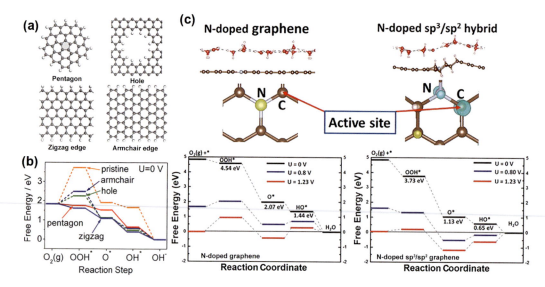

FIGURE 4.4 (a) Structural models of various carbon defects. (b) Calculated free energy diagrams for different defects. (Adapted with permission [21], Copyright (2015), American Chemical Society.) (c) Atomic structure of N-doped graphene and N-doped sp³/sp² hybrid, and corresponding Gibbs free energy diagrams for ORR. (Adapted with permission [22], Copyright (2015), John Wiley and Sons.)

Another example is the identification of active sites in the defective feroxyhyte nanosheet for OER [23]. The Fe atom in the pristine FeOOH surface, the first neighboring Fe atoms to the Fe vacancy, and the second neighboring Fe atoms to the Fe vacancies have been theoretically investigated. Based on the Gibbs free energy diagrams, the second neighboring Fe atoms have been recognized as the active sites with the lowest theoretical overpotential.

4.3.1.2 Reaction Mechanisms

DFT calculations based on the CHE model can quantitatively estimate the adsorption Gibbs free energy change of intermediates on a given electrocatalyst in each elementary step, while the magnitudes of Gibbs free energy changes help identify the potential RDS for OER and ORR. Such information can be used to evaluate the reactivity of the active sites on electrocatalysts and then guide the mechanistic hypothesis. For example, Bao's group calculated the Gibbs free energy diagrams of N-doped graphene for ORR applications [24]. It was found that the energy barrier for the hydrogenation of adsorbed O_2 to OOH* in the first step was 0.51 eV, much lower than that for direct O_2* dissociation into O* (1.56 eV). Therefore, the more energetically favorable associative mechanism has been generally accepted for ORR on N-doped carbon materials.

However, the associate mechanism is not always favored for ORR/OER from the experiments. For example, some recent experimental studies on Ag nanoparticles have also suggested the existence of the [2e⁻ + 2e⁻] dissociative pathway for ORR [25]. The 4e⁻ associative mechanism for ORR has been experimentally and theoretically demonstrated on Ag(111) [26]. Our recent DFT results confirmed that the dissociative ORR pathway was energetically preferred on Ag(322) surface by using the advanced SCAN-rVV10 DFT method (Figure 4.5a) [27]. The existence of low-coordination number surface atoms on the Ag(322) surface facilitated the dissociation of O_2, highlighting the importance of the surface morphology and coordination numbers for the catalyst design.

Different from the conventional adsorption evolution mechanism (AEM) described as OH* → O* → OOH* → O_2 for OER, lattice oxygen mechanism (LOM) and dual-site mechanism (DSM) have been proposed to additionally elucidate the OER mechanism on metal oxide catalysts (Figure 4.5b) [28]. In a representative LOM, the lattice oxygen on the catalyst surface participates in oxygen

Mathematical Modeling for Enhanced Electrochemical Properties

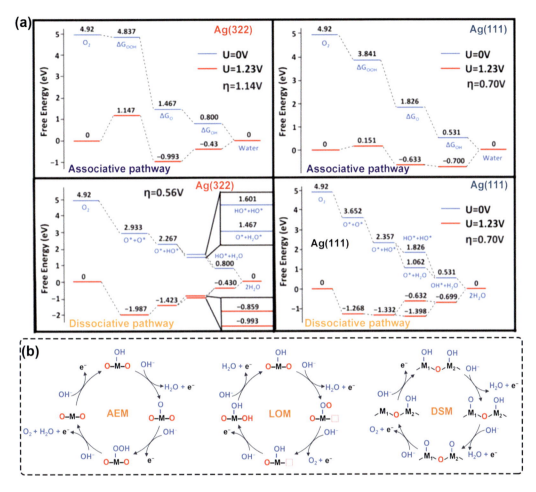

FIGURE 4.5 (a) Gibbs free energy diagrams of ORR on Ag(322) and Ag(111) surfaces calculated from associative and dissociative pathways. (Adapted with permission [27], Copyright (2022), MDPI.) (b) Reaction pathways for OER via the AEM, LOM, and DSM.

production via the reversible formation of surface oxygen vacancies (V_O) [29]. When the distance between two neighboring metal sites is short enough, DSM can become dominant. In DSM, dual-metal sites connected by one oxygen can be occupied by OH$^-$ simultaneously. Then, O−O is directly coupled, and O_2 is released from two adjacent metal sites instead of forming OOH*. The LOM and DSM mechanisms are supported by theoretical and experimental reports, providing thermodynamic reasoning that the LOM and DSM can be energetically preferred over the AEM [30, 31]. In order to clarify the properties that govern the preferred OER mechanism, Kolpak and co-workers used ab initio computational schemes to investigate the surface structure OER mechanism on ABO_3 perovskite oxides (A = La, La$_{0.5}$Sr$_{0.5}$, Sr; B = Mn, Fe, Co, and Ni) [31]. The newly identified intermediates of H* and −OO with V_O in LOM led to a new scaling relationship between the binding energy of intermediates apart from those identified between O*, OH*, and OOH*. It was found that ΔG linearly correlates with V_O formation enthalpy ($\Delta E_f^{V_O}$) and N—V descriptor, where N was the number of unpaired electrons on the isolated B atom and V was the nominal valence charge of B in the stoichiometric bulk ABO_3. It was predicted that OER occurred via AEM for large N−V and switched to LOM after $LaCoO_3$. The exploration of LOM and DSM provides an essential understanding for designing next-generation OER catalysts. Moreover, electrocatalysts that catalyze the

OER via LOM or DSM could avoid the formation of the OOH* intermediate. Consequently, the scaling relation between OOH* and OH* can be avoided, thus achieving a lower overpotential than the theoretical limitation of AEM [30].

4.3.1.3 Solvent Effect

Electrochemical ORR/OER processes are associated with the electrode-electrolyte interface. Solvent molecules can interact with reactive intermediates adsorbed on the catalyst surface, thus affecting their adsorption strength. For a more reliable and accurate description of the electrocatalytic process, the solvation effects should be a common practice and included in computational electrocatalysis calculations.

Few methods are commonly used to model the electrolyte. One is the explicit model with the introduction of solvent molecules and ions into the catalytic system for energy calculations. This approach typically adopts a few solvent molecules or multiple solvent layers into the periodic slab model. The explicit model allows the specific interaction between solvent and electrode surface to be accurately described in energetic calculations, i.e., adsorption configuration of adsorbates and free energy with long-range electrostatic interactions. For example, Wang et al. investigated the ORR catalyzed by the $Mn-N_4$ single-atom catalyst with explicit water molecules [32]. It was found that the solvent environment transformed the adsorption of O_2 from the end-on configuration to the more stable side-on configuration. The hydrogen bonds formed between intermediates and surrounding water molecules elongated the Mn–O bond, thus facilitating the protonation for O* and OH* intermediates. Our group constructed explicit water layer structures comprised of hexagonal water arrangements on the monolayer graphene embedded with atomic metal sites [33]. When the electrocatalytic reactions were investigated, one water molecule was replaced by the reaction intermediates, i.e., OOH*, O*, and OH*. Under the effect of explicit water, the ORR on $Fe-N_4$ and $Ni-N_4$ moieties presented a different free energy profile compared to that without explicit water, showing improved agreement with experiments and a reasonable estimation of the ORR overpotential. Using ab initio molecular dynamics simulations and DFT calculations, Wang et al. revisited the ORR on the Fe–N–C catalyst in the aqueous medium [34]. They proposed that the first electron transfer step involved the release of a hydroxide anion instead of forming OOH*. The released hydroxide anion would be confined in a solvated "pseudo-adsorption" state near the catalyst surface. The formation of a pseudo-adsorption state lowered the energy of the whole system. At the same time, the solvent water was found to dynamically coordinate to the metal center adaptively, adjusting its binding strength of intermediates. The RDS was turned to the protonation of the pseudo-adsorbed hydroxide with the reduction of metal centers in the last PCET step.

However, the explicit models require a considerable number of solvent molecules and large supercells to converge the average structure of the liquid and avoid artificial electron work functions and molecular arrangements caused by the size effect, which will result in huge computational costs. Moreover, it is difficult to determine the global minimum energy structure of solvent molecules. Another approach is the implicit solvation model, which treats solvent as a polarizable continuum with certain dielectric properties [35]. Several widely used first-principle calculation software, such as Adaptive Poisson-Boltzmann Solver (APBS) [36], Quantum ESPRESSO [37], CP2K [38], and VASP (VASPsol) [39], integrate implicit solvation models for calculating the solvation energy. In contrast to explicit models, implicit models are much computationally cheaper. However, the calculated stabilizations for reaction intermediates are different from the results obtained by explicit solvent models. This is probably because implicit models neglect granular structure of the solvent, and cannot describe all the solvent-adsorbate interactions, such as the hydrogen bonding between solvent and adsorbates [40].

4.3.2 Design of Electrocatalysts

Theoretical models can provide reasonable strategies to promote the performance of the catalyst. As discussed earlier, the electrocatalytic activity is actually determined by the binding strength of the key intermediates and underlying electronic structures of electrocatalysts. Thus, the general guideline is to achieve the optimal electronic configurations that deliver decent intermediate binding. The established volcano plot based on the Sabatier principle enables the mapping of catalytic activity using key descriptors, facilitating the catalyst optimization within a family of compounds. One can enhance or weaken the binding strength of electrocatalysts toward reaction intermediates via proper material engineering, such as alloying, multisite functionalization, and heteroatom-doping.

4.3.2.1 Alloying

Previous studies have demonstrated optimizing the catalytic activity of catalysts upon broadening or narrowing the d-band position. For example, Pt and Pd stand out as the nearest metal to the top of the activity volcano plot for ORR [9]. However, the binding energy of O* on Pt and Pd is stronger than the optimal value [41, 42]. By alloying Pt/Pd with other elements (especially transition metals), the O* binding could be weakened, thus maximizing the ORR activity of Pt/Pd-based catalysts. Nørskov et al. constructed the volcano plot based on the experimentally measured ORR activity of various Pt_3X (X = Ti, Sc, Fe, Co, and Ni) alloys as a function of the adsorption energy of O* relative to Pt ($\Delta E_O - \Delta E^{Pt}_O$). Several Pt_3X alloys (i.e., Pt_3Ni, Pt_3Co, Pt_3Fe, Pt_3Sc, Pt_3Ti) with weakened O* binding should exhibit better ORR activity than pure Pt [41]. Liu et al. investigated the electronic structure of MPd_{12} (M = Fe, Co, Ni, Cu, Zn) nanoparticles dispersed on graphene substrates (MPd_{12}/SVG) and their reactivity toward O* adsorption [42]. It was found a linear relationship between ΔE_{ads} and ε_d (Figure 4.6a). Therefore, the ORR kinetics hindered by strong adsorption of O* could be promoted by adjusting the ε_d of the catalysts. Recently, Lim et al. probed the oxygen dissociation pathway on Pt(111), Pt_3M(111), and PtM(111) surfaces (M = Co, Ni, Mn, and Ir) [43]. The calculated ORR activity of Pt-M surfaces was inferior to that of Pt(111) except for the Pt_3Ir(111) surface, implying that binary Pt-M alloys have insufficient ORR activity to replace Pt catalysts. When an additional single Pt skin layer was introduced at the top of Pt-M alloys (denoted as Pt/Pt-M(111), the adsorption strengths of key intermediates (O* and OH*) were decreased to values below those observed for the Pt(111), thus resulting in enhanced ORR performance. Further analyses showed that the d-band center position of Pt/Pt-M(111) surfaces was more negative than that of Pt(111) and Pt-M(111), indicating that the promoted ORR activity could originate from the downshift of d-band centers.

Similarly, engineering strain is an effective method to optimize intermediate binding on Pt catalysts. The tensile strain can generally lead to an up-shift of the d-band center, while a compressive strain can induce a down-shift [44]. For example, the lattice mismatch between the Pt shell and the Pt-Cu core could cause a reduced Pt-Pt interatomic distance in the shell and, hence, induce more strain. Moderate compressive lattice strain was predicted to enhance the ORR catalysis [45].

For the aprotic Li-O_2 battery, the adsorption strength of LiO_2 on the metal surfaces strongly correlates with the calculated charging overpotentials and can be adjusted via metal alloying. Choi et al. calculated the free energy diagram on the Co(0001), Pt(111), and Pt_3Co(111) surfaces for OER in Li-O_2 battery [46]. The calculated OER overpotential of Pt_3Co(111) was 0.49 V lower than pure Pt, thus contributing to a more efficient decomposition of discharging reaction products.

4.3.2.2 Multisite Functionalization

The major obstacle in optimizing the catalyst is the interdependence of the binding of the intermediates. Scaling relations among the Gibbs free energy changes of reactive intermediates in ORR/OER make it difficult to tune the adsorption strength of one intermediate on the active site without affecting the others. Therefore, breaking or circumventing the adsorption-energy scaling relations to achieve superior performance is the main goal for catalyst design and engineering. The multisite functionalization can offer several active sites or local binding environments to stabilize different intermediates. It is possible to decouple the ΔG_{OOH^*} from the ΔG_{OH^*} via a dynamic switch of active sites. As such, the binding energy of each intermediate can be optimized independently, thus breaking the limit of energy scaling relations toward the lower overpotential. For example, Shen et al. utilized dual active sites on SnO_x/Pt-Cu-Ni heterojunction catalyst for ORR (Figure 4.6b) [47].

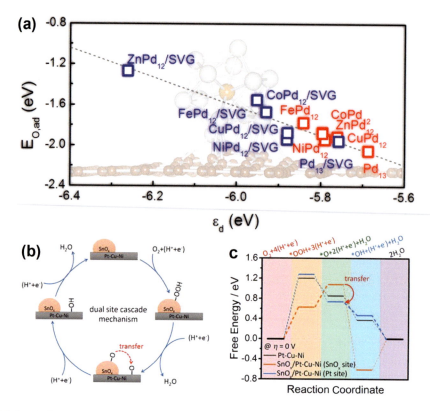

FIGURE 4.6 (a) Calculated O adsorption energies over various MPd_{12} composites plotted as a function of ε_d. (Adapted with permission [42], Copyright (2013), American Chemical Society.) (b) Scheme of the dual-site cascade mechanism on SnO_x/Pt-Cu-Ni. (c) ORR free energy diagrams for Pt-Cu-Ni and SnO_x/Pt-Cu-Ni. (Adapted with permission [47], Copyright (2019), American Chemical Society.)

Theoretical studies demonstrated that the RDS was the formation of OOH* on Pt-Cu-Ni with an energy barrier of 1.20 eV. On SnO_x/Pt-Cu-Ni, the first two elementary ORR steps occurred on a SnO_x site, followed by barrierless transfer of O* intermediate to an adjacent Pt site for the remaining steps. The dynamic switch of reactive sites reduced the overall reaction energy barrier to 0.63 eV (Figure 4.6c). Du et al. integrated NiO nanoparticles on nickel-iron layered double hydroxide (NiFe LDH) for OER [48]. The intersection between NiO and NiFe LDH nanosheet provided a tridimensional adsorption site. In contrast to traditional single-site adsorption, the adsorption configurations differed for each intermediate and varied dynamically during the OER process. Consequently, the binding energy of each intermediate could be adjusted independently, and the scaling relationship could be circumvented by the resultant NiO/NiFe LDH composite.

4.3.2.3 Heteroatom-Doping

Via doping heteroatom or functional moieties, the local geometric structure of catalysts and reactivity of active sites around the dopants can be altered, thus influencing the reaction pathways. For instance, the incorporation of Zn into CoOOH gave rise to oxygen non-bonding states with different local configurations. The formed $Zn-O_2-Co-O_2-Zn$ configuration enabled a switch of the OER mechanism from the AEM pathway to the LOM, thus achieving a lower overpotential than the theoretical limitation of AEM [30]. For transition metal-based catalysts, the established adsorption-energy scaling relations are associated with the corresponding d-band states. It has been proposed that introducing p-block elements could influence the metal d states via hybridization of d- and p-orbitals, thus circumventing the original scaling relations. For example, Tong et al. modified the Pt surface with sulfur, a typical surface poison for metal catalysts [49]. The adsorption strength of OOH* and OH* deviated from the original scaling relationship of pure metals, resulting in enhanced ORR activity. The scaling relations can also be broken by introducing a proton donor/acceptor to stabilize adsorbed species. For example, by introducing a hydrogen acceptor on the metal oxide surface, the hydrogen bond formed between OOH* and acceptor facilitates its stabilization, thus effectively lowering the OER potential needed for breaking bonds to the surface [50]. Busch et al. modified M–N_4 (M = transition metal, i.e., Mn, Fe, Co, Ni, etc.) sites embedded graphene with functional groups such as –COOH, –OH, and –NH_2 as the proton donor/acceptor [51]. DFT calculations revealed the improved ORR and OER activity beyond the theoretical limitation originating from the scaling relations.

4.3.3 Machine Learning-Based High-Throughput Screening

High-throughput screening is emerging as a powerful tool to use mathematical models and theoretical insights discussed earlier to accelerate the design and discovery of efficient catalyst materials. Conventionally, studies on electrocatalysts usually involve time-consuming and trial-and-error procedures. A systematic high-throughput computational screening with well-developed activity descriptors is promising for assessing and screening a large number of candidate catalysts in a time-efficient manner. Shui's group investigated 180 types of single-atom catalysts with different structure configurations for ORR via DFT [15]. The calculated theoretical overpotentials and ΔG_{OH^*} enabled a high forecast precision. Notably, the Co-doped metal-organic frameworks, Ir-doped phthalocyanine, Co-doped N-coordination graphene, Co-doped graphdiyne, and Rh-doped phthalocyanine stood out as promising candidates due to their low overpotentials.

However, the DFT computations are still unaffordable for large-scale high-throughput screening. The recent efforts in theoretical and experimental investigations of electrocatalysts have generated enormous data. A data-driven framework can then be developed to determine the structure-performance relationship of active sites under reaction conditions. Sets of key features/descriptors D_{ik} of known electrocatalysts M_i, and an inference model F will be used to calculate the overpotential η_{ij} of M_i for the electrocatalysis in operando conditions C_j. Generally, the databases will be randomly divided into two parts: a training set and a test set. The training set will be used to determine the key features of electrocatalysts related to their performance and build an inference model. The test data set will be used for cross-validation of the inference model to overcome overfitting and/or underrepresented class problems of machine learning. Additionally, F must possess a solid physical basis supported by computational and experimental results.

Recently, Xin et al. developed an adaptive machine learning strategy in search of high-performance ABO_3-type cubic perovskites for catalyzing OER [52]. They concluded that the e_g orbital electronic structure characteristics of the metal B-site were the underlying factors that govern the OER activity, providing physical insights from data-driven models. Moreover, reasonable expectations for promising candidate catalysts can be provided by ML methods. The output of the results can be further validated by experiments. Li's group systematically examined a series of dual-metal-site embedded graphene catalysts (DMSCs) via DFT computations with ML [14]. Electron affinity, electronegativity, and radii of the embedded transition metal atoms were fundamental parameters to determine the intrinsic ORR activity. Based on the free energy diagram and microkinetic simulations, several DMSCs (i.e., Ni/Cu) presented good ORR catalytic performances in terms of calculated U_L and simulated half-wave potential, even superior to the Pt benchmark. The excellent performances of DMSCs have also been verified in experimental studies.

High-entropy alloys (HEAs) have drawn considerable interest as ORR/OER electrocatalysts [53]. Their complex compositions and atomic arrangements provide innumerable possibilities in microstructures and properties. Meanwhile, this also imposes a great challenge for discovering HEAs with desirable electrocatalytic activity via the conventional experiment-verification process.

By employing DFT computations and machine learning algorithms, it is possible to establish an accurate correlation between the adsorption energies of reaction intermediates and surface structures of HEAs [53, 54]. The cost of exploring the properties of unknown compounds is reduced. It should be noted that the accuracy and performance of the computational methods depend heavily on the quality and scale of the database since unbalanced data would lead to underfitting problems and biased predictions.

4.4 CONCLUSIONS AND OUTLOOK

This chapter summarizes the DFT-based mathematical modeling and calculation methods in investigating the electrocatalytic ORR/OER process. Via these approaches, an elementary reaction mechanism and the corresponding energetic information on reaction barriers are extracted from first-principle computations. Computational analyses can reveal the origin of the ORR/OER overpotential and help understand the intrinsic limitation of electrocatalysts due to scaling relations. Finally, we have reviewed the advances in theoretical modeling for catalysis applications, emphasizing the mechanistic revealing and research efforts for breaking scaling relations toward enhanced electrochemical properties. ML methods and data-driven techniques integrated with computational methods will significantly enhance the efficiency in designing and discovering new electrocatalysts. It is believed that the understanding of the detailed calculation methods as well as the underlying activity origin can facilitate the development of electrocatalysts in heterogeneous ORR and OER.

Challenges are also faced by using computational approaches. One of the issues is that the typically used CHE model neglects the kinetic barriers and dynamic changes of the micro-reaction environment that correlate to reaction performances. The static vacuum calculations cannot reflect the electrochemical environments in a real catalysis system, possibly leading to discrepancies between DFT-predicted activity and experimentally measured activity. Although the computational studies with the explicit or implicit models indeed provide insights into the catalytic behavior at the electrochemical interface, it should be noted that both methods only represent an approximation to the actual system due to the challenge of considering the dynamic change (i.e., pH, concentration, surface charges of the catalyst, and electrode potentials) during the electrocatalysis process within a DFT model [55]. In addition, while the d-band model is widely applicable for many adsorbate/transition metal systems, there still exist outliers and exceptions where its accuracy is unsatisfactory. Moreover, current studies mainly focus on thermodynamics to estimate the catalytic activity of catalyst materials, while kinetic information plays a decisive role in evaluating the overall catalytic performance. Also, the selectivity and stability of catalysts lack reliable evaluation approaches and descriptors in computational catalysis. These call for more advanced methods for modeling the full details of the complex electrochemical system, which are essential to bridge the gap between theoretical computations and real catalysis systems.

There is no doubt that exciting applications of the computational and data science approach will continue to flourish in accelerating catalyst design and mechanistic understanding. The concepts and insights discussed in this chapter will be redeveloped along with a deeper understanding of electrochemistry in MABs unveiled by the combination of theory and experiment, thus further advancing this field.

ACKNOWLEDGMENTS

The authors gratefully acknowledge financial support from the Australia Research Council Discovery Projects (DP210103266) of Australia. This work was also supported by computational resources provided by the Australian Government through the National Computational Infrastructure (NCI) under the National Computational Merit Allocation Scheme and the Pawsey Supercomputing Centre with funding from the Australian Government and the Government of Western Australia.

REFERENCES

[1] Z.L. Wang., D. Xu., J.J. Xu., X.B. Zhang., Oxygen electrocatalysts in metal-air batteries: From aqueous to nonaqueous electrolytes, Chem. Soc. Rev. 43 (2014) 7746–7786.

[2] M. Wu., G. Zhang., M. Wu., J. Prakash., S. Sun., Rational design of multifunctional air electrodes for rechargeable Zn-air batteries: Recent progress and future perspectives, Energy Storage Mater. 21 (2019) 253–286.

[3] Y. Tian, L, Xu., J. Qiu., X. Liu., S. Zhang., Rational design of sustainable transition metal-based bifunctional electrocatalysts for oxygen reduction and evolution reactions, Sustain. Mater. Technol. 25 (2020) e00204.

[4] P. Hohenberg, W. Kohn, Inhomogeneous electron gas, Phy. Rev. 136 (1964) B864–B871.

[5] H. Ooka, R. Nakamura, Shift of the optimum binding energy at higher rates of catalysis, J. Phys. Chem. Lett. 10 (2019) 6706–6713.

[6] H. Ooka, J. Huang, K.S. Exner, The sabatier principle in electrocatalysis: Basics, limitations, and extensions, Front. Energy Res. 9 (2021) 654460.

[7] A.H. Motagamwala, J.A. Dumesic, Microkinetic modeling: A tool for rational catalyst design, Chem. Rev. 121 (2021) 1049–1076.

[8] T. Risthaus, S. Grimme, Benchmarking of London dispersion-accounting density functional theory methods on very large molecular complexes, J. Chem. Theory Comput. 9 (2013) 1580–1591.

[9] J.K. Nørskov, J. Rossmeisl, A. Logadottir, L. Lindqvist, J.R. Kitchin, T. Bligaard, H. Jónsson, Origin of the overpotential for oxygen reduction at a fuel-cell cathode, J. Phys. Chem. B. 108 (2004) 17886–17892.

[10] A.A. Peterson, F. Abild-Pedersen, F. Studt, J. Rossmeisl, J.K. Nørskov, How copper catalyzes the electroreduction of carbon dioxide into hydrocarbon fuels, Energy Environ. Sci. 3 (2010) 1311–1315.

[11] A. Kulkarni, S. Siahrostami, A. Patel, J.K. Nørskov, Understanding catalytic activity trends in the oxygen reduction reaction, Chem. Rev. 118 (2018) 2302–2312.

[12] H. Xu, D. Cheng, D. Cao, X.C. Zeng., A universal principle for a rational design of single-atom electrocatalysts, Nat. Catal. 1 (2018) 339–348.

[13] Z.W. Seh, J. Kibsgaard, C.F. Dickens, I. Chorkendorff, J.K. Nørskov, T.F. Jaramillo, Combining theory and experiment in electrocatalysis: Insights into materials design, Science. (2017) 355. DOI: 10.1126/science.aad4998.

[14] X. Zhu, J. Yan, M. Gu, T. Liu, Y. Dai, Y. Gu, Y. Li, Activity origin and design principles for oxygen reduction on dual-metal-site catalysts: A combined density functional theory and machine learning study, J. Phys. Chem. Lett. 10 (2019) 7760–7766.

[15] Y. Wang, R. Hu, Y. Li, F. Wang, J. Shang, J. Shui, High-throughput screening of carbon-supported single metal atom catalysts for oxygen reduction reaction, Nano Res. 15 (2021) 1054–1060.

[16] M. García-Mota, M. Bajdich, V. Viswanathan, A. Vojvodic, A.T. Bell, J.K. Nørskov, Importance of correlation in determining electrocatalytic oxygen evolution activity on cobalt oxides, J. Phys. Chem. C. 116 (2012) 21077–21082.

[17] B. Hammer, J.K. Nørskov, Electronic factors determining the reactivity of metal surfaces, Surface Science. 343 (1995) 211–220.

[18] J.K. Nørskov, F. Abild-Pedersen, F. Studt, T. Bligaard, Density functional theory in surface chemistry and catalysis, Proc. Natl. Acad. Sci. U.S.A. 108 (2011) 937–943.

[19] J.K. Nørskov, T. Bligaard, J. Rossmeisl, C.H. Christensen, Towards the computational design of solid catalysts, Nat. Chem. 1 (2009) 37–46.

[20] J.S. Hummelshoj, J. Blomqvist, S. Datta, T. Vegge, J. Rossmeisl, K.S. Thygesen, A.C. Luntz, K.W. Jacobsen, J.K. Nørskov, Communications: Elementary oxygen electrode reactions in the aprotic Li-air battery, J. Chem. Phys. 132 (2010) 071101.

[21] Y. Jiang, L. Yang, T. Sun, J. Zhao, Z. Lyu, O. Zhuo, X. Wang, Q. Wu, J. Ma, Z. Hu, Significant contribution of intrinsic carbon defects to oxygen reduction activity, ACS Catal. 5 (2015) 6707–6712.

[22] J. Gao, Y. Wang, H. Wu, X. Liu, L. Wang, Q. Yu, A. Li, H. Wang, C. Song, Z. Gao, M. Peng, M. Zhang, N. Ma, J. Wang, W. Zhou, G. Wang, Z. Yin, D. Ma, Construction of a sp^3/sp^2 carbon interface in 3D N-doped nanocarbons for the oxygen reduction reaction, Angew. Chem. Int. Ed. 58 (2019) 15089–15097.

[23] B. Liu, Y. Wang, H.Q. Peng, R. Yang, Z. Jiang, X. Zhou, C.S. Lee, H. Zhao, W. Zhang, Iron vacancies induced bifunctionality in ultrathin feroxyhyte nanosheets for overall water splitting, Adv. Mater. 30 (2018) 1803144.

[24] L. Yu, X.L. Pan, X.M. Cao, P. Hu, X.H. Bao, Oxygen reduction reaction mechanism on nitrogen-doped graphene: A density functional theory study, J. Catal. 282 (2011) 183–190.

[25] Q. Wang, X. Cui, W. Guan, L. Zhang, X. Fan, Z. Shi, W. Zheng, Shape-dependent catalytic activity of oxygen reduction reaction (ORR) on silver nanodecahedra and nanocubes, J. Power. Sources. 269 (2014) 152–157.

[26] J.M. Linge, H. Erikson, J. Kozlova, J. Aruväli, V. Sammelselg, K. Tammeveski, Oxygen reduction on electrodeposited silver catalysts in alkaline solution, J. Solid. State. Electrchem. 22 (2017) 81–89.

[27] J.J. Hinsch, J. Liu, J.J. White, Y. Wang, The role of steps on silver nanoparticles in electrocatalytic oxygen reduction, Catalysts. 12 (2022) 576.

[28] C. Ling, Y. Cui, S. Lu, X. Bai, J. Wang, How computations accelerate electrocatalyst discovery. Chem. 8 (2022) 1575.

[29] W.G. Hardin, D.A. Slanac, X. Wang, S. Dai, K.P. Johnston, K.J. Stevenson, Highly active, nonprecious metal perovskite electrocatalysts for bifunctional metal-air battery electrodes, J. Phys. Chem. Lett. 4 (2013) 1254–1259.

[30] Z.F. Huang, J. Song, Y. Du, S. Xi, S. Dou, J.M.V. Nsanzimana, C. Wang, Z.J. Xu, X. Wang, Chemical and structural origin of lattice oxygen oxidation in Co—Zn oxyhydroxide oxygen evolution electrocatalysts, Nat. Energy. 4 (2019) 329–338.

[31] X. Rong, J. Parolin, A.M. Kolpak, A fundamental relationship between reaction mechanism and stability in metal oxide catalysts for oxygen evolution, ACS Catal. 6 (2016) 1153–1158.

[32] H. Cao, G.J. Xia, J.W. Chen, H.M. Yan, Z. Huang, Y.G. Wang, Mechanistic insight into the oxygen reduction reaction on the Mn—N_4/C single-atom catalyst: The role of the solvent environment, J. Phys. Chem. C. 124 (2020) 7287–7294.

[33] Z. Zhu, H. Yin, Y. Wang, C.H. Chuang, L. Xing, M. Dong, Y.R. Lu, G. Casillas-Garcia, Y. Zheng, S. Chen, Y. Dou, P. Liu, Q. Cheng, H. Zhao, Coexisting single-atomic Fe and Ni sites on hierarchically ordered porous carbon as a highly efficient ORR electrocatalyst. Adv. Mater. 32 (2020) 2004670.

[34] J.W. Chen, Z. Zhang, H.M. Yan, G.J. Xia, H. Cao, Y.G. Wang, Pseudo-adsorption and long-range redox coupling during oxygen reduction reaction on single atom electrocatalyst. Nat. Commun. 13 (2022) 1734.

[35] K. Mathew, V.S.C. Kolluru, S. Mula, S.N. Steinmann, R.G. Hennig, Implicit self-consistent electrolyte model in plane-wave density-functional theory, J. Chem. Phys. 151 (2019) 234101.

[36] N.A. Baker, D. Sept, S. Joseph, M.J. Holst, J.A. McCammon, Electrostatics of nanosystems: Application to microtubules and the ribosome, Proc. Natl. Acad. Sci. U.S.A. 98 (2001) 10037–10041.

[37] P. Giannozzi, O. Andreussi, T. Brumme, O. Bunau, M. Buongiorno Nardelli, M. Calandra, R. Car, C. Cavazzoni, D. Ceresoli, M. Cococcioni, N. Colonna, I. Carnimeo, A. Dal Corso, S. de Gironcoli, P. Delugas, R.A. DiStasio, Jr., A. Ferretti, A. Floris, G. Fratesi, G. Fugallo, R. Gebauer, U. Gerstmann, F. Giustino, T. Gorni, J. Jia, M. Kawamura, H.Y. Ko, A. Kokalj, E. Kucukbenli, M. Lazzeri, M. Marsili, N. Marzari, F. Mauri, N.L. Nguyen, H.V. Nguyen, A. Otero-de-la-Roza, L. Paulatto, S. Ponce, D. Rocca, R. Sabatini, B. Santra, M. Schlipf, A.P. Seitsonen, A. Smogunov, I. Timrov, T. Thonhauser, P. Umari, N. Vast, X. Wu, S. Baroni, Advanced capabilities for materials modelling with Quantum ESPRESSO, J. Phys. Condens. Matter. 29 (2017) 465901.

[38] J. VandeVondele, F. Mohamed, M. Krack, J. Hutter, M. Sprik, M. Parrinello, The influence of temperature and density functional models in ab initio molecular dynamics simulation of liquid water, J. Chem. Phys. 122 (2005) 14515.

[39] K. Mathew, R. Sundararaman, K. Letchworth-Weaver, T.A. Arias, R.G. Hennig, Implicit solvation model for density-functional study of nanocrystal surfaces and reaction pathways, J. Chem. Phys. 140 (2014) 084106.

[40] M. Reda, H.A. Hansen, T. Vegge, DFT study of stabilization effects on N-doped graphene for ORR catalysis, Catal. Today. 312 (2018) 118–125.

[41] J. Greeley, I.E. Stephens, A.S. Bondarenko, T.P. Johansson, H.A. Hansen, T.F. Jaramillo, J. Rossmeisl, I. Chorkendorff, J.K. Nørskov, Alloys of platinum and early transition metals as oxygen reduction electrocatalysts, Nat. Chem. 1 (2009) 552–556.

[42] X. Liu, C. Meng, Y. Han, Defective graphene supported MPd_{12}(M = Fe, Co, Ni, Cu, Zn, Pd) nanoparticles as potential oxygen reduction electrocatalysts: A first-principles study, J. Phys. Chem. C. 117 (2013) 1350–1357.

[43] D.Y. Shin, Y.J. Shin, M.S. Kim, J.A. Kwon, D.H. Lim, Density functional theory-based design of a Pt-skinned PtNi catalyst for the oxygen reduction reaction in fuel cells, Appl. Surf. Sci. 565 (2021) 150518.

[44] Y. Jiao, Y. Zheng, M. Jaroniec, S.Z. Qiao, Design of electrocatalysts for oxygen- and hydrogen-involving energy conversion reactions, Chem. Soc. Rev. 44 (2015) 2060–2086.

[45] P. Strasser, S. Koh, T. Anniyev, J. Greeley, K. More, C. Yu, Z. Liu, S. Kaya, D. Nordlund, H. Ogasawara, M.F. Toney, A. Nilsson, Lattice-strain control of the activity in dealloyed core-shell fuel cell catalysts, Nat. Chem. 2 (2010) 454–460.

[46] B.G. Kim, H.J. Kim, S. Back, K.W. Nam, Y. Jung, Y.K. Han, J.W. Choi, Improved reversibility in lithium-oxygen battery: Understanding elementary reactions and surface charge engineering of metal alloy catalyst, Sci. Rep. 4 (2014) 4225.

[47] X. Shen, T. Nagai, F. Yang, L.Q. Zhou, Y. Pan, L. Yao, D. Wu, Y.S. Liu, J. Feng, H. Guo, H. Jia, Z. Peng, Dual-site cascade oxygen reduction mechanism on SnO_x/Pt-Cu-Ni for promoting reaction kinetics, J. Am. Chem. Soc. 141 (2019) 9463–9467.

[48] Z.W. Gao, J.Y. Liu, X.M. Chen, X.L. Zheng, J. Mao, H. Liu, T. Ma, L. Li, W.C. Wang, X.W. Du, Engineering NiO/NiFe LDH intersection to bypass scaling relationship for oxygen evolution reaction via dynamic tridimensional adsorption of intermediates, Adv. Mater. 31 (2019) 1804769.

[49] Y.Y. Wang, D.J. Chen, T.C. Allison, Y.J. Tong, Effect of surface-bound sulfide on oxygen reduction reaction on Pt: Breaking the scaling relationship and mechanistic insights, J. Chem. Phys. 150 (2019) 041728.

[50] R. Frydendal, M. Busch, N.B. Halck, E.A. Paoli, P. Krtil, I. Chorkendorff, J. Rossmeisl, Enhancing activity for the oxygen evolution reaction: The beneficial interaction of gold with manganese and cobalt oxides, Chemcatchem. 7 (2015) 149–154.

[51] M. Busch, N.B. Halck, U.I. Kramm, S. Siahrostami, P. Krtil, J. Rossmeisl, Beyond the top of the volcano? A unified approach to electrocatalytic oxygen reduction and oxygen evolution, Nano Energy. 29 (2016) 126–135.

[52] Z. Li, L.E.K. Achenie, H. Xin, An adaptive machine learning strategy for accelerating discovery of perovskite electrocatalysts, ACS Catal. 10 (2020) 4377–4384.

[53] J. Hwang, R.R. Rao, L. Giordano, Y. Katayama, Y. Yu, Y. Shao-Horn, Perovskites in catalysis and electrocatalysis, Science. 358 (2017) 751–756.

[54] T.A.A. Batchelor, J.K. Pedersen, S.H. Winther, I.E. Castelli, K.W. Jacobsen, J. Rossmeisl, High-entropy alloys as a discovery platform for electrocatalysis, Joule. 3 (2019) 834–845.

[55] F. Calle-Vallejo, A. Krabbe, J.M. Garcia-Lastra, How covalence breaks adsorption-energy scaling relations and solvation restores them, Chem. Sci. 8 (2017) 124–130.

5 Materials and Technologies of Mg-Air Primary Batteries

Xingrui Chen, Qichi Le, Jeffrey Venezuela, and Andrej Atrens

CONTENTS

5.1 Introduction 65
5.2 The Discharge Nature of Mg Anode in Aqueous Solution 66
5.3 The Development of the High-Performance Mg Anodes 67
    5.3.1 The Intrinsic Discharge Performance of Pure Mg 67
    5.3.2 The Alloying Elements 68
    5.3.3 The Influence of the Production Method 70
5.4 Electrolyte and Additives 72
5.5 The Air Cathode 73
5.6 Outlook 73
References 74

5.1 INTRODUCTION

The lightest structural materials until now are Mg and its alloys, which have been widely used in several aspects, such as aerospace, transportation, and electronic devices. Mg and alloys are also promising energy storage materials due to their appealing qualities, including a relatively negative electrode potential (−2.37 V_{SHE}) and high volumetric capacity (3832 mAh/cm³). The Mg-air battery is one of the most typical energy storage systems, as it has a high theoretical voltage of 3.1 V and an energy density of 6800 Wh/kg, which is regarded as the next-generation battery. In addition, Mg is the 5th most abundant element on earth, ensuring the low cost of this battery system. Mg-air primary batteries possess simple structures and undemanding operating conditions. By using O_2 from the air as the cathode of the magnesium-air battery, it is possible to reduce the weight of the battery and increase its energy density. Electrolytes used in these batteries are simple saline solutions such as NaCl solution. All the components of Mg-air primary batteries are therefore nontoxic to humans and nature, thus meeting the requirements of a green lifestyle. On the anode, the oxidation reaction produces Mg^{2+} and two electrons, as shown in equation (1). At the air cathode, the oxygen in the air is reduced to OH^- through a cathodic reaction, which consumes four electrons, as shown in equation (2).

$$Mg \rightarrow Mg^{2+} + 2e^- \qquad (1)$$

$$O_2 + 2H_2O + 4e^- \rightarrow 4OH^- \qquad (2)$$

Mg-based batteries are currently faced with the following scientific and technological challenges. Because of the high rate of self-corrosion of the Mg anode at the overpotential of the Mg anode, the Mg anode has a low utilization efficiency. The battery voltage is much lower than the theoretical value (i) because the accumulation of discharge products on the anode and cathode surface increases the internal resistance of the battery, and (ii) because of the high charge transfer resistance on the anode surface.

In recent years, there have been an increasing number of research studies focusing on Mg-air primary batteries. This chapter contains the state-of-the-art of Mg-air batteries, including the fundamentals of Mg dissolution in aqueous electrolytes, the development of the Mg-based anode, the electrolyte and additives, and the air cathode. This chapter discusses recent developments, current research opportunities, and future directions for research. We expect this chapter to guide new studies that will advance the development of Mg-based primary batteries in the future.

5.2 THE DISCHARGE NATURE OF MG ANODE IN AQUEOUS SOLUTION

The primary Mg-based batteries undergo the same electrochemical reaction on the anode, but their cathode reactions differ according to the cathode material and battery type [1]. Figure 5.1 schematically presents the discharge nature of an Mg anode in an aqueous solution. In a theoretical reaction of the Mg anode, the Mg loses two electrons and becomes an Mg^{2+} ion, as shown in Eq. (1). However, recent results have confirmed that Mg loses one electron at a time [2]. Furthermore, quantum mechanics prohibits the simultaneous transfer of two electrons [3]. As a result, the first step in the Mg anode reaction of Mg-based batteries is:

$$Mg \rightarrow Mg^+ + e^- \tag{3}$$

Due to the high chemical activity of the uni-positive Mg+ ions, gas-phase studies have demonstrated that Mg^+ reacts with water within milliseconds [4], suggesting that their lifespan on the anode is very short. As part of the battery reaction, some of them lose their outermost electron to Eq. (4)

$$Mg^+ \rightarrow Mg^{2+} + e^- \tag{4}$$

Other Mg^+ ions chemically split water using Eq. (5) to produce H_2 and $Mg(OH)_2$. As a result, electrons cannot be supplied to the external circuit in this reaction, which contributes to the loss of the anodic efficiency of the Mg anode in the Mg-air battery.

$$Mg^+ + H_2O \rightarrow 1/2\,Mg(OH)_2 + 1/2\,H_2 \tag{5}$$

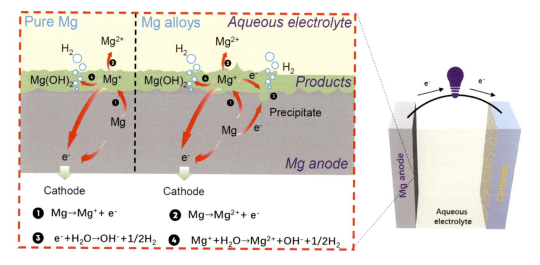

FIGURE 5.1 The discharge nature of an Mg anode in an aqueous electrolyte.

As a result of reactions (3) and (4), the ideal consumption of metal Mg will result in energy conversion and electrons for the external circuit. Alternatively, Mg+ going through reaction (4) is considered a wasteful process since it does not produce electrons for external use. The apparent phenomena of this reaction are the anodic hydrogen evolution and the negative difference effect (NDE) [1, 5]. According to the results by Chen et al. [6], the wasteful consumption of Mg via reactions (5) is the only cause of the energy loss of an ultra-high-purity Mg anode in the Mg-air battery.

Mg alloys present more complicated discharge behavior than pure Mg due to the presence of precipitates (second phases and impurities). Mg normally has lower standard electrode potential than these precipitates [7, 8]. During the discharge process, these precipitates have a lower priority of dissolution than the Mg matrix. Mg surrounding precipitate particles dissolve preferentially due to micro galvanic corrosion, which goes through reactions (3) and (4) as well as the production of electrons. Nevertheless, these electrons do not provide current to the external circuit. They split water via Eq. (5) as the cathodic reaction that balances the anodic partial reactions Eq. (3) and (4).

$$e^- + H_2O \rightarrow OH^- + 1/2H_2 \qquad (6)$$

Due to this process, the Mg consumption is not available to provide electrons in the external circuit, and therefore this is also considered a wasteful consumption of Mg. As the discharge progresses, precipitates may lose support from the surrounding Mg and then drop out from the anode. Supporting evidence has been provided by the lamellar surface of Mg anodes containing the LPSO phases after discharge [9, 10], and the detection of $Mg_{17}Al_{12}$ particles in the discharge products of an Mg-Al anode [11]. Exfoliation of precipitates can remove some magnesium from the anode, resulting in excessive consumption of magnesium, which is also known as the chunk effect [12].

A second problem associated with Mg-based primary batteries is their low discharge voltage. There are usually lower reported discharge voltages than 1.6 V, which is significantly lower than the theoretical value of 3.1 V. This issue is associated with three aspects: (i) the high charge transfer resistance of Mg anode, as reported by Chen et al. [6]; (ii) the resistance from the accumulation of discharge products on the anode and cathode surface; and (iii) the decreased anodic reaction potential due to the addition of alloying elements in the Mg anode. However, research on the discharge voltage of Mg-air batteries is sparse, which is indeed an aspect for future researchers.

5.3 THE DEVELOPMENT OF THE HIGH-PERFORMANCE MG ANODES

As one of the major parts of the Mg-air battery, the Mg anode plays a key role in the discharge performance. Therefore, the development of the high-performance Mg anode has attracted increasing attention during the past decades. An overview of the intrinsic discharge performance of pure magnesium is presented in this section, which serves as a benchmark for the Mg-based anode. This section also presents some efforts to enhance the discharge properties of a magnesium anode using alloying methods and advanced manufacturing approaches.

5.3.1 THE INTRINSIC DISCHARGE PERFORMANCE OF PURE MG

The impurities (i.e. Fe, Ni, Cu) are the most common factors for the corrosion of pure Mg as well as the discharge performance for Mg anode. As mentioned in section 5.2, these impurities can cause the wasteful consumption of Mg. Therefore, it is important to keep impurities below the tolerance limit. The tolerance limit for cast Mg and alloys is about 180 ppm for Fe, 1000 ppm for Cu, and 5 ppm for Ni [13]. Specific research on the discharge behavior of ultra-high-purity (UHP) Mg anode for Mg-air battery was reported by Chen et al. using both experiment and modeling methods [6]. The results reveal that (i) instead of the resistance caused by discharge products, the main anodic resistance is the charge transfer resistance; (ii) for the UHP Mg anode, the only reason behind the drop in anodic efficiency is the anodic hydrogen reaction; and (iii) during discharge of UHP Mg, a strong dissolution orientation occurs at low current densities, which disappears with increasing current densities. The UHP Mg shows a peak discharge capacity and anodic efficiency of 1391 mAh/g and 63 % at 20 mA/cm^2, respectively, which should be regarded as the benchmark of the intrinsic discharge properties of Mg.

5.3.2 THE ALLOYING ELEMENTS

Alloying other elements into Mg can change the phase composition to improve such properties as strength and formability. Alloying to produce improved Mg anodes has been an active research area since 1967. With the development in these decades, several Mg-based alloy systems have been used in the Mg-air battery, including Mg-Al, Mg-Zn, Mg-Li, Mg-Sn, and Mg-Ca.

Al is the most popular alloying element to Mg because of its cheap price, relatively low density, and solubility, which produces the most widely used AZ series (Mg-Al-Zn) alloys. The maximum solubility of Al in Mg is 12 wt.%, but it decreases significantly with decreasing temperature, and as a consequence, the β phase ($Mg_{17}Al_{12}$) usually appears in the Mg-Al alloy when the Al content is higher than 3 wt.%. The Mg-Al-based anodes generally present relatively high anodic efficiency and discharge capacity (i.e. an anode efficiency of 78.9% and a discharge capacity of 1826.5 mWh/g for an Mg-9Al anode [14]), which is attributed to the suppressed self-corrosion by Al. However, the higher content of Al may cause passivation, decreasing the discharge voltage. Among all AZ series alloys, AZ31 is usually used as the control group in many studies. Anode performance of AZ31 anode was investigated by Nakatsugawa et al., who concluded that a higher discharge current promoted uniform dissolution and increased anode efficiency [15]. Based on a comparative study conducted by Ma et al., the AZ61 alloy showed the best electrochemical performance in terms of discharge voltage and anodic capability of Mg-air batteries [11]. Mg-Al-Pb is another popular anodic material for Mg-air batteries. Wang and co-authors investigated the discharge performance of Mg-6Al-5Pb and concluded as follows [16]. The precipitation of Pb^{2+} from the surface of the alloy as different forms of lead oxides (PbO_x) during the discharge process, which in turn aids in the precipitation of Al^{3+} as $Al(OH)_3$. The $Mg(OH)_2$ film can therefore be easily removed by combining the $Mg(OH)_2$ with the $Al(OH)_3$, thereby increasing the anodic activity of the Mg-Al-Pb anode. This anode system was optimized by Wen et al., who found that Mg-6Al-7Pb-0.5Zn provided the best discharge performance [17].

Mg-Zn-based alloy is the second most popular alloying system that also has attracted increasing attention as the anode for Mg-air batteries. The effect of Zn content (from 2 wt.% to 15 wt.%) on the discharging performance of Mg-Zn binary alloy as anode for Mg-air batteries was investigated by Tong et al., which indicated that with the increase of Zn content, the volume fraction of Mg-Zn phases increased, thereby increasing corrosion rates and reducing discharging performance [18]. The $Ca_2Mg_6Zn_3$ phase tends to be formed in an Mg-2Zn-1Ca anode, which is beneficial to decreasing the chunk effect and then improving the discharge performance [19]. A quasicrystal phase containing Mg-Zn-Y alloy was reported by Chen et al. as the anode of the Mg-air battery, which had a better open circuit potential, electrochemical activity, corrosion resistance, and discharge performance than the ZK60 anode [20]. Controlling the Zn/Y ratio can create another unique structure named the long period stacking ordered (LPSO) phase, which helps improve the self-corrosion resistance and discharge performance at a low volume fraction [21].

Li has attracted attention as an additional element to the Mg anodes because of its high electrochemical activity and ability to modify the crystal structure. Liu et al. provided a comprehensive investigation of the discharge properties of Mg-Li anodes with different crystal structures, namely the α-Mg phase only, the β-Li phase only, and the α-Mg + β-Li dual-phase [22, 23]. The results indicate that: (i) the increase of β phase is beneficial to the higher electrochemical activity and better discharge performance; (ii) the discharge product of β-Li based anode is dissolvable LiOH which peels away easily, resulting in the stable discharge potential; and (iii) as a result of the prior dissolution of the β-Li phase and the dissolvable reaction product, the hydrogen evolution during discharge is suppressed. Due to the ability to increase the anode activity, Sn is added to the Mg-based anode [24]. The Sn content plays a key role in the electrochemical properties and discharge performance,

Materials and Technologies of Mg-Air Primary Batteries

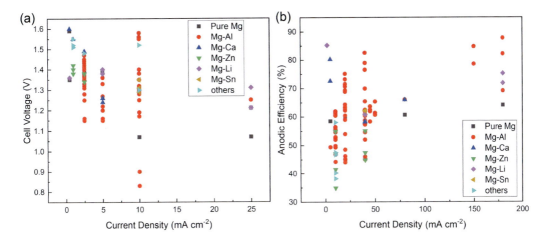

FIGURE 5.2 The best discharge properties of different Mg-based anodes: (a) discharge voltage and (b) anodic efficiency.

as demonstrated by Gu et al. [7]. Their results showed that appropriate Sn content presented a better discharge performance while high Sn content caused passivation during the discharge.

In recent years, Ca has been used to modify a magnesium-based anode, since it has a more negative electrode potential than Mg, and therefore, the addition of Ca into Mg usually shifts the open circuit potential to more negative values. Deng et al. systematically studied the discharge performance of Mg-Ca binary alloys (Ca content 0.1, 0.5, 1, 2, 4 wt. %) as anodes for primary Mg-air batteries [25]. Their results revealed that the micro-alloyed Mg-Ca anode (i.e., 0.1 wt.%) had a higher self-corrosion resistance and a better discharge performance due to the high electrochemical activity of Ca, while a high content of Ca produced Mg_2Ca particles, which were harmful to the anodic efficiency. However, Liu et al. believed that Mg_2Ca particles were helpful for the uniform consumption of the Mg anode during discharge for an as-extruded Mg-1.5 wt.% Ca anode [26]. Figure 5.2 presents the best discharge performance (discharge voltage and anodic efficiency) of different Mg-based anodes. The Mg-Al-based alloys are the most common anodic materials for Mg-air batteries, as shown in the number of red dots in Figure 5.2. Other alloy systems have received relatively little research.

According to Figure 5.2a, there is usually a higher discharge voltage at a lower current density. The discharge voltage tends to decrease with the increase of current density, which is generally attributed to the increased IR drop. So far, the highest discharge voltage is possessed by the Mg-0.1Ca with the value of 1.6 V at 0.5 mA/cm² [25], which is slightly higher than the value of 1.57 V for the UHP Mg anode [6]. The extruded and annealed Mg-6Al-1Sn anode presents a relatively high discharge voltage of 1.55 V at 10 mA/cm², which is the result of the function of self-peeling of discharge products because of the second phases and the fine-grained structure [27].

Indeed, the exfoliation of discharge products plays a key role in the discharge voltage, which is considered the major reason for the voltage drop of the Mg-air battery. However, recent results from Chen et al. present different opinions [6]. The anodic reaction voltage (−2.69 V) and the cathodic reaction voltage (0.4 V) compose the theoretical voltage of the Mg-air battery (3.1 V). Therefore, the operating discharge voltage (E) of the Mg-air battery can be described as Eq. (7).

$$E = E_{theo} - i(R_{anode} + R_{cathode} + R_{electrolyte}) \qquad (7)$$

where E_{theo} is the theoretical voltage, i is the applied current density, R_{anode}, $R_{cathode}$, and $R_{electrolyte}$ is the resistance from the anode, cathode, and electrolyte, respectively. The $R_{cathode}$ results from the cathode reaction, which involves the diffusion and catalysis of O_2 from the air into the electrolyte. The $R_{electrolyte}$ is normally less than 5 Ω cm^2 in 3.5 wt.% NaCl solution (the common electrolyte), which should be steady because the anode is positioned at the same position in a specific battery testing system. The R_{anode} is composed of the R_p, the resistance caused by the growth of discharge products, and R_{ct}, the charge transfer resistance due to anodic reaction. Recent results from Chen et al., reveal that the R_p is quite lower than the R_{ct} [6]. Additionally, due to the impact of the hydrogen bubble, the discharge products can exfoliate from the anode, which may reach a dynamic balance between the accumulation and exfoliation. The impurities and some alloying elements may also decrease the OCP value of Mg to reduce the theoretical anodic reaction voltage, which is another reason for the difference between the operating voltage and theoretical voltage.

A typical method used to calculate anodic efficiency is the mass loss method, which is based on the anode's ability to convert energy into electricity. In general, anodic efficiency increases with an increase in applied current density. Nevertheless, some studies suggest that anodic efficiency achieves a maximum value at a specific current density and decreases with increasing current density. According to Figure 5.2b, Mg-Zn-based anodes commonly presents a low anodic efficiency from 35% to 55%. Mg-Li alloy systems, on the other hand, have high anodic efficiency with values exceeding 60%. The Mg-Al alloy system presents a relatively wide range of 45% to 85%. Existing anodes based on Mg range in anodic efficiency between 30% and 90%. In this area of research, the primary objective is to develop an anode with 100% anodic efficiency. Unfortunately, this objective appears to be difficult to achieve due to NDE or anodic hydrogen reactions, as shown in section 5.2. Because of the very close relationship between the anodic hydrogen evolution and anodic efficiency, it is therefore recommended for future researchers to collect the hydrogen during the discharge.

5.3.3 The Influence of the Production Method

Mg anodes do not need to possess high mechanical properties, such as strength and ductility, to serve as an energy source, which makes the most of Mg-based anodes being as-cast alloys. Even so, the production process has an impact on the microstructure and the properties of the magnesium alloy.

Casting is the most common method used to manufacture magnesium anodes. The advantages of casting are easy customization and cheap. Chen et al. produced the AZ80 anodes using the direct-chill casting coupled with different external fields, as shown in Figure 5.3a [28]. During the casting process, the author employed several external physical fields, including the low-frequency electromagnetic field (LEF), the fixed-frequency ultrasonic field (FUF), and the variable-frequency ultrasonic field (VUF). The results showed that external fields refined both a-Mg grains and $Mg_{17}Al_{12}$ phases (as shown in Figure 5.3b and 3c) due to the positive function during the solidification such as ultrasonic cavitation and ultrasonic streaming by ultrasonic fields and forced convection by LEF [29, 30]. The discharge performance of AZ80 anode was significantly improved through such microstructure modification (Figure 5.3d).

Hot deformation is another useful method to modify the microstructure of Mg anodes, especially for the texture and grain orientation. There are two main methods for producing magnesium anodes: hot extrusion and hot rolling. Hot extrusion provides strong three-dimensional compressive stress during the deformation, which can smash grains and some second-phase particles to refine the microstructure. Therefore, the hot extrusion is used to produce fine-grained Mg anodes. Wang et al. investigated the influence of extrusion ratio on microstructure (Figure 5.4a) and discharge performance (Figure 5.4b) of the Mg-Al-Pb-Ce-Y anode [31]. Anodes produced with a small extrusion ratio had the finest grain size and the strongest basal texture, which improved corrosion resistance, however, some dense dislocations and nano-sized $Mg_{0.78}Pb_{0.22}$ phases adversely affected anodic efficiency and discharge voltage. The increased die exit temperature caused coarsening of the grains and weakening of the basal texture as the extrusion ratio was increased, but it also removed the nano-sized $Mg_{0.78}Pb_{0.22}$ phases and reduced the amount of dislocation, enhancing the discharge properties.

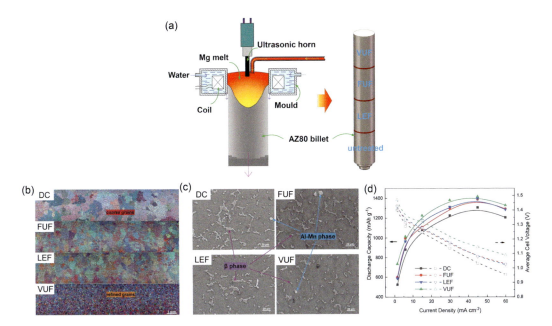

FIGURE 5.3 (a) Schematic of the casting process for chill-direct casting process with the participation of external fields. (b) The influence of different external fields on grain size. (c) The distribution and morphology of $Mg_{17}Al_{12}$ in Mg anodes and (d) the discharge performance of AZ80 anodes. DC (direct-chill casting), FUF (fixed-frequency ultrasonic field), LEF (low-frequency ultrasonic field), and VUF (variable-frequency ultrasonic field). (Adapted with permission [28]. Copyright (2020), Elsevier.)

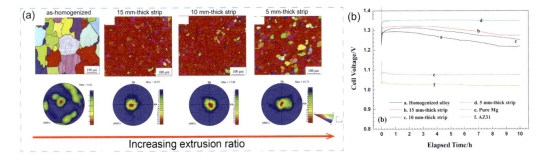

FIGURE 5.4 (a) BESD images of as-extruded Mg-Al-Pb-Ce-Y anodes with different extrusion ratios, showing the grain size and texture evolution, and (b) the discharge curves at 10 mA/cm² in NaCl electrolyte of Mg-air batteries. (Adapted with permission [31]. Copyright (2019), Elsevier.)

Hot rolling is the common method to manufacture the Mg sheets, which is very suitable for Mg anode. However, due to the HCP structure of Mg and the inhomogeneous deformation during hot rolling, the rolling surface (RS) and the cross-section surface (CS) generally present different crystallographic orientations. For instance, Zhao et al. reported that compared to the RS surface of the AZ31 anode, the CS surface had better corrosion resistance and anodic efficiency [32]. Results from Wang et al. indicated that the (0001) basal crystal plane was the dominant texture of the RS of the as-rolled AP65 anode, which was beneficial to the anode efficiency but decreased the discharge voltage [33].

Anodes of Mg containing soluble elements can be subject to heat treatment, which can modify the phase composition, distribution, and morphology. For example, heat treatment changed the phase composition for the Mg-Gd-Zn anode, as reported by Chen et al. [9]. They employed T4 and T6 heat

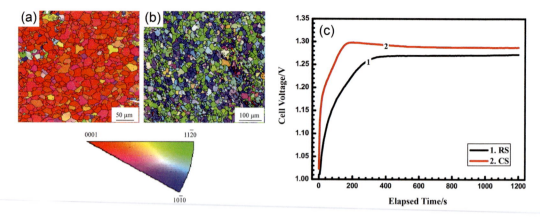

FIGURE 5.5 EBSD images of as-rolled AP65 anode on (a) RS and (b) CS, showing the grain size and crystallographic orientation. (c) the discharge curves at 10 mA cm^{-2} in 3.5 wt. % NaCl as anode for Mg-air battery. (Adapted with permission [33]. Copyright (2018), Elsevier.)

treatment to fabricate the $Mg_{95.28}Gd_{3.72}Zn_{1.00}$ anode with a 14 H LPSO phase. The results showed that the T6-treated anode performed better than the cast anode due to the LPSO phase. A recent study indicates that the heat treatment also eliminates the harmful effect of Mg-Al particles and Mg-Zn particles for Mg-8Al and Mg-6Zn anodes, significantly increasing the anodic efficiency [34].

5.4 ELECTROLYTE AND ADDITIVES

The electrolyte is one of the three main parts of Mg-air batteries, which directly influence the discharge performance of the battery. Typically, the electrolyte of Mg-based primary batteries is an aqueous solution, i.e. NaCl solution. Other reported aqueous electrolytes included $NaNO_3$, $NaPO_4$, $MgSO_4$, and $Mg(NO_3)_2$ [35, 36]. Although some anions can influence the discharge performance somehow, they cannot essentially solve the issue of low anodic efficiency due to parasitical self-corrosion, as stated in section 2. The early attempt of electrolyte additive for Mg-based batteries was back in 2015. Dinesh et al. [37] added water-soluble graphene (WSG) to 3.5 wt. % NaCl solution as the electrolyte for Mg-air battery. The WSG improved discharge capacity via the decreased self-corrosion due to deposited graphene on the anode surface as a protection layer. However, the deposition hindered the direct contact between the electrolyte and the anode, resulting in a decrease in the discharge voltage.

As mentioned earlier, the two major issues of the Mg-air battery are low energy conversion rate and low discharge voltage. To solve these problems, using electrolyte additives has gradually become a feasible method, which attracts increasing attention in this aspect. A comprehensive work was done by Höche et al. [38] in 2018, who studied the influence of the addition of several iron complexing agents to 0.5 wt. % NaCl solution on discharge performance of Mg-based primary batteries. The theoretical basis was the redeposition of Fe [39, 40]. The concept was to use additives to consume the Fe and Mg ions, which weakened the side effects from impurities to improve the anodic efficiency and prohibited the formation of $Mg(OH)_2$ on the anode surface. Despite the theoretical controversy, their results provided basic data on the functions of these additives, which were classified as follows: (i) suppress self-corrosion and remove discharge products: glycolate and salicylate; (ii) suppress self-corrosion but decrease discharge voltage: dodecylbenzenesulfonic acid sodium salt (DBS) and oxalate; (iii) remove discharge products but accelerate the self-corrosion: Tiron, ethylenediaminetetraacetic acid dipotassium salt dehydrate (K_2-EDTA) and nitrilotriacetic acid disodium salt (NTA).

One of the advantages of electrolyte additive is that it can be customized according to the target Mg-based anode. For example, Wang et al. customized the dedicated Mg^{2+} complexing agents for

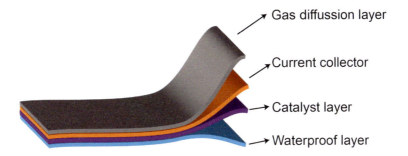

FIGURE 5.6 The conventional four-layer structure of the air cathode.

an Mg-Ca anode, which increased the discharge voltage and specific energy of respective Mg-air batteries [41]. To avoid the usage of the aqueous solution, gel electrolyte was also an option for an Mg-based primary battery. A poly(ethylene oxide) (PEO) organic gel and a crosslinked polyacrylamide (PAM) hydrogel composed a dual-layer gel electrolyte [42]. This gel electrolyte dramatically increased the anodic efficiency to 99.3%, which was the highest value in all reported batteries. This battery shows promising application potential due to its high flexibility and deformability.

5.5 THE AIR CATHODE

According to the discussion in section 5.3.2, the air cathode is also a key factor in the performance of the Mg-air battery. The cathodic reaction is generally the oxygen reduction reaction (ORR), which is shown in Eq. (2). Typically, the air cathode consists of a gas diffusion layer, a current collector, a catalyst layer, and a waterproof layer as shown in Figure 5.6. The waterproof layer separates the electrolyte from the air. The current controller is usually made of metal mesh (i.e. Ni or steel). The catalyst contains the effective catalysts to accomplish the cathodic reaction. Therefore, the gas diffusion layer and the catalysts are the most important parts of the performance of the air cathode. Studies on the gas diffusion layer are not quite rare, while research on the catalysts is more thriftily. Pt is one of the ideal ORR catalysts due to its high efficiency. However, the high price hinders its commercialization in Mg-air batteries. Seeking an economical low-Pt or non-Pt catalyst has then become a topic. One pathway is to add other metals to Pt, including Fe, Ni, Co, Cu, and Ag. To further decrease the cost, some researchers used non-metal materials, such as carbon-based materials, transition metal oxides (MnO_2), and perovskite-based oxides [43]. For example, Li et al. reported a mixed-phase mullite ($SmMn_2O_5$) electrocatalyst, which had a similar catalytic activity but superior stability to Pt/C in 1 M NaCl solutions [44]. Moreover, the support material also plays a role in the cathode performance, particularly for those oxide catalysts because of the low electric conductivity. The common support materials are usually carbon-based, including graphene, carbon nanofibers (CNFs), and the aforementioned multi-walled carbon nanotubes (MWCNTs) [45].

5.6 OUTLOOK

This chapter covered several aspects of the fast-growing Mg-air primary battery, including the discharge nature of Mg anode in aqueous solution, the effects of alloying and manufacturing methods for Mg anode, the development of electrolyte and additives, and the progress in the air cathode. Several fundamental research studies have already proven that Mg-air batteries have significant potential as a next-generation, advanced primary battery. Consequently, it attracts an increasing number of researchers to participate. However, unsatisfied battery performance still hinders its commercialization and wide application. To this end, more efforts should be inputted to promote the better performance of different components.

The development of advanced Mg-based anode requires more investigation of broad alloy systems and manufacturing technologies. It is still unknown whether any allotting elements can reduce the HER and NDE. Additionally, it would be very interesting to observe the usage of other advanced manufacturing methods to improve the discharge properties of Mg anodes. The improvement of air cathode can be conducted by structure modification, catalyst optimization, and appropriate choice of substrate materials. For electrolytes, two aspects of the electrolyte additives and nonaqueous electrolytes are both important. It is expected to see the discovery of new electrolyte additives and the development of a synergistic mixture strategy.

REFERENCES

[1] M. Deng, L. Wang, B. Vaghefinazari, W. Xu, C. Feiler, S.V. Lamaka, D. Höche, M.L. Zheludkevich, D. Snihirova, High-energy and durable aqueous magnesium batteries: Recent advances and perspectives, Energy Storage Materials. 43 (2021) 238–247.

[2] X. Chen, Y. Jia, Z. Shi, Q. Le, J. Li, M. Zhang, M. Liu, A. Atrens, Understanding the discharge behavior of an ultra-high-purity Mg anode for Mg-air primary battery, J. Mater. Chem. A. 9(37) (2021) 21387–21401.

[3] J. Bockris, A. Reddy, Modern electrochemistry, 6th printing. Plenum Press, New York, 1977.

[4] A. Atrens, G.L. Song, M. Liu, Z. Shi, F. Cao, M.S. Dargusch, Review of recent developments in the field of magnesium corrosion, Adv. Eng. Mater. 17 (2015) 400–453.

[5] X. Liu, J. Xue, D. Zhang, Electrochemical behaviors and discharge performance of the as-extruded Mg-1.5 wt% Ca alloys as anode for Mg-air battery, Journal of Alloys and Compounds. 790 (2019) 822–828.

[6] X. Chen, Y. Jia, Z. Shi, Q. Le, J. Li, M. Zhang, M. Liu, A. Atrens, Understanding the discharge behavior of an ultra-high-purity Mg anode for Mg-air primary battery, J. Mater. Chem. A. 9 (2021) 21387–21401.

[7] X.-J. Gu, W.-L. Cheng, S.-M. Cheng, H. Yu, Z.-F. Wang, H.-X. Wang, L.-F. Wang, Discharge behavior of Mg—Sn—Zn—Ag alloys with different Sn contents as anodes for Mg-air batteries, Journal of the Electrochemical Society, 167 (2020) 020501.

[8] X. Chen, Q. Zou, Q. Le, S. Ning, C. Hu, X. Li, A. Atrens, Two distinct roles of Al2Sm and Al11Sm3 phases on the corrosion behavior of the magnesium alloy Mg-5Sm-xAl, Progress in Natural Science: Materials International. 31 (2021) 599–608.

[9] X. Chen, H. Wang, Q. Zou, Q. Le, C. Wen, A. Atrens, The influence of heat treatment on discharge and electrochemical properties of Mg-Gd-Zn magnesium anode with long period stacking ordered structure for Mg-air battery, Electrochimica Acta. 367 (2021) 137518.

[10] X. Chen, H. Wang, Q. Le, Y. Jia, X. Zhou, F. Yu, A. Atrens, The role of long-period stacking ordered phase on the discharge and electrochemical behaviors of magnesium anode Mg-Zn-Y for the primary Mg-air battery, International Journal of Energy Research. 44 (2020) 8865–8876.

[11] J. Ma, G. Wang, Y. Li, F. Ren, A.A. Volinsky, Electrochemical performance of Mg-air batteries based on AZ series magnesium alloys, Ionics. 25 (2019) 2201–2209.

[12] M. Deng, L. Wang, D. Höche, S.V. Lamaka, D. Snihirova, B. Vaghefinazari, M.L. Zheludkevich, Clarifying the decisive factors for utilization efficiency of Mg anodes for primary aqueous batteries, Journal of Power Sources. 441 (2019) 227201.

[13] R. McNulty, J. Hanawalt, Some corrosion characteristics of high purity magnesium alloys, Transactions of the Electrochemical Society. 81 (1942) 423.

[14] X. Li, H. Lu, S. Yuan, J. Bai, J. Wang, Y. Cao, Q. Hong, Performance of Mg—9Al—1In alloy as anodes for Mg-air batteries in 3.5 wt% NaCl solutions, Journal of the Electrochemical Society, 164 (2017) A3131–A3137.

[15] I. Nakatsugawa, Y. Chino, Performance of AZ31 alloy as anodes for primary magnesium-air batteries under high current discharge, Materials Transactions. 61 (2020) 200–205.

[16] N.-G. Wang, R.-C. Wang, C.-Q. Peng, F. Yan, X.-Y. Zhang, Influence of aluminium and lead on activation of magnesium as anode, Transactions of Nonferrous Metals Society of China. 20 (2010) 1403–1411.

[17] L. Wen, K. Yu, H. Xiong, Y. Dai, S. Yang, X. Qiao, F. Teng, S. Fan, Composition optimization and electrochemical properties of Mg-Al-Pb-(Zn) alloys as anodes for seawater activated battery, Electrochimica Acta. 194 (2016) 40–51.

[18] F. Tong, X. Chen, Q. Wang, W. Gao, Hypoeutectic Mg—Zn binary alloys as anode materials for magnesium-air batteries, Journal of Alloys and Compounds. (2020) 157579.

[19] C. Gong, X. He, D. Fang, B. Liu, X. Yan, Effect of second phases on discharge properties and corrosion behaviors of the Mg-Ca-Zn anodes for primary Mg-air batteries, Journal of Alloys and Compounds. 861 (2021) 158493.

[20] X. Chen, Q. Zou, Q. Le, J. Hou, R. Guo, H. Wang, C. Hu, L. Bao, T. Wang, D. Zhao, The quasicrystal of Mg—Zn—Y on discharge and electrochemical behaviors as the anode for Mg-air battery, Journal of Power Sources. 451 (2020) 227807.

[21] X. Chen, H. Wang, Q. Le, Y. Jia, X. Zhou, F. Yu, A. Atrens, The role of long period stacking ordered phase on the discharge and electrochemical behaviors of magnesium anode Mg-Zn-Y for the primary Mg-air battery, International Journal of Energy Research. 44(11) (2020) 8865–8876.

[22] X. Liu, J. Xue, S. Liu, Discharge and corrosion behaviors of the α-Mg and β-Li based Mg alloys for Mg-air batteries at different current densities, Mater. Des. 160 (2018) 138–146.

[23] X. Liu, S. Liu, J. Xue, Discharge performance of the magnesium anodes with different phase constitutions for Mg-air batteries, Journal of Power Sources, 396 (2018) 667–674.

[24] B. Graver, A.M. Pedersen, K. Nisancioglu, Anodic activation of aluminum by trace element tin, ECS Transactions, 16 (2009) 55–69.

[25] M. Deng, D. Höche, S.V. Lamaka, D. Snihirova, M.L. Zheludkevich, Mg-Ca binary alloys as anodes for primary Mg-air batteries, Journal of Power Sources. 396 (2018) 109–118.

[26] X. Liu, J. Xue, D. Zhang, Electrochemical behaviors and discharge performance of the as-extruded Mg-1.5 wt%Ca alloys as anode for Mg-air battery, Journal of Alloys and Compounds. 790 (2019) 822–828.

[27] H. Xiong, K. Yu, X. Yin, Y. Dai, Y. Yan, H. Zhu, Effects of microstructure on the electrochemical discharge behavior of Mg-6wt% Al-1wt% Sn alloy as anode for Mg-air primary battery, Journal of Alloys and Compounds. 708 (2017) 652–661.

[28] X. Chen, S. Ning, Q. Le, H. Wang, Q. Zou, R. Guo, J. Hou, Y. Jia, A. Atrens, F. Yu, Effects of external field treatment on the electrochemical behaviors and discharge performance of AZ80 anodes for Mg-air batteries, J. Mater. Sci. Technol. 38 (2020) 47–55.

[29] Y. Jia, J. Hou, H. Wang, Q. Le, Q. Lan, X. Chen, L. Bao, Effects of an oscillation electromagnetic field on grain refinement and Al8Mn5 phase formation during direct-chill casting of AZ31B magnesium alloy, Journal of Materials Processing Technology, 278 (2020) 116542.

[30] X. Chen, Y. Jia, Q. Liao, W. Jia, Q. Le, S. Ning, F. Yu, The simultaneous application of variable frequency ultrasonic and low frequency electromagnetic fields in semi continuous casting of AZ80 magnesium alloy, Journal of Alloys and Compounds, 774 (2019) 710–720.

[31] N. Wang, W. Li, Y. Huang, G. Wu, M. Hu, G. Li, Z. Shi, Wrought Mg-Al-Pb-RE alloy strips as the anodes for Mg-air batteries, Journal of Power Sources. 436 (2019) 226855.

[32] Z. Yanchun, H. Guangsheng, Z. Cheng, C. Lin, H. Ting-zhuang, P. Fusheng, Effect of texture on the performance of Mg-air battery based on rolled Mg-3Al-1Zn alloy sheet, Rare Metal Mat. Eng., 47 (2018) 1064–1068.

[33] N. Wang, Y. Mu, W. Xiong, J. Zhang, Q. Li, Z. Shi, Effect of crystallographic orientation on the discharge and corrosion behaviour of AP65 magnesium alloy anodes, Corrosion Science. 144 (2018) 107–126.

[34] X. Chen, Q. Zou, Q. Le, M. Zhang, M. Liu, A. Atrens, Influence of heat treatment on the discharge performance of Mg-Al and Mg-Zn alloys as anodes for the Mg-air battery, Chemical Engineering Journal. (2021) 133797.

[35] J. Xu, Q. Yang, C. Huang, M.S. Javed, M.K. Aslam, C. Chen, Influence of additives fluoride and phosphate on the electrochemical performance of Mg—MnO_2 battery, Journal of Applied Electrochemistry. 47 (2017) 767–775.

[36] F.W. Richey, B.D. McCloskey, A.C. Luntz, Mg anode corrosion in aqueous electrolytes and implications for Mg-Air batteries, Journal of the Electrochemical Society. 163 (2016) A958–A963.

[37] M.M. Dinesh, K. Saminathan, M. Selvam, S. Srither, V. Rajendran, K.V. Kaler, Water soluble graphene as electrolyte additive in magnesium-air battery system, Journal of Power Sources. 276 (2015) 32–38.

[38] D. Höche, S.V. Lamaka, B. Vaghefinazari, T. Braun, R.P. Petrauskas, M. Fichtner, M.L. Zheludkevich, Performance boost for primary magnesium cells using iron complexing agents as electrolyte additives, Scientific Reports. 8 (2018) 1–9.

[39] D. Höche, C. Blawert, S.V. Lamaka, N. Scharnagl, C. Mendis, M.L. Zheludkevich, The effect of iron re-deposition on the corrosion of impurity-containing magnesium, Physical Chemistry Chemical Physics. 18 (2016) 1279–1291.

[40] S. Lamaka, D. Höche, R. Petrauskas, C. Blawert, M. Zheludkevich, A new concept for corrosion inhibition of magnesium: Suppression of iron re-deposition, Electrochemistry Communications. 62 (2016) 5–8.

[41] L. Wang, D. Snihirova, M. Deng, B. Vaghefinazari, S.V. Lamaka, D. Höche, M.L. Zheludkevich, Tailoring electrolyte additives for controlled Mg-Ca anode activity in aqueous Mg-air batteries, Journal of Power Sources. 460 (2020) 228106.

[42] L. Li, H. Chen, E. He, L. Wang, T. Ye, J. Lu, Y. Jiao, J. Wang, R. Gao, H. Peng, High-energy-density magnesium-air battery based on dual-layer gel electrolyte, Angewandte Chemie International Edition. 60 (2021) 15317–15322.

[43] F. Cheng, J. Chen, Metal-air batteries: From oxygen reduction electrochemistry to cathode catalysts, Chemical Society Reviews. 41 (2012) 2172–2192.

[44] Y. Li, X. Zhang, H.-B. Li, H.D. Yoo, X. Chi, Q. An, J. Liu, M. Yu, W. Wang, Y. Yao, Mixed-phase mullite electrocatalyst for pH-neutral oxygen reduction in magnesium-air batteries, Nano Energy. 27 (2016) 8–16.

[45] P. Yue, Z. Li, S. Wang, Y. Wang, MnO_2 nanorod catalysts for magnesium-air fuel cells: Influence of different supports, International Journal of Hydrogen Energy. 40 (2015) 6809–6817.

6 Materials and Technologies of Al-Air Batteries

Weng Cheong Tan, Lip Huat Saw, Ming Chian Yew, and Ming Kun Yew

CONTENTS

- 6.1 Introduction 77
- 6.2 Aluminum-Air Battery 78
- 6.3 Aluminum Anode 79
  - 6.3.1 Pure Aluminum 79
  - 6.3.2 Aluminum Alloy 79
- 6.4 Electrolyte 81
  - 6.4.1 Aqueous Electrolyte 81
    - 6.4.1.1 Alkaline Electrolyte 81
    - 6.4.1.2 Saline Electrolyte 82
    - 6.4.1.3 Acid Electrolyte 82
    - 6.4.1.4 Dual Electrolyte 83
  - 6.4.2 Non-Aqueous Electrolyte 84
    - 6.4.2.1 Gel Electrolyte (Solid Electrolyte) 84
    - 6.4.2.2 Ionic Liquid Electrolyte 85
- 6.5 Cathode 85
- 6.6 Application of Aluminum-Air Battery 86
- 6.7 Conclusion 87
- Acknowledgments 87
- References 87

6.1 INTRODUCTION

Industrial revolution 4.0 pushes the agenda on promoting renewable energy globally to combat the issues of global warming. More research work needs to be done to achieve a sustainable energy system. The development of renewable energy is crucial in reducing the dependency on generating electricity through burning fossil fuels, which contribute to releasing harmful gas into the environment. Renewable energy such as solar energy, wind energy, and tidal energy are introduced to harvest energy from nature. However, these energy harvesting systems suffer common drawbacks in which the power output is highly dependent on the weather, time, and location. This will cause fluctuation in power output and affects the operation of the industry and daily routines. Hence, there is a need to introduce new technology to solve the problems.

Metal-air battery is a new type of energy storage system in which electricity is generated through the consumption of a metal anode. Metals such as aluminum, magnesium, zinc, lithium, sodium, and potassium are suitable to be used as the anode for electrochemical reactions. In general, the metal anode is consumed through the oxidation process while the oxygen in the atmosphere undergoes the reduction process at the cathode to generate electricity. Depending on the types of metal-air batteries, different types of by-products will be generated during the electrochemical process.

However, most of the by-products involve their corresponding metal oxide. Among all metal-air battery candidates, aluminum provides huge potential, such as a high specific energy density of 8.1 kWh/kg and a theoretical potential voltage of 2.71 V in the alkaline solution. Besides that, it is abundant in nature and easily available in the market. The recyclability of aluminum is considered the highest among all the metals as it can reach 100 % recycle value.

Although aluminum provides many advantages when used in metal-air batteries, aluminum-air batteries also suffer from some limitations in practical application such as a protective oxide layer forming on the surface of the aluminum anode will limit the rate of electrochemical reaction. On the other hand, aluminum is highly reactive when immersed in an alkaline solution. The alkaline electrolyte used in the aluminum-air battery will induce self-corrosion on the aluminum anode and reduce the coulombic efficiency. Both phenomena affect the electrical performance of the aluminum-air battery and require further work before the commercialization of aluminum-air batteries.

This chapter provides insight into the latest development in the aluminum-air battery, especially on the anode, electrolyte, and cathode. Besides, using different methods and technologies to suppress or reduce the parasitic reaction of the aluminum anode in the aluminum-air battery is discussed.

6.2 ALUMINUM-AIR BATTERY

Among many types of metal-air batteries, the aluminum-air battery with its theoretical specific energy of 8140 Wh/kg, the theoretical gravimetric capacity of 2980 mAh/g, a theoretical voltage of 2.7 V, low cost, high abundance, low energy spent on production, and environmental friendliness has received sustainable interest from worldwide researchers. For a comparison, aluminum-air batteries show higher specific energy which is about 20 times higher than that of the commercial lithium-ion batteries in the market [1]. Besides that, the aluminum-air battery is environmentally friendly due to the natural recyclability of aluminum metal. It can be recycled with minimal losses and possesses low pollution to the environment. Most of the aluminum-air batteries available on the market or undergoing research are primary batteries. In other words, the battery needs to be replaced when the aluminum anode is fully consumed. On the other hand, there is also research on the development of a secondary aluminum-air battery in which the aluminum anode can be recharged [2]. However, the idea is new, and it is still in the infant stage. Works dealing with secondary aluminum-air batteries are limited and most of them are on the laboratory scale. Extensive work and research are needed before commercializing a secondary aluminum-air battery is made possible. In general, the electrochemical reactions that occur within the batteries are as shown next:

Cathode Reaction in an Alkaline Electrolyte

$$O_2 + 2H_2O + 4e^- \rightarrow 4OH^- \qquad E = 0.2 \text{ V vs. Ag/AgCl} \tag{1}$$

Anode Reaction in an Alkaline Electrolyte

$$Al + 3OH^- \rightarrow Al(OH)_3 + 3e^- \qquad E = -2.55 \text{ V vs. Ag/AgCl} \tag{2}$$

The Overall Reaction of a Single Electrolyte System

$$4Al + 3O_2 + 6H_2O \rightarrow 4Al(OH)_3 \qquad E = 2.75 \text{ V} \tag{3}$$

Side Reaction (Parasitic Reaction)

$$2Al + 6H_2O \rightarrow 2Al(OH)_3 + 3H_2 \qquad (4)$$

From the electrochemical reactions, it can be seen that the aluminum anode reacts with hydroxyl ions (OH−) and produces aluminum hydroxide $Al(OH)_3$ as a final product. In this process, electrons are released, which is crucial in electricity generation. On the other hand, the oxygen at the cathode undergoes reduction and receives the electrons emitted by the anode. This completes a cycle and generates electricity in the process. Unfortunately, a side reaction also occurs in which the aluminum reacts with water to form aluminum hydroxide and release hydrogen gas. This reaction is unfavorable and reduces the coulombic efficiency of the aluminum-air battery as some of the aluminum is consumed by the side reaction. If the side reaction is left intact, the aluminum anode will undergo severe corrosion. The side reaction does not release electrons and hence, does not contribute to electricity production. The main obstacle that affects the performance of the aluminum-air battery is the side reaction. It can reduce the specific capacity and shorten the aluminum-air battery's life cycle. Many approaches have been proposed to tackle the side reaction, such as alloying the aluminum anode and adding an electrocatalyst to the electrolyte. In the aluminum-air battery, the aluminum anode will be gradually consumed to generate electricity when the aluminum anode is in contact with the electrolyte in the presence of an air cathode. The battery stops supplying electricity when all the aluminum anode is fully consumed and converted to aluminum hydroxide. The aluminum hydroxide produced can be recycled back to aluminum through the Hall-Heroult process [3]. The recyclability of aluminum is high, indicating that the aluminum-air battery is green and has less pollution.

6.3 ALUMINUM ANODE

An aluminum anode is a primary component in an aluminum-air battery. The selection of aluminum anode is important as it can decide the overall electrical performance of the aluminum-air battery. The feasibility of using aluminum as an anode in a battery started in the 1850s. Many studies have been involved throughout the years that bring aluminum-air batteries to today's technology.

6.3.1 Pure Aluminum

Pure aluminum anode shows good electrochemical properties in the aluminum-air battery as it can produce high energy density and high voltage [4, 5]. However, the main drawback of using pure aluminum anode is the corrosion issues. Pure aluminum anode shows a strong reaction with water which promotes aluminum corrosion. The undesirable product such as $Al(OH)_3$ is formed during the process. The $Al(OH)_3$ layer will be deposited along the surface of the aluminum anode, preventing fresh aluminum from being consumed for electricity generation, and shorten the life cycle of the aluminum-air battery [6]. Another issue that arises is forming a protective oxide layer of the aluminum anode [7]. Pure aluminum tends to form aluminum oxide along the surfaces, reducing the active aluminum available for electrochemical reactions. Although pure aluminum shows good electrochemical properties, it is not recommended to be used as an anode in the aluminum-air battery due to its low electrochemical activity and high self-corrosion rate. Hence, some researchers used aluminum alloy to reduce the effect of self-corrosion.

6.3.2 Aluminum Alloy

The disadvantages of the pure aluminum anode can be minimized by alloying the aluminum anode. The main purpose is to reduce the corrosion rate so that the operating time of the aluminum-air battery can be prolonged. The selection of additives is important as the performance of the aluminum-air battery is highly dependent on the anode and the corrosion properties [8]. Some of the metals that are suitable for metal alloying include Zinc (Zn), Indium (In), Tin (Sn), Gallium (Ga), and Magnesium (Mg). The additional elements can help to reduce the side reaction, which enhances the utilization of aluminum. Indium can control the generation of hydrogen by positively shifting the anode potential. Furthermore, tin can improve the dissolution rate of aluminum in an aqueous solution which in turn reduces the corrosion rate. As for the issues dealing with the oxide film on the aluminum surface, they can be resolved through the introduction of gallium. Gallium helps to keep the oxide film inactive. Although additives seem like a promising solution to improve the aluminum-air battery's performance, not all additives are suitable to be used as alloy material for the aluminum anode. For example, the addition of iron (Fe) will enhance the corrosion rate, reducing the aluminum-air battery's performance. On the other hand, zinc is an interesting additive that can affect the performance of the aluminum-air battery in positive and negative ways. Zinc can reduce hydrogen production and limits the corrosion of aluminum anode. Nevertheless, it can also form an oxide film on the layer. The impact on the performance is debatable depending on which factor is dominant.

Different grades of aluminum alloy also show different impacts on the performance of the aluminum-air battery. Aluminum alloy 1070, a pure aluminum alloy, is known for high corrosion resistance and performed even better with calcium (Ca) as additives. It helps to lower the corrosion rate and shows good corrosion inhibition characteristics. The weight percent of calcium plays an important factor in which the discharge voltage improves with respect to the increase of calcium content in the aluminum alloy. Copper (Cu) is also an excellent element in combating corrosion. Aluminum alloy 7075-T7351 shows good corrosion inhibition properties due to the presence of copper. Copper will form a barrier and shield the anode from corrosion. When the corrosion rate is reduced, it causes an improvement in the specific capacity of the aluminum-air battery.

Comparing high purity and low purity aluminum alloy on the performance of the aluminum-air battery, it is believed that low purity aluminum alloy can perform better in reducing the corrosion rate. For example, the aluminum alloy 1085 (high purity) shows more significant corrosion than aluminum alloy 7475 (low purity). This is because, in the high purity aluminum alloy, the weight percentage of aluminum is higher and with fewer additives. This causes the aluminum to be more readily available and subject to corrosion. On the other hand, low purity aluminum alloy contains a higher weight amount of impurities that help to reduce the corrosion rate. As a result, the specific capacity of aluminum-air batteries using low purity aluminum alloy is always higher than high purity aluminum alloy under the same operating conditions. The study has also revealed that low purity aluminum alloy can also outperform high purity aluminum alloy in terms of power density and energy density [9].

The grain size of the aluminum alloy is also an important factor that affects the performance of the aluminum-air battery. A highly polydisperse grain size aluminum alloy shows poor performance in corrosion inhibition. For example, an aluminum alloy 6061 that contains magnesium and silicon shows a higher corrosion rate than a pure aluminum plate. This is due to the highly dispersed grain size of aluminum alloy 6061, which provides more sites for aluminum corrosion.

Aluminum alloy has been shown to perform superior as a corrosion inhibitor in aluminum-air batteries. The electrical performance of the aluminum-air battery can be improved by replacing the pure aluminum anode with an aluminum alloy anode in the aluminum-air battery. However, three factors need to be considered in determining the effectiveness of aluminum alloy's corrosion inhibition capabilities: types of additives, grain size, and purity of an aluminum alloy. These three factors are mutually exclusive, and proper analysis is needed to determine which aluminum alloy performs the best as a corrosion inhibitor in the aluminum-air battery.

Materials and Technologies of Al-Air Batteries

6.4 ELECTROLYTE

An electrolyte is an important medium in the aluminum-air battery that facilitates ions' movement for electrochemical reactions. The electrolyte can be aqueous, non-aqueous, or ionic liquid in an aluminum-air battery. Different types of electrolytes have their advantages and disadvantages.

6.4.1 Aqueous Electrolyte

An aqueous electrolyte is a type of electrolyte in which solute is dissolved in water to produce ions. Depending on the types of solute, an aqueous electrolyte can be acid, alkaline, or neutral. Since water is involved in preparing the aqueous electrolyte, it is usually associated with a high corrosion rate in the aluminum-air battery. An aqueous electrolyte is the most common use and basic electrolyte in the aluminum-air battery.

6.4.1.1 Alkaline Electrolyte

Alkaline aqueous electrolyte shows a good electrical performance in the aluminum-air batteries, with the main drawback of the short life cycle of the aluminum-air battery. This is because aluminum is very reactive in alkaline solutions. The electrochemical reaction is vigorous and leads to a high electron generation rate and is subjected to a high corrosion reaction. Since both reactions are rapid in the aluminum-air battery, the aluminum-air battery will be consumed rapidly and eventually covered by aluminum hydroxide. The corrosion issue in the alkaline electrolyte is the worst among all the other types of electrolytes. The most popular alkaline electrolytes in the aluminum-air battery are potassium hydroxide (KOH) and sodium hydroxide (NaOH). They are well known for high electrical performance and notorious for corrosion issues. The KOH and NaOH electrolytes contain a high amount of OH- ions which help to increase the electrical performance. The performance can be boosted by increasing the concentration of the electrolyte. The discharge voltage and discharge current are enhanced and positively correlated with the electrolyte concentration. Nevertheless, this leads to a higher corrosion rate and causes a reduction in the aluminum-air battery-specific capacity as more aluminum anode is consumed due to the parasitic reaction. On top of that, the performance of KOH is generally better than NaOH at the same concentration.

Additives also work well in alkaline aqueous solutions to reduce the corrosion rate. Inorganic inhibition such as components that contain polar groups of nitrogen, sulfur, and oxygen can help to reduce corrosion. This is because polar groups can form a protective layer through the binding of the aluminum surface and lead to a slower corrosion rate. Some inhibitors such as sodium stannate (Na_2SnO_3), potassium stannate (K_2SnO_3), methanol, aromatic carboxylic acids, zinc oxide, tin, and indium show good corrosion inhibition performance when added to KOH and NaOH. Although the corrosion rate of the aluminum anode shows a reduction through the introduction of the corrosion inhibitor, it does not mean that the electrical performance of the aluminum-air battery is improved. The battery will tend to operate at a lower discharge voltage and discharge current when a corrosion inhibitor is added to the alkaline electrolyte.

Apart from introducing a single corrosion inhibitor, adding a hybrid addictive to the alkaline electrolyte can also help to further reduce the corrosion of the aluminum-air battery. There are a few combinations of additives such as zinc oxide together with 8-hydroxyquinoline (8-HQ), carboxymethyl (CMC) with zinc oxide, and amino acid with cerium in an alkaline electrolyte that show good corrosion inhibition capabilities. Hybrid additives show good corrosion inhibition as compared to single additives while ensuring better anode utilization efficiency.

Instead of improving the corrosion issues through additives, special designs on the aluminum-air battery structure can also help to suppress the corrosion. The oil displacement method is a method in which a layer of oil is introduced in between the aluminum anode and separator [10]. The aluminum-air battery can generally perform with the aqueous alkaline solution when it is in use. When it is not in use, a pump will be used to circulate the oil and wash away all the alkaline electrolytes. Hence, the electrochemical reactions stop and prevent the aluminum anode from self-discharging. However, this method requires extra components such as oil and a closed-loop pumping system, increasing the operating cost and design complexity.

A new design integrating a hydrogen fuel cell system with an aluminum-air battery is also proposed [11]. This design utilizes the generated hydrogen gas due to the parasitic reaction and converts it into electricity. A hydrogen reservoir is introduced to gather the hydrogen gas generated from the aluminum-air battery. Next, the hydrogen gas is channeled towards a fuel cell system. Since the hydrogen fuel cell receives hydrogen fuel from the aluminum-air battery, the fuel cell can generate electricity. The performance of the fuel cell tends to improve if the corrosion at the aluminum anode in the battery is severe. By integrating both designs, the total power output is generally greater than that of the aluminum-air battery alone. Since this method involves a new design on the aluminum-air battery, it introduces extra cost, and the design suffers from limitations due to the complexity.

6.4.1.2 Saline Electrolyte

Saline electrolytes are also used as an electrolyte in aluminum-air batteries. However, this is a less common selection due to the electrical performance being generally poorer than the alkaline electrolyte. A common example of a saline electrolyte is sodium chloride (NaCl). The aluminum-air battery shows weak performance when employing a pure aluminum anode with sodium chloride as the electrolyte. This is because an oxide layer is formed on the surface of the pure aluminum anode. The sodium chloride is unable to break down the film and no aluminum dissolution occurs on the aluminum surface. The sodium chloride electrolyte will work if the pure aluminum anode is replaced with aluminum alloy. Depending on the types of aluminum alloy, the corrosion rate behaves differently. The main factor determining the corrosion rate of the aluminum alloy in the saline electrolyte is the elements in the aluminum alloy. For example, aluminum alloy that contains magnesium, tin, and gallium shows superior corrosion inhibition capabilities in saline solution [12]. However, in the absence of magnesium, the corrosion occurs just in one day. The presence of indium in an aluminum alloy shows good performance as well. The addition of indium offers a huge leap in performance. With indium as an additive, the electrical performance of a NaCl electrolyte is increased as compared to a NaOH electrolyte [13].

6.4.1.3 Acid Electrolyte

The acid electrolyte can help to minimize the corrosion of the aluminum anode with the trade-off in the power output. Acid electrolytes such as sulphuric acid (H_2SO_4) and hydrochloric acid (HCl) have shown exemplary performance in hindering dendrite formation on the aluminum anode. Thus, reducing the formation of by-products in the electrochemical reaction. The discharge duration of an acid electrolyte is generally longer than an alkaline electrolyte. One of the benefits of using acid electrolytes is enhancing the oxygen reduction rate (ORR) at the cathode. However, it causes another issue in which the oxidation rate at the anode does not catch up with the oxygen reduction reaction. Hence, causing wastage of the energy output. Besides that, using an acidic electrolyte also shows a lower open-circuit voltage (OCV) compared to the alkaline electrolyte. The OCV tends to reduce with a reduction in pH value of the electrolyte [14]. The performance of an acidic electrolyte can be further improved through an additive. An additive such as Xathan hydrogel works well in the acidic electrolyte.

In general, the power output of an aluminum-air battery is the highest for alkaline electrolytes, followed by acid electrolytes, and performed the worst for saline electrolytes. On the other hand, the corrosion inhibition capabilities are reduced in the order of saline electrolyte, acid electrolyte, and alkaline electrolyte. Since there exists no consensus on which electrolyte performs the best in an aluminum-air battery, a new idea is needed to find a solution to compromise the benefits and drawbacks of the electrolyte to produce an aluminum-air battery with better performance.

6.4.1.4 Dual Electrolyte

A dual electrolyte system in an aluminum-air battery is developed to utilize the benefits of alkaline electrolyte and acidic electrolyte. In a dual electrolyte system, an alkaline electrolyte is used as the anode, while the acidic electrolyte is used as the cathode. A layer of a separator is introduced to isolate the electrolyte and prevent mixing of the electrolyte while ensuring a transfer of ions within the battery. The advantage of high-performance alkaline electrolytes and the advantage of high oxygen reduction reaction of acidic electrolytes are combined in the dual electrolyte aluminum-air battery.

The electrochemical reactions that occur in a dual electrolyte system are as shown here:

Cathode Reaction in an Acidic Electrolyte

$$O_2 + 4H^+ + 4e^- \rightarrow 2H_2O \qquad E = 1.03 \text{ V vs. Ag/AgCl} \tag{5}$$

Anode Reaction in an Alkaline Electrolyte

$$Al + 3OH^- \rightarrow Al(OH)_3 + 3e^- \qquad E = -2.55 \text{ V vs. Ag/AgCl} \tag{6}$$

The Overall Reaction of a Dual Electrolyte System

$$4Al + 16OH^- + 3O_2 + 12H^+ \rightarrow 4Al(OH)_4^- + 6H_2O \qquad E = 3.58 \text{ V} \tag{7}$$

The main difference between the single electrolyte and dual electrolyte systems is the cathode reaction. The ORR reaction at the cathode side is greatly improved from a voltage of 0.2 V vs. Ag/AgCl to 1.03 V vs. Ag/AgCl. The theoretical voltage of the overall reaction is improved by 30% for a dual electrolyte system as compared to a single electrolyte. Hence, it is deduced that a dual electrolyte system aluminum-air battery has higher OCV. The dual electrolyte system mainly consists of an alkaline electrolyte at the anode and an acid electrolyte at the cathode. Similar to a single electrolyte system, the most commonly used alkaline electrolytes are KOH and NaOH while the acid electrolytes are HCl and H_2SO_4. Additives discussed previously in the single electrolyte system are also applicable in the dual electrolyte system. By combining two electrolytes, the OCV of the aluminum-air battery can be improved. It is observed that the power density and current density are also improved. Since there is an introduction of separator in dual electrolyte system, the selection of separator is also important. It suggests that an anion exchange membrane is a more suitable separator than a bipolar membrane. Using a bipolar membrane does not show good performance on the battery. Besides that, the thickness of the electrolyte and separator are also important factors affecting the performance of the battery. By reducing the thickness of both the anolyte and catholyte and the separator, the power density can be improved, but it does not show any changes to the OCV of the aluminum-air battery.

The dual electrolyte system aluminum-air batteries offer an interesting concept in which different combinations of electrolytes are possible in the system. For example, developing an aluminum-air battery with saline electrolyte (anolyte) and alkaline electrolyte (catholyte) is possible. Still, the performance is generally poorer than the alkaline/acid electrolyte combination. Another interesting combination is using an aqueous electrolyte and a gel electrolyte as the catholyte. The overall performance is better than an aqueous single electrolyte system. Dual electrolyte system aluminum-air batteries provide different possibilities for study in the future.

In general, a dual electrolyte system aluminum-air battery offers better electrical performance due to improving the ORR reaction at the cathode. This is a good approach as it can break through the limits of a single electrolyte system as the maximum OCV of a single electrolyte system is only

1.8 V up to date. Using a dual electrolyte system with alkaline and acid electrolytes can boost the OCV to more than 2.2 V. Hence, it is good to implement a dual electrolyte system in the aluminum-air battery. However, the effects of the separator play an important role in a dual electrolyte aluminum-air battery. This is because the life cycle will be greatly affected due to the mixing of the anolyte and electrolyte. Although high discharge current and high voltage are achieved, dual electrolyte system aluminum-air batteries generally have a shorter operating life cycle and required extension works before applying in real-life applications.

6.4.2 Non-Aqueous Electrolyte

The main purpose of introducing non-aqueous electrolytes is to solve the corrosion issues faced by the aqueous electrolyte. Most of the non-aqueous electrolytes eliminate the presence of water which is responsible for aluminum corrosion. A secondary aluminum-air battery is made possible with a non-aqueous electrolyte, opening up possibilities of a rechargeable aluminum-air battery in the future.

6.4.2.1 Gel Electrolyte (Solid Electrolyte)

Gel electrolyte is produced using an alkaline solution, gelling agent, and polymerization initiator mixed with distilled water. Additives can be added to the mixture depending on the application. Additives can be added to improve the performance of the aluminum-air batteries. Gel electrolytes remove the electrolyte leakage which is one of the problems in aqueous aluminum-air batteries. Besides that, gel electrolyte shows higher chemical stability and better mechanical strength. Moreover, gel electrolyte shows flexibility in different shapes. It can be made into different shapes for different applications, offering different interesting designs for the aluminum-air battery. Some examples of gel electrolytes are polyacrylic acid (PAA), KOH-based hydroponics gel, KOH-based polyacrylamide, NaOH-based agarose, and polyvinyl alcohol (PVA) based gel electrolyte. The performance of the aluminum-air batteries is affected by the ionic conductivity of the gel electrolyte. Hence, developing a high ionic conductivity gel electrolyte is important to ensure good performance. The gel electrolyte can be based on alkaline electrolytes such as KOH and NaOH, providing good performance as an aqueous electrolyte. In general, gel electrolytes based on alkaline electrolytes show higher ionic conductivity values. The gel electrolyte can be customized depending on the ingredients. A high concentration of gel electrolyte is made possible if a high concentration of KOH or NaOH is used as an ingredient. Gel electrolyte has shown good corrosion inhibition capabilities while improving the aluminum anode's specific capacity compared to an aqueous aluminum-air battery. However, the peak power density is generally lower than an aqueous aluminum-air battery.

It is also possible to combine two different types of gel electrolytes and integrate them to form a single electrolyte. For example, combining the (PVA) and (PAA) gel electrolyte to form a PVA/PAA/KOH matrix electrolyte shows even better performance than a single PVA/KOH electrolyte. The electrical performance is better as it can generate higher power density while offering higher strength and ductility.

A rechargeable aluminum-air battery is made possible by the gel electrolyte. Hydrogel-based gel electrolyte shows rechargeable capabilities but low anodic efficiency. This technology is relatively new, and the understanding of the recharge capability performance is still limited. On the other hand, a gel electrolyte based on an acidic medium shows rechargeability as well. It is produced by mixing $AlCl_3$ (EMlmCl-$AlCl_3$) complexes acrylamide with ethyl-3-methylimidazolium (EMlmCl) via radical polymerization. The gel electrolyte can be electroplated on the aluminum surface, reducing the contact resistance and increasing the battery performance. Adding aluminum chloride ($AlCl_3$) to poly(vinylidene fluoride) (PVDF) helps to provide a stable electrochemical window while enabling rechargeability of the aluminum-air battery. When the $AlCl_3$ is present in the electrolyte, it is possible to develop a secondary aluminum-air battery.

In general, gel electrolytes can help to solve the leakage issues, maintaining a low corrosion rate in the aluminum-air battery while providing flexibility in the battery design. However, the performance is still not on par with the aqueous electrolyte. The main limitation of gel electrolyte is its low ionic conductivity. Hence, extensive study is required to apply gel electrolytes in the aluminum-air battery application.

6.4.2.2 Ionic Liquid Electrolyte

To develop a secondary aluminum-air battery, it is necessary to use ionic liquids electrolytes [15]. Usually, the ionic liquid electrolyte is produced from aluminum chloride ($AlCl_3$) with other organic salts such as urea, acetamide, and methylimidazolium chloride (EMImCl). A chloroaluminate melts ionic liquids electrolytes that allow the aluminum electrodeposition which makes it possible to develop a rechargeable aluminum-air battery [2]. It has the general formulae of MCl-$AlCl_3$, where M consists of a cation such as Na^+, Li^+, or K^+. Organic cation such as imidazolium or pyrrolidinium is also a suitable candidate. The drawback of using chloroaluminate melts ionic liquids electrolyte is its sensitivity to moisture in which water will reduce the anodic and cathodic potential window. Based on the amounts of chloroaluminate melts in $AlCl_3$, the electrolyte can be alkaline, acidic, and even neutral. When the amount of $AlCl_4^-$ is high, the electrolyte is neutral. In an acidic melt, the formation of $Al_2Cl_7^-$ occurs and allows the deposition of aluminum to happen. The equations explaining the pH medium are shown below:

In Basic or Neutral Medium

$$AlCl_4^- + 3e^- \rightarrow Al + 4Cl^- \tag{8}$$

In Acidic Medium

$$4Al_2Cl_7^- + 3e^- \xleftrightarrow{\text{reversible}} Al + 7AlCl_4^- \tag{9}$$

The preceding equation shows that the acidic medium is a reversible reaction indicating that the aluminum can undergo dissolution and deposition which suggests that a secondary aluminum-air battery is possible. The aluminum consumed in the electrochemical reaction can be charged back into the aluminum anode. Since acidic medium allows the development of secondary aluminum-air batteries, it is widely studied and investigated. Some other examples of ionic liquid electrolytes include 1-ethyl-3-methylimidazolium oligo-fluorohydrogenate, and 1-butyl-3-methylimidazolium chloride. Although $AlCl_3$-based ionic liquids electrolytes allow the possibility of developing rechargeable aluminum-air batteries, the performance is limited due to the deposition of Al_2O_3 and $Al(OH)_3$ at both the anode and cathode. Hence, it is important to choose a proper cathode to reduce the accumulation of the by-products. Titanium nitride (TiN) and Titanium carbide (TiC) are good candidates for the air cathode.

6.5 CATHODE

Air cathode is an important component in the aluminum-air battery to provide oxygen for the ORR. It is important to enhance the ORR at the air cathode to achieve good electrical performance. The cathode should possess porous properties so that oxygen can diffuse through it for the electrochemical reaction. The air cathode consists of three main components: the gas diffusion layer (GDL), active catalytic layer, and current collector. A layer of polytetrafluoroethylene (PTFE) is coated on the GDL to prevent the diffusion of water while maintaining permeability for the oxygen to diffuse. The active catalytic layer plays an important role in enhancing the ORR. There are four main issues associated with the air cathode. Firstly, the energy and power density is limited due to the sluggish oxygen reaction limits. Next, the alkaline electrolyte tends to react with carbon dioxide to form carbonate on the cathode. Thirdly, the electrolyte will dry out due to water evaporation happening at the cathode. Lastly, the air cathode electrolyte suffers from water flooding and reduces the diffusion of oxygen.

Platinum (Pt) is a good cathode choice as it can greatly improve the ORR but it is also expensive and limits its usage in real-life applications. Hence, other types of air cathode are introduced. The most commonly used cathode material is activated carbon. Activated carbon is synthesized from coal, coconut, or any carbon-based materials. It possesses many advantages such as low cost, excellent chemical stability, good electrocatalytic activity, and high surface area. Large surface area is an important parameter because it allows more binding sites for the electrochemical reaction. Despite the mentioned advantages, the performance of the activated carbon is limited by the growth of cathode biofilms after a long-term operation. The performance of the battery will reduce due to the deposition of biofilm after some time. However, the performance can be partially restored if the biofilm is removed from the air cathode. Graphene is an expensive material that shows superior performance as an air cathode. It consists of different mesopores (2–50 nm), which provide excellent contact area while also reducing the diffusion path. The ORR is greatly improved due to the graphene's unique structure.

A cathode diffusion layer is responsible for controlling water leakage. The cathode should be designed to contain the water to avoid any form of leakage which will cause the electrolyte to dry out. Polyvinylidene fluoride (PVDF) is an excellent candidate for retaining water but at the expense of reducing the electrical performance of the aluminum-air battery. This is due to the dense polymer network that prevents oxygen diffusion at the cathode. Hence, reducing the ORR leads to lower voltage and power output.

Since the ORR and water flooding problems depend on each other, it is necessary to solve both issues by introducing an electrocatalyst at the cathode. The catalytic activity of the cathode can be improved by using different catalysts such as copper oxide (Cu_2O), copper (Cu), cobalt oxide (Co_3O_4), nickel cobalt oxide ($NiCo_2O_4$), cobalt (Co), and iron (Fe). All these catalysts have shown a promising performance in enhancing the electrical performance due to their capabilities of improving the catalytic activities which leads to better ORR. The electrocatalyst can be combined for even better battery performance. For instance, Fe/N/C and Cu/N/C have shown superior performance as a catalyst. Manganese oxide (MnO_2) is a good catalyst as it has high oxygen adsorption and reduction rate. The performance even can be boosted by adding silver (Ag) to the MnO_2. This can greatly improve the specific area and leads to a more binding site for the electrochemical process. The performance can be even better than Pt/C, which is notorious for being expensive but good in performance.

6.6 APPLICATION OF ALUMINUM-AIR BATTERY

Alcoa and Phinergy have developed an aluminum-air battery and applied it in an electric vehicle (EV) which provides a travel distance of about 1600 km. The aluminum-air battery consists of 50 aluminum plates with each plate able to sustain a travel distance of 32 km. To prevent the electrolyte from drying out, a water refill is needed when 10 aluminum plates are consumed. When all the aluminum plates are consumed, the battery needs to be replaced. RiAlAir Group is another company developing aluminum-air batteries. The company aims to develop an electric boat with an aluminum-air battery as fuel for their project. The main application of the aluminum-air battery is in the marine field or as a standby energy application. Austin Electric planned to develop an EV with an aluminum-air battery as the power source. It is claimed that the battery can replace the commercially used lithium-ion battery in the market. Based on the study, replacing the lithium-ion battery from the Tesla Model S with an aluminum-air battery will boost the total travel distance to 4345 km as compared to 595 km for a single charge. This improvement is huge as it can greatly improve the feasibility of long-distance travel of the EV. There is also some invention that involves aluminum-air batteries that have been patented to date. For example, a hybrid cell that combined the aluminum-air battery and an accumulator was proposed and patented in 2015. The anode can be made from aluminum or aluminum alloy with an $AlCl_3$-based electrolyte. Controlling the molar ratio of $AlCl_3$ helps to minimize aluminum corrosion. Another patent in the year 2013 proposed a new method to remove the sediment along the aluminum surface. This can help to ensure the aluminum surface contains no impurities so that fresh aluminum is readily available for electrochemical reaction.

6.7 CONCLUSION

In general, the main issue that remains unsolved regarding the aluminum anode is corrosion. The corrosion cannot be eliminated based on current technology. Most of the approach aims to slow down the corrosion rate. The corrosion of the aluminum anode depends on the chemical components of the aluminum anode and the electrolyte. Alloying the aluminum anode shows promising results in suppressing corrosion. However, electrolyte selection becomes an important factor as there is no conclusion on which electrolyte works the best in the aluminum-air battery. Usually, if the electrolyte shows good corrosion inhibition properties, it will come with a penalty of low electrical performance, which limits the power output. Extensive work is required to develop an electrolyte that can minimize the impact on the electrical performance while securing good corrosion stability. On the other hand, the main purpose of the air cathode is to deliver oxygen to the anode. The oxygen reduction reaction is important to ensure sufficient oxygen is supplied to the anode and no energy will be wasted due to the vigorous reaction at the anode side. Numerous approaches and air cathode materials that can improve the oxygen reduction reaction have been proposed, and most of them behave positively in the aluminum-air battery. Although many studies have been conducted to study the nature of aluminum-air batteries, their performance is still limited compared to lithium-ion batteries.

ACKNOWLEDGMENTS

This project is supported by Research Environment Links Grant No. MIGHT/CEO/NUOF/1–2022 (2) from the British Council and Malaysia Industry-Government Group for High Technology, as part of the British Council's Going Global Partnerships program. The program builds stronger, more inclusive, internationally connected higher education and TVET systems.

REFERENCES

[1] Y.G. Li, J. Lu, Metal air batteries: Will they be the future electrochemical energy storage device of choice?, ACS Energy Letter. 2 (2017) 1370–1377.
[2] R. Mori, Recent developments for aluminum-air batteries, Electrochemical Energy Reviews. 3(2) (2020 June) 344–369.
[3] T. Husband, Recycling aluminum, a way of life or a lifestyle?, Chemmatters. (2012 April) 15–17.
[4] J. Bernard, M. Chatenet, F. Dalard, Understanding aluminum behaviour in aqueous alkaline solution using coupled techniques. Part I: Rotating ring-disk study, Electrochimica Acta. 52(1) (2006) 86–93.
[5] A.A. Mohamad, Electrochemical properties of aluminum anodes in gel electrolyte-based aluminum-air batteries, Corrosion Science. 50(12) (2008) 3475–3479.
[6] M.L. Doche, J.J. Rameau, R. Durand, F. Novel-Cattin, Electrochemical behaviour of aluminum in concentrated NaOH solutions, Corrosion Science. 41(4) (1999) 805–826.
[7] L. Fan, H. Lu, The effect of grain size on aluminum anodes for Al-air batteries in alkaline electrolytes, Journal of Power Sources. 284 (2015) 409–415.
[8] D.D. Macdonald, K.H. Lee, A. Moccari, D. Harrington, Evaluation of alloy anodes for aluminum-air batteries: corrosion studies, Corrosion. 44(9) (1988) 652–657.

[9] L. Fan, H. Lu, J. Leng, Z. Sun, C. Chen, The study of industrial aluminum alloy as anodes for aluminum-air batteries in alkaline electrolytes, J. Electrochem. Soc. 163(2) (2016) A8–A12.

[10] B.J. Hopkins, Y. Shao-Horn, D.P. Hart, Suppressing corrosion in primary aluminum—air batteries via oil displacement, Science. 362(6415) (2018 November 9) 658–661.

[11] L. Wang, W. Wang, G. Yang, D. Liu, J. Xuan, H. Wang, M.K.H. Leung, F. Liu, A hybrid aluminum/hydrogen/air cell system, International Journal of Hydrogen Energy. 38 (2013) 14801–14809.

[12] M. Nestoridi, D. Pletcher, R.J.K. Wood, S. Wang, R.L. Jones, K.R. Stokes, I. Wilcock, The study of aluminum anodes for high power density Al/air batteries with brine electrolytes, Journal of Power Sources. 178(1) (2008) 445–455.

[13] H. Xiong, X. Yin, Y. Yan, Y. Dai, S. Fan, X. Qiao, K. Yu, Corrosion and discharge behaviors of Al-Mg-Sn-Ga-In in different solutions, Journal of Materials Engineering and Performance. 25(8) (2016) 3456–3464.

[14] T.M. Di Palma, F. Migliardini, M.F. Gaele, P. Corbo, Aluminum-air batteries with solid hydrogel electrolytes: Effect of pH upon cell performance, Analytical Letters. 54(1–2) (2021) 28–39.

[15] M. Kar, T.J. Simons, M. Forsyth, D.R. MacFarlane, Ionic liquid electrolytes as a platform for rechargeable metal-air batteries: A perspective, Physical Chemistry Chemical Physics. 16(35) (2014) 18658–18674.

7 Materials and Technologies of Other Metal-Air Batteries

Dhavalkumar N. Joshi and Vinod Kumar

Department of Mechanical Engineering, Gyeongsang National University, Jinju 52828, South Korea

Department of Physics, University of the West Indies, St. Augustine, Trinidad and Tobago

CONTENTS

7.1 Introduction 89
7.2 Types of Metal-Air Batteries 90
    7.2.1 Role of Electrodes in Aqueous Metal-Air Battery 91
    7.2.2 Role of Electrodes in Non-Aqueous Metal-Air Battery 92
7.3 Research Prospective for Different Research Batteries 93
    7.3.1 Fe-Air Batteries 93
    7.3.2 Na-Air Battery 95
    7.3.3 K-Air Battery 99
    7.3.4 Mg-Air Battery 99
7.4 Conclusions 102
References 102

7.1 INTRODUCTION

Batteries have a broad range of applications, including grid energy storage and electric vehicle propulsion. Numerous factors, such as the rapid development of electric vehicles, the Internet of Things (IoT), and the promotion of renewable energies, have augmented the demand for batteries with greater energy densities without compromising safety [1–4]. The future of electric vehicles depends entirely on developing safe and efficient electrochemical energy storage devices. Lithium-ion (Li-ion) batteries have been the focus of the majority of battery technology research, and these batteries have reached nearly every household committed to the scientific community's endeavors. Traditional Li-ion batteries, which use intercalation chemistry, are only expected to achieve 30% efficiency, and Li is also a scarce resource on the planet [1, 5]. Consequently, it is impossible to meet the increasing energy demand with these traditional Li batteries. For electric vehicles to be marketable, they must have a minimum range of 500 kilometers on a single charge, which is currently not possible to achieve with Li-ion batteries [2]. To achieve the desired energy density and meet the ever-increasing energy demand, new materials and technologies incorporating new chemistry must be continuously developed. This motivates researchers to seek a more effective, environmentally friendly, and secure energy storage solution.

Theoretically, metal-air batteries (MABs) have a significantly higher energy density than Li-ion batteries, which makes them a viable option for the next generation of electronic devices. Although interest in MABs began in 1970, this technology did not advance due to many obstacles, most of which were related to the battery's efficiency and the stability of its components, which affected its life. The development of nanotechnology has enabled superior electrode design through the creation of nano-sized metals (such as Fe, Li, Zn, Al, and Sn), and carbon-based materials (such as graphene, rGO, and CNTs) have reignited the scientific community's interest in these battery systems [6]. MABs include a metal anode (positive electrode), an air-breathing cathode (negative electrode), and a suitable electrolyte. The metal anode can be composed of electrochemically equivalent alkali metals (such as Li, Na, and K), alkaline earth metals (such as Mg), or first-row transition metals (such as Fe and Zn). MABs combine standard battery and fuel cell characteristics. As it is possible to design the electrodes of MABs with earth-abundant, inexpensive, and non-toxic materials such as Fe, Zn, Al, etc., and as oxygen or air is the primary fuel for the reaction, the use of these batteries will be an ideal way to meet the energy demand with renewable and inexpensive sources of energy. As Li, Zn, and Al MABs were covered in the preceding chapter, here we examine the most recent developments in materials and technologies of electrodes (i.e. anode and cathode) of other MABs, such as Fe, Mg, Na, and K MABs.

7.2 TYPES OF METAL-AIR BATTERIES

As discussed, the core components of MABs are the anode, cathode, and electrolyte. Fe-air, Na-air, K-air, Sn-air, and Mg-air batteries are named according to the active metal (such as Fe, Na, K, Sn, and Mg) used as the anode in the MABs. While the cathode has an open porous structure that allows for a constant oxygen supply from the surrounding air or oxygen. Aqueous or non-aqueous electrolytes can be employed depending on the nature of the anode. Often, MABs are categorized into two groups based on the electrolyte employed: (a) aqueous MABs and (b) non-aqueous MABs. The schematic representing the structure and the working principle of MABs is depicted in Figure 7.1. In MABs, the metals (at the anode) and oxygen (at the cathode) engage in the electrochemical processes. The generic reaction formula distinguishing aqueous and non-aqueous batteries can be presented as shown in equations 1 to 7. The discharge results in aqueous MABs are generally hydroxides and oxides of metals. These discharge products are typically easy to disassociate in an aqueous electrolyte. However, in non-aqueous MABs the discharge products are oxides or superoxide and are difficult to disassociate by the non-aqueous electrolyte.

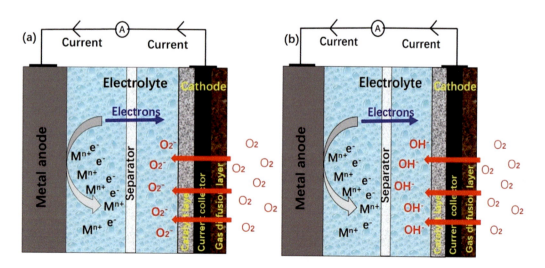

FIGURE 7.1 Schematic representation of working mechanism for (a) non-aqueous, and (b) aqueous metal-air batteries. (Adapted with permission [7]. Copyright The Authors, some rights reserved; exclusive licensee MDPI. Distributed under a Creative Commons Attribution License 4.0 (CC BY).)

- **Reaction at the metal anode in aqueous MABs**

$$M + nOH^- \leftrightarrow M(OH)_2 + ne^- \quad (1)$$

$$M(OH)_2 + nOH^- \leftrightarrow M-O + xH_2O + ne^- \quad \text{(reaction is not balanced)} \quad (2)$$

- **Reaction at the cathode in aqueous MABs**

$$O_2 + 2H_2O + 4e^- \leftrightarrow 2OH^- \quad (3)$$

- **Reaction at the anode in non-aqueous MABs**

$$M \leftrightarrow M^{+n} + ne^- \quad (4)$$

- **Reaction at the cathode in non-aqueous MABs**

$$O_2 + e^- \leftrightarrow O_2^- \quad (5)$$

$$O_2^- + M^{+n} \leftrightarrow MO_2 \quad (6)$$

and/or

$$MO_2 + M^{+n} + e^- \leftrightarrow M_2O_2 \quad (7)$$

To maximize the potential of these batteries, significant material and technological advancements are required to overcome the obstacles posed by these core components. All three mentioned core components must work in perfect harmony to design highly functional MABs, and materials and technologies must continue to evolve to improve energy and power density as well as the safety index of the batteries. Any loophole in one of the battery's core components will result in poor performance.

7.2.1 Role of Electrodes in Aqueous Metal-Air Battery

Even though the metals employed in aqueous MABs are thermodynamically unstable in aqueous media, their surface can be passivated by the matching oxides or hydroxides to make them compatible with the aqueous electrolyte. During the discharge cycle, the anode oxidizes the metal while the catalyst-cathode reduces the O_2 from the air and vice versa during the charging cycle. Thus, MABs are recognized as the only metal-fuel-operating fuel cell. The oxygen reduction reaction (ORR) at the air-cathode contact is analogous to the ORR in the hydrogen fuel cell. So, many operating concepts and designs are shared between these two systems. Although the theoretical energy density of MABs is relatively high, the practical energy density has only reached one-third of the theoretical value. The inefficiency of the air cathode catalyst and the poor energy efficiency due to the overpotential of the air cathode are significant obstacles that restrict their use in low-power applications. Due to the absence of robust bifunctional electrocatalysts and a cyclable metal anode, MABs have a concise life cycle aggravated by repeated cycling at high currents. The engineering and logical design of cathode (and anode) materials in conjunction with the usage of suitable electrodes can give answers to these problems. Because ORR is a prolonged reaction, the air cathode in MABs is the performance-limiting electrode. Consequently, cathode design modification is crucial for improving battery performance. This reaction happens mainly at the triple-phase boundary at the air cathode, where the solid electrode concurrently interfaces with liquid electrolyte and gaseous O_2. Therefore, the growth of active air catalysts to speed up the ORR kinetics and the design of suitable electrode architecture to extend the triple-phase boundary would have an enormously favorable effect on the battery discharge performance.

Alternatively, the bifunctionality of the air cathode can be obtained by combining suitable materials. Various operating components such as precious metals and non-precious metal oxides (and carbonaceous materials, have been explored as the air catalyst [6]. When alkaline electrolyte used in aqueous MABs, due to its high surface coverage of OH species, the ORR activity of Pt, which is often used as reference material, is hindered and surpassed by other non-precious metal-based materials [6]. For rechargeable MABs, the air cathode must successfully catalyze oxygen reduction and evolution (ORR and OER). Therefore, the development of bifunctional oxygen electrocatalysts is necessary [8, 9]. Alternately, the bifunctionality of the air cathode may be realized with the proper mix of materials with various operational sections. The second technique offers greater design freedom for the air cathode since the ORR and OER catalysts may be adjusted independently and combined. These are substantial drawbacks of MABs compared to Li-ion technology. During discharge-charge cycles, the air cathode of rechargeable MABs has a limited tolerance, particularly for the ORR active component, for alternating reductive-oxidative conditions. The severe oxidative potential can reason permanent damage to ORR active sites at the cathode and electrochemical corrosion of the carbon support during battery recharge, and there are only a handful of bifunctional air cathodes with the requisite high activity and cycle stability [1].

Recently, researchers have used a bi-cathode design to address the issue by separating ORR and OER catalysts and loading them onto two separate cathodes for discharge and charge. Moreover, the metallic anode must be constructed to avoid corrosion, non-uniform dissolving, and deposition. The long-term cycle stability of rechargeable MABs cannot be attained without a suitable answer to this issue; nevertheless, this subject has been comparatively overlooked. This problem can be rectified by alloying several metals and increasing the metals' purity [10]. In addition to material improvements for a superior anode, the design modification may be helpful for mechanical recharging and the design of flow batteries. Almost all aqueous metal batteries have similar issues. Only a small portion of the potential energy density of Al-air and Mg-air batteries can be utilized because of severe parasitic corrosion of their metallic anodes in contact with the aqueous electrolyte. Moreover, these batteries are only capable of mechanical charge.

7.2.2 Role of Electrodes in Non-Aqueous Metal-Air Battery

Since metals like K and Na are excessively reactive in aqueous solutions, non-aqueous electrolytes are typically used. ORR in this battery's non-aqueous electrolytes occurs via an entirely different process than in batteries with aqueous electrolytes. In short, On the catalyst surface, O_2 is first reduced by one electron to create superoxide anion O^{-2}, which combines with alkali metal cation to generate peroxide MO_2 (M = Na/K). Despite this, the reaction changes slightly according to the metal and electrolyte selected. Larger cations (Na^+ and K^+) are better able to stabilize the superoxide anion, according to the hard-soft acid-base hypothesis. Therefore, the discharge product of Na-air batteries is often a combination of Na_2O_2 and NaO_2, whereas the discharge product of $K-O_2$ batteries is primarily KO_2. Due to the restricted solubility of these superoxides or peroxides in the electrolyte, they are prepared on the air cathode. Their collection gradually obstructs the cathode's accessible surface area, shutting down the battery in the end. Therefore, the discharge capacity of these non-aqueous MABs is governed by the storage capacity of the air cathode in relation to the discharge product and is much lesser than the theoretical value. Non-aqueous MABs are in their infancy and face much bigger challenges than aqueous MABs. The anode of a non-aqueous battery suffers the same difficulties as the anode of an aqueous battery, such as dendrite growth and metal corrosion [11, 12]. Despite possessing the highest potential energy density, the Mg-air (and Al-air) batteries have received the least amount of study. This is because the presence of hydrogen causes the metal to corrode parasitically, reducing its potential practical values. The major issues for MAB's anodes (electrodes) are metal dendrites, corrosion, surface passivation, and side reactions.

In addition to dendrites, it is possible to observe changes in electrode thickness and surface area during the cycling of metal-based batteries. The shape change of the metal electrode is attributed to the non-uniform distribution of current and reaction zones on the electrode, which causes densification of the electrode and a decrease in capacity. Hydrogen accumulation on the metal electrode is the second challenge. In aqueous MABs, the hydrogen evolution process (HER) is the most significant side reaction on the metal electrode. Creating hydrogen using a part of the energy given for charging would result in low Coulombic efficiency. Due to the connection between hydrogen evolution and zinc oxidation, metal electrodes may corrode while inactive. This self-discharge mechanism is also a major driver of capacity decline. According to the preceding description, the fundamental goal of boosting the performance of metal electrodes is to maintain their starting state after extended discharge-charge cycles. Efforts to improve the reversibility of metal electrodes centered on altering the electrode structure and using electrode additives. Metal electrode batteries frequently experience dendrites, metal oxide deposition, and electrode degradation. Therefore, identical design concepts for the metal electrode in other types of metal-based batteries are also applicable to MABs.

7.3 RESEARCH PROSPECTIVE FOR DIFFERENT RESEARCH BATTERIES

7.3.1 Fe-Air Batteries

Due to the high cost of Li, scientists are continually exploring substitutes such as Fe, Na, and K-based MABs, even though Li-air batteries are the subject of the most study. As Na and K are equally as explosive as Li, safety concerns remain. In this context, the Fe-air battery is attractive for study because of its low production cost, long life cycle, enhanced rechargeability, and lower safety risk. As Fe-air batteries are resistant to the formation of undesirable dendrites during repeated charge-discharge cycles (a significant constraint in most MABs) and have superior charge-discharge characteristics, they have a great deal of potential for the development of anode electrodes for MABs. The energy density of rechargeable Fe-air batteries is as low as 60–80 Wh/kg, which is scarcely comparable to Li-ion technology. Due to their extended life (>1000 cycles), cost-effectiveness (~$100/kWh), and environmental friendliness, however, they are suited for stationary energy storage. However, it cannot be considered for use in pure EV applications [13]. Fe-air batteries are mostly aqueous batteries. The redox reaction at the metallic anode during the charge-discharge cycle depends primarily on the electrochemical reaction of Fe in an alkaline solution. The initial step involves the oxidation of Fe metal into $Fe(OH)_2$, and the subsequent step is the oxidation of $Fe(OH)_2$ to form Fe_3O_4 and $FeOOH$, followed by an H_2 evolution reaction.

Typically, the anode of a Fe-air battery is constructed of Fe or Fe oxide powders covered with a current collector based on Ni. These components enhance the active surface area and conductivity at the anode surface, speeding redox processes. Due to carbon's reduced weight and greater conductivity, current research focuses on carbon-based electrodes instead of Ni. This enables the design and production of lightweight energy systems suitable for portable and wearable devices. To rationally design the anode for Fe-air batteries, the most important factors are (a) to maximize utilization of the Fe/Fe oxides, (b) to occupy the Fe passivation, and (c) to suppress the hydrogen evolution reaction during charging; and numerous methods have been implemented to improve these factors. Optimizing the loading quantity of Fe/Fe oxide in the electrode is crucial, since a larger loading amount results in a higher discharge capacity but a lower specific energy density, and vice versa. Increasing the surface area permits more active sites and greater accessibility of the electrolyte to the anode materials, which is the most efficient method for enhancing activity. As an active anode material for Fe-air batteries, submicron Fe/Fe oxide nanoparticles are employed for this purpose. To increase the efficiency and discharge capacity of aqueous and solid-state Fe-air batteries, simple and quick deposition of submicron-sized Fe-oxide coated on carbon paper was utilized. Numerous ways for changing anode materials to maximize performance and provide simpler components for large-scale, low-cost production have been published. Co-precipitation and calcination, as outlined, can satisfy market needs for low-cost, large-scale production. It has also been reported that FeS/C composite powder was used as an anode by calcination and co-precipitation [13, 14]. By introducing 3% Bi_2O_3 into the FeS/C composite, the reversible discharge capacity was enhanced to 333 mAh/g with a discharge current of 60 mA.g^{-1} and a faradic efficiency of 91.9%. At a higher discharge rate of 1500 mAh.g^{-1}, a specific capacity of 230 mAh/g was recorded, demonstrating that 3% Bi_2O_3 in FeS/C displays a high discharge rate and improved cycle potential [15]. Figuerdo et al. presented a unique 3D printing method for the production of Fe-air batteries and achieved an energy density of 453 Wh/kg [16]. To create the anode, they utilized the molten salt method, which consisted of 95 wt.% Fe/carbon paste, 85 wt.% Fe_2O_3 and 4 wt.% Bi_2S_3, and 5 wt.% polytetrafluoroethylene (PTFE) polymer, which was hot pressed at 200°C and 12Mpa for one hour. Their work presented a novel quick electrode design method that can speed up the development of electrodes for future scientific research. In rechargeable MABs, the active materials (Fe and its oxide in Fe-air) undergo a significant volume expansion (200 to 500 percent) during the oxidation reaction, making it essential to design a porous electrode to prevent cavity blockage that prevents electrolyte diffusion to the active material.

Hayashi et al. studied the design of porous electrodes to enhance redox reactions by increasing their active surface area [17]. They combined micrometer-sized Fe powder with pore-forming polyvinyl alcohol (PVA). By calcining a mixture at 1120 °C for 20 minutes in a nitrogen atmosphere, an interconnected porous Fe anode for rechargeable Fe-air batteries was developed. Despite claiming a lower discharge capacity of 120 mAh/g, this work permits the large-scale, low-cost manufacture of Fe electrodes. This approach also permits porosity regulation with a modification in the molecular weight and amount of PVA employed during the atomization process. In addition, the atomization process allows adding the proper component to form a suitable metal alloy, opening the door to further optimization. Wai et al. synthesized Fe_3O_4 embedded rGO composite using a simple chemical precipitation method and used it as an anode for rechargeable Fe-air batteries. Fe-oxide was found to be distributed homogeneously in the exfoliated graphene sheet as shown in the schematic (Figure 7.2). 60 wt% of Fe-Oxide in rGO, found to have promising electrochemical behavior and prolonged charge-discharge cycle performance with better capacity retention, was achieved [18]. The same group also has demonstrated Fe-oxide decorated carbon paper for all-solid-state rechargeable Fe-air batteries.

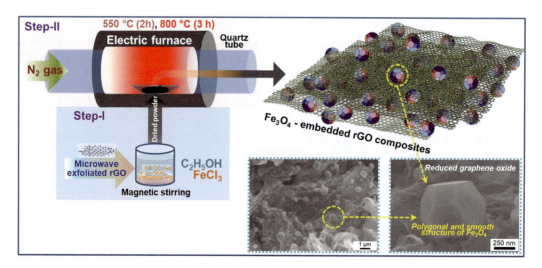

FIGURE 7.2 Schematic representation of Fe_3O_4 decorated rGO used in Fe-air battery. (Adapted with permission [18]. Copyright (2022), Elsevier.)

The impact of adding sulfur to atomized Fe powder was examined using a similar procedure. Fe powder containing sulfur was shown to improve the charge/discharge characteristics of rechargeable Fe-air batteries by around three times more than bare Fe powder. In the aforementioned cases Bi_2O_3, Bi_2S_3, FeS are used as an additive to suppress the HER and enhance the charge efficacy of the anode [13]. The electrode can also be fabricated without a Ni/C current collector; however, in this case, the performance of the device will depend on the thickness of the Fe pellet, which can be adjusted by varying the amount of material used to create the pellet. If the battery's charging capacity is fixed, the thickness does not affect its discharge performance. Due to the hydrophobic binder, however, the thicker electrode may result in less material usage within the electrode. However, when the alkaline solution is used in an aqueous Fe-air battery, the progressive wetting of the pores and partial displacement of the hydrophobic binder molecules permit the penetration and activate the initially inaccessible inactive sites for the redox reactions. Additionally, the particle coarsening caused by repeated charge/discharge cycles overcomes the initial wetting limitation, resulting in direct contact between the Fe atom and electrolyte. The addition of Bi to carbonyl Fe further reduced the HER rate by a factor of 10 and demonstrated a charge efficiency of 96%. The decomposition of hydrated α-$Fe_2C_2O_4$-PVA composite at 600°C under vacuum, to obtain Fe and Fe-oxide with a prominent presence of α-Fe, was also reported as an in-situ method for carbon-grafted Fe anodes in Fe-air batteries. Fe-air batteries fabricated with an anode rich in Fe demonstrated a specific charge capacity of 400 mAh/g at the current design of 100 mAh/g with a faradic efficiency of 80%, unaffected by low or high discharge [13]. Numerous groups have also studied core-shell structures with Fe cores interconnected with graphitic networks or intercalated with reduced graphene oxide, which provides improved cyclic performance. The Fe encapsulation prevents the active materials' detachment and desorption, thereby enhancing the electrode's stability. Additionally, it provides the conductive path for the electron and increases the system's robustness. Recently MOS_2 was proven to be a good additive in the Fe anode, due to its higher activity for anodic and cathodic reactions, suppressing Fe passivation as well as HER at the anode [19].

During ORR and OER reactions, the redox reaction of oxygen from the air at the air cathode is transformed into OH- (oxidation) or H_2O (reduction) and vice versa. In MABs, the cathode electrode is also bifunctional because it operates both ORR and OER. As low overpotential is required for ORR and OER generally noble metals are used as cathode materials. The ORR and OER performance of MABs including noble metal alloys (e.g., Ag, Au, Cu) was examined. Numerous researchers have used non-noble materials and distinct electrocatalyst systems for ORR and OER, using the same electrode configuration, to substitute costly noble metals. The search for an inexpensive non-noble metal catalyst capable of ORR and OER reactions continues. Another critical trait the positive electrode must exhibit is resilience to alkaline conditions (aqueous 8M KOH in Fe-air batteries). Some perovskite materials share bifunctional characteristics with platinum and palladium. To increase the conductivity of this material, conductive support materials are typically added or combined. Layered double hydroxides (LDH) are also utilized as the cathode in Fe-air batteries. Kubo et al. reported superior performance using Ni-Fe CO_3^{-2} LDH as an air electrode compared to other LDHs. Ni-Fe-based LDH formed a favorable triple phase boundary in the catalyst's layer [19]. Metal inorganic frameworks (MOFs) and their derivatives, which have unique crystalline porosity and modifiable chemical properties, are also used as bifunctional positive electrodes [20]. These catalysts have a larger surface area and facilitate electron and mass transfer for a more effective catalytic process due to their metal framework.

7.3.2 Na-Air Battery

Even though Li-air batteries had garnered great attention within the category of MABs, the significant overpotential, inferior round-trip energy storage efficiency, and limited Li supplies continue to prevent widespread applications of MABs [21]. The earth-abundant Na metal, approximately five times more abundant than Li, is also used in metal-air battery systems as a substitute for Li. Due to its promising high energy density of 3164 Wh/kg and high specific capacity of 1166 mAh/g, the Na-air battery can also be a viable option for meeting rising energy demand [22]. Additional advantages of Na-air batteries include their environmental friendliness and inexpensive precursor supply (about 1/3 of the cost of Li-air batteries). These advantages had stimulated important research interest in developing efficient Na-air battery systems [23, 24]. However, low cycle performance, carbon substrate and electrolyte degradation during the discharge-charge process, and specific capacity loss under high current density are the primary difficulties that must be tackled before the widespread adoption of Na-air batteries.

In 2012, Sun et al. reported the first room-temperature battery with a discharge capacity of 1058 mAh/g, 20 cycles of life, and 85% of CE [23]. Typical Na-air batteries contain a Na metal anode and a porous cathode that are separated in the electrolyte by a separator. Depending on the type of electrolyte employed, these batteries are either aqueous or non-aqueous. In these systems, the electrochemical processes and discharge products are different (Figure 7.3). During discharge, the Na metal on the anode is oxidized to produce Na^+, which goes to the cathode through the electrolyte, interacts with an electron from the external circuit, and reduces oxygen (non-aqueous)/water (aqueous) to make NaO_2 (or Na_2O_2)/NaOH, and vice versa during the charging process. The use of porous carbon as the cathode in non-aqueous rechargeable Na-air batteries facilitates the supply of O_2/air required for the battery's electrochemical reactions. The principal stumbling block in the functioning of Na-air batteries is the clogging of the cathode's porous structure by insoluble discharge products (i.e., NaO_2 and Na_2O_2), which hinders the continuous supply of air/O_2 and Na+ ions and reduces the batteries' performance to a cyclic level. Hence, decomposition and reversible formation of these discharge products are crucial to improving the Na-air battery performance. By managing the growth and shape of these oxides, the researchers are still investigating the development of materials and methods for effective Na-air batteries with appropriate air cathodes and

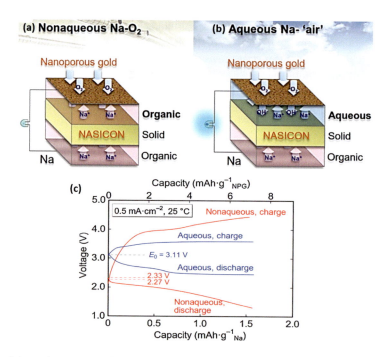

FIGURE 7.3 Schematic structure of Na-air batteries, (a) non-aqueous and (b) aqueous; and (c) first discharge charge curves of non-aqueous and aqueous batteries at 0.5 mA/cm² up to 7 Ma. (Adapted with permission [35]. Copyright (2022), Elsevier.)

electrolytes to mitigate these difficulties. This section will examine various carbon materials, metals and metal complexes, and redox mediators to build suitable cathode and anode materials [21].

Developing an appropriate cathode electrode that improves ion and air mobility requires adjusting the thickness, pore size, pore volume, density, surface area, and electrical characteristics. Carbon-based materials have been the focus of extensive research due to their good conductivity and huge surface area. Sun et al. used a carbon electrode to achieve a discharge capacity of 1884 mAh/g in the first cycle [23]. Using graphene nanosheets as a cathode resulted in a discharge capacity of 6208 mAh/g at 300 mA/g, double that of a thin-film carbon electrode. However, due to the accumulation of discharge products in the graphene sheets, it was discovered that the battery life cycle was only around 10 cycles [25]. When carbon nanotubes (CNT) were employed as a carbon cathode instead of a carbon electrode, the discharge capacity at 500 mA/g increased fourfold in the first cycle [26]. Nitrogen doping of carbon materials was discovered to increase the cathode activity of Na-air batteries. It was discovered that nitrogen-doped graphene sheets have a greater double discharge capacity than undoped graphene sheets [27]. This study reveals that N-doping in graphene increases the cathode's ORR electrocatalytic activity. Additionally, the surface area of the materials and their porosity have a substantial effect on the performance of these batteries. Due to their increased surface area, N-doped CNTs were found to have a higher discharge capacity than N-rich carbon materials [28]. Using multiwalled CNTs as the cathode resulting in a 400 mV reduction in charge-discharge polarization [29]. Possessing a linear relationship between discharge capacity and surface area, a significant range of highly ordered porous networks with a large surface area have been studied to reduce charge-discharge polarization due to insoluble Na-oxides [28]. In the long run, however, all substances exhibit capacitive fading, and Na-oxide cannot decompose. Significant effort must be made in the modification of carbon-based materials to improve the catalytic performance of discharge products.

Using different metal and metal oxide composites, the performance of the Na-air battery is also investigated. $NiCo_2O_4$ nanosheets/Ni foam was discovered to be a stable electrocatalyst for non-aqueous Ni-air batteries [29]. On the initial cycle, the cathode containing this composite produced discharge and charge capacities of 1185 mAh/g and 1112 mAh/g, respectively, and after 10 cycles, up to 34% of the initial discharge capacity was maintained. α-MnO_2 nanowires retained > 85% of their initial specific capacity after 55 cycles [30]. Various metal/metal oxide composites with carbon materials, like Pt@GNSs and $CaMnO_3$/C, are utilized to boost coulombic efficiency and decrease overpotentials [21]. It was established that the decomposition of nonconductive oxides by carbon materials and metals was insufficient, leading to the degeneration of the cells and necessitating additional research for developing an appropriate cathode for non-aqueous Na-air batteries. It was discovered that adding soluble mediators such as NaI and ferrocene to the electrolyte increased the breakdown of Na-oxides, hence improving cycle performance [31]. NaI increases the charge-discharge capacity to 150 cycles with a 1000 mAh/g cut-off capacity, whereas ferrocene increases the Na-air cycle life to 230 cycles [21, 31].

The research on aqueous Na-air batteries is still in its infancy, with cathode modifications serving as the primary emphasis. The formation of a highly soluble discharge product is a fundamental benefit of aqueous Na-air batteries (NaOH). These batteries have low overpotentials for charge and discharge processes and excellent EE and cyclic stability. The various characteristics of aqueous and non-aqueous MABs are compared in Table 7.1.

Using carbon paper coated with Vulcan XC72R and Pt/C as the air electrode, Sun et al. created the first rechargeable Na-air battery (aqueous) in 2015 [32]. With a steady discharge voltage of 2.85 V, the voltage efficiency of the Pt/C-coated cathode was 84.3%, whereas the voltage efficiency of uncoated carbon paper was 72.43%. Hashimoto et al. developed an aqueous Na-air battery utilizing $NaClO_4$ or an aqueous NaOH solution as the catholyte, a porous gold cathode, and a NASICON separator (Figure 7.3) [32]. In addition, Mn_3O_4-catalyzed activated carbon has been studied for the development of compact aqueous Na-air batteries. Also, Pt/C coated carbon as a cathode was discovered to improve the number of cycles [31, 33]. Cheon et al. reported the

TABLE 7.1
Comparison of Characteristics of Aqueous and Non-Aqueous Batteries

	Non-aqueous Na-air battery	Aqueous Na-air battery
Electrolyte	Non-aqueous organic	Aqueous $NaClO_4$ or NaOH
Separator	Polymers	NASICON
Discharge products	NaO_2, Na_2O_2 (Insoluble)	NaOH (Soluble)
Overpotential	High	Low
Energy efficiency	High	Low
Safety	Good	Average
Cyclic stability	Poor	Good

lowest polarization of 115mV for a cathode composed of "graphitic nanoshell/mesoporous carbon nanohybrid." However, corrosion of the electrodes in the aqueous media is a significant problem for these batteries [34].

The performance of a Na-air cell also depends on the anode. Sodium metal has the lowest electrochemical potential of all sodium anode materials. Low redox potential results in enhanced chemical reactivity, promoting HER side reactions and material deterioration. In the Na-air battery, Na metal anodes are challenged by highly oxidative contaminants, such as dissolved O_2, O_2^-, and singlet oxygen ($^1O_2$). Therefore, metallic Na is more chemically reactive and orientation growth is simpler than metallic Li. Similar to Li metal anodes, the stripping and plating of Na metal anodes result in the production of damaging dendrites and dead sodium. Dendrite formation and separator permeation hinder the charging process of Na-air batteries prematurely, reducing their cycle life to a few cycles. These dendrites induce a severe internal short circuit, leading to thermal runaway. As Na dendrites expand, they can fracture and detach, resulting in the formation of sodium that has died. In addition, the uneven anode surface and amount of dead sodium will increase the diffusion routes and resistance of Na^+ ions and electrons, leading to a significant rise in polarization and discharging and charging overpotentials. New SEI contacts decrease the amount of active Na anode material and increase electrolyte consumption. Dendritic development in hybrid Na-air cells can be inhibited by facilitating the formation of stable solid electrolyte interphase (SEI) layer on the Na anode or by purposely creating the metallic Na anode [36]. Peled et al. explored in depth the formation of SEI on freshly deposited sodium metal and the faradaic efficiency of the deposition/dissolution processes involving sodium [37]. They discovered that the solid electrolyte interphase (SEI), which forms on the anode during battery operation, is the most sensitive factor affecting the safety and cycle life of Na-air batteries, making the production of a suitable SEI a crucial step. Liang et al. presented a modified approach employing a liquid anode to circumvent Na dendrite development [38]. The solid electrolyte separated the sodium-containing liquid anode from the air electrode. Due to the strong ionic and adequate electronic conductivity of the liquid anode as well as the low interfacial resistance between the liquid anode and solid electrolyte, the cell exhibited the smallest overpotential gap (0.14 V) and the highest round-trip efficiency (95.9%), with exceptional stability. It has also been claimed that bismuth-containing Na anodes are suitable for Na-air batteries. The bismuth coating on the Na anode hinders both the chemical reaction of electrolytes and Na metal anodes and the production of Na dendrites [39]. The Na/Bi anode demonstrates enhanced cycling stability in both symmetric and Na-O_2 batteries. A Na/rGO composite anode exhibits enhanced chemical stability and reduced dendrite development in Na-air batteries. High electric conducting matrices such as 3D conducting Cu substrates are used to lower the surface current density, allowing for Na stripping and deposition at high current densities without dendrite development [40].

7.3.3 K-Air Battery

K-air batteries offer a higher theoretical energy density (935 Wh/kg) and greater energy efficiency and durability than other alkalies MABs [38]. In the K-air battery, the quick and reversible O_2/KO_2 single-electron reaction has better redox kinetics than Li-O_2 redox chemistry, hence eliminating the need for redox analyzers or mediators. Furthermore, the abundance of K on earth mitigates the worldwide deficit of Li. These features of the K-air system make it an excellent contender for low-cost, large-scale energy storage devices [41]. However, the development of a K-air battery is still in its infancy, and round-trip efficiency is still poor. Also, the cycle life and the energy efficiency must be improved before its practical implementation. The operating environment, degradation process, and electrode design are cathode reversibility and stability drivers in K-air batteries [41]. As K-air batteries employ a single-electron reaction, it mitigates the slow kinetics resulting in increased energy efficiency compared to Li-air (two-electron reaction) batteries. In other MABs, such as in Na-air batteries, the spontaneous decomposition of NaO_2 into $Na_2O_2/Na_2O_2.2H_2O$ restricts the cell's reversibility, whereas, KO_2, the discharge product of the K-air battery, is kinetically favorable and thermodynamically stable [38]. The quick O_2/KO_2 redox couple assures low polarization and completes recharging well below 3.0 V versus K/K^+, which substantially reduces the formation of singlet oxygen and parasitic reaction caused by high potential [38]. Compared to Li-air, Zn-air, and Na-air batteries, substantially less research has been conducted on the influence of electrodes on K-O_2 batteries. Ren and Wu presented the first demonstration of reversible electrochemical production and breakdown of KO_2 in K-air batteries in 2013 [11]. The morphology of the discharge product (KO_2), which is determined by its growth mechanism, greatly influences the discharge capacity of K-air batteries. Both potassium peroxide (K_2O_2-unfavorable) and potassium oxide (K_2O-favorable) are observed to develop and oxidize in K-air batteries [42, 43]. K_2O_2 decreased K-air batteries' Coulombic and energy efficiency (CE and EE). Consequently, optimizing the cathode design is essential to maximize the air-electrode surface area and increase rate performance. The search for appropriate cathode materials continues to obtain high capacity and long-term cycle stability in K-air batteries. Due to their high theoretical abilities and good rate performance, layered structured transition metal oxides have been studied extensively as potential cathode candidates [44]. Similar to Na metal, the highly reactive K metal anodes offer substantial battery safety and life challenges. Electrolyte deterioration and K dendrite formation are found to be critical challenges with K metal. Therefore, stability and modification of the anode, as well as enhancement of the solid electrolyte interface at the anode-electrolyte interface, are essential for boosting the performance of the K-air battery. To reduce K-dendrite formation and increase the charge-discharge cycle of K-air batteries, numerous hybrid anode materials such as K_3Sb, Na-K alloy, and KC_8 have been studied [45–47]. Due to their high porosity, diverse topologies, and abundant presence of heteroatoms, different carbon materials derived from metal-organic frameworks (MOFs) are also exploited in potassium-ion batteries. The following graphic depicts the structural advantages of carbon derived from MOFs as an anode material [48], as shown in Figure 7.4.

7.3.4 Mg-Air Battery

Typical configuration of Mg-air batteries, comprised of Mg or Mg alloy anode (negative electrode), air cathode loaded with catalysts (positive electrode), and electrolyte (aqueous or non-aqueous), together with the working mechanism and reactions involved, are depicted in Figure 7.5. In Mg-air batteries, Mg anode oxidizes by sacrificing two electrons during discharge, and an air cathode uses these electrons to convert O_2 to OH^- in the presence of H_2O. Mg-air batteries are attractive electrochemical devices because of their high theoretic voltage (3.1 V) and specific energy density (6.8 kWh/kg), as well as the abundance of Mg, low toxicity, and high reaction activity.

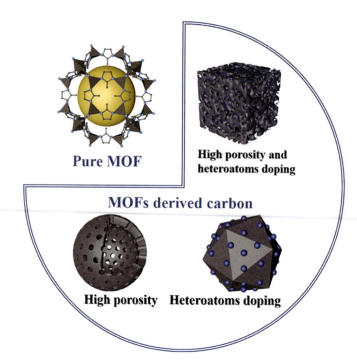

FIGURE 7.4 Advantages of the MOFs derived carbon as an anode material. (Adapted with permission [48]. Copyright (2022), Elsevier.)

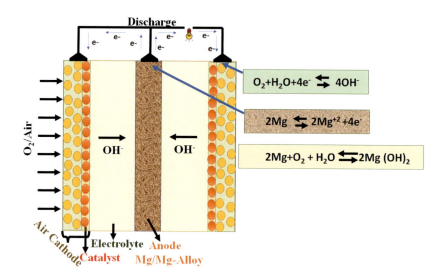

FIGURE 7.5 Schematic representing structure and working mechanism of Mg-air battery.

Mg-air batteries are among the least investigated MABs. To create an effective rechargeable Mg-air battery that can support reversible oxygen reduction and evolution reactions, it is necessary to conduct extensive research and development into the progress of materials and technology. The two greatest constraints of Mg-air batteries are their high level of polarization and weak coulombic efficiency. The actual specific energy is often less than 10% of the theoretical quantity, and the working voltage is frequently less than 1 to 2 V. Two aspects contribute to this issue are the sluggish kinetics of the oxygen reduction process in the air cathode and the corrosion of the magnesium anode brought on by the interaction between magnesium and the electrolyte. For Mg-air batteries, current breakthroughs in anode materials comprised of magnesium or its alloys and popular categories of air cathode catalysts. The design of the anodes and cathodes for Mg-air batteries has also been researched to further increase performance [10]. It has been demonstrated that bifunctional catalysts with reversible oxygen reduction and evolution processes pose obstacles to the creation of rechargeable Mg-air batteries. Hence, further study and understanding are needed to address the existing problems with Mg-air batteries.

The Mg anode is crucial to the performance of Mg-air batteries. At the anode, the hydrogen evolution process (HER), one of these side reactions, is what promotes magnesium corrosion [48–52]. One issue is the Mg anode's side reaction, namely the corrosion of Mg. The negative difference effect is another mechanism causing the Mg anode to corrode (NDE) [49]. A slower HER response rate, a smaller NDE, and a lack of impurities are necessary for a good Mg anode material. There are two approaches applied to address these issues. The first is alloying magnesium with other metals to stop HER, and the second is enhancing the properties of magnesium itself. Recently, there has been a huge interest in alloying magnesium with other metals like aluminum, manganese, or zinc since they inhibit the HER process at the anode site. The Al added to the Mg in the Mg-Al alloy anode not only blocks the HER but also improves the physical strength of materials, protecting the anode from corroding itself [10, 53]. For Mg-Al-Zn alloys, are also widely explored as an anode material for Mg-air batteries. They are often created by combining Mg-Al alloys with zinc [10, 47]. The operating voltage of these batteries during discharge testing in a neutral, 3.5 to 7 wt.% NaCl solution at a current density of ~5 mA/cm^2 is 1.12 to 1.28 V, and with a specific discharge capacity is ~1125 mAh/g [54]. Alloying with manganese was also found to be an appropriate choice. The component particles of these alloys were found to have their particle sizes reduced by manganese. Owing to higher specific energy, higher Faradic capacity, and stronger negative standard potential, Mg-Li alloys have recently been looked into as prospective battery anodes for Mg-air batteries [50]. Altering the inherent features of Mg by manipulating its purity and shape was also utilized as a potential method for enhancing the performance of the Mg anode. In terms of lowering corrosion rate, a 99.99% pure Mg anode has been found to perform better than a number of the alloys described earlier [52]. Among the numerous examined morphologies, sea-urchin-like Mg was shown to have the highest rate performance and energy density. The porous network in the sea-urchin structure was discovered to contain a greater number of active sites, which speed up the urchin's growth. It was discovered that the sedimentation of discharge products such as $Mg(OH)_2$ improves the performance of the Mg-air battery [50].

Similar to the anode, the cathode considerably impacts the performance of Mg-air batteries. The overpotential of the air cathode, created by the slow kinetics of the ORR, is responsible for the Mg-air battery's poor coulombic efficiency. Also, the discharged product ($Mg(OH)_2$) should be simple to remove to supply fresh active sites, to avoid overpotential concerns. A typical air cathode is made up of four layers: (1) a current collector layer with a (2) catalyst layer (active of ORR like Ag, Pt, Pd), (3) a gas diffusion layer (high conductive very porous materials such as polytetrafluoroethylene (PTFE)), and (4) a waterproof layer (water repellent porous substance like paraffin or wax). The catalysts are a crucial aspect to increase the Mg-air performance by improving the ORR activity of the cathode materials. Many catalysts appropriate for acidic, alkaline, and neutral solutions have been explored in this aspect. Powerful catalysts are necessary to ease the recurring problem of overpotential in Mg-air batteries. Among all catalysts, Pt was shown to be the most suitable electrochemical performance. Carbon-supported Pt clusters with exposed (111) phases were found to have good ORR activity. However, as Pt is among rare materials on earth and also not cost-effective there is considerable focus on producing non-Pt catalysts. In this aspect, alloying Pt (such as Pt_3Y, Pt-O, Pt_3Co) with other transition metal have proven to have superior activity than intrinsic Pt. Closed packed binding of the intermediates such as OH^- and O^{-2} on the surface is shown to be the main reason for increasing the ORR activity of these alloys [10]. As ORR catalysts, other noble metals such as Pd, Ag, and Cu as well as their alloys have garnered significant interest. Also employed as catalysts in Mg-air batteries include carbon materials (CNTs, porous carbon, graphene, and rGO), transition metal oxides (MnO_x, $CaMnO_3$, and $CaMn_2O_4$), metal/carbon composites (Co_3O_4/r-GO and MnCoO), and nitrogen. Numerous kinds of research indicate that N-doped high-surface-area carbon materials, such as graphene and CNTs, are a possible alternative to platinum. It was discovered that the presence of N imposes a significant positive charge on nearby C atoms, which is responsible for the activity boost. Pyridinic N was discovered to boost the Mg-air battery's onset potential. Nanostructured transition metal oxides (such as MnO_2) have also been examined as a cathode catalyst for the Mg-air batteries. When utilized as a cathode, the ORR activity of several phases of the same material was found to be significantly different. The addition of noble metals to these transition metal oxides was also found to increase the ORR activity of the cathode materials [55]. The composite of metal oxides (such as Co_3O_4) with porous carbon materials was discovered to have higher stability than the Pt electrodes [10]. However, a substantial study is required to produce suitable cathode and anode materials for this category of MABs.

7.4 CONCLUSIONS

This chapter focuses on the problems with the cathode and anode of many understudied MABs. Issues such as dendrite development, electrode corrosion, and blockage by discharge products must be addressed in all kinds of batteries to increase battery performance and life cycle. The many strategies investigated to address the aforementioned problems are reviewed in relation to Na-air, Mg-air, Fe-air, and K-air batteries.

REFERENCES

[1] Y. Li, J. Lu, Metal-air batteries: Will they be the future electrochemical energy storage device of choice?, ACS Energy Lett. 2 (2017) 1370–1377.
[2] J. Lu, Z. Chen, Z. Ma, F. Pan, L.A. Curtiss, K. Amine, The role of nanotechnology in the development of battery materials for electric vehicles, Nat. Nanotechnol. 11 (2016) 1031–1038.
[3] B. Dunn, H. Kamath, J.-M. Tarascon, Electrical energy storage for the grid: A battery of choices, Science (80-.). 334 (2011) 928–935.
[4] A. Raj, D. Steingart, Power sources for the internet of things, J. Electrochem. Soc. 165 (2018) B3130.
[5] N.R. Van, The rechargeable revolution: A better battery, Nature. 507 (2014) 26–28.
[6] A. Kraytsberg, Y. Ein-Eli, The impact of nano-scaled materials on advanced metal—air battery systems, Nano Energy. 2 (2013) 468–480.
[7] C. Wang, Y. Yu, J. Niu, Y. Liu, D. Bridges, X. Liu, J. Pooran, Y. Zhang, A. Hu, Recent progress of metal-air batteries: A mini review, Appl. Sci. 9 (2019) 2787.
[8] J. Zhang, Z. Zhao, Z. Xia, L. Dai, A metal-free bifunctional electrocatalyst for oxygen reduction and oxygen evolution reactions, Nat. Nanotechnol. 10 (2015) 444–452.
[9] Y. Liang, Y. Li, H. Wang, J. Zhou, J. Wang, T. Regier, H. Dai, Co3O4 nanocrystals on graphene as a synergistic catalyst for oxygen reduction reaction, Nat. Mater. 10 (2011) 780–786.
[10] T. Zhang, Z. Tao, J. Chen, Magnesium-air batteries: From principle to application, Mater. Horizons. 1 (2014) 196–206.
[11] X. Ren, Y. Wu, A low-overpotential potassium-oxygen battery based on potassium superoxide, J. Am. Chem. Soc. 135 (2013) 2923–2926.
[12] P. Hartmann, C.L. Bender, M. Vračar, A.K. Dürr, A. Garsuch, J. Janek, P. Adelhelm, A rechargeable room-temperature sodium superoxide (NaO2) battery, Nat. Mater. 12 (2013) 228–232.
[13] W.K. Tan, G. Kawamura, H. Muto, A. Matsuda, Current progress in the development of Fe-air batteries and their prospects for next-generation batteries, Sustain. Mater. Next Gener. Energy Devices. (2021) 59–83.
[14] X. Zhao, Y. Gong, X. Li, N. Xu, K. Huang, Performance of solid oxide iron-air battery operated at 550 C, J. Electrochem. Soc. 160 (2013) A1241.

[15] E. Shangguan, F. Li, J. Li, Z. Chang, Q. Li, X.-Z. Yuan, H. Wang, FeS/C composite as high-performance anode material for alkaline nickel-iron rechargeable batteries, J. Power Sources. 291 (2015) 29–39.

[16] H.A. Figueredo-Rodríguez, R.D. McKerracher, M. Insausti, A.G. Luis, C.P. de León, C. Alegre, V. Baglio, A.S. Aricò, F.C. Walsh, A rechargeable, aqueous iron air battery with nanostructured electrodes capable of high energy density operation, J. Electrochem. Soc. 164 (2017) A1148.

[17] T. Tsuda, N. Ando, K. Matsubara, T. Tanabe, K. Itagaki, N. Soma, S. Nakamura, N. Hayashi, T. Gunji, T. Ohsaka, Improvement of high-rate charging/discharging performance of a lithium ion battery composed of laminated LiFePO4 cathodes/graphite anodes having porous electrode structures fabricated with a pico-second pulsed laser, Electrochim. Acta. 291 (2018) 267–277.

[18] W.K. Tan, K. Asami, K. Maegawa, R. Kumar, G. Kawamura, H. Muto, A. Matsuda, Fe3O4-embedded rGO composites as anode for rechargeable FeOx-air batteries, Mater. Today Commun. 25 (2020) 101540.

[19] E.O. Aremu, K.-S. Ryu, Performance and degradation behavior of carbonyl Fe-MoS2 composite as anode material in Fe-air batteries, Electrochim. Acta. 313 (2019) 468–477.

[20] T. Tsuneishi, T. Esaki, H. Sakamoto, K. Hayashi, G. Kawamura, H. Muto, A. Matsuda, Iron composite anodes for fabricating all-solid-state iron-air rechargeable batteries, in: Key Eng. Mater., Trans Tech Publ. (2014) 114–119.

[21] W. Yin, Z. Fu, The potential of Na-Air batteries, ChemCatChem. 9 (2017) 1545–1553.

[22] N. Chawla, M. Safa, Sodium batteries: A review on sodium-sulfur and sodium-air batteries, Electronics. 8 (2019) 1201.

[23] P. Hartmann, C.L. Bender, J. Sann, A.K. Dürr, M. Jansen, J. Janek, P. Adelhelm, A comprehensive study on the cell chemistry of the sodium superoxide (NaO 2) battery, Phys. Chem. Chem. Phys. 15 (2013) 11661–11672.

[24] C.L. Bender, P. Hartmann, M. Vračar, P. Adelhelm, J. Janek, On the thermodynamics, the role of the carbon cathode, and the cycle life of the sodium superoxide (NaO2) battery, Adv. Energy Mater. 4 (2014) 1301863.

[25] W. Liu, Q. Sun, Y. Yang, J.-Y. Xie, Z.-W. Fu, An enhanced electrochemical performance of a sodium—air battery with graphene nanosheets as air electrode catalysts, Chem. Commun. 49 (2013) 1951–1953.

[26] Z. Jian, Y. Chen, F. Li, T. Zhang, C. Liu, H. Zhou, High capacity Na-O2 batteries with carbon nanotube paper as binder-free air cathode, J. Power Sources. 251 (2014) 466–469.

[27] Y. Li, H. Yadegari, X. Li, M.N. Banis, R. Li, X. Sun, Superior catalytic activity of nitrogen-doped graphene cathodes for high energy capacity sodium-air batteries, Chem. Commun. 49 (2013) 11731–11733.

[28] M.N. Banis, H. Yadegari, Q. Sun, T. Regier, T. Boyko, J. Zhou, Y.M. Yiu, R. Li, Y. Hu, T.K. Sham, Revealing the charge/discharge mechanism of Na—O2 cells by in situ soft X-ray absorption spectroscopy, Energy Environ. Sci. 11 (2018) 2073–2077.

[29] G.A. Elia, I. Hasa, J. Hassoun, Characterization of a reversible, low-polarization sodium-oxygen battery, Electrochim. Acta. 191 (2016) 516–520.

[30] S. Rosenberg, A. Hintennach, In situ formation of α-MnO2 nanowires as catalyst for sodium-air batteries, J. Power Sources. 274 (2015) 1043–1048.

[31] W.-W. Yin, Z. Shadike, Y. Yang, F. Ding, L. Sang, H. Li, Z.-W. Fu, A long-life Na—air battery based on a soluble NaI catalyst, Chem. Commun. 51 (2015) 2324–2327.

[32] S.H. Sahgong, S.T. Senthilkumar, K. Kim, S.M. Hwang, Y. Kim, Rechargeable aqueous Na—air batteries: Highly improved voltage efficiency by use of catalysts, Electrochem. Commun. 61 (2015) 53–56.

[33] F. Liang, K. Hayashi, A high-energy-density mixed-aprotic-aqueous sodium-air cell with a ceramic separator and a porous carbon electrode, J. Electrochem. Soc. 162 (2015) A1215.

[34] J.Y. Cheon, K. Kim, Y.J. Sa, S.H. Sahgong, Y. Hong, J. Woo, S. Yim, H.Y. Jeong, Y. Kim, S.H. Joo, Graphitic nanoshell/mesoporous carbon nanohybrids as highly efficient and stable bifunctional oxygen electrocatalysts for rechargeable aqueous Na—air batteries, Adv. Energy Mater. 6 (2016) 1501794.

[35] T. Hashimoto, K. Hayashi, Aqueous and nonaqueous sodium-air cells with nanoporous gold cathode, Electrochim. Acta. 182 (2015) 809–814.

[36] S.M. Hwang, J. Park, Y. Kim, W. Go, J. Han, Y. Kim, Y. Kim, Rechargeable seawater batteries: From concept to applications, Adv. Mater. 31 (2019) 1804936.

[37] E. Peled, D. Golodnitsky, H. Mazor, M. Goor, S. Avshalomov, Parameter analysis of a practical lithium- and sodium-air electric vehicle battery, J. Power Sources. 196 (2011) 6835–6840.

[38] B. Lee, E. Paek, D. Mitlin, S.W. Lee, Sodium metal anodes: Emerging solutions to dendrite growth, Chem. Rev. 119 (2019) 5416–5460.

[39] Z. Khan, M. Vagin, X. Crispin, Can hybrid Na—air batteries outperform nonaqueous Na-O2 batteries?, Adv. Sci. 7 (2020) 1902866.
[40] X. Lin, Q. Sun, K. Doyle Davis, R. Li, X. Sun, The application of carbon materials in nonaqueous Na-O2 batteries, Carbon Energy. 1 (2019) 141–164.
[41] N. Xiao, X. Ren, M. He, W.D. McCulloch, Y. Wu, Probing mechanisms for inverse correlation between rate performance and capacity in K-O2 batteries, ACS Appl. Mater. Interfaces. 9 (2017) 4301–4308.
[42] L. Qin, N. Xiao, S. Zhang, X. Chen, Y. Wu, From K-O2 to K-Air batteries: Realizing superoxide batteries on the basis of dry ambient air, Angew. Chemie Int. Ed. 59 (2020) 10498–10501.
[43] N. Xiao, X. Ren, W.D. McCulloch, G. Gourdin, Y. Wu, Potassium superoxide: A unique alternative for metal-air batteries, Acc. Chem. Res. 51 (2018) 2335–2343.
[44] Z. Liu, H. Su, Y. Yang, T. Wu, S. Sun, H. Yu, Advances and perspectives on transitional metal layered oxides for potassium-ion battery, Energy Storage Mater. 34 (2021) 211–228.
[45] L. Qin, S. Zhang, J. Zheng, Y. Lei, D. Zhai, Y. Wu, Pursuing graphite-based K-ion O_2 batteries: A lesson from Li-ion batteries, Energy Environ. Sci. 13 (2020) 3656–3662.
[46] W.D. McCulloch, X. Ren, M. Yu, Z. Huang, Y. Wu, Potassium-ion oxygen battery based on a high capacity antimony anode, ACS Appl. Mater. Interfaces. 7 (2015) 26158–26166.
[47] W. Yu, K.C. Lau, Y. Lei, R. Liu, L. Qin, W. Yang, B. Li, L.A. Curtiss, D. Zhai, F. Kang, Dendrite-free potassium-oxygen battery based on a liquid alloy anode, ACS Appl. Mater. Interfaces. 9 (2017) 31871–31878.
[48] Y. Zhang, M. Sha, Q. Fu, H. Zhao, Y. Lei, An overview of metal-organic frameworks derived carbon as anode materials for sodium- and potassium-ion batteries, Mater. Today Sustain. (2022) 100156.
[49] R. Balasubramanian, A. Veluchamy, N. Venkatakrishnan, R. Gangadharan, Electrochemical characterization of magnesium/silver chloride battery, J. Power Sources. 56 (1995) 197–199.
[50] Y. Ma, N. Li, D. Li, M. Zhang, X. Huang, Performance of Mg—14Li—1Al—0.1 Ce as anode for Mg-air battery, J. Power Sources. 196 (2011) 2346–2350.
[51] Y.D. Milusheva, R.I. Boukoureshtlieva, S.M. Hristov, A.R. Kaisheva, Environmentally-clean Mg-air electrochemical power sources, Bulg. Chem. Commun. 43 (2011) 42–47.
[52] G. Song, A. Atrens, Understanding magnesium corrosion: A framework for improved alloy performance, Adv. Eng. Mater. 5 (2003) 837–858.
[53] B.A. Shaw, Corrosion resistance of magnesium alloys, ASM Handb. 13 (2003) 692–696.
[54] R.P. Hamlen, E.C. Jerabek, J.C. Ruzzo, E.G. Siwek, Anodes for refuelable magnesium-air batteries, J. Electrochem. Soc. 116 (1969) 1588.
[55] F. Cheng, Y. Su, J. Liang, Z. Tao, J. Chen, MnO2-based nanostructures as catalysts for electrochemical oxygen reduction in alkaline media, Chem. Mater. 22 (2010) 898–905.

8 Novel Architectural Designs for Improved Performance

*José Béjar, Minerva Guerra-Balcázar,
Lorena Álvarez-Contreras, and Noé Arjona*

CONTENTS

8.1 Introduction 105
8.2 Types of Metal-Air Batteries.. 105
    8.2.1 Primary Metal-Air Battery 105
    8.2.2 Secondary Metal-Air Battery 106
    8.2.3 Flexible Metal-Air Battery 106
8.3 Wearable Technology and Metal-Air Batteries 107
8.4 Architectures of Metal-Air Batteries.. 109
    8.4.1 Cable-Like Metal-Air Batteries.. 109
    8.4.2 Sandwich-Like Metal-Air Batteries 111
    8.4.3 Stacked Metal-Air Batteries 112
8.5 Improvement of Power Density through Optimizing the Battery Design........ 114
8.6 Conclusions.. 117
References.. 117

8.1 INTRODUCTION

Metal-air batteries are a promising technology for the development of portable applications. Metal-air batteries (MABs) have a higher energy density than other batteries with the advantage of being safer and cheaper. MABs use oxygen from the environment in the cathode, which further reduces its cost, weight, and volume [1]. MABs are mainly composed of three components: a metal anode, an electrolyte which can be an aqueous alkaline solution using salts like potassium hydroxide, and an air electrode. The anode can be constructed of Zn, Fe, Li Al, Mg, and those presented in Figure 8.1, while their theoretical specific energy densities range between 1900 and 10300 Wh/kg [2]. MABs have a higher theoretical energy density than other batteries including those commercially available Li-ion batteries. In addition, metal-air batteries are safer, environmentally friendly, and less expensive, which makes them very attractive for various applications. Metal-air batteries can be simply classified into primary and secondary. In this chapter, different designs for primary and secondary batteries are discussed, and improvements for higher performances are highlighted.

8.2 TYPES OF METAL-AIR BATTERIES

8.2.1 PRIMARY METAL-AIR BATTERY

Primary MABs are characterized by being non-rechargeable energy storage devices. And, it can be attributed to the difficulty of electrochemically regenerating the metal ion to its reduced form. Among primary MABs, Al-air batteries present good theoretical energy properties; however, it presents the disadvantage of their high surface passivation and the presence of parasitic reactions such as the hydrogen evolution reaction (HER). On the other hand, Mg-air batteries have an energy

DOI: 10.1201/9781003295761-8

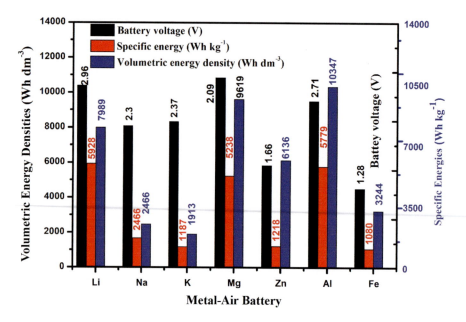

FIGURE 8.1 Electrochemical characteristics of selected metal-air batteries. (Adapted with permission [2]. Copyright (2019), MDPI.)

density of 6.8 kWh/kg, Mg being one of the most affordable metals. In most cases, these anodic materials are prepared in the form of alloys with other metals to decrease the aforementioned issues. Because of these issues, the Zn-air batteries have preferably worked, displaying many advantages over the previous ones [3].

8.2.2 Secondary Metal-Air Battery

Secondary batteries are those which can be electrically recharged, involving charging and discharging processes. In this regard, the Zinc-air battery (ZAB) entails different challenges to primary batteries, mainly in relation to the reversibility of the processes, and the bifunctionality of the cathode. Some of the important challenges in the development of metal-air batteries are the need for bifunctional materials and changes in the shape and composition of the anode during cycling.

There are some successful strategies to improve the charging and discharging processes of a metal-air battery. In the case of the cathode, it is necessary to develop bifunctional materials that are efficient for the oxygen reduction reaction (ORR) and the oxygen evolution reaction (OER). In general, bifunctional materials must have a high surface area, high charge transfer, environmental friendliness, high abundance, and availability to ensure a low cost. Some cobalt oxides with or without spinel-like structures have been used for MABs due to their high performance and stability. Some cobalt oxides with doped structures such as CO_3O_4/NPC have onset potentials of 0.9 and 0.3 V for ORR and OER respectively. This electrocatalyst showed to have a high amount of oxygen vacancies [4]. However, because of the scarcity of cobalt, new materials must arise.

8.2.3 Flexible Metal-Air Battery

Flexible MABs are great candidates to be used as flexible energy sources mostly because of their safety. In literature, there are some strategies to obtain flexible batteries which will be mentioned during this chapter. A flexible primary Al-air battery was built based on paper saturated with aqueous electrolyte; the effect of bending angle was evaluated, and found that the flexible battery works between 0 and 180° bindings angles [5]. There are several configurations for flexible batteries like cable-type batteries, sandwich-type, and multi-array types. For cable-type batteries, it was found no difference in bending conditions using a material composed of polypyrrole fibers, and ZIF 67 calcined, this material allowed to obtain a flexible cathode. Bending tests exhibited no difference during the discharge process at 0.1 mA/cm^2 under the bending and nonbinding conditions [6].

8.3 WEARABLE TECHNOLOGY AND METAL-AIR BATTERIES

Regarding applications for flexible MABs, wearable technology (WT) arises as the most suitable. The WT is an area that is projected to grow over the next few years in such a way that it has impacted important aspects such as health care and the global economy, due to this technology being used for solving concrete problems in different fields such as medicine, leisure, sports, etc. The terms wearables, wearable devices, and wearable technology nowadays refer to small electronic and portable devices, or computers with wireless communication capabilities that are incorporated into gadgets, accessories, or clothing, etc., generally, technology that is carried on or near the human body, or even invasive versions such as microchips or smart tattoos [7]. These devices make it possible to monitor human activity and behavior, thus they provide essential data for dealing with specific human needs. To have a clearer idea of the interaction with the body, four levels have been suggested, which are summarized in Table 8.1. More broadly, other classification factors are related to the location of the wearable device on the human body and based on the energy-consumption profile (Figure 8.2) [8].

On the other hand, WT could be subdivided into two categories: primary, those that operate independently and function as central connectors for other devices, and secondary, monitoring specific actions or performance of measurement and sending the measurement to a primary wearable device for analysis [9].

Portable devices typically use the user's smartphone as an entry to connect to the Internet. Compared to current smartphones and tablets, the main added value is that wearable devices can provide continuous monitoring and scanning functions, including biofeedback or other physiological sensory functions, such as those related to biometrics [10]. Therefore, the operational efficiency depends on the battery or connection options of the link bridge. Some portable devices nowadays have a long-range wireless radio. However, continuous use is still not recommended due to high power consumption [11]. In this sense, the architecture of wearable devices has evolved from independent portable devices, which could monitor and collect information from users, then connect by cable to more powerful external devices for data processing. These architectures restrict user mobility, which is considered an obstacle to expanding portable solutions. Consequently, they have been replaced by portable wireless systems using smartphones as a portable cloud to transmit data, either directly or with pre-processing in custom applications protecting information with the Internet of Things (IoT), which allows a direct connection between the user and a variety of smart devices, which builds a powerful, broad, and robust system that adapts to the needs of the user.

TABLE 8.1
Levels of the Interaction of Wearable Technology

Level	Borderline	Example
Embedded	Implanted inside the body	Pacemaker
Intimate	Attached to the body in such a way that it could be considered indistinguishable	Contact lens, prosthesis
Mounted	Attached to the body, not covered	Smart watch, head-mounted display
Carried	A device that is carried and used close to the body	Smart phone

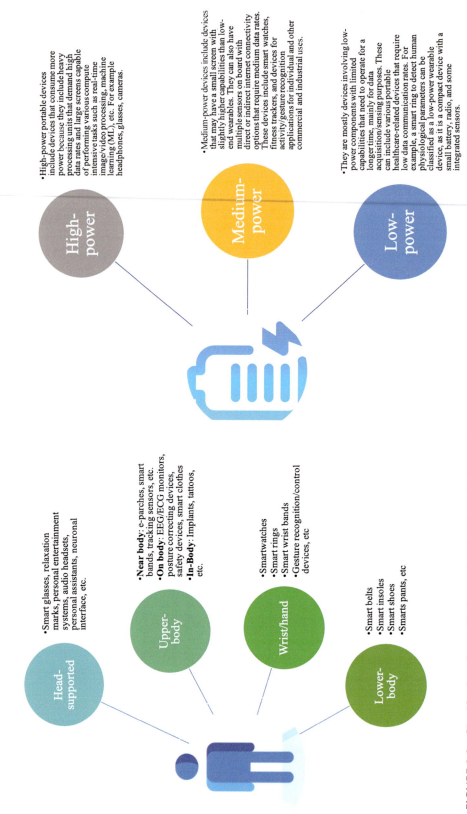

FIGURE 8.2 Classification of wearable devices based on the on-body location and based on their energy-consumption profile.

In the next step in the evolution of wearables, device manufacturers are considering equipping wearables with short-range and long-range connectivity chipsets, so these wearables can function independently of other devices. The challenges with this architecture range from network design and sizing, to battery life. The stage of information collection and processing carried out by portable devices will depend on the particular functionality, through a series of stages that include a collection of data generated by the user, preprocessing, filtering, structured, cleaned and validated to improve data quality, in order to reduce the data processing time, improving the output information, getting a better criteria of decision. The data transfer phase is an essential part of the portable device and, finally, the data processing stage. Adding to the challenges in processing, connectivity, functionality, aesthetics, privacy, portability (lightness), and security, an extremely important aspect is the energy consumption of portable devices.

The power supply system is the crucial component that guarantees continuous, uninterrupted, and long-term operation, which is required to have high flexibility, high power density, long time, durability, and high security for wearable use. In this sense, battery systems must be designed that preserve the viewpoint of wearable devices in terms of flexibility and safety, which demands the development of flexible energy supply devices to portable self-powered devices based on intrinsically flexible self-healing materials of perovskite composites, metals, alloys, carbon nanostructures, etc., and conductive polymers. In summary, to address the aforementioned challenges, future research should focus on developing reliable and flexible materials with high electrochemical performance, including innovative elements such as elasticity, self-healing, transparency, hydrophobicity, biocompatibility, and breathability [12]. It should explore processing techniques that enable low-cost scale-up and design electrode materials with improved mechanical properties.

8.4 ARCHITECTURES OF METAL-AIR BATTERIES

Metal-air batteries possess the versatility to be manufactured in different shapes, allowing them to satisfy the energy demand for many technologies. For example, the rapid development of portable electronic devices and flexible technology can be covered by cable-like and sandwich-like flexible batteries, which have become a worldwide trend in recent years. On the other hand, due to their high energy density, metal-air batteries can be implemented in high-power technology, such as electric vehicles or as energy storage systems in renewable energies sources (sun and wind), this can be carried out in manufacturing stacks with higher storage capacities.

8.4.1 CABLE-LIKE METAL-AIR BATTERIES

Cable-like MABs arise as a response to the development of new medical devices, wearable technologies, or artificial intelligence, which demand flexible energy storage systems that can withstand deformations in three dimensions, and are comfortable for use in textile technology. Due to the low-energy density displayed by cable Li-ion batteries, MABs have emerged as a better alternative, because the geometry allows a greater oxygen interaction with the air electrode, being their potential use in textile technologies, since the cloth always is in contact with ambient air [13]. In this sense, different cable-type-like designs have been proposed (all of them initially applied for Lithium-ion batteries), where the most important designs are the coaxial structure, twisted structure, and stretchable Li-ion batteries. However, for metal-air systems, the coaxial structure is the most studied alternative, since its structure maximizes the interaction with air [13, 14].

Regarding the coaxial architecture, there are two different configurations according to the electrode layout: metal outside and metal inside (Figure 8.3a). In the metal outside arrangement (Figure 8.3a, left scheme), the battery possesses internal ventilation through its structure, enabling it to drive air inside the battery. Then, the air electrode is covered by the electrolyte and the flexible metal; this configuration allows the construction of a battery network that could be fed through itself by pumping pure oxygen into the tubular structure. Regarding the metal inside configuration

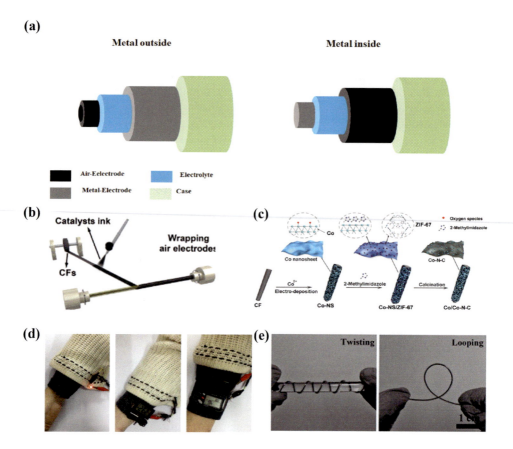

FIGURE 8.3 (a) Schematic presentation of the two coaxial cable-like metal-air battery configurations: metal outside and metal inside. (b) The automatic spray-coating process to prepare air-electrode for cable-like Zn-air battery. (Adapted with permission [15]. Copyright (2018), Wiley.) (c) Schematic representation of in-situ Co/Co-N-C synthesis. (Adapted with permission [16]. Copyright (2019), Wiley.) (d) Photographs of different electronic devices powered by knittable Zn-air batteries. (Adapted with permission [17]. Copyright (2018), Wiley.) (e) Photographs of Silicon-oxygen battery under different deformations. (Adapted with permission [18]. Copyright (2017), Wiley.)

(Figure 8.3a, right scheme), the metal is covered by the electrolyte and the air-electrode, and contrary to the internal cathode, these batteries favor the use of air as an oxygen source because the catalyst is exposed to the ambient. This aspect makes this configuration the most promising alternative for wearable technology applications. Additionally, since the metal is in the battery´s center, it is protected from oxidation generated by CO_2 in the air [13]. The new developments in the construction of cable-like batteries are focused on the three principal components: air-electrode, metal-electrode, and electrolyte, and thus, relevant aspects of these topics are going to be analyzed.

For the air-electrode construction, there are two main methods: the spray/cast-coating and the *in-situ* approach. In the first case, the ORR/OER catalyst is dispersed until getting a catalytic ink which is brushed on a flexible substrate (like carbon cloth). In this way, this method allows a continuous fabrication of flexible air-electrodes [13] as was reported by Jun Lu and co-workers [15], where they synthesized Co_3O_4/N-rGO as a bifunctional electrocatalyst, and it was spray-coated onto Toray carbon fibers using a semi continue to process (Figure 8.3b). This allowed them to obtain batteries with several centimeters in length (35 cm) with high volumetric energy density (36.1 mWh/cm³). However, despite the coating's easiness, the spray/casting-coating methods require the use of additives (like Nafion) to prepare the catalytic ink [13], forming different solid interfaces, diminishing the electrocatalytic activity [19, 20]. The advantage of *in-situ* methods is that binders are eliminated because the catalytic material is synthesized directly onto a substrate surface. Additionally, the obtained catalyst usually exhibits a highly homogeneous layer, improving the ORR/ER kinetic, besides a better electron/ionic transport (Figure 8.3c) [16]. The typical anodic shapes for these batteries include wires and strips because they exhibit certain malleability; however, the mechanical properties of cable-like batteries demand high flexibility and stretchable capability. Therefore, in recent reports wires in a spiral shape to form a spring-like anode have been reported, allowing to increase the stretchable up to 30% [21]. Besides, spring-like anodes enhance the contact area with the electrolyte, improving the battery stability.

Another important issue in this type of anode is the formation of dendrites, the electrode corrosion, and the shape-memory effect, which decreases the lifetime of a MAB, and because of this, different metal alloys have been reported to solve that drawback [13]. Concerning mechanical properties, the use of a flexible and lightweight substrate has been an important alternative, since the anodic material can be deposited in this substrate, maintaining its flexibility while it reduces the metal excess (Figure 8.3d) [17]. Finally, a special anode is used in Li-air batteries, because, unlike Zn and Al anodes, Li is highly reactive, which is difficult for battery fabrication. Therefore, the principal strategy to solve fabrication issues is the replacement of these metals with alloyed compounds like M_xLi (M:Si, Ge, Sn, Al) to decrease problems like dendrites formation maintaining the anode flexibility [13]. An important work that supports the previously mentioned, is the novel method to obtain a Li electrode with low metal content reported by H. Peng and collaborators [18]. They introduced Si nanoparticles into carbon nanotubes followed by a lithiation process performed by electrochemical methods. The hybrid fiber not only avoids the dendrite formation but also showed great flexibility (Figure 8.3e), displaying an excellent battery performance with an energy density of 512 Wh/kg.

The electrolyte is the medium where the ions and oxygen are transported, therefore, the electrolyte must possess high ionic conductivity, electrochemical stability, oxygen solubility, and for cable-like batteries, high flexibility and stretchability. For this purpose, polymer-based gel electrolytes (GPEs) are a promising alternative to be applied in this kind of battery, since GPEs avoid the volatilization issues generated by aqueous electrolytes. Additionally, its quasi-solid state inhibits the dendrite growth and due to hydrophobicity properties, they prevent the metal corrosion generated by the CO_2 in ambient air, while possessing excellent mechanical properties to support the stress condition during battery operation.

8.4.2 Sandwich-Like Metal-Air Batteries

Sandwich-like metal-air batteries have been studied due to their simple configuration and feasible operation because the typical arrangement of a sandwich-like battery consists of an anode, electrolyte, and the cathode assembled layer by layer in parallel form (Figure 8.4a). Unlike cable-like batteries, where an important goal is the possibility to be used like a thread in textile devices, in sandwich-like batteries, the goal is its application like a patch sewn on the cloth, which looks like a closer technology. This type of electrode layout is compatible with battery production systems used nowadays [22].

For sandwich-like battery construction, some procedures are also used in cable-like batteries as the air-electrode construction, because in spray/cast-coating and in-situ methods the key component is the employed substrate, which can be applied to both systems. Thus, special attention is given to the metal-electrode and electrolyte components. For example, excluding the pure metal sheets, it is possible to produce anodes by different methods, like zinc coating on conductive and flexible substrates such as carbon cloth, conductive papers, or even polymers with superior mechanical properties [23, 24]. Mitra and co-workers [23] used a polyethylene terephthalate (PET) substrate covered by silver ink (which worked like a collector), and then, they mixed Zn powder, multi-walled carbon

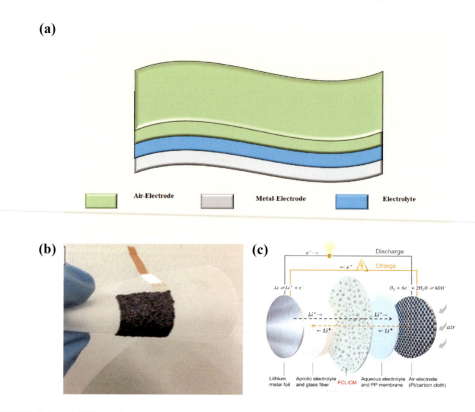

FIGURE 8.4 (a) Schematic representation of a sandwich-like battery. (b) Flexible Zn-air battery with a Zn composite casted on PET. (Adapted with permission [23]. Copyright (2017), Elsevier.) (c) Schematic representation of a hybrid Li-air battery. (Adapted with permission [24]. Copyright (2021), Elsevier.)

nanotubes (MWCNTs), Bismuth (III) oxide (90–210 nm), and the polymer binder to form a slurry which was cast onto the substrate (Figure 8.4b).

During sandwich-like battery construction, the electrolyte must maintain its laminar structure, and because of this, hydrogels are the most studied alternatives, being polyvinyl alcohol, polyethylene oxide, and polyacrylic acid the most used polymers, which provide structural integrity, maintaining a good ionic conductivity [22]. These polymers have been reported mostly for Al-air and Zn-air batteries. For Li-air batteries, it is necessary to use organic electrolytes to avoid the presence of water in the battery. In this sense, recent reports incorporate polymer matrixes, solid inorganic fillers, and flexible organic polymers [24]. A flexible composite Li-ion conducting membrane (FCLICM) is presented in Figure 8.4c, where this membrane is used as a barrier to water permeation in a hybrid Li-air battery, which was assembled with two kinds of electrolytes, an aprotic electrolyte on the metal side and aqueous electrolyte in the air-electrode side. The FCLICM was synthesized using a $Li_{1.3}Al_{0.3}Ti_{1.7}(PO_4)_3$ (LATP) as a Li-ion conductor and a PVDF-HFP as a polymer matrix, then LATP powders were encapsulated into the PVDF-HFP matrix, cast and dried.

8.4.3 Stacked Metal-Air Batteries

Cable-like and sandwich-like metal-air batteries are focused on the miniaturization and portability of electronic devices; however, because of their high theoretical energy densities, metal-air batteries can be used in the electric grid, green energy, and high-power supplies, through the battery's stacks manufacturing. In this sense, the most promising systems are the Zn-air

Novel Architectural Designs for Improved Performance

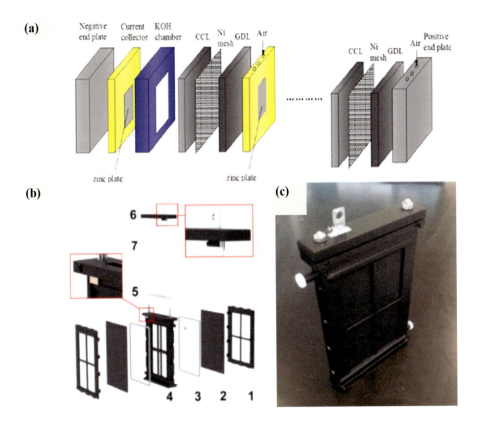

FIGURE 8.5 (a) Schematic configuration of conventional Zinc-air battery stack. (Adapted with permission [27]. Copyright (2014), MDPI.) (b) Schematic representation of BCZAB components and (c) Assembled BCZAB. (Adapted with permission [25]. Copyright (2020), Elsevier.)

batteries, because of the stability and electrochemical reversibility of Zn. There are few works about the battery configuration in stacked MABs, being the coin-like Zn-air battery is the only optimized system and the only battery today commercialized for hearing aids. The construction of optimized devices is an unexplored topic limiting the realistic applications of MABs in electric vehicles or power stations [25].

Some developments in stack designs are focused on demonstrating the application of Zn-air batteries through a sort battery arrangement connected in series [26]. Ma and co-workers [27] reported a rechargeable battery stack using a novel bipolar plate structure. However, these works use the same conventional design (Figure 8.5), which has challenges to overcome to achieve realistic applications [25]: (1) the low-effective volume generated by the use of sealing plates with large dimensions; (2) the uneven forces during stack assembly, which can involve electrolyte leakage; and (3) the stack length variability with the number of used batteries, which is inconvenient in terms of standardization because accessories are going to be different in function of the stack size. To solve these problems, C. Zhong and co-workers [25] proposed a novel boltless compact Zn-air battery (BCZAB), which was compared against the conventional design (labeled B#CC in their work). The stack construction consisted in developing three principal components: the catalyst layer on the electrolyte side, a waterproof diffusion layer on the gas side, and a current collector (Ni mesh). After 70 h, the BCZA did not exhibit electrolyte losses, while displaying a volumetric capacity cell almost four times higher than that of the B#CC (100.6 vs 27.6 Ah/L).

8.5 IMPROVEMENT OF POWER DENSITY THROUGH OPTIMIZING THE BATTERY DESIGN

The battery design is fundamental to increasing the performance and durability of metal-air batteries. Many designs have been proposed for metal-air batteries, including those designs in which water and air-stable metals (Al, Zn) are used as anodes, those batteries operated with liquid electrolytes, and for those all-solid metal-air batteries.

For instance, for Al-air batteries (AlAB) operated with liquid electrolytes, the use of a continuous flow AlAB allows for an increase in the capacity in terms of discharge time, and times improvements from 10 h for static to almost 35 h for a continuous flow-AlAB (Figure 8.6a) have been reported [28]. Additionally, the maximum power density was 302.8 mW/cm^2 (OCP = 1.5 V) using Al alloyed anode, a MnOOH@CeO$_2$ cathode, and 6 M KOH containing corrosion inhibitors as electrolyte.

The main problem of continuous-flow AlAB is the use of external pumps, which increase the weight and volume of the device, and at the same time, it can consume a significant amount of the produced energy. Because of this, non-continuous (passive flow) batteries are preferred.

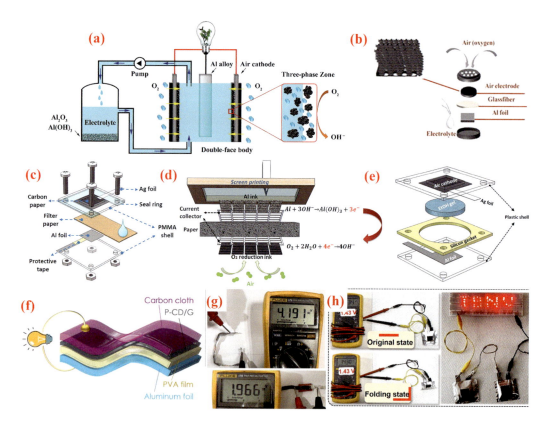

FIGURE 8.6 Battery designs for Aluminum-air batteries. (a) Continuous flow AlAB. (Adapted with permission [28], Copyright (2021), Elsevier.) (b) Coin-like AlAB. (Adapted with permission [29]. Copyright (2020), Elsevier.) (c) Passive-flow AlAB. (Adapted with permission [30]. Copyright (2019), Elsevier.) (d) Printable AlAB. (Adapted with permission [32]. Copyright (2019), Elsevier.) (e) Gel-based AlAB. (Adapted with permission [33]. Copyright The Authors, some rights reserved; exclusive licensee Elsevier. Distributed under a Creative Commons Attribution License 4.0 (CC BY).) (f) All-solid-state AlAB. (Adapted with permission [34]. Copyright (2020), Wiley.) (g) Flexible AlAB. (Adapted with permission [35]. Copyright (2019), Elsevier.) (h) Stacked AlAB. (Adapted with permission [36]. Copyright (2022), Elsevier.)

Among them, the coin-like battery, which is highly known in Li-ion batteries has been used for AlAB (Figure 8.6b). Because of the internal resistances of using glass fiber and the limited mass transport, these batteries usually show low power densities, e.g., an AlAB operated with Co_2O_4/CNTs/3D graphene as cathode displayed a maximum power density of 4.88 mW/cm^2 [29]. The substitution of glass fiber with paper-based technology allows to obtain low-cost paper-based AlAB (Figure 8.6c). Filter (cellulose) paper is usually employed to drive the electrolyte by capillarity, and it can decrease the Al self-corrosion [30], improving the power density (21 mW/cm^2) and the specific capacity (1273 mAh/g) of a liquid electrolyte AlAB in contrast to the coin AlAB. The same authors tested the paper-based AlAB using a gel electrolyte, where the higher viscosity and thus, lower ionic mobility decreased the power density to 3.5 mW/cm^2 (specific capacity = 900 mAh/g) [31].

Printable Al-air batteries are a low-cost alternative to fabricating components (Figure 8.6d). The printable AlAB using an Al-based ink allowed to obtain a maximum power density of 6.6 mW/cm^2 and a specific capacity of 951 mAh/g [32]. It is found from these works that the gel polymer electrolytes allow to decrease anodic issues like passivation, corrosion, and dendrite formation, but they require further modifications to improve the power density. And, these modifications can be made from a materials point of view and a battery design perspective. From the latter, a coin-battery-inspired ethanol gel electrolyte AlAB was constructed (Figure 8.6e) using silicon as a gasket. The ethanol electrolyte prevented the Al corrosion, maintaining 96.6% of its initial weight after being submerged for 30 days [33]. Despite this advantage, the maximum power density was found to be 4 mW/cm^2 with a specific capacity of 2546 mAh/g. The change in the battery design to a sandwich-like architecture (Figure 8.6f) together with the use of polyvinyl alcohol (PVA) gel electrolyte allowed an increase in the power density of a non-aqueous AlAB to 157.3 mW/cm^2 [34].

All-solid-state AlAB is ideal for wearable applications. As earlier mentioned, gel polymers like PVA allow for an increase in the performance of batteries with sandwich-like designs. There are some works related to increasing the performance in this similar architecture, but with a materials science perspective. The use of cobalt nanoparticles in hollow carbon nanotubes (Figure 8.6g) allows it to obtain open-circuit potentials as high as 2.04 V with power densities from 28 to 29.23 mW/cm^2 [35]. In similar conditions (maintaining PVA gel electrolyte), the power density can be achieved by decreasing the sluggish oxygen kinetics by designing single-atom cathodic materials, boosting the power density until achieving power densities up to 184 mW/cm^2 [36], while the stacking of AlAB can provide the required energy to turn on devices like displays (Figure 8.6h).

Battery designs for Zinc-air systems (Figure 8.7) are similar to those reported for Al-air (Figure 8.6); however, the relatively easy electrochemical reduction of Zn ions to metallic Zn brings a rechargeable character to ZABs, and thus, materials science and battery designs are focused on increasing not only the performance and capacity but also to increase the rechargeability. ZABs have been tested in a continuous flow in a micro-confined channel (Figure 8.7a), in which the intrinsic phenomena allow to form a natural interface between the anodic and cathodic streams, removing the need for a solid membrane [37]. OCP values between 1.39 and 1.47 V were obtained using MnO_2 and Pt as cathodes, while a maximum power density of 308 mW/cm^2 and a specific capacity of 758 mAh/g were achieved. For passive-flow ZABs, the battery design presented in Figure 8.7b is the most characteristic when aqueous solutions with zinc acetates are used as electrolytes [38]. In this regard, the most suitable battery design is that in which the thickness of the electrolyte reservoir is minimized to decrease ohmic losses from electrode separation, thus, increasing the current/power density. Concerning the sandwich-like configuration (Figure 8.7c) [39], most of the efforts are made to improve the activity via better bifunctional materials and to improve the mechanical and electrochemical characteristics of gel polymer electrolytes to decrease anode issues (passivation, corrosion, shape changes, dendritic growth during charge, and the HER as a parasitic reaction). The sandwich-like ZAB reported by J. Fu et al. [39], displayed a voltage near 1.2 V with a specific capacity of almost 450 mAh/g.

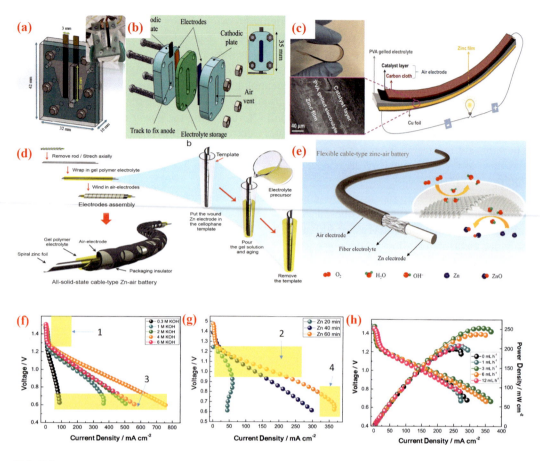

FIGURE 8.7 Battery designs for Zn-air batteries: (a) continuous flow battery. (Adapted with permission [37]. Copyright (2020), American Chemical Society.) (b) Rechargeable ZAB operated with liquid electrolyte. (Adapted with permission [38]. Copyright (2020), American Chemical Society.) (c) Sandwich-like ZAB. (Adapted with permission [39]. Copyright (2015), Wiley.) (d) Cable-like architecture with gelatin-based electrolyte. (Adapted with permission [40]. Copyright (2015), Wiley.) (e) Cable-like architecture with fiber electrolyte. (Adapted with permission [41]. Copyright (2022), Elsevier.) Polarization curves for a continuous flow ZAB (also called Zn-air fuel cell) analyzing the effect of (f) KOH concentration, (g) deposition time of Zn, and (h) polarization and power density curves analyzing the effect of the flow rate. (Adapted with permission [37]. Copyright (2020), American Chemical Society.)

In the case of flexible cable-like ZABs (Figures 8.7d–e), their functionality strongly depends on the gel polymer electrolyte. In the first attempts to develop this kind of device, gelatin was analyzed as an electrolyte, but a maximum voltage near 1 V and short duration times were found [40]. Recently, the use of a fiber electrolyte allowed users to obtain a highly functional flexible cable-like ZAB, achieving a voltage near 1.5 V, and a specific capacity close to 800 mAh/g [41]. To finalize the discussion about the design effect on the performance, the author from reference [37] analyzed the KOH effect, the effect of the deposition time of Zn, and the flow rate effect (Figures 8.7f–h). These polarization curves can be used to highlight specific design and materials science modifications to increase the performance of Zn-air batteries to guide the reader to understand the motivations behind some reported designs (Figure 8.7a–e), and to propose new ones.

Four zones were marked in yellow squares in *Figures 8.7f–g*. The first zone is the polarization zone; at this zone, the battery design has a low contribution, and uniquely it must ensure a proper seal, which in aqueous ZABs is easily achieved with screws. Most of the polarization losses (observed as decreasing of the battery voltage from the thermodynamic values) are associated with the overpotentials to start the oxygen reduction/evolution reactions, and thus, these can be diminished only through modifications to electrocatalysts aiming to decrease energy barriers. The second zone is concerning to the ohmic losses region (Figure 8.7g). The ionic conductivity of the electrolyte (aqueous, gel, or solid) strongly affects this region. However, from a battery design viewpoint, the reduction of the reservoir dimensions or the thickness of the polymer can greatly decrease this zone and as a consequence, higher current densities should be observed at lower overpotentials, the current density increase with the proper modifications.

The third region is related to the mass transport losses, and two cases are highlighted in the figures and marked as 3 and 4. For active-flow batteries, the proper ionic conductivity of the electrolyte and flow rate should result in discharge curves with a linear behavior as observed in Figure 8.7f, and the increase of the flow rate should remove species from the anodic interface (like zincates or ZnO particles), increasing the power density (Figure 8.7h). However, the low ionic conductivity and low mobility in all-solid-state MABs promote a behavior as is highlighted as "4" in Figure 8.7g, where a curved behavior is observed at high overpotentials (near to 0.4/0 V). In this way, to decrease this effect, a plausible solution is to increase the intrinsic properties of the electrolyte, but from a design viewpoint, the increase of the exposed area for air-breathing could significantly decrease this effect.

8.6 CONCLUSIONS

The aspects reviewed in the chapter make it clear that metal-air batteries are a proven technology because of their specific capacity, diversity of potential applications, and the variety of used metals. However, to achieve its extensive use as an energy storage system, it is necessary to consider the development of functional designs, because most of the reported works on MABs are focused on a specific battery component, and, in most of those works, the practicality and viability of the reported systems to be used in real life is dismissed. For MABs operated with liquids electrolytes (like Al and Zn-based), the sandwich-like architecture is the most widely employed. Even though electrode separation plays an important role, most of the works reporting this configuration are centered on the catalytic effect of nanomaterials. An opposite case occurs when all-solid-state MABs are designed; the coin-like, sandwich-like and cable-like architectures follow a smart design enhancing the air electrode exposure to the ambient to enhance the air-breathing. In addition, gel/solid polymer electrolytes are designed to decrease electrode separation by decreasing the membrane thickness. Moreover, the battery design is highly influenced by the application. For wearable technology, the cable-like architecture is the most promising one. For grill applications, as presented, the architectures promoting an active flow without any membrane are promising.

REFERENCES

[1] A.G. Olabi, E.T. Sayed, T. Wilberforce, A. Jamal, A.H. Alami, K. Elsaid, S. Mohammod, A. Rahman, S.K. Shah, Metal-air batteries: A review, Energies. 14 (2021) 7373 (1–46).

[2] C. Wang, Y. Yu, J. Niu, Y. Liu, D. Bridges, X. Liu, J. Pooran, Y. Zhang, A. Hu, Recent progress of metal-air batteries-a review, Appl. Sci. 9 (2019) 2787.

[3] X. Chen, X. Liu, Q. Le, M. Zhang, M. Liu, A. Atrens, A comprehensive review of the development of magnesium anodes for primary batteries, J. Mater. Chem. A. 9 (2021) 12367–12399.

[4] J. Sun, Y. Yang, J. Wang, Z. Zhang, J. Guo, In-situ construction of cobalt oxide/nitrogen-doped porous carbon compounds as efficient bifunctional catalysts for oxygen electrode reactions, J. Alloys Compd. 827 (2020) 154308.

[5] Y. Wang, H.Y.H. Kwok, W. Pan, Y. Zhang, H. Zhang, X. Lu, D.Y.C. Leung, Combining Al-air battery with paper-making industry, a novel type of flexible primary battery technology, Electrochim. Acta. 319 (2019) 947–957.

[6] F. Meng, H. Zhong, D. Bao, J. Yan, X. Zhang, In situ coupling of strung Co_4N and intertwined N-C fibers toward free-standing bifunctional cathode for robust, efficient, and flexible Zn-Air batteries, J. Am. Chem. Soc. 138 (2016) 10226–10231.

[7] T. Luczak, R. Burch, E. Lewis, H. Chander, J. Ball, State-of-the-art review of athletic wearable technology: What 113 strength and conditioning coaches and athletic trainers from the USA said about technology in sports, Int. J. Sport. Sci. Coach. 15 (2020) 26–40.

[8] A. Ometov, V. Shubina, L. Klus, J. Skibińska, S. Saafi, P. Pascacio, L. Flueratoru, D.Q. Gaibor, N. Chukhno, O. Chukhno, A. Ali, A. Channa, E. Svertoka, W. Bin Qaim, R. Casanova-Marqués, S. Holcer, J. Torres-Sospedra, S. Casteleyn, G. Ruggeri, G. Araniti, R. Burget, J. Hosek, E.S. Lohan, A survey on wearable technology: History, state-of-the-art and current challenges, Comput. Networks. 193 (2021).

[9] A. Godfrey, V. Hetherington, H. Shum, P. Bonato, N.H. Lovell, S. Stuart, Maturitas from A to Z : Wearable technology explained, Maturitas. 113 (2018) 40–47.

[10] S. Khan, S. Parkinson, L. Grant, N. Liu, S. Mcguire, Biometric systems utilising health data from wearable devices, ACM Comput. Surv. 53 (2021) 1–29.

[11] W. Bin Qaim, A. Ometov, A. Molinaro, I. Lener, C. Campolo, E.S. Lohan, J. Nurmi, Towards energy efficiency in the internet of wearable things: A systematic review, IEEE Access. 8 (2020) 175412–175435.

[12] J. Gao, K. Shang, Y. Ding, Z. Wen, Material and configuration design strategies towards flexible and wearable power supply devices: A review, J. Mater. Chem. A. 9 (2021) 8950–8965.

[13] L. Ye, Y. Hong, M. Liao, B. Wang, D. Wei, H. Peng, L. Ye, Y. Hong, M. Liao, B. Wang, D. Wei, H. Peng, Recent advances in flexible fiber-shaped metal-air batteries, Energy Storage Mater. 28 (2020) 364–374.

[14] X. Xu, S. Xie, Y. Zhang, H. Peng, The rise of fiber electronics, Angew. Chemie-Int. Ed. 58 (2019) 13643–13653.

[15] Y. Li, C. Zhong, J. Liu, X. Zeng, S. Qu, X. Han, Y. Deng, W. Hu, J. Lu, Atomically thin mesoporous Co_3O_4 layers strongly coupled with N-rGO nanosheets as high-performance bifunctional catalysts for 1D Knittable Zinc—air batteries, Adv. Mater. 30 (2018) 1703657.

[16] P. Yu, L. Wang, F. Sun, Y. Xie, X. Liu, J. Ma, X. Wang, C. Tian, J. Li, H. Fu, Co nanoislands rooted on Co—N—C nanosheets as efficient oxygen electrocatalyst for Zn—air batteries, Adv. Mater. 31 (2019) 1901666.

[17] X. Chen, C. Zhong, B. Liu, Z. Liu, X. Bi, N. Zhao, X. Han, Y. Deng, J. Lu, W. Hu, Atomic layer Co_3O_4 nanosheets: The key to knittable Zn—air batteries, Small. 14 (2018) 1702987.

[18] Y. Zhang, Y. Jiao, L. Lu, L. Wang, T. Chen, H. Peng, An ultraflexible silicon-oxygen battery fiber with high energy density, Angew. Chemie. 129 (2017) 13929–13934.

[19] L. An, Y. Li, M. Luo, J. Yin, Y.Q. Zhao, C. Xu, F. Cheng, Y. Yang, P. Xi, S. Guo, Atomic-level coupled interfaces and lattice distortion on CuS/NiS_2 nanocrystals boost oxygen catalysis for flexible Zn-air batteries, Adv. Funct. Mater. 27 (2017) 1–9.

[20] A. Kushima, T. Koido, Y. Fujiwara, N. Kuriyama, N. Kusumi, J. Li, Charging/discharging nanomorphology asymmetry and rate-dependent capacity degradation in Li-oxygen battery, Nano Lett. 15 (2015) 8260–8265.

[21] Y. Xu, Y. Zhao, J. Ren, Y. Zhang, H. Peng, An all-solid-state fiber-shaped aluminum—air battery with flexibility, stretchability, and high electrochemical performance, Angew. Chemie—Int. Ed. 55 (2016) 7979–7982.

[22] T. Li, X. Peng, P. Cui, G. Shi, W. Yang, Z. Chen, Y. Huang, Y. Chen, J. Peng, R. Zou, X. Zeng, J. Yu, J. Gan, Z. Mu, Y. Chen, J. Zeng, J. Liu, Y. Yang, Y. Wei, J. Lu, Recent progress and future perspectives offlexible metal-air batteries, SmartMat. 2 (2021) 519–553.

[23] Z. Wang, X. Meng, Z. Wu, S. Mitra, Development of flexible zinc—air battery with nanocomposite electrodes and a novel separator, J. Energy Chem. 26 (2017) 129–138.

[24] S.H. Lu, H.C. Lu, Pouch-type hybrid Li-air battery enabled by flexible composite lithium-ion conducting membrane, J. Power Sources. 489 (2021) 229431.

[25] Z. Zhao, B. Liu, X. Fan, X. Liu, J. Ding, W. Hu, C. Zhong, An easily assembled boltless zinc—air battery configuration for power systems, J. Power Sources. 458 (2020) 228061.

[26] T. An, X. Ge, N.N. Tham, A. Sumboja, Z. Liu, Y. Zong, Facile one-pot synthesis of CoFe alloy nanoparticles decorated N-doped carbon for high-performance rechargeable zinc-air battery stacks, ACS Sustain. Chem. Eng. 6 (2018) 7743–7751.

[27] H. Ma, B. Wang, Y. Fan, W. Hong, Development and characterization of an electrically rechargeable zinc-air battery stack, Energies. 7 (2014) 6549–6557.

[28] D. Liu, J. Tian, Y. Tang, J. Li, S. Wu, S. Yi, X. Huang, D. Sun, H. Wang, High-power double-face flow Al-air battery enabled by CeO_2 decorated MnOOH nanorods catalyst, Chem. Eng. J. 406 (2021) 126772.

[29] Y. Liu, L. Yang, B. Xie, N. Zhao, L. Yang, F. Zhan, Q. Pan, J. Han, X. Wang, J. Liu, J. Li, Y. Yang, Ultrathin Co_3O_4 nanosheet clusters anchored on nitrogen doped carbon nanotubes/3D graphene as binder-free cathodes for Al-air battery, Chem. Eng. J. 381 (2020) 122681.

[30] Y. Wang, H.Y.H. Kwok, W. Pan, H. Zhang, X. Lu, D.Y.C. Leung, Parametric study and optimization of a low-cost paper-based Al-air battery with corrosion inhibition ability, Appl. Energy. 251 (2019) 113342.

[31] Y. Wang, W. Pan, H.Y.H. Kwok, H. Zhang, X. Lu, D.Y.C. Leung, Liquid-free Al-air batteries with paper-based gel electrolyte: A green energy technology for portable electronics, J. Power Sources. 437 (2019) 226896.

[32] Y. Wang, H.Y.H. Kwok, W. Pan, Y. Zhang, H. Zhang, X. Lu, D.Y.C. Leung, Printing Al-air batteries on paper for powering disposable printed electronics, J. Power Sources. 450 (2020) 227685.

[33] Y. Wang, W. Pan, K.W. Leong, S. Luo, X. Zhao, D.Y.C. Leung, Solid-state Al-air battery with an ethanol gel electrolyte, Green Energy Environ. (2021) In Press.

[34] M. Wang, Y. Li, J. Fang, C.J. Villa, Y. Xu, S. Hao, J. Li, Y. Liu, C. Wolverton, X. Chen, V.P. Dravid, Y. Lai, Superior oxygen reduction reaction on phosphorus-doped carbon dot/graphene aerogel for all-solid-state flexible Al—air batteries, Adv. Energy Mater. 10 (2020) 1902736.

[35] C. Zhu, Y. Ma, W. Zang, C. Guan, X. Liu, S.J. Pennycook, J. Wang, W. Huang, Conformal dispersed cobalt nanoparticles in hollow carbon nanotube arrays for flexible Zn-air and Al-air batteries, Chem. Eng. J. 369 (2019) 988–995.

[36] T.H. Nguyen, P.K.L. Tran, D.T. Tran, T.N. Pham, N.H. Kim, J.H. Lee, Single (Ni, Fe) atoms and ultrasmall Core@shell Ni@Fe nanostructures Dual-implanted CNTs-Graphene nanonetworks for robust Zn- and Al-Air batteries, Chem. Eng. J. 440 (2022) 135781.

[37] E. Ortiz-Ortega, L. Díaz-Patiño, J. Bejar, G. Trejo, M. Guerra-Balcázar, F. Espinosa-Magaña, L. Álvarez-Contreras, L.G. Arriaga, N. Arjona, A flow-through membraneless microfluidic Zinc-air cell, ACS Appl. Mater. Interfaces. 12 (2020) 41185–41199.

[38] J. Béjar, F. Espinosa-Magaña, M. Guerra-Balcázar, J. Ledesma-García, L. Álvarez-Contreras, N. Arjona, L.G. Arriaga, Three-dimensional-order macroporous AB_2O_4 spinels (A, B =Co and Mn) as electrodes in Zn-air batteries, ACS Appl. Mater. Interfaces. 12 (2020) 53760–53773.

[39] J. Fu, D.U. Lee, F.M. Hassan, Z. Bai, M.G. Park, Z. Chen, Flexible high-energy polymer-electrolyte-based rechargeable Zinc-air batteries, Adv. Mater. 27 (2015) 5617–5622.

[40] J. Park, M. Park, G. Nam, J. Lee, J. Cho, All-solid-state cable-type flexible Zinc—air battery, Adv. Mater. 27 (2015) 1396–1401.

[41] P. Zhang, K. Wang, Y. Zuo, M. Wei, P. Pei, J. Liu, H. Wang, Z. Chen, N. Shang, A flexible zinc-air battery using fiber absorbed electrolyte, J. Power Sources. 531 (2022) 231342.

9 Phase-Engineered Materials as Efficient Electrocatalysts for Metal-Air Batteries

Mengyang Dong, Huajie Yin, Porun Liu, and Huijun Zhao

CONTENTS

9.1 Introduction 121
9.2 Phase-Engineered Noble Metal-Based Nanomaterials Electrocatalysts 122
9.3 Phase-Engineered Transition Metal-Based Nanomaterials Electrocatalysts 123
    9.3.1 Transition Metal-Alloy Electrocatalysts 123
    9.3.2 Transition Metal Oxides Electrocatalysts 123
    9.3.3 Optimized-Phase Spinels Electrocatalysts 124
    9.3.4 Perovskites Electrocatalysts 125
9.4 Phase-Engineered Transition Metal Chalcogenides/Dichalcogenides-Based Electrocatalysts 126
9.5 Phase-Engineered Other Transition Metal-Based Electrocatalysts 129
    9.5.1 Transition Metal Carbides Electrocatalysts 129
    9.5.2 Transition Metal Nitrides Electrocatalysts 129
    9.5.3 Transition Metal Phosphides Electrocatalysts 129
9.6 Conclusion and Prospects 131
References 131

9.1 INTRODUCTION

As a new type of high-efficiency energy conversion device, metal-air batteries (MABs) perfectly fit the current development direction of energy devices toward higher functionality, stability, and safety. As the most crucial part of MABs, breakthroughs in the research on bifunctional catalysts, the "heart" of driving batteries, are urgently needed. Metal nanomaterials, including noble metal-based nanomaterials and non-precious metal-based nanomaterials, are a class of highly efficient bifunctional catalysts for oxygen reduction reaction (ORR)/oxygen evolution reaction (OER) with broad application prospects. As a parameter describing the regularity of atomic arrangement in metal materials, phase is a key structural parameter that affects the physicochemical properties and application performance of metal nanomaterials. Therefore, phase-engineered nanomaterials have become an important method of controlling the properties and functions of metal-based nanomaterials. However, to date, most studies have been limited to bifunctional metal nanocatalysts with conventional thermodynamically stable phases, which results in the inability to improve the performance of bifunctional catalysts. Therefore, it is of great significance to explore new metal nanocatalysts with unconventional crystal phases and to study the effect of phase engineering on their ORR/OER performance. In this chapter, studies on different types of MABs (Zn-air batteries (ZABs), lithium-air batteries (LiABs), magnesium-air batteries (MgABs), and sodium-oxygen batteries (SOBs)) are described, and different kinds of bifunctional catalysts are classified. The current research results in this field can be obtained intuitively with emphasis on the impact of phase engineering. Visually, it is shown that the performance of the MABs prepared under the phase

DOI: 10.1201/9781003295761-9

engineering adjustment has been improved to a relevant degree, indicating its application significance. Microscopically, through characterization techniques and theoretical calculations, the phase transition process under the effect of phase engineering and the origins for the improvement of bifunctional catalytic performance caused by the corresponding phase transition can be established.

9.2 PHASE-ENGINEERED NOBLE METAL-BASED NANOMATERIALS ELECTROCATALYSTS

As bifunctional electrocatalysts for ORR/OER, noble metal-based materials such as Pt/Ru/Ir have been tentatively commercialized. Research on bifunctional noble metal catalysts has mostly focused on maintaining or even improving the performance of electrocatalysts while minimizing the loading of noble metals. Among them, strategies such as alloying and even multi-phase alloying, building carbon phase carriers, and constructing nanoscale phase structures have been developed. Shao et al. reported that ordered Pt-Fe inter-phases with ultra-low Pt loadings can be easily realized by thermal treatment of Pt nanocrystals anchored on Fe, N co-doped surface functional carbons derived from zeolitic imidazolate frameworks [1]. The optimized samples have higher maximum power density and better specific capacity when applied to the ZAB system configuration. Additionally, Goodenough et al. demonstrate the ordered Fe_3Pt inter-phases nanoalloys supported by porous metal nickel-iron nitride (Ni_3FeN) [2]. The ordered Fe_3Pt nanoalloys mainly donate to the high activity of ORR, while the bimetallic nitride Ni_3FeN donates to the superb activity of OER. ZAB constructed by Fe_3Pt/Ni_3FeN catalyst achieved highly efficient long-term cycling.

Palladium (Pd) metal-based catalysts are cheaper, more abundant than Pt metal-based catalysts, and also stable. Recently, Kang et al. reported that controlling the crystal phase of palladium-copper nanoparticles (PdCu NPs) to tune their surface atomic arrangement, thereby controlling the catalytic performance in non-aqueous lithium-oxygen ($Li-O_2$) batteries [3], theoretically and experimentally demonstrated how crystal phase modulation modulates the nucleation behavior and growth kinetics of ORR/OER discharge products (Figure 9.1). Furthermore, the coupling of Pd nanoparticles with $M(OH)_2$ (M = Ni, Co) can greatly enhance the electrocatalytic activity [4]. Pd nanoparticles supported on cobalt hydroxide nanosheets ($Pd/Co(OH)_2$) were developed as air cathodes for MABs. The ZAB composed of $Pd/Co(OH)_2$ exhibits stable charge-discharge voltage and durable cycling stability, also making it a high-efficiency air electrode for LiOB.

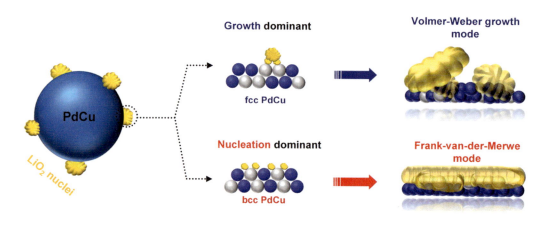

FIGURE 9.1 Schematic diagram of the growth of discharge products on the surface of PdCu. (Adapted with permission [3]. Copyright (2020) WILEY-VCH Verlag GmbH & Co. KGaA, Weinheim.)

9.3 PHASE-ENGINEERED TRANSITION METAL-BASED NANOMATERIALS ELECTROCATALYSTS

Transition metal oxide (TMO)-based composites evince lavish activity and stability for ORR/OER, with better potential than noble metal-based catalysts. This paragraph succinctly launches the recent progress of bifunctional TMO-based electrocatalysts for MABs, pertaining to metal alloys, metal oxides (including cobalt and manganese-based oxides), polymetallic oxides (mainly spinels and perovskites) were scrutinized.

9.3.1 Transition Metal-Alloy Electrocatalysts

Transition metal alloy catalysts, biphase FeCo, NiCo, NiFe, and multi-phase metal alloys, etc., are characterized in that different metals can undergo electronic transitions between different valence states, and have excellent electrical conductivity, which is an important factor for the adsorption and adsorption of O_2. Activation provides surface redox centers that improve ORR/OER electrocatalytic activity. Alloy nanoparticles alone tend to agglomerate, which leads to the reduction of catalytically active sites. On the other hand, when working for a long time in a strong acid or strong alkali environment, problems such as corrosion will occur, and the surface state will change, which will reduce its electrocatalytic activity. Therefore, the dispersibility and stability of alloy nanoparticles can be improved by compounding with carbon supports with good chemical stability, and the application of alloy nanocatalysts in the field of ORR/OER electrocatalysis can be promoted. For instance, Chen et al. reported a strategy of bimetallic-phase Co/CoFe alloys supported on N-doped graphitic carbon [5]. The aqueous ZAB presented a slight voltage drop of 60 mV after 360 hours. Similarly, Liu et al. developed *in situ* growth reduced crystal size of CoFe-ZIF phase on hollow polypyrrole spheres (CoFe/N-HCSs) [6]. Interestingly, ZAB with CoFe/N-HCSs showed eminent stable cycling and energy density.

9.3.2 Transition Metal Oxides Electrocatalysts

Specifically, manganese oxides (MnO_x) have attracted attention due to their low cost, abundance, environmental friendliness, and affirmative bifunctional catalytic activity. Studies of manganese oxides have mainly emphasized the effect of morphology and crystallography, which further affect catalytic activity. For example, Zhang et al. presented a chemical bath deposition method for *in situ* growth of MnO_x on carbon cloth (MnO_x-CC) [7]. MnO_x-CC as a catalyst for solid-state ZAB displays high round-trip efficiency, long cycle life, and high capacity. Moreover, Jiang et al. developed NiFeO@MnO_x NPs in porous layers, their bifunctional activity benefit from the synergistic effect between core/shell [8]. The NiFeO@MnO_x constructed cathode of ZAB conveys better durability than its counterparts. Likewise, Lee et al. reported poly-phase uniform MnO_2 and Fe_2O_3 nanorods on logically designed porous CNT microspheres [9]. MnO_2-Fe_2O_3/CNT for ZAB discloses a power density of 253 mW/cm^2, a low polarization voltage gap, and long-cycling stability. In addition, Tang et al. also reported an *in situ* poly-phase Co/MnO NPs immobilization in N, S co-doped nanotube and nanofiber (Co/MnO@N, S-CNT/CNF) [10]. Benefiting from the compositional synergy and structural advantages, Co/MnO@N, S-CNT/CNFs indicate terrific ORR activity, excellent long-term stability, and good methanol tolerance.

In addition to the above studies, many other TMOs have been extensively studied as the main components of bifunctional catalysts. Such as nickel oxides, ferric oxides, and cobalt oxides-based nanomaterials, etc., For instance, Wang et al. synthesized Ni nanoparticles embedded into porous NiO nanosheets (Ni/NiO) [11]. Profiting from the subsidy of Ni/NiO phases and pore structures, the ZAB with porous Ni/NiO nanosheets showed a specific capacity of 853 mAh/g. Especially, the Mott-Schottky electrocatalyst poly-phase Cu nanodots/Fe_2O_3-NPCs were inaugurated by heterojunction of Cu NDs/Fe_2O_3 nano-islands which enables electron transfer between metallic Cu/semiconducting

Fe$_2$O$_3$ phase [12]. Consequently, the bifunctional catalytic activities were enhanced appreciably as the enhanced adsorption of O$_2$ and OH- species. Furthermore, the Cu NDs/Fe$_2$O$_3$-NPCs assembled ZAB manifested a power density of 138 mW/cm^2. As for the study of cobalt oxide-based nanomaterials, Zhang et al. proposed a governable vacuum calcination approach for converting Co(OH)$_2$ into oxygen-deficient amorphous crystalline CoO (ODAC-CoO) [13]. The cathode of ODAC-CoO for ZAB showed sensational power density, specific capacity, and matchless cycling stability. In addition, Zhang et al. proposed poly-phase CoS/CoO nanocrystals on N-doped graphene (CoS/CoO@NGNs) [14]. The as-prepared CoS/CoO@NGNs for ZAB convey a high power density of 137.8 mW/cm^2 and exceptional cycling stability.

Although the current MgAB is a primary battery, it can be charged by changing the old magnesium anode and electrolyte with a new magnesium anode and electrolyte, so the research on MgAB cathode material is whether electrochemical charging can be finally realized. MgAB typically works in pH-neutral aqueous electrolytes because the generation of Mg(OH)$_2$ in alkaline electrolytes passivates the Mg surface, preventing further reaction at the anode. Nonetheless, it is a major challenge to develop high-efficient and economic pH-neutral electrocatalysts. Yao et al. have reported that poly-phase mullite (SmMn$_2$O$_5$) electrocatalysts showed similar catalytic activity but better stability in 1M NaCl solution than Pt/C [15]. Density functional theory simulations indicate that the moderate bonding strength between O atoms in the mullite structure and the pyramidal Mn dimer sites is beneficial to the catalytic activity.

9.3.3 Optimized-Phase Spinels Electrocatalysts

Spinel-type (M$_3$O$_4$ or AB$_2$O$_4$) oxides with multiple cations, the flexibility of A or B-site cation composition control make them highly potential as bifunctional catalytic materials. It is worth noting that an efficient electrocatalytic process needs to consider both the macroscopic and microstructure effects on the catalyst during mass transfer and charge transfer, while unmodified spinel oxides are used as electrocatalysts due to their low activity density and low density. Low activity intensity cannot meet the actual requirements of device service. When it comes to famous studies on spinel oxides, for bifunctional electrocatalyst of MAB, which the studies by Chen et al. have to be mentioned: the prepared Co$_x$Mn$_{3-x}$O$_4$ NPs have been reported to exhibit exceptional catalytic activity for ORR/OER due to their high surface area and numerous defects [16]. Lately, they also synthesized Co-Mn ultraminiature cobalt-manganese spinels with tailored structural symmetry and composition [17]. As shown in Figure 9.2, nanocrystalline spinels exhibit properties that are strongly related to phase transition and composition during the catalytic ORR/OER process, which presents a comparable activity to Pt/C and superior durability, and can be used as an effective tool for the construction of rechargeable ZABs and LiABs.

Similarly, Liu et al. developed a structure of nanocarbon with poly-phase spinel MnCo$_2$O$_4$ nanocrystals [18]. The MnCo$_2$O$_4$/CNT as cathode for ZAB has been flourishingly manipulated for 768 hours. Moreover, Qiu et al. introduced multi-phase Mn$_3$O$_4$-based and FeCo-based spinel nanocomposites by dealloying a sequence of Al-rich precursor alloys [19]. ZAB with multi-phase (FeCoNi)$_3$O$_4$/Mn$_3$O$_4$ nanocomposite displayed the open-circuit voltage of the battery reached 1.44 V, and the charge/discharge of ZAB could run stably for 400 hours at a certain current density. Moreover, Zhang et al. demonstrated the controllable phase synthesis of NiFe$_2$O$_4$/FeNi$_2$S$_4$ heterostructured nanosheets (HNSs) [20], the multi-phase NiFe$_2$O$_4$/FeNi$_2$S$_4$ HNSs exhibited exceptional bifunctional electrocatalytic activity in 0.2 M phosphate solution. The potential of multi-phase NiFe$_2$O$_4$/FeNi$_2$S$_4$ HNSs of ORR and OER performance is much lower than that of single-phase FeNi$_2$S$_4$ or NiFe$_2$O$_4$. Additionally, Zhong et al. introduced a novel poly-phase NiCo$_2$O$_4$/Co, N-CNTs heterogeneous nanocages (NCs) transferred from the internal/external self-etching of ZIF-67 [21]. The power density of NiCo$_2$O$_4$/Co, N-CNTs as cathode for ZAB reached 173.7 mW/cm^2. Similarly, Wan et al. reported a poly-phase CoFe alloy and CoFe$_2$O$_4$ spinel phase [22], denoted as CoFe-CoFe$_2$O$_4$-NCNT, the catalyst not only surpassed Pt-Ru/C in ORR, but simultaneously, the activity

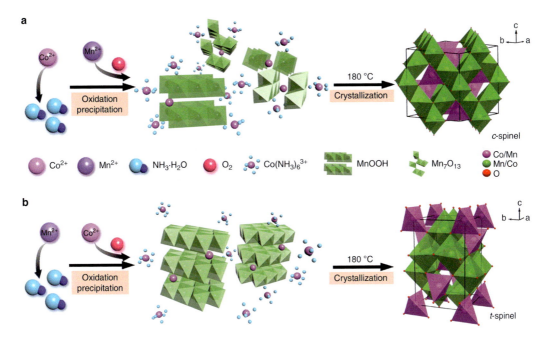

FIGURE 9.2 Synthesis of CoMnO spinels. Adapted with permission [17]. Copyright (2015) Nature Publishing Group.

also showed low over-potential at 10 mA/cm² in OER. Furthermore, as a Co_3O_4-based material, Chen et al. proposed a poly-phase metallic Co and spinel Co_3O_4 "Janus" nanoparticles capable of interpenetrating in a porous graphitized shell (Co/Co_3O_4@PGS) [23]. This heterogeneous mixture catalyst can provide excellent charge transport channels and excellent adsorption capacity for oxygen. The ZAB using Co/Co_3O_4@PGS in a rechargeable air electrode exhibits excellent stability at 10 mA/cm² for as long as 800 hours.

9.3.4 Perovskites Electrocatalysts

Perovskite oxides with the canonical molecular formula ABO_3 (where A: an alkaline earth/rare earth element; B: a transition metal) have recently received numerous follows of interest as bifunctional electrocatalysts for ORR/OER due to their distinctive and well-built crystal structure. Improving the catalytic ability of perovskites can be achieved by engineering morphologies, tailoring defects, building hybrids, etc. Shao et al. have done deep research in this field, for instance, they developed biphase Pt-Sr$(Co_{0.8}Fe_{0.2})_{0.95}P_{0.05}O_{3-\delta}$ (SCFP)/C-12 exhibits outstanding bifunctional activity for ORR/OER with a potential difference ΔE (between the overpotential of OER and the half-wave potential for ORR) = 0.73 V [24]. Characterization suggests that the excellent capability is due to the strong electronic interrelation between Pt and SCFP, such that electrons are rapidly transferred through Pt-O-Co bonds and high concentrations of surface oxygen vacancies. At the same time, more active sites can be increased by decreasing the energy barrier on the catalyst surface. In the next year, they developed an active inter-phase between perovskite $LaCoO_{3-\delta}$ (LC) and nanostructured Co-MOF [25]. Notably, the charge potential of Co-MOF/LC-0.5 for ZAB reached a low value of 2.03 V and exhibited superior cycling stability. Separately, due to the existence of electron defects and oxygen vacancies, Azad et al. reported a biphase perovskite material $LnBa_{0.5}Sr_{0.5}Co_{1.5}Fe_{0.5}O_6$ (LnBSCF) [26] with a variety of indicators that represent a good catalyst. Furthermore, Huang

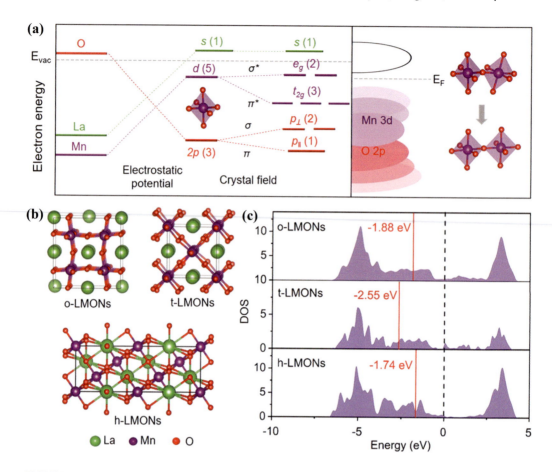

FIGURE 9.3 Electronic and crystal structures of LMONs. (Adapted with permission [27]. Copyright (2021) WILEY-VCH Verlag GmbH & Co. KGaA, Weinheim.)

et al. have developed LaMnO$_3$ surfaces with multiple crystalline phases [27], ultrathin o-LMONs achieved high OER performance with suitable surface binding energies (from e.g. filling, oxygen vacancies on the surface and the degree of hybridization of Mn), in very high OER activity and excellent durability are attained while obtained by co-tuning the crystallographic and electronic structures (Figure 9.3).

9.4 PHASE-ENGINEERED TRANSITION METAL CHALCOGENIDES/ DICHALCOGENIDES-BASED ELECTROCATALYSTS

Metal sulfides have animated attention as emerging bifunctional catalysts, which can be classified into monometallic sulfides and bimetallic sulfides. Most of the current research focuses on composite metal sulfides. Phase and structure optimization is more strenuous to achieve in polymetallic sulfides than in monometallic counterparts due to different kinetic and thermodynamic parameters during the crystallization of different phases. Notably, Yang et al. proposed a nickel sulfide (NiS$_x$) free-standing porous membrane (FHF) [28]. Due to its exceptional bifunctional catalytic performance and free-standing properties, NiS$_x$ FHF provides an ideal gettering cathode for use in flexible ZABs. Moreover, Liu et al. reported poly-phase colloidal nickel sulfide (NiS$_x$) with a high surface area as a bifunctional catalyst [29]. The rechargeable ZAB with NiS$_x$/S-rGO catalyst provides a low charge/discharge voltage difference of 1 V over 590 cycles.

Equally, Guo et al. demonstrated an oxygen vacancy-dominated multi-phase nickel-cobalt sulfide interfacial porous nanowire (NiS_2/CoS_2-O NWs) to promote OER catalysis through *in situ* electrochemical reactions of interfacial NWs [30], using the catalyst as the constructed ZAB of the air cathode has an extremely high open-circuit voltage of 1.49 V, which can be conserved for more than 30 hours. Correspondingly, Yang et al. synthesized poly-phase CoS_x@ g-C_3N_4 (PCN)/rGO catalyst after heat treatment in a sulfur atmosphere [31]. CoS_x@PCN/rGO as an air electrode in ZAB system also exhibits long cyclability, superior to Pt and other noble metal electrocatalysts, ascribed to the internally approachable nitrogen sites, and promoted conveyance of intermediates. Recently, Shao et al. showed the phase transformation from cobalt-copper oxide nanosheets (Co_2Cu_1-ONS) into the structure of stacked-shape porous sulfide networks (Co_2Cu_1-S) [32], and the S thermal annealing process can lead to faster mass transfer of reactants and intrinsic activity, in Figure 9.4, phase transition effect based on DFT calculation reveals which makes the ORR/OER catalytic performance of Co_2Cu_1-S significantly higher.

Likewise, Lee et al. developed a uniformly poly-phase multi-elements disulfides/N, S-co-doped graphene network (CoS_x@Cu_2MoS_4-MoS_2/NSG) [33]. ZABs using this catalyst as a cathode can provide a high cell voltage of 1.44 V and a peak power density of 40 mW/cm^2. Furthermore, Yu et al. reported that poly-phase $FeCo_8S_8$ nanosheets (NS) screened on rGO were derived from a colloidal strategy to prepare FeCo-based bimetallic-phase sulfides on reduced graphene oxide (rGO) by appropriate customizing of Co_9S_8 with Fe at the atomic level [34], showed efficient oxygen electrocatalytic performance. Specifically, the air electrode for ZAB exhibits exceptional stability after 200 cycles with a low voltage gap (0.77 V) at 10 mA/cm^2.

Recently, molybdenum disulfide (MoS_2) has emanated as an attractive candidate for the construction of highly efficient catalysts. Surprisingly, however, most of the research on molybdenum disulfide focuses on HER. The following studies reveal the ORR/OER performance of molybdenum disulfide. For instance, Mu et al. reported a poly-phase Mo-N/C framework encapsulated by MoS_2 nanosheets with interfacial Mo-N coupling centers [35], which was able to exhibit a high power density of 196.4 mW/cm^2 and high voltaic efficiency with excellent stability. Correspondingly, Cao et al. fabricated uniform biphase Co_9S_8@MoS_2 based on the core-shell heterostructure [36]. During the reaction process of OER, Co_9S_8@MoS_2 exhibits low overpotential, while Co_9S_8@MoS_2 exhibits strong ORR performance, excellent stability when tested as a component of ZAB. Moreover, the excellent catalytic performance is derived from the abundant catalytic active sites brought by the core-shell structure, which can also show the synergistic effect of Co_9S_8@MoS_2 and MoS_2.

As an alternative to lithium-ion batteries, LiOB has become the darling of many hot industries using power batteries due to its high theoretical specific density. However, how to make LiOB exert its excellent energy density in practical applications and how to design a high-efficiency ORR/OER bifunctional catalyst to avoid the insufficiency of kinetic reactions have become hot research topics. Recently, for the first time, Kim et al. reported two-dimensional (2D) triangular MoS_2 (1T-phase MoS_2) nanosheets as highly active electrocatalysts for LiOB, taking advantage of the excellent catalytically active surface of 1T-phase MoS_2 for its excellent absorption of Li^+ ions and oxygen [37]. Both the experimental results and DFT calculations confirm the bifunctional catalytic activity of 1T-phase MoS_2 for ORR/OER compared to the semiconducting hexagonal phase MoS_2 (2H-phase MoS_2) which has catalytic activity only at the edge.

Among rechargeable metal-air batteries, sodium-air batteries are comparable to lithium-air batteries. However, the cubic NaO_2 particles generated on the cathode after discharge of the sodium-air rechargeable battery can further hinder the cycling efficiency of the battery and greatly reduce its performance. Gu et al. reported a bifunctional catalyst using CuS as a cathode in a Na-O_2 battery (SOB) [38] and adopted a series of *in situ* methods to observe and judge the changes in products and catalysts phases during the ORR/OER reaction. With the change of potential, Na_2O_2 is the main ORR product and presents a state of uniform coverage at the cathode. During the following OER process, Na_2O_2 is converted into NaO_2, and decomposed into Na ions and O_2 gas. This process occurs on the surface of the CuS electrode, where the extruded Cu nanocrystals serve as an effective

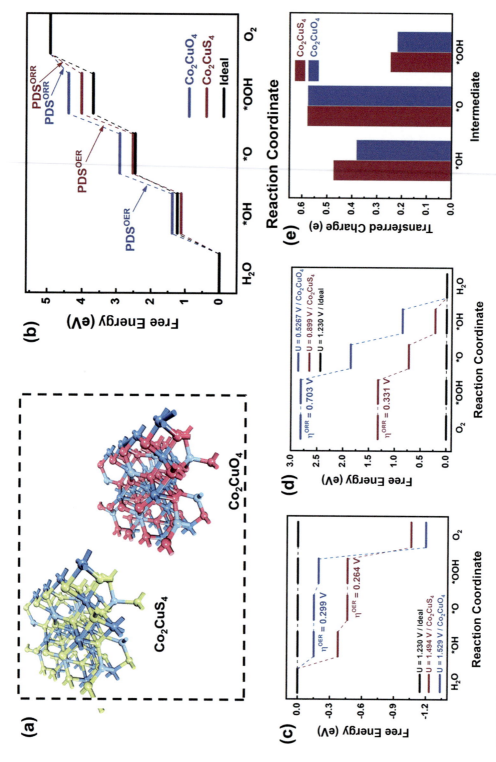

FIGURE 9.4 Research on the mechanism of phase transition effect based on DFT calculation. (Adapted with permission [32]. Copyright (2021) Royal Society of Chemistry.)

catalyst for the subsequent ORR and OER. In the assembled SOB battery, it exhibits high stability and the maximum capacity can exceed 3 mAh/cm^2.

9.5 PHASE-ENGINEERED OTHER TRANSITION METAL-BASED ELECTROCATALYSTS

9.5.1 Transition Metal Carbides Electrocatalysts

Transition metal carbides (TMCs) exhibit promising electrochemical performance during the bifunctional electrocatalytic reaction of ORR/OER. TMC has excellent electronic properties, good chemical, and thermal stability, and is not easily corroded by dilute acids (except oxidizing acids and hydrofluoric acid) or alkaline solutions. It has great advantages in cost and strong resistance to catalyst poisoning, so it is expected to be a high-quality material to replace Pt-based catalysts. Recently, Hu et al. fabricated poly-phase N-doped carbon nanoboxes decorated with interfacial Co/MoC nanoparticles (Co@IC/MoC@PC) [39]. Benefiting from the good charge transfer of the interfacial phase, the ZAB assembled with the Co@IC/MoC@PC catalyst possesses a large peak power density of 221 mW/cm^2, a high specific capacity of 728 mAh/g and a low charge/discharge voltage difference of 0.41 V. Correspondingly, Wang et al. described poly-phase Fe/Fe$_3$C@C (Fe@C) NPs embedded in 3D N-doped graphene/CNTs (Fe@C—NG/NCNTs) [40]. Using Fe@C-NG/NCNTs as the cathode catalyst for ZAB, the peak power density can reach 101.2 mW/cm^2, and the energy density can reach 764.5 Wh/kg. Besides, Jia et al. reported the synthesis of ultra-small poly-phase Fe$_7$C$_3$/N-doped carbon hybrids (u-Fe$_7$C$_3$@NC) utilizing Fe^{3+}, polydopamine, and ZnO nanospheres [41]. The charge transfer during O$_2$ adsorption and ORR is facilitated due to the strong interaction between the u-Fe$_7$C$_3$ nanoparticle phase and the NCs phase.

9.5.2 Transition Metal Nitrides Electrocatalysts

Transition metal nitrides (TMNs) retain metal-like physical properties and chemical properties with unique electronic structures. As determined, the introduction of the N element in transition metals can increase the d-band electron density and shrink the d-band of transition metals, leading to electronic structures and cross-Fermi-level state densities similar to those of noble Pt metals. Therefore, TMNs can be used as efficient electrocatalysts in energy-related processes. For example, Fan et al. reported a composite of multi-phase iron-nickel nitride/alloy nanospheres with carbon (FeNi$_3$-N), displaying the phase transformation of FeNi$_3$-N (Figure 9.5) [42]. The OER catalytic process of this catalyst in the alkaline medium can exhibit an overpotential as low as 222 mV and the Tafel slope as low as 41.53 mV/dec. DFT calculations indicate that the extremely high electrochemical performance of the dual-phasic nitride originates from the lowered energy barrier and the adjustment of the d-band center to balance the absorption of the intermediate. Equally, Lin et al. not only investigated the layered MAX phase MoSi$_2$N$_4$ as ZAB cathode and anode but also revealed the zinc storage mechanism [43]. As the phase transition changes from the state I to state III, the Zn loading in MoSi$_2$N$_4$ increases accordingly, and the maximum theoretical capacity can reach 257 mAh/g.

9.5.3 Transition Metal Phosphides Electrocatalysts

Transition metal phosphides (TMPs) have great potential to replace PGMs as catalysts for ZABs. Wang et al. developed tri-phase structural Co$_2$P/FeCo nanoparticles anchored on multi elements-codoped bamboo-like CNTs (Co$_2$P/FeCo/MnNP-BCNTs) [44]. The prepared catalyst exhibits good bifunctional catalytic activity for the ORR/OER process. The ORR half-wave potential can reach 0.88 V, and the OER process has an overpotential of 324 mV at 10 mA/cm². The ZAB with this catalyst shows exceptional performance, conveying cycle stability, and the power density can reach 220 mW/cm^2. Similarly, Fu et al. developed a biphase 2D In-doped CoO/CoP heterostructure

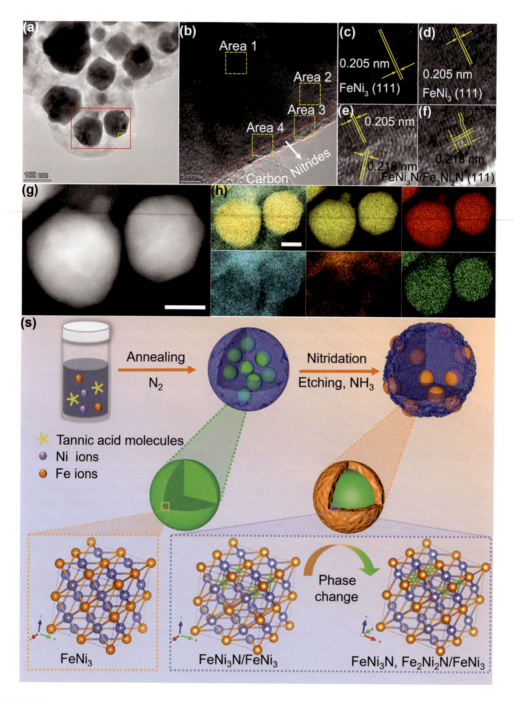

FIGURE 9.5 (a–h) Morphology of FeNi$_3$-N (FeNi$_3$-N-350–8). (s) Synthesis scheme of catalyst FeNi$_3$-N. (Adapted with permission [42]. Copyright (2021) WILEY-VCH Verlag GmbH & Co. KGaA, Weinheim.)

nanosheet (In-CoO/CoP FNS) rich in oxygen vacancies due to the abundant oxygen vacancies and porous heterostructures [45]. The In-CoO/CoP FNS catalyst was used as the cathode material of ZAB, which exhibited high cycling stability, high power density (139.4 mW/cm^2), and energy density (938 Wh/kg).

9.6 CONCLUSION AND PROSPECTS

To conclude, phase engineering has emerged as an effective venue to fabricate robust, cost-effective bifunctional catalysts for MABs. In the future, the improvement of the performance of various MABs should be considered as a whole, not only to improve the catalytic activity toward ORR/OER but also to optimize the structure of the battery configuration, so that the "battery kit-electrodes-electrolyte" complete system can work together to boost the overall battery performances. There are many challenges in the following areas:

1. Different phases of ORR products are deposited on the surface of the air electrode, which will hinder the contact between the catalyst and oxygen, and the blockage of the active site will affect the stability of the catalyst, resulting in poor reversibility of the electrode.
2. The high potential generated during OER can cause corrosion of the carbon support, making it more susceptible to degradation.
3. The catalytic activity of carbon-based materials in non-aqueous electrolyte systems involves the synergistic effect of the intrinsic catalyst activity and its structural design, which leads to a large gap between theoretical catalytic activity and practical performance in MABs.
4. The cost of preparing heterogeneous composite catalysts is high, and the exact reaction mechanism is not clear. Therefore, the structure of OER/ORR electrode catalyst should be optimized, combined with theoretical calculations to understand the electrocatalytic reaction mechanism. The phase of the catalysts is to be engineered to increase the exposure of active sites for improving the catalytic activity and effectively promoting the overall performance of MABs.

REFERENCES

[1] L. Zhou, X. Deng, Q. Lu, G. Yang, W. Zhou, Z. Shao, Zeolitic imidazolate framework-derived ordered Pt—Fe intermetallic electrocatalysts for high-performance Zn-air batteries, Energy Fuels. 34(9) (2020) 11527–11535.

[2] Z. Cui, G. Fu, Y. Li, J.B. Goodenough, Ni_3FeN-supported Fe_3Pt intermetallic nanoalloy as a high-performance bifunctional catalyst for metal—air batteries, Angewandte Chemie International Edition. 56(33) (2017) 9901–9905.

[3] M. Park, C. Liang, T.H. Lee, D.A. Agyeman, J. Yang, V.W. h Lau, S.I. Choi, H.W. Jang, K. Cho, Y.M. Kang, Regulating the catalytic dynamics through a crystal structure modulation of bimetallic catalyst, Advanced Energy Materials. 10(8) (2020) 1903225.

[4] S. Hyun, A. Saejio, S. Shanmugam, Pd nanoparticles deposited on $Co(OH)_2$ nanoplatelets as a bifunctional electrocatalyst and their application in Zn-air and $Li-O_2$ batteries, Nanoscale. 12(34) (2020) 17858–17869.

[5] Y. Niu, X. Teng, S. Gong, M. Xu, S.-G. Sun, Z. Chen, Engineering two-phase bifunctional oxygen electrocatalysts with tunable and synergetic components for flexible Zn—air batteries, Nano-Micro Letters. 13(1) (2021) 1–16.

[6] J. Li, Y. Kang, W. Wei, X. Li, Z. Lei, P. Liu, Well-dispersed ultrafine CoFe nanoalloy decorated N-doped hollow carbon microspheres for rechargeable/flexible Zn-air batteries, Chem. Eng. J. 407 (2021) 127961.

[7] S. Chen, X. Shu, H. Wang, J. Zhang, Thermally driven phase transition of manganese oxide on carbon cloth for enhancing the performance of flexible all-solid-state zinc—air batteries, Journal of Materials Chemistry A. 7(34) (2019) 19719–19727.

[8] Y. Cheng, S. Dou, J.-P. Veder, S. Wang, M. Saunders, S.P. Jiang, Efficient and durable bifunctional oxygen catalysts based on $NiFeO@MnO_x$ core—shell structures for rechargeable Zn—air batteries, ACS Applied Materials & Interfaces. 9(9) (2017) 8121–8133.

[9] J.H. Hong, J.H. Kim, G.D. Park, J.Y. Lee, J.-K. Lee, Y.C. Kang, A strategy for fabricating three-dimensional porous architecture comprising metal oxides/CNT as highly active and durable bifunctional oxygen electrocatalysts and their application in rechargeable Zn-air batteries, Chem. Eng. J. 414 (2021) 128815.

[10] Q. Zhou, S. Hou, Y. Cheng, R. Sun, W. Shen, R. Tian, J. Yang, H. Pang, L. Xu, K. Huang, Interfacial engineering Co and MnO within N,S co-doped carbon hierarchical branched superstructures toward high-efficiency electrocatalytic oxygen reduction for robust Zn-air batteries, Applied Catalysis B: Environmental. 295 (2021) 120281.

[11] P. Liu, J. Ran, B. Xia, S. Xi, D. Gao, J. Wang, Bifunctional oxygen electrocatalyst of mesoporous Ni/NiO nanosheets for flexible rechargeable Zn—air batteries, Nano-Micro Letters. 12(1) (2020) 1–12.

[12] Q. Dong, H. Wang, J. Ren, X. Wang, R. Wang, Activating Cu/Fe_2O_3 nanoislands rooted on N-rich porous carbon nanosheets via the Mott-Schottky effect for rechargeable Zn-air battery, Chem. Eng. J. 442 (2022) 136128.

[13] Y. Tian, X. Liu, L. Xu, D. Yuan, Y. Dou, J. Qiu, H. Li, J. Ma, Y. Wang, D, Su, Engineering crystallinity and oxygen vacancies of Co(II) oxide nanosheets for high performance and robust rechargeable Zn—air batteries, Advanced Functional Materials 31(20) (2021) 2101239.

[14] Y. Tian, L. Xu, M. Li, D. Yuan, X. Liu, J. Qian, Y. Dou, J. Qiu, S. Zhang, Interface engineering of CoS/CoO@N-doped graphene nanocomposite for high-performance rechargeable Zn—air batteries, Nano-Micro Letters. 13(1) (2021) 1–15.

[15] Y. Li, X. Zhang, H.-B. Li, H.D. Yoo, X. Chi, Q. An, J. Liu, M. Yu, W. Wang, Y. Yao, Mixed-phase mullite electrocatalyst for pH-neutral oxygen reduction in magnesium-air batteries, Nano Energy. 27 (2016) 8–16.

[16] F. Cheng, J. Shen, B. Peng, Y. Pan, Z. Tao, J. Chen, Rapid room-temperature synthesis of nanocrystalline spinels as oxygen reduction and evolution electrocatalysts, Nature Chemistry. 3(1) (2011) 79–84.

[17] C. Li, X. Han, F. Cheng, Y. Hu, C. Chen, J. Chen, Phase and composition controllable synthesis of cobalt manganese spinel nanoparticles towards efficient oxygen electrocatalysis, Nature Communications. 6(1) (2015) 1–8.

[18] X. Ge, Y. Liu, F.T. Goh, T.A. Hor, Y. Zong, P. Xiao, Z. Zhang, S.H. Lim, B. Li, X. Wang, Dual-phase spinel $MnCo_2O_4$ and spinel $MnCo_2O_4$/nanocarbon hybrids for electrocatalytic oxygen reduction and evolution, ACS Applied Materials & Interfaces. 6(15) (2014) 12684–12691.

[19] S. Li, X. Zhou, G. Fang, G. Xie, X. Liu, X. Lin, H.-J. Qiu, Multicomponent spinel metal oxide nanocomposites as high-performance bifunctional catalysts in Zn—air batteries, ACS Applied Energy Materials. 3(8) (2020) 7710–7718.

[20] L. An, Z. Zhang, J. Feng, F. Lv, Y. Li, R. Wang, M. Lu, R.B. Gupta, P. Xi, S. Zhang, Heterostructure-promoted oxygen electrocatalysis enables rechargeable zinc–air battery with neutral aqueous electrolyte, Journal of the American Chemical Society. 140(50) (2018) 17624–17631.

[21] J. Li, S. Lu, H. Huang, D. Liu, Z. Zhuang, C. Zhong, ZIF-67 as continuous self-sacrifice template derived $NiCo_2O_4$/Co,N-CNTs nanocages as efficient bifunctional electrocatalysts for rechargeable Zn—air batteries, ACS Sustainable Chemistry & Engineering. 6(8) (2018) 10021–10029.

[22] I. Kone, Z. Ahmad, A. Xie, L. Kong, Y. Tang, Y. Sun, Y. Chen, X. Yang, P. Wan, In situ-grown nitrogen-doped carbon-nanotube-embedded two phases of bimetal CoFe alloy and $CoFe_2O_4$ spinel oxide as highly efficient bifunctional catalyst for oxygen reduction and evolution reactions in rechargeable zinc—air batteries, Energy Technology. 9(4) (2021) 2001117.

[23] Y. Jiang, Y.P. Deng, J. Fu, D.U. Lee, R. Liang, Z.P. Cano, Y. Liu, Z. Bai, S. Hwang, L. Yang, Interpenetrating triphase cobalt-based nanocomposites as efficient bifunctional oxygen electrocatalysts for long-lasting rechargeable Zn—air batteries, Advanced Energy Materials. 8(15) (2018) 1702900.

[24] X. Wang, J. Sunarso, Q. Lu, Z. Zhou, J. Dai, D. Guan, W. Zhou, Z. Shao, High-performance platinum-perovskite composite bifunctional oxygen electrocatalyst for rechargeable Zn—air battery, Advanced Energy Materials. 10(5) (2020) 1903271.

[25] X. Wang, L. Ge, Q. Lu, J. Dai, D. Guan, R. Ran, S.-C. Weng, Z. Hu, W. Zhou, Z. Shao, High-performance metal-organic framework-perovskite hybrid as an important component of the air-electrode for rechargeable Zn-Air battery, Journal of Power Sources. 468 (2020) 228377.

[26] S. Kumar, M. Singh, R. Pal, U.P. Azad, A.K. Singh, D.P. Singh, V. Ganesan, A.K. Singh, R. Prakash, Lanthanide based double perovskites: Bifunctional catalysts for oxygen evolution/reduction reactions, Int. J. Hydrogen Energy. 46(33) (2021) 17163–17172.

[27] Q. Li, J. Wu, T. Wu, H. Jin, N. Zhang, J. Li, W. Liang, M. Liu, L. Huang, J. Zhou, Phase engineering of atomically thin perovskite oxide for highly active oxygen evolution, Advanced Functional Materials. 31(38) (2021) 2102002.

[28] K. Marcus, K. Liang, W. Niu, Y. Yang, Nickel sulfide freestanding holey films as air-breathing electrodes for flexible Zn–air batteries, The Journal of Physical Chemistry Letters. 9(11) (2018) 2746–2750.

[29] A. Sumboja, J. Chen, Y. Ma, Y. Xu, Y. Zong, P.S. Lee, Z. Liu, Sulfur-rich colloidal nickel sulfides as bifunctional catalyst for all-solid-state, flexible and rechargeable Zn-air batteries, ChemCatChem. 11(4) (2019) 1205–1213.

[30] J. Yin, Y. Li, F. Lv, M. Lu, K. Sun, W. Wang, L. Wang, F. Cheng, Y. Li, P. Xi, Oxygen vacancies dominated NiS_2/CoS_2 interface porous nanowires for portable Zn—air batteries driven water splitting devices, Advanced Materials. 29(47) (2017) 1704681.

[31] W. Niu, Z. Li, K. Marcus, L. Zhou, Y. Li, R. Ye, K. Liang, Y. Yang, Surface-modified porous carbon nitride composites as highly efficient electrocatalyst for Zn-air batteries, Advanced Energy Materials. 8(1) (2018) 1701642.

[32] L. Zhuang, H. Tao, F. Xu, C. Lian, H. Liu, K. Wang, J. Li, W. Zhou, Z. Xu, Z. Shao, Phase and morphology engineering of porous cobalt—copper sulfide as a bifunctional oxygen electrode for rechargeable Zn—air batteries, Journal of Materials Chemistry A. 9(34) (2021) 18329–18337.

[33] D.C. Nguyen, D.T. Tran, T.L.L. Doan, D.H. Kim, N.H. Kim, J.H. Lee, Rational design of core@shell structured CoS_x@Cu_2MoS_4 hybridized MoS_2/N,S-codoped graphene as advanced electrocatalyst for water splitting and Zn-air battery, Advanced Energy Materials. 10(8) (2020) 1903289.

[34] S. Lu, J. Jiang, H. Yang, Y.J. Zhang, D.-N. Pei, J.-J. Chen, Y. Yu, Phase engineering of iron—cobalt sulfides for Zn—air and Na—ion batteries, ACS Nano. 14(8) (2020) 10438–10451.

[35] I.S. Amiinu, Z. Pu, X. Liu, K.A. Owusu, H.G.R. Monestel, F.O. Boakye, H. Zhang, S. Mu, Multifunctional Mo—N/C@MoS_2 electrocatalysts for HER, OER, ORR, and Zn—air batteries, Advanced Functional Materials. 27(44) (2017) 1702300.

[36] J. Bai, T. Meng, D. Guo, S. Wang, B. Mao, M. Cao, Co_9S_8@MoS_2 core—shell heterostructures as trifunctional electrocatalysts for overall water splitting and Zn—air batteries, ACS Applied Materials & Interfaces. 10(2) (2018) 1678–1689.

[37] Z. Sadighi, J. Liu, L. Zhao, F. Ciucci, J.-K. Kim, Metallic MoS_2 nanosheets: Multifunctional electrocatalyst for the ORR, OER and $Li-O_2$ batteries, Nanoscale. 10(47) (2018) 22549–22559.

[38] S. Han, C. Cai, F. Yang, Y. Zhu, Q. Sun, Y.G. Zhu, H. Li, H. Wang, Y. Shao-Horn, X. Sun, Interrogation of the reaction mechanism in a $Na-O_2$ battery using in situ transmission electron microscopy, ACS Nano. 14(3) (2020) 3669–3677.

[39] L. Zhang, Y. Zhu, Z. Nie, Z. Li, Y. Ye, L. Li, J. Hong, Z. Bi, Y. Zhou, G. Hu, Co/MoC nanoparticles embedded in carbon nanoboxes as robust trifunctional electrocatalysts for a Zn—air battery and water electrocatalysis, ACS Nano. 15(8) (2021) 13399–13414.

[40] Q. Wang, Y. Lei, Z. Chen, N. Wu, Y. Wang, B. Wang, Y. Wang, Fe/Fe_3C@C nanoparticles encapsulated in N-doped graphene—CNTs framework as an efficient bifunctional oxygen electrocatalyst for robust rechargeable Zn—air batteries, Journal of Materials Chemistry A. 6(2) (2018) 516–526.

[41] L. Chen, Y. Zhang, X. Liu, L. Long, S. Wang, W. Yang, J. Jia, Strongly coupled ultrasmall-Fe_7C_3/N-doped porous carbon hybrids for highly efficient Zn—air batteries, Chemical Communications. 55(39) (2019) 5651–5654.

[42] Q. Chen, N. Gong, T. Zhu, C. Yang, W. Peng, Y. Li, F. Zhang, X. Fan, Surface phase engineering modulated iron-nickel nitrides/alloy nanospheres with tailored d-band center for efficient oxygen evolution reaction, Small. 18(4) (2022) 2105696.

[43] X.-M. Li, Z.-Z. Lin, L.R. Cheng, X. Chen, Layered $MoSi_2N_4$ as electrode material of Zn—air battery, Physica Status Solidi (RRL)—Rapid Research Letters. 16(5) (2022) 2200007.

[44] Z. Han, J.-J. Feng, Y.-Q. Yao, Z.-G. Wang, L. Zhang, A.-J. Wang, Mn, N, P-tridoped bamboo-like carbon nanotubes decorated with ultrafine Co_2P/FeCo nanoparticles as bifunctional oxygen electrocatalyst for long-term rechargeable Zn-air battery, J. Colloid Interface Sci. 590 (2021) 330–340.

[45] W. Jin, J. Chen, B. Liu, J. Hu, Z. Wu, W. Cai, G. Fu, Oxygen vacancy: Rich in-doped CoO/CoP heterostructure as an effective air cathode for rechargeable Zn—air batteries, Small. 15(46) (2019) 1904210.

10 Noble Metal-Based Electrocatalysts for Metal-Air Batteries

Jun Mei

CONTENTS

10.1 Introduction 135
10.2 Noble Metal-Based Electrocatalysts for MOBs 136
    10.2.1 Ru-Based Electrocatalysts 137
    10.2.2 Ir-Based Electrocatalysts 138
    10.2.3 Ag-Based Electrocatalysts 139
    10.2.4 Other Noble Metal-Based Electrocatalysts 141
10.3 Noble Metal-Based Electrocatalysts for MCBs 142
10.4 Electrocatalyst Optimization for Metal-Air Batteries 143
    10.4.1 Alloying with Non-Noble Metals 144
    10.4.2 Modification by Introducing Non-Metal Species 145
10.5 Conclusion 145
References 146

10.1 INTRODUCTION

High-density energy storage systems (ESSs) are in high demand to meet the ever-growing requirements for high-mileage electric vehicles (EVs) [1]. As one of the representative ESSs, metal-air batteries possess attractive energy densities without serious safety issues, particularly when the low-cost metal anodes (*e.g.* Zn, Mg, and Al) are used as anodes. As for cathodes in metal-air batteries, air electrodes in the presence of efficient electrocatalysts toward oxygen redox chemistry are often used [2]. An ideal cathode catalyst for rechargeable metal-air batteries should be favorable for accelerating reaction kinetics and promoting the reversibility of oxygen-involved electrocatalytic reactions, thus reducing the overpotentials and increasing round-trip efficiency [3, 4]. Exploring suitable cathode catalysts for high-performance metal-air batteries has received extensive attention in recent years.

Currently, it remains challenging to directly use ambient air consisting of O_2, H_2O, CO_2, and other environmental contaminants [5]. Based on the specific gaseous component used in the cathodes, metal-air batteries can be described as meta-O_2 (MOBs), meta-N_2 (MNBs), or metal-CO_2 (MCBs) batteries. In the rechargeable MOBs, the primary oxygen-involved reactions in cathodes include oxygen reduction reaction (ORR) and oxygen evolution reaction (OER). To date, various types of cathode catalysts including noble metals (*e.g.* Ru, Ag, Ir, Pd, and Pt), transition metals (*e.g.* Fe, Co, V, Cu, and Ni), metal-free materials (*e.g.* C_3N_4), and their derived alloys or hybrids, have been explored as bifunctional electrocatalysts for OER and ORR. Currently, the dominant catalysts are based on noble metals, particularly for large-scale practical applications, which are primarily due to their superior catalytic performance and stable reactivity under different reaction conditions. Despite many endeavors on low-cost transition metals and metal-free catalysts, the overall performance is still inferior to noble metals, far away from practical applications at this stage.

To overcome the major shortcomings of natural scarcity and high costs of noble-metal catalysts, in recent years, some effective strategies have been proposed to achieve the cost-effective goal. Some common approaches for reducing the utilization of noble metals without the expense of catalytic performance include coating/doping low-percentage noble metals into non-precious metals, alloying noble metals with low-cost metals, and hybridizing noble metal nanostructures with other natural abundant components [2, 6]. These noble metal derivations, such as alloys and hybrids, exhibit great potential as alternative cathode catalysts for MOBs. With respect to MNBs and MCBs using the intrinsically inert nitrogen and carbon dioxide, respectively, the advantages of noble metal-based catalysts are more obvious.

In this chapter, different types of noble-metal electrocatalysts, including Ru, Ir, Ag, Pd, Pt, and their derivations, are comprehensively reviewed for metal-air batteries, particularly the widely studied MOBs, and the emerging MCBs. Then, some intensively used strategies for structural optimization of these noble-metal electrocatalysts are discussed for different types of metal-air batteries. Finally, the current challenges on noble-metal electrocatalysts for metal-air batteries are critically analyzed, and some possible solutions are proposed for offering some guidelines for future research.

10.2 NOBLE METAL-BASED ELECTROCATALYSTS FOR MOBs

Based on different metal anodes, MOBs can be categorized as $Li-O_2$ (LOBs), $Na-O_2$ (NOB), $Zn-O_2$ (ZOBs), $Mg-O_2$ (MOBs), $Al-O_2$ (AOBs), and so on. Also, based on the electrolyte systems, these can be divided into aqueous, non-aqueous, solid-state, and hybrid batteries. Generally, an aqueous electrolyte can be employed when the metal anode counterpart is not sensitive to moisture, such as Zn, Mg, and Al. Otherwise, for these sensitive metal anodes, such as Li and Na, an aprotic or organic electrolyte should be used.

Non-aqueous LOBs possess a high theoretical energy density of around 3500 Wh/kg, which is much higher than that of Li-ion batteries (200–400 Wh/kg). As for the battery configuration, there are three main parts in a non-aqueous LOB, including the lithium anode, the electrolyte, and the air cathode. In principle, on discharge, gaseous oxygen is reduced at the cathode side (oxygen reduction reaction, ORR) accompanied by the formation of lithium peroxide (Li_2O_2) via intermediate superoxide lithium (LiO_2). On charge, Li_2O_2 decomposes back into Li and O_2 (oxygen evolution reaction, OER) [7, 8]. To achieve high-performance LOBs, one of the major challenges is to overcome the sluggish kinetics of the ORR and OER. Besides, the highly reactive superoxide and peroxide species often result in the undesired decomposition of electrolytes and electrodes, which require effective solutions for achieving rational control of these side reactions [9]. For non-aqueous NOBs, compared to Li-ions, Na-ions are less electrophilic and have a weaker Lewis base, the stable sodium superoxide (NaO_2) appears as the main product during ORR. Moreover, the oxidization of NaO_2 in NOBs requires lower overpotentials than that of Li_2O_2 in LOBs. It should be noteworthy that various other types of discharge products, such as sodium peroxide (Na_2O_2), sodium hydroxide (NaOH), sodium peroxide hydrate ($Na_2O_2\ 2H_2O$), and deficient sodium peroxide (Na_2xO_2), may be produced under different conditions [10].

Aqueous ZOBs have received much attention due to their low costs, good reversibility, and excellent safety. A typical ZOB includes an oxygen cathode, a Zn anode, and an alkaline electrolyte. To ensure electrically rechargeable of ZOBs, a bifunctional electrocatalyst for OER and ORR is highly required as cathode catalysts. Considering that most catalysts show good catalytic activity towards a single reaction, either ORR or OER, one commonly used strategy for achieving bifunctional catalysts used in ZOBs is by coupling these ORR-favorable species and OER-active components into an integrated composite catalyst, such as noble metals/carbon hybrids [11]. If the anode was replaced by Mg or Al, aqueous MOBs or AOBs will be obtained with similar working principles [12].

In the following section, three common types of noble metal-based electrocatalysts, Ru, Ir, and Ag, for MOBs are primarily reviewed. The most widely studied two types of MOBs include non-aqueous LOBs and aqueous ZOBs. The ideal cathode electrocatalyst is expected to achieve the desired compromise between the utilization ratio of noble metals and battery performance, thus promoting the commercial applications of ZOBs.

10.2.1 Ru-Based Electrocatalysts

The most widely studied catalyst for LOBs is Ru-based ones, including Ru supported onto or hybridized with conductive species, such as Ru/porous carbon [15] and Ru/graphene [16], and RuO_2 supported onto or hybridized with conductive species, such as dendrimer-encapsulated RuO_2 nanoparticles [17] and Ru/RuO_2-supported on reduced graphene oxide [18]. As early in 2015, a multiwalled carbon nanotube sponge (MWCNT) functionalized with Ru or RuO_2 particles (Figure 10.1a) was compared for LOBs, concluding that the MWCNT/Ru cathode exhibited a reduced charge overpotential of ~0.3 V than that of MWCNT/RuO_2 after the first discharge step (Figure 10.1b) [13]. Further investigations verified that the appearance of solid electrolyte interphase (SEI) layer on the cathode surfaces after the initial cycle, which remained stable after cycling, was crucial for preserving reversibility at the interface between the cathode and the electrolyte, thus promoting the long-time operation of dimethyl sulfoxide (DMSO)-based LOBs [13].

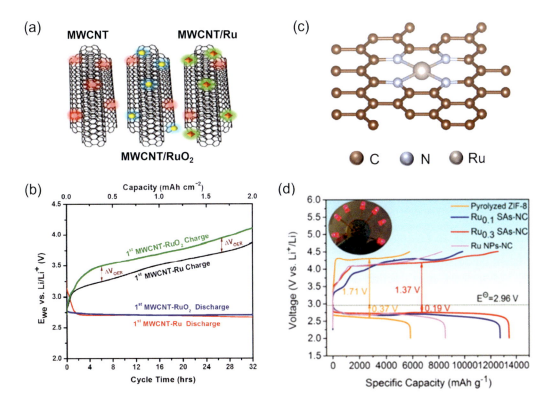

FIGURE 10.1 (a) Schematic illustration of the multiwalled carbon nanotube sponge (MWCNT) functionalized with either Ru or RuO_2 for LOBs, and (b) comparison of the initial charge/discharge curves. (a, b) (Adapted with permission [13]. Copyright (2015), American Chemical Society.) (c) Schematic illustration of Ru single atoms within N-doped porous carbon (Ru SAs-NC) used for LOBs, and (d) comparison of the first galvanostatic discharge/charge profiles at 0.02 mA/cm². ((c, d) Adapted with permission [14]. Copyright (2020), American Chemical Society.)

To reduce the overall costs and maximize atomic utilization efficiency, single-atom catalysts have received much attention in recent years, and it is expected that the desired catalysts should possess superior redox efficiency and good reversibility for LOBs. For example, in a LOB cathode, Ru single atoms within N-doped porous carbon (Ru SAs-NC, Figure 10.1c) on carbon cloth, were synthesized by employing the MOF-assisted confinement strategy and the ion substitution process from Ru^{3+} to Zn^{2+} [14]. Figure 10.1d compares the initial discharge/charge profiles at a voltage window between 2.0 and 4.5 V at 0.02 mA/cm², showing that the $Ru_{0.3}$SAs-NC-based LOB exhibits a capacity as high as 13,424 mAh/g, which outperforms other batteries based on the $Ru_{0.1}$SAs-NC (12,724 mAh/g), Ru nanoparticles on N-doped carbon (Ru NPs-NC, 8,511 mAh/g), or pyrolyzed zeolitic imidazolate frameworks (ZIF-8) (5,822 mAh/g) [14]. In comparison to the pyrolyzed ZIF-8 catalyst, the cathode catalyzed by $Ru_{0.3}$SAs-NC manifested a lower overpotential, suggesting that the modification of Ru could improve ORR and OER activities. Furthermore, the developed LOB based on the optimized $Ru_{0.3}$SAs-NC electrocatalyst showed the lowest overpotential of 0.55 V with a fixed capacity of 1,000 mAh/g at 0.02 mA/cm². Besides, theoretical calculations confirmed that the $Ru-N_4$ acted as the primary driving force center, and the rate-limiting steps of ORR and OER were the generation and oxidization of Li_2O_2, respectively [14].

Apart from the common Li_2O_2, lithium hydroxide (LiOH) or lithium carbonate (Li_2CO_3), has been identified as one of the discharge products in some LOBs in the presence of water, such as together with a soluble LiI additive or a solid Ru/MnO_2 on carbon black Super P catalyst [19–21]. Ru was proposed to catalyze the formation and decomposition reactions of the LiOH product and the decomposition of the Li_2CO_3 by-product. It has been evidenced that, the formation of LiOH follows a $4e^-$ ORR, and the hydrogen in LiOH originates from the added water and the oxygen in LiOH from both gaseous oxygen and water, upon discharge [22]. When charging, the oxidization of LiOH occurs at 3.1 V, accompanied by the trapping of oxygen in a form of dimethyl sulfone ($DMSO_2$) in the electrolyte instead of O_2 evolution. Compared to Li_2O_2, the formation of LiOH over Ru brings about much fewer side reactions, which is a critical advantage for developing long-life LOBs [22].

Besides, Ru-based electrocatalysts have been applied for other types of MOBs. For example, Ru nanoparticles on carbon nanotubes substrates were introduced for a NOB, in which the Na_2xO_2 with less reactivity than the NaO_2, was identified as the main discharge product [23]. The cathode catalyst exhibited bifunctional catalytic effects towards OER and ORR, thus remarkably increasing the cycle life of NOBs [23]. In addition to NOBs, RuO_2-coated ordered mesoporous carbon nanofiber arrays [24] have been reported for ZOBs.

10.2.2 Ir-Based Electrocatalysts

Ir-based catalysts have been regarded as promising multifunctional electrocatalysts for OER, ORR, and hydrogen evolution reaction (HER) [27, 28], which can be also used as cathode catalysts in LOBs. Various Ir-containing cathode electrocatalysts, including Ir-incorporated into deoxygenated hierarchical graphene [29], and Ir-decorated N-doped carbon spheres [30], manifest good activities towards OER and ORR for rechargeable LOBs. Besides, the introduction of Ir-based catalysts is verified to well address the negative effects of the main by-products, such as Li_2CO_3. As demonstrated in Figures 10.2a and 2b, the Ir-decorated boron carbide (Ir/B_4C) electrode could decompose nearly all preloaded Li_2CO_3 at a charge voltage below 4.37 V in an ether-based electrolyte, however, the B_4C electrode with no Ir catalyst exhibited a decompose efficiency as low as 4.7% [25]. Further analysis confirmed the role of synergistic effects, in which Ir showed a high affinity towards oxygen-containing species that could decrease the energy barrier for the oxidation of preloaded Li_2CO_3, and B_4C delivered high catalytic activity and good stability for $Li-O_2$ reactions [25]. Another typical example is to protect the surface of carbon nanofibers by using a thin Ir layer (Figure 10.2c) for effectively decomposing Li_2CO_3 by-products and achieving fast electron transfer between the carbon and the Li_2O_2 layer (Figure 10.2d) [26]. Compared with bare carbon nanofiber

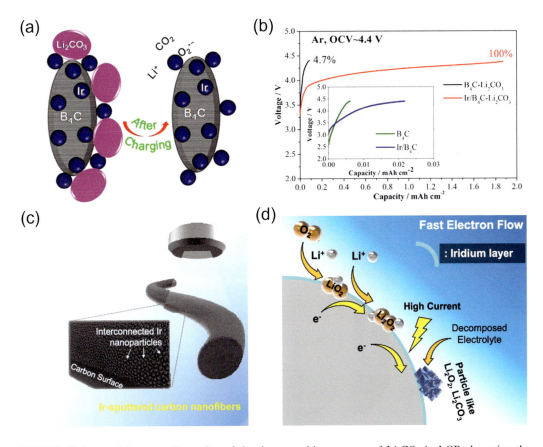

FIGURE 10.2 (a) Schematic illustration of the decomposition process of Li_2CO_3 in LOBs by using the Ir-modified boron carbide (Ir/B_4C) catalyst, and (b) comparison of the charge voltage profiles in Ar. (a, b) (Adapted with permission [25]. Copyright (2017), American Chemical Society.) (c) Schematic illustration of the Ir-sputtered carbon nanofibers for LOBs, and (d) the proposed mechanism of electrochemical reactions. ((c, d) Adapted with permission [26]. Copyright (2020), American Chemical Society.)

cathode for LOBs, the Ir-coated cathode presented two times longer operation lifespan, together with 0.2 V_{Li} lower overpotentials for OER [26].

10.2.3 Ag-Based Electrocatalysts

In comparison to Pt, Ag possesses a much lower cost and better stability in alkaline solutions. Different Ag micro-/nano-structures [33], and their derived composites, such as Co_3O_4/Ag [34], AgPd/Pd [35], and Ag/perovskite [36], have been studied for non-aqueous or aqueous LOBs. As a typical example, Lu et al. explored the effect of the size-selective Ag clusters on the morphologies of electrochemically grown Li_2O_2 in LOBs by depositing sub-nanometer Ag clusters with different sizes and atoms on the passivated carbon [31]. It was identified that there is obvious dependence relation between the morphologies of the Li_2O_2 and the size of Ag clusters, which was due to different formation mechanisms, as illustrated in Figure 10.3. In these two proposed mechanisms, the initial three steps, including the oxygen adsorption, the electron transfer, and the formation of LiO_2 are the same. In the first mechanism, further solvation occurs accompanied by the formation of supersaturated LiO_2 solution, and then the nucleation and further growth of LiO_2 occur at surface sites. In this case, a fast electron transfer rate is favorable for the formation of the solvated LiO_2,

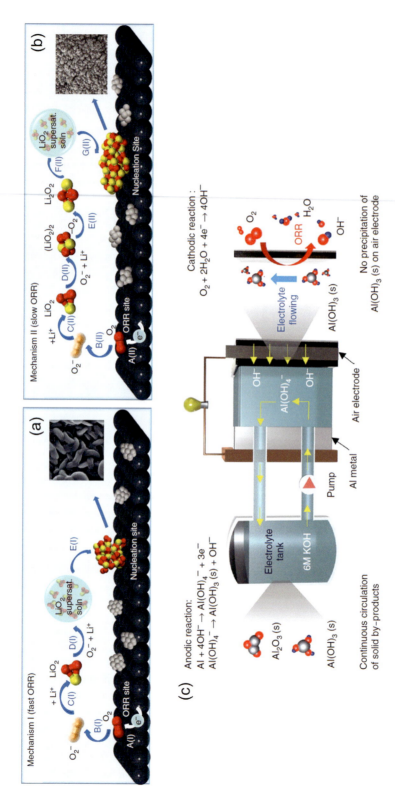

FIGURE 10.3 (a, b) Schematic illustration of two possible discharge mechanisms by using different sized Ag clusters for LOBs. (Adapted with permission [31]. Copyright (2014), Nature Publishing Group.) (c) Schematic illustration of the Al-air flow battery system. (Adapted with permission [32]. Copyright (2018), The Authors, some rights reserved; exclusive license Nature Publishing Group. Distributed under a Creative Commons Attribution License 4.0 (CC BY).)

thus resulting in the heterogeneous nucleation of LiO_2, which can be identified by the spherical-like morphology of the discharge product by using the Ag-15-atoms catalyst. In the second mechanism, the further reaction of the solvated LiO_2 with a superoxide anion and a Li^+ cation occurs, resulting in the appearance of a LiO_2 dimer and then disproportionation to Li_2O_2. When the supersaturated state reaches, the Li_2O_2 starts to nucleate and grow on surface sites. In this case, a slow electron transfer rate causes a slow formation rate of LiO_2, enabling the subsequent disproportionation step and thus leading to heterogeneous nucleation of Li_2O_2, which could be found in the consistent discharge products based on Ag-3-atoms (plates) and Ag-9-atoms (nanorods) catalysts [31].

Except for LOBs, Ag-based electrocatalysts have been extensively explored for AOBs and ZOBs [37, 38]. For AOBs, one of the representative Ag-based catalysts is the Ag/MnO_2 hybrid, which is largely due to its attractive ORR activity [39–41]. As depicted in Figure 10.3c, silver manganate nanoplates loaded onto air electrodes were introduced into an Al-air flow battery, in which the electrolytes are continuously circulated through an additional pump to avoid the undesired precipitation of by-products, resulting in an energy density of ~2552Wh kg_{Al}^{-1} at a high rate of 100 mA/cm^2 [32]. As an example of ZOBs, through *in situ* integration of Ag nanowire and graphene oxide, an attractive discharge rate as high as 300 mA/cm^2 was achieved, implying the promising potential for rechargeable ZOBs [42].

10.2.4 Other Noble Metal-Based Electrocatalysts

Other noble metal-based electrocatalysts, such as Au-based materials (*e.g.* Au-Pt [46]), Pt-based composites (*e.g.* Pt/graphene [47], and Pt/perovskite [48]), and Pd-containing hybrids (*e.g.* Pd/carbon [49], PdO/carbon [50] and Pd/carbon nanotubes [51]), have been reported for different types of MOBs. For example, an anisotropic Pt catalyst exposed with high-index {411} facets was rationally synthesized for significantly improving ORR and OER activities [43]. As shown in Figure 10.4a,

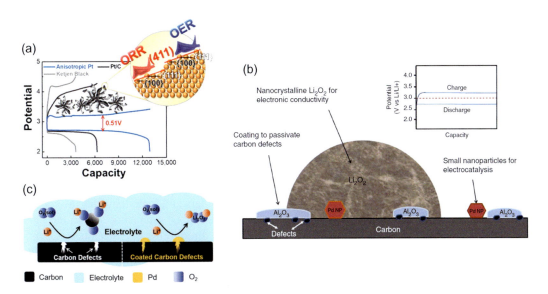

FIGURE 10.4 (a) Comparison of galvanostatic discharge/charge profiles of Pt and carbon catalysts for LOBs and schematic illustration of the anisotropic Pt catalyst with high-index {411} facets. (Adapted with permission [43]. Copyright (2018), American Chemical Society.) (b) Schematic illustration of the nanostructured cathode consisting of Al_2O_3 coating, Pd nanoparticles, and the nanocrystalline Li_2O_2 for LOBs. (Adapted with permission [44]. Copyright (2013), Nature Publishing Group.) (c) Schematic illustration of selective growth of a Pd film on carbon defects for LOBs. (Adapted with permission [45]. Copyright (2019), American Chemical Society.)

the resultant anisotropic Pt-based LOB delivered a large capacity of 12,985 mAh/g_{carbon}, much superior to those based on the commercial Pt/carbon (6,272 mAh/g_{carbon}) and carbon (3,622 mAh/g_{carbon}). Moreover, the anisotropic Pt-based LOB exhibited the first round-trip efficiency of 87% at a current density of 200 mA/g_{carbon}.[43]. Figure 10.4b proposed a new cathode architecture for addressing the charge overpotential problem of LOBs, in which a protective Al_2O_3 coating was implanted for passivating the porous carbon, which was used for passivating carbon defect sites and thus preventing electrolyte decomposition, in the presence of Pd nanoparticles as effective catalysts for the decomposition of nanocrystalline Li_2O_2 [44]. Based on this cathode architecture, the charge overpotential of the resultant LOBs was greatly reduced to about 0.2 V [44]. Then, a wet impregnation chemistry method was employed to selectively produce a thin Pd layer on carbon defect sites (Figure 10.4c), which was used for suppressing the undesired side reactions and promoting the Li_2O_2 decomposition at a low charge potential [45]. These designs on the cathode catalysts could improve the overall efficiency and extend the lifespan of LOBs.

10.3 NOBLE METAL-BASED ELECTROCATALYSTS FOR MCBs

A natural plant can capture CO_2 gas from the atmosphere for producing organic species via photosynthesis to support life. Similarly, MCBs (M = Li/Na/K/Mg/Al) can utilize the captured CO_2 as a cathode fuel for energy storage. In Li-CO_2 batteries (LCBs), the primary discharge products are Li_2CO_3 and carbon [52]. However, a higher overpotential is always required for electrochemical oxidization on the Li_2CO_3 product, a wide band gap insulator, upon charge of the LCB, compared to the Li_2O_2 product in a LOB. This will greatly increase the charge overpotential and result in poor cycling performance for LCBs. Similar to LCBs, the discharge products of Na-CO_2 batteries (NCBs) were Na_2CO_3 and carbon, and one of the major challenges is to promote the reversibility of the Na_2CO_3 product [53].

In 2017, Ru particles deposited on the Super P (Ru@Super P) were proposed as the cathode in a Li-CO_2 battery, which delivered a reduced charge potential below 4.4 V and stably worked for 80 cycles with a cut-off capacity of 1,000 mAh/g at the current densities of 100–300 mA/g [54]. As illustrated in Figure 10.5a, upon discharge, Li-ions transfer to the Ru@Super P porous cathode and then react with gaseous CO_2 accompanied by the formation of the Li_2CO_3 product, which can be further decomposed back into Li-ions and CO_2 on the subsequent charging step. Further studies confirmed that the enhanced LCB performance was largely associated with the selective catalytic characteristic of Ru particles, in which the desired charge reaction between Li_2CO_3 and carbon was considerably promoted, and meanwhile, the unfavorable self-decomposition of Li_2CO_3 product was significantly inhibited (Figure 10.5b) [54]. Later, a flexible LCB was rationally designed by using chemically treated wood as the cathode supporter for loading Ru nanocatalyst [55]. Figure 10.5c elucidates the unique advantages of the wood-derived porous cathode architecture for the LCB without a transport barrier. The microchannels are very beneficial to the efficient CO_2 gas flow while the nanochannels offer sufficient space for electrolyte fillers and ionic transport. Moreover, in the presence of Ru catalyst, the deposition of the discharge Li_2CO_3 product occurs on the interior walls of the microchannels as the CO_2 gas stream flows through the microchannels. During recharge, rapid transport for the as-formed lithium ions is achieved along the channel walls and the resultant CO_2 gas can be rapidly transported through the microchannels [55]. Thus, this LCB manifests excellent rechargeability with stable cycling over 200 times with a low overpotential of 1.5 V as well as a discharge capacity as high as 11 mAh/cm^2 using a 2-mm-thick cathode [55]. Further performance optimization can be achieved depending on the synergistic effect of electrolytes and catalysts, for example, a quinary molten salt electrolyte and Ru particles on the carbon cathode, for high-power-density LCBs [57]. The nitrate-based molten salt electrolyte is much more favorable for capturing CO_2 and reducing the gap of discharge or charge overpotentials. As a result, the as-obtained LCB delivered a maximum power density of 33.4 mW/cm^2 in the presence of the Ru catalyst [57]. Besides, Ru nanoparticles dispersed onto activated carbon nanofibers or carbon nanotubes

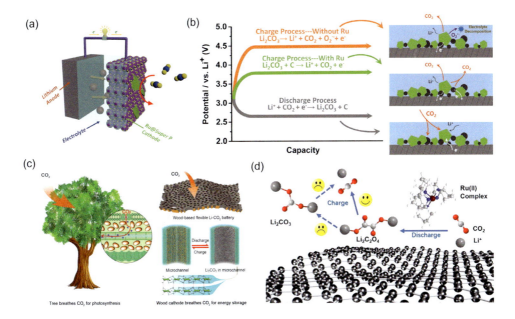

FIGURE 10.5 (a) Schematic illustration of the battery configuration of a LCB battery by using the Ru deposited onto Super P carbon (Ru@ Super P) as cathode, and (b) the reaction mechanisms with or without Ru electrocatalyst. ((a, b) Adapted with permission [54]. Copyright (2017), Royal Society of Chemistry.) (c) Schematic illustration of nature plant absorbing CO_2 for photosynthesis, and the flexible LCB based on the wood cathode in the presence of Ru particles. (Adapted with permission [55]. Copyright (2018), Royal Society of Chemistry.) (d) Schematic illustration of the function for the $Ru(bpy)_3Cl_2$ catalyst in the LCB. (Adapted with permission [56]. Copyright (2021), Wiley-VCH.)

were synthesized for LCBs [58,59], and Ru nanoparticles supported on ketjen black carbon were also explored for NCBs [60].

Apart from the single-component Ru, noble metal Ru-derived catalysts, such as RuRh alloy on Vulcan XC72R carbon [61], Ru-Cu nanoparticles on graphene [62], and RuP_2 nanoparticles on N, P-doped carbon [63], have been studied as efficient cathodes in LCBs. Xing *et al.* synthesized the triangular RuRh alloy ultrathin sheets used for LCBs [61]. It was confirmed that alloying Ru with Rh could enlarge the ability of electron transfer for these surface Ru atoms, which could facilitate both electroreduction and Li_2CO_3 decomposition reactions, sustaining the resultant LCB up to 180 cycles at a current density of 1.0 A/g [61]. Recently, a Ru-complex-based mobile catalyst, tris(2,2'-bipyridyl)-dichloro-ruthenium(II) ($Ru(bpy)_3Cl_2$) was developed for LCBs, in which the Ru^{II} center interacted with the dissolved CO_2 and stabilized the discharge intermediate ($Li_2C_2O_4$), thus promoting the reduction reaction and delaying the undesired transformation reaction into carbonate (Figure 10.5d) [56]. Consequently, the charge potential of the assembled LCB was only 3.86 V, which was 540 mV lower than that of the LCB in the absence of Ru^{II} catalyst, and the stable cycles were maintained for over 60 times with a fixed capacity of 1,000 mAh/g at 300 mA/g [56]. Additionally, other noble metal-based catalysts, such as Ir/carbon nanofibers, have been developed as cathode catalysts for rechargeable LCBs.

10.4 ELECTROCATALYST OPTIMIZATION FOR METAL-AIR BATTERIES

As mentioned earlier, the goal for electrocatalyst optimization is to achieve the desired compromise between the utilization ratio of expensive noble metals and overall electrochemical performance, thus accelerating the development of high-performance metal-air batteries for commercial applications. The simplest and widely studied strategy is to hybridize with other components with different catalytic features, as discussed earlier. Herein, two strategies are highlighted for noble-metal catalysts, alloying with low-cost metals and modification by non-metal species,

10.4.1 Alloying with Non-Noble Metals

Alloying the noble metal (e.g. Pt, Pd, Ru, Ag, or Ir) with a transition metal (*e.g.* Co, Ni, Fe, Cu) is an effective strategy for various MOBs, particularly for LOBs (*e.g.* Pt_3Co, Pt_3Ni, Pt_3V, Pt_3Nb, Pt_3Ta, Pd_3Co, Pd_3Fe, PtCu, PdFe, PdCo, PdNi) and ZOBs (*e.g.* PtRuCu, RuCo, Ni/Ni_3Pt, PtCo) [66–70]. For example, alloying Pt with low-cost Fe to form an intermetallic phase is beneficial to considerably improving catalytic activities and meanwhile reducing the utilization of high-cost Pt catalysts. As displayed in Figure 10.6a, a bifunctional oxygen electrocatalyst, in which the ordered Fe_3Pt intermetallic alloy as a cost-effective ORR catalyst supported on the porous nickel-iron nitride (Ni_3FeN) as an efficient OER catalyst, brings about a long-term cycling life of over 480 h for ZOBs [64]. Recently, a CO_2-bubble template strategy was proposed to expose Pt active sites in the PtFeNi porous films, leading to high catalytic properties towards ORR and OER, in which a half-wave potential of 0.87 V for ORR and an onset overpotential of 288 mV for OER at a current density

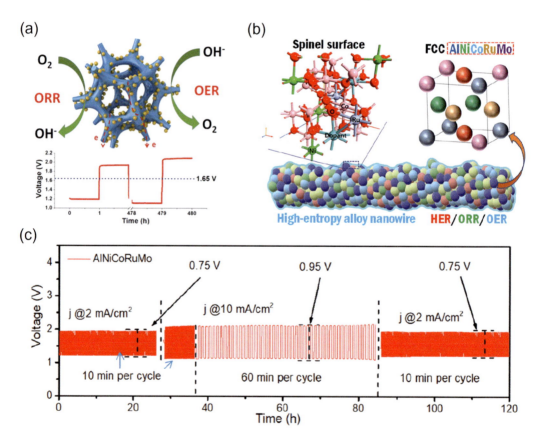

FIGURE 10.6 (a) Schematic illustration of a bifunctional oxygen electrocatalyst composed of Fe_3Pt alloy supported on the porous nickel-iron nitride (Ni_3FeN) for ZOBs and the corresponding discharge/charge voltage profiles. (Adapted with permission [64]. Copyright (2017), Wiley-VCH.) (b) Schematic illustration of a high-entropy AlNiCoRuMo alloy nanowire as a trifunctional electrocatalyst, and (c) the corresponding cycling performances for ZOBs at different current densities. (Adapted with permission [65]. Copyright (2020), American Chemical Society.)

of 10 mA/cm^2 were achieved at a Pt loading as low as 0.01 mg/cm^2 [71]. Moreover, the assembled ZOB delivered a maximum power density of 175.0 mW/cm^2 and only 0.64 V of the voltage gap was identified after long cycling tests over 500 h [71].

Noble metal-containing high-entropy alloys have been regarded as promising electrocatalysts for MOBs. These mechanically and chemically stable alloy catalysts can greatly reduce the consumption of noble metals and meanwhile possess excellent flexibility on electronic structures to adapt for different catalytic reactions. Recently, a general dealloying approach was proposed to synthesize a series of high-entropy AlNiCoRu-X (X = Mo, Cu, V, or Fe) alloy nanowires as trifunctional electrocatalysts for HER, OER, and ORR [65]. Particularly, the resultant AlNiCoRuMo nanowires exhibit a similar HER activity to Pt/C, an enhanced OER activity compared to RuO$_2$, and an increased ORR activity compared to Pt/C (Figure 10.6b). Furthermore, the stable oxidized surfaces on these alloy nanowires are favorable for continuous operation cycles for rechargeable ZOBs. Specifically, the AlNiCoRuMo-based ZOB exhibited a constant potential gap of ~0.75 V for over 500 h at 2 mA/cm^2 and a stable gap of 0.95 V for nearly 80 h at a large current density of 10 mA/cm^2 (Figure 10.6c) [65].

10.4.2 Modification by Introducing Non-Metal Species

For electrocatalytic reactions occurred in metal-air batteries, surface states act as crucial roles in affecting catalytic kinetics and stability. Further surface modification on these noble-metal-containing electrocatalysts is another optimization strategy to adapt for OER and/or ORR, which is much favorable for promoting reversibility of metal-air batteries. For example, a phosphate-ion ($H_2PO_4^-$) functionalization strategy was applied on a bismuth rhodium oxide ($Bi_2Rh_2O_{6.8}$) pyrochlore electrocatalyst for electrocatalysis [72]. During ORR (Figure 10.7a) and OER (Figure 10.7b), the functionalized $H_2PO_4^-$ ion can considerably improve surface reactivities for accelerating electron/charge transfer, facilitate the cationic oxidation, and promote the electron donation processes via improving electron transport. When the modified catalyst (P-$Bi_2Rh_2O_{6.8}$) catalyst was implanted into an aqueous Na-air battery (Figure 10.7c), the lowest overpotential gap of 0.17 V and the highest round trip efficiency of 94.9% were achieved compared to these air cathodes by using Pt/C (0.21 V, 93.6%) or Ir/C (0.32 V, 90.2%) (Figure 10.7d) [72]. This strategy was also used for other pyrochlores, such as thallium ruthenium oxide ($Tl_2Ru_2O_7$) [73].

10.5 CONCLUSION

In summary, metal-air batteries deliver obvious advantages on high energy densities, which show great potential in large power storage applications, such as EVs. For LOBs, most attention has been paid to the nonaqueous systems, and the direct utilization of air remains challenging, primarily due to the presence of moisture and other environmental contaminations. For other MOBs, such as ZOBs, MOBs, and AOBs, the battery configurations are relatively simple, and the practical operation is easy to be achieved in aqueous systems, resulting in low manufacturing costs without serious safety issues. Some major progress in the development of the efficient bifunctional electrocatalysts used in ZOBs has been achieved in recent years, however, research on MOBs and AOBs is still required to promote the reaction reversibility as long-life rechargeable batteries. For other metal-air batteries, such as MCBs and MNBs, the current study is only at the very initial stage, and more work is needed in the future. In terms of noble metal-based electrocatalysts, alloy and hybrid catalysts have been intensively explored for different types of MOBs, and these derived catalysts show a good compromise on the reduced consumption of expensive noble metals and the overall battery performance to meet the requirements for the practical applications. Alloying of noble metals with these low-cost transition metals is one of the effective solutions for addressing the issues on manufacture costs. For hybrid catalysts, conductive and porous carbonaceous materials, such as porous carbon, graphene gel, and carbon fiber/tube networks, are the most widely used supporter for loading noble

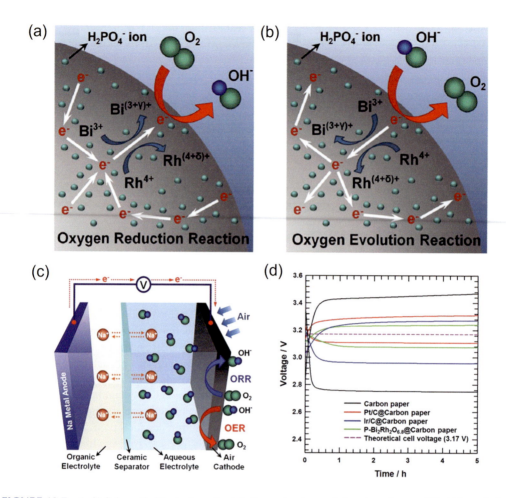

FIGURE 10.7 (a, b) Schematic illustration of catalytic mechanisms of phosphate-ion-modified bismuth rhodium oxide (P-$Bi_2Rh_2O_{6.8}$) catalyst for (a) ORR and (b) OER. (c) Schematic illustration of the assembled aqueous Na-air batteries, and (d) comparison of charge/discharge curves. (Adapted with permission [72]. Copyright (2018), Royal Society of Chemistry.)

metal-based electrocatalysts in the cathodes of metal-air batteries. Furthermore, to enlarge the catalytic activity and meet the requirements on rate capability or cycling stability, these electrocatalysts can be further modified by surface and interface engineering, such as anion-functionalized surfaces on noble metal-containing catalysts for promoting ORR and OER in MOBs. It is expected that the development of efficient cathode electrocatalysts will greatly boost the innovation of metal-air batteries and accelerate the commercialization of large-scale applications.

REFERENCES

[1] P.G. Bruce, S.A. Freunberger, L.J. Hardwick, J.-M. Tarascon, $Li-O_2$ and Li-S batteries with high energy storage, Nat. Mater. 11 (2012) 19–29.
[2] J. Mei, T. Liao, J. Liang, Y. Qiao, S.X. Dou, Z. Sun, Toward promising cathode catalysts for nonlithium metal-oxygen batteries, Adv. Energy Mater. 10 (2020) 1901997.
[3] F. Cheng, J. Chen, Metal—air batteries: From oxygen reduction electrochemistry to cathode catalysts, Chem. Soc. Rev. 41 (2012) 2172–2192.

[4] Y.C. Lu, H.A. Gasteiger, Y. Shao-Horn, Catalytic activity trends of oxygen reduction reaction for non-aqueous Li-air batteries, J. Am. Chem. Soc. 133 (2011) 19048–19051.

[5] G. Girishkumar, B. McCloskey, A.C. Luntz, S. Swanson, W. Wilcke, Lithium-air battery: Promise and challenges, J. Phys. Chem. Lett. 1 (2010) 2193–2203.

[6] J. Mei, Q. Zhang, H. Peng, T. Liao, Z. Sun, Phase engineering activation of low-cost iron-containing sulfide minerals for advanced electrocatalysis, J. Mater. Sci. Technol. 111 (2022) 181–188.

[7] F. Li, T. Zhang, H. Zhou, Challenges of non-aqueous Li-O_2 batteries: Electrolytes, catalysts, and anodes, Energy Environ. Sci. 6 (2013) 1125–1141.

[8] D. Aurbach, B.D. McCloskey, L.F. Nazar, P.G. Bruce, Advances in understanding mechanisms underpinning lithium-air batteries, Nat. Energy. 1 (2016) 16128.

[9] K.X. Wang, Q.C. Zhu, J.S. Chen, Strategies toward high-performance cathode materials for lithium-oxygen batteries, Small. 14 (2018) 1800078.

[10] H. Yadegari, Q. Sun, X. Sun, Sodium-oxygen batteries: A comparative review from chemical and electrochemical fundamentals to future perspective, Adv. Mater. 28 (2016) 7065–7093.

[11] J. Fu, Z.P. Cano, M.G. Park, A. Yu, M. Fowler, Z. Chen, Electrically rechargeable zinc—air batteries: Progress, challenges, and perspectives, Adv. Mater. 29 (2017) 1604685.

[12] M.A. Rahman, X. Wang, C. Wen, High energy density metal-air batteries: A review, J. Electrochem. Soc. 160 (2013) A1759–A1771.

[13] M.A. Schroeder, A.J. Pearse, A.C. Kozen, X. Chen, K. Gregorczyk, X. Han, A. Cao, L. Hu, S.B. Lee, G.W. Rubloff, M. Noked, Investigation of the cathode-catalyst-electrolyte interface in aprotic Li-O_2 batteries, Chem. Mater. 27 (2015) 5305–5313.

[14] X. Hu, G. Luo, Q. Zhao, D. Wu, T. Yang, J. Wen, R. Wang, C. Xu, N. Hu, Ru single atoms on N-doped carbon by spatial confinement and ionic substitution strategies for high-performance Li–O_2 batteries, J. Am. Chem. Soc. 142 (2020) 16776–16786.

[15] C. Shu, C. Wu, J. Long, H. Guo, S.X. Dou, J. Wang, Highly reversible Li-O_2 battery induced by modulating local electronic structure via synergistic interfacial interaction between ruthenium nanoparticles and hierarchically porous carbon, Nano Energy. 57 (2019) 166–175.

[16] Y.S. Jeong, J.B. Park, H.G. Jung, J. Kim, X. Luo, J. Lu, L. Curtiss, K. Amine, Y.K. Sun, B. Scrosati, Y.J. Lee, Study on the catalytic activity of noble metal nanoparticles on reduced graphene oxide for oxygen evolution reactions in lithium-air batteries, Nano Lett. 15 (2015) 4261–4268.

[17] P. Bhattacharya, E.N. Nasybulin, M.H. Engelhard, L. Kovarik, M.E. Bowden, X.S. Li, D.J. Gaspar, W. Xu, J.G. Zhang, Dendrimer-encapsulated ruthenium oxide nanoparticles as catalysts in lithium-oxygen batteries, Adv. Funct. Mater. 24 (2014) 7510–7519.

[18] H.G. Jung, Y.S. Jeong, J.B. Park, Y.K. Sun, B. Scrosati, Y.J. Lee, Ruthenium-based electrocatalysts supported on reduced graphene oxide for lithium-air batteries, ACS Nano. 7 (2013) 3532–3539.

[19] T. Liu, M. Leskes, W. Yu, A.J. Moore, L. Zhou, P.M. Bayley, G. Kim, C.P. Grey, Cycling Li-O_2 batteries via LiOH formation and decomposition, Science. 350 (2015) 530–533.

[20] F. Li, S. Wu, D. Li, T. Zhang, P. He, A. Yamada, H. Zhou, The water catalysis at oxygen cathodes of lithium-oxygen cells, Nat. Commun. 6 (2015) 7843.

[21] B. Sun, L. Guo, Y. Ju, P. Munroe, E. Wang, Z. Peng, G. Wang, Unraveling the catalytic activities of ruthenium nanocrystals in high performance aprotic Li-O_2 batteries, Nano Energy. 28 (2016) 486–494.

[22] T. Liu, Z. Liu, G. Kim, J.T. Frith, N. Garcia-Araez, C.P. Grey, Understanding LiOH chemistry in a ruthenium-catalyzed Li-O_2 battery, Angew. Chem. Int. Ed. 56 (2017) 16057–16062.

[23] J.H. Kang, W.J. Kwak, D. Aurbach, Y.K. Sun, Sodium oxygen batteries: One step further with catalysis by ruthenium nanoparticles, J. Mater. Chem. A. 5 (2017) 20678–20686.

[24] Z. Guo, C. Li, W. Li, H. Guo, X. Su, P. He, Y. Wang, Y. Xia, Ruthenium oxide coated ordered mesoporous carbon nanofiber arrays: A highly bifunctional oxygen electrocatalyst for rechargeable Zn-air batteries, J. Mater. Chem. A. 4 (2016) 6282–6289.

[25] S. Song, W. Xu, J. Zheng, L. Luo, M.H. Engelhard, M.E. Bowden, B. Liu, C.M. Wang, J.G. Zhang, Complete decomposition of Li_2CO_3 in Li-O_2 batteries using Ir/B_4C as noncarbon-based oxygen electrode, Nano Lett. 17 (2017) 1417–1424.

[26] J.S. Nam, J.W. Jung, D.Y. Youn, S.H. Cho, J.Y. Cheong, M.S. Kim, S.W. Song, S.J. Kim, I.D. Kim, Free-standing carbon nanofibers protected by a thin metallic iridium layer for extended life-cycle Li-oxygen batteries, ACS Appl. Mater. Interfaces. 12 (2020) 55756–55765.

[27] Q. Zhang, Z. Duan, Y. Wang, L. Li, B. Nan, J. Guan, Atomically dispersed iridium catalysts for multifunctional electrocatalysis, J. Mater. Chem. A. 8 (2020) 19665–19673.

[28] J. Mei, T. He, J. Bai, D. Qi, A. Du, T. Liao, G.A. Ayoko, Y. Yamauchi, L. Sun, Z. Sun, Surface-dependent intermediate adsorption modulation on iridium-modified black phosphorus electrocatalysts for efficient pH-universal water splitting, Adv. Mater. 33 (2021) 2104638.

[29] W. Zhou, Y. Cheng, X. Yang, B. Wu, H. Nie, H. Zhang, H. Zhang, Iridium incorporated into deoxygenated hierarchical graphene as a high-performance cathode for rechargeable Li-O_2 batteries, J. Mater. Chem. A. 3 (2015) 14556–14561.

[30] J. Shen, H. Wu, W. Sun, J. Qiao, H. Cai, Z. Wang, K. Sun, In-situ nitrogen-doped hierarchical porous hollow carbon spheres anchored with iridium nanoparticles as efficient cathode catalysts for reversible lithium-oxygen batteries, Chem. Eng. J. 358 (2019) 340–350.

[31] J. Lu, L. Cheng, K.C. Lau, E. Tyo, X. Luo, J. Wen, D. Miller, R.S. Assary, H.H. Wang, P. Redfern, H. Wu, J.B. Park, Y.K. Sun, S. Vajda, K. Amine, L.A. Curtiss, Effect of the size-selective silver clusters on lithium peroxide morphology in lithium-oxygen batteries, Nat. Commun. 5 (2014) 4895.

[32] J. Ryu, H. Jang, J. Park, Y. Yoo, M. Park, J. Cho, Seed-mediated atomic-scale reconstruction of silver manganate nanoplates for oxygen reduction towards high-energy aluminum-air flow batteries, Nat. Commun. 9 (2018) 3715.

[33] J.B. Park, X. Luo, J. Lu, C.D. Shin, C.S. Yoon, K. Amine, Y.K. Sun, Improvement of electrochemical properties of lithium-oxygen batteries using a silver electrode, J. Phys. Chem. C. 119 (2015) 15036–15040.

[34] R. Gao, Z. Yang, L. Zheng, L. Gu, L. Liu, Y. Lee, Z. Hu, X. Liu, Enhancing the catalytic activity of Co_3O_4 for Li-O_2 batteries through the synergy of surface/interface/doping engineering, ACS Catal. 8 (2018) 1955–1963.

[35] W. Bin Luo, X.W. Gao, S.L. Chou, J.Z. Wang, H.K. Liu, Porous AgPd-Pd composite nanotubes as highly efficient electrocatalysts for lithium-oxygen batteries, Adv. Mater. 27 (2015) 6862–6869.

[36] Y. Cong, Q. Tang, X. Wang, M. Liu, J. Liu, Z. Geng, R. Cao, X. Zhang, W. Zhang, K. Huang, S. Feng, Silver-intermediated perovskite $La_{0.9}FeO_{3-\delta}$ toward high-performance cathode catalysts for nonaqueous lithium-oxygen batteries, ACS Catal. 9 (2019) 11743–11752.

[37] J. Ryu, M. Park, J. Cho, Advanced technologies for high-energy aluminum-air batteries, Adv. Mater. 31 (2019) 1804784.

[38] Y. Li, H. Dai, Recent advances in zinc-air batteries, Chem. Soc. Rev. 43 (2014) 5257–5275.

[39] S. Sun, H. Miao, Y. Xue, Q. Wang, S. Li, Z. Liu, Oxygen reduction reaction catalysts of manganese oxide decorated by silver nanoparticles for aluminum-air batteries, Electrochim. Acta. 214 (2016) 49–55.

[40] S. Sun, H. Miao, Y. Xue, Q. Wang, Q. Zhang, Z. Dong, S. Li, H. Huang, Z. Liu, High electrocatalytic activity of silver-doped manganese dioxide toward oxygen reduction reaction in aluminum-air battery, J. Electrochem. Soc. 164 (2017) F768–F774.

[41] B. Xu, H. Lu, W. Cai, Y. Cao, Y. Deng, W. Yang, Synergistically enhanced oxygen reduction reaction composites of specific surface area and manganese valence controlled α-MnO_2 nanotube decorated by silver nanoparticles in Al-air batteries, Electrochim. Acta. 305 (2019) 360–369.

[42] S. Hu, T. Han, C. Lin, W. Xiang, Y. Zhao, P. Gao, F. Du, X. Li, Y. Sun, Enhanced electrocatalysis via 3D graphene aerogel engineered with a silver nanowire network for ultrahigh-rate zinc-air batteries, Adv. Funct. Mater. 27 (2017) 1700041.

[43] K. Song, J. Jung, M. Park, H. Park, H.J. Kim, S. Il Choi, J. Yang, K. Kang, Y.K. Han, Y.M. Kang, Anisotropic surface modulation of Pt catalysts for highly reversible Li-O_2 batteries: High index facet as a critical descriptor, ACS Catal. 8 (2018) 9006–9015.

[44] J. Lu, Y. Lei, K.C. Lau, X. Luo, P. Du, J. Wen, R.S. Assary, U. Das, D.J. Miller, J.W. Elam, H.M. Albishri, D.A. El-Hady, Y.K. Sun, L.A. Curtiss, K. Amine, A nanostructured cathode architecture for low charge overpotential in lithium-oxygen batteries, Nat. Commun. 4 (2013) 2383.

[45] T. Zhang, B. Zou, X. Bi, M. Li, J. Wen, F. Huo, K. Amine, J. Lu, Selective growth of a discontinuous subnanometer Pd film on carbon defects for Li-O_2 batteries, ACS Energy Lett. 4 (2019) 2782–2786.

[46] Y.C. Lu, Z. Xu, H.A. Gasteiger, S. Chen, K. Hamad-Schifferli, Y. Shao-Horn, Platinum-gold nanoparticles: A highly active bifunctional electrocatalyst for rechargeable lithium-air batteries, J. Am. Chem. Soc. 132 (2010) 12170–12171.

[47] F. Wu, Y. Xing, X. Zeng, Y. Yuan, X. Zhang, R. Shahbazian-Yassar, J. Wen, D.J. Miller, L. Li, R. Chen, J. Lu, K. Amine, Platinum-coated hollow graphene nanocages as cathode used in lithium-oxygen batteries, Adv. Funct. Mater. 26 (2016) 7626–7633.

[48] X. Wang, J. Sunarso, Q. Lu, Z. Zhou, J. Dai, D. Guan, W. Zhou, Z. Shao, High-performance platinum-perovskite composite bifunctional oxygen electrocatalyst for rechargeable Zn-air battery, Adv. Energy Mater. 10 (2020) 1903271.

[49] Y. Lei, J. Lu, X. Luo, T. Wu, P. Du, X. Zhang, Y. Ren, J. Wen, D.J. Miller, J.T. Miller, Y.K. Sun, J.W. Elam, K. Amine, Synthesis of porous carbon supported palladium nanoparticle catalysts by atomic layer deposition: Application for rechargeable lithium-O_2 battery, Nano Lett. 13 (2013) 4182–4189.

[50] H. Cheng, K. Scott, Selection of oxygen reduction catalysts for rechargeable lithium-air batteries-metal or oxide?, Appl. Catal. B Environ. 108–109 (2011) 140–151.

[51] N. Chawla, A. Chamaani, M. Safa, B. El-Zahab, Palladium-filled carbon nanotubes cathode for improved electrolyte stability and cyclability performance of Li-O_2 batteries, J. Electrochem. Soc. 164 (2017) A6303–A6307.

[52] Y. Liu, R. Wang, Y. Lyu, H. Li, L. Chen, Rechargeable Li/CO_2-O_2 (2:1) battery and Li/CO_2 battery, Energy Environ. Sci. 7 (2014) 677–681.

[53] X. Hu, J. Sun, Z. Li, Q. Zhao, C. Chen, J. Chen, Rechargeable room-temperature Na-CO_2 batteries, Angew. Chem. Int. Ed. 55 (2016) 6482–6486.

[54] S. Yang, Y. Qiao, P. He, Y. Liu, Z. Cheng, J.J. Zhu, H. Zhou, A reversible lithium-CO_2 battery with Ru nanoparticles as a cathode catalyst, Energy Environ. Sci. 10 (2017) 972–978.

[55] S. Xu, C. Chen, Y. Kuang, J. Song, W. Gan, B. Liu, E.M. Hitz, J.W. Connell, Y. Lin, L. Hu, Flexible lithium-CO_2 battery with ultrahigh capacity and stable cycling, Energy Environ. Sci. 11 (2018) 3231–3237.

[56] Z. Zhang, W.L. Bai, Z.P. Cai, J.H. Cheng, H.Y. Kuang, B.X. Dong, Y.B. Wang, K.X. Wang, J.S. Chen, Enhanced electrochemical performance of aprotic Li-CO_2 batteries with a ruthenium-complex-based mobile catalyst, Angew. Chem. Int. Ed. 60 (2021) 16404–16408.

[57] K. Baek, W.C. Jeon, S. Woo, J.C. Kim, J.G. Lee, K. An, S.K. Kwak, S.J. Kang, Synergistic effect of quinary molten salts and ruthenium catalyst for high-power-density lithium-carbon dioxide cell, Nat. Commun. 11 (2020) 456.

[58] Y. Qiao, S. Xu, Y. Liu, J. Dai, H. Xie, Y. Yao, X. Mu, C. Chen, D.J. Kline, E.M. Hitz, B. Liu, J. Song, P. He, M.R. Zachariah, L. Hu, Transient, in situ synthesis of ultrafine ruthenium nanoparticles for a high-rate Li-CO_2 battery, Energy Environ. Sci. 12 (2019) 1100–1107.

[59] C.J. Chen, J.J. Yang, C.H. Chen, D.H. Wei, S.F. Hu, R.S. Liu, Improvement of lithium anode deterioration for ameliorating cyclabilities of non-aqueous Li-CO_2 batteries, Nanoscale. 12 (2020) 8385–8396.

[60] L. Guo, B. Li, V. Thirumal, J. Song, Advanced rechargeable Na-CO_2 batteries enabled by a ruthenium@porous carbon composite cathode with enhanced Na_2CO_3 reversibility, Chem. Commun. 55 (2019) 7946–7949.

[61] Y. Xing, K. Wang, N. Li, D. Su, W.T. Wong, B. Huang, S. Guo, Ultrathin RuRh alloy nanosheets enable high-performance lithium-CO_2 battery, Matter. 2 (2020) 1494–1508.

[62] Z. Zhang, C. Yang, S. Wu, A. Wang, L. Zhao, D. Zhai, B. Ren, K. Cao, Z. Zhou, Exploiting synergistic effect by integrating ruthenium-copper nanoparticles highly Co-dispersed on graphene as efficient air cathodes for Li-CO_2 batteries, Adv. Energy Mater. 9 (2019) 1802805.

[63] Z. Guo, J. Li, H. Qi, X. Sun, H. Li, A.G. Tamirat, J. Liu, Y. Wang, L. Wang, A highly reversible long-life Li—CO_2 battery with a RuP_2-based catalytic cathode, Small. 15 (2019) 1803246.

[64] Z. Cui, G. Fu, Y. Li, J.B. Goodenough, Ni_3FeN-supported Fe_3Pt intermetallic nanoalloy as a high-performance bifunctional catalyst for metal-air batteries, Angew. Chem. Int. Ed. 56 (2017) 9901–9905.

[65] Z. Jin, J. Lyu, Y.L. Zhao, H. Li, X. Lin, G. Xie, X. Liu, J.J. Kai, H.J. Qiu, rugged high-entropy alloy nanowires with in situ formed surface spinel oxide as highly stable electrocatalyst in Zn-air batteries, ACS Mater. Lett. 2 (2020) 1698–1706.

[66] J. Liu, T. Zhang, G.I.N. Waterhouse, Complex alloy nanostructures as advanced catalysts for oxygen electrocatalysis: From materials design to applications, J. Mater. Chem. A. 8 (2020) 23142–23161.

[67] Y.J. Kang, S.C. Jung, H.J. Kim, Y.K. Han, S.H. Oh, Maximum catalytic activity of Pt_3M in Li-O_2 batteries: M=group V transition metals, Nano Energy. 27 (2016) 1–7.

[68] S.M. Cho, S.W. Hwang, J.H. Yom, W.Y. Yoon, Pd_3Co/MWCNTs composite electro-catalyst cathode material for use in lithium-oxygen batteries, J. Electrochem. Soc. 162 (2015) A2236–A2244.

[69] Z. Cui, L. Li, A. Manthiram, J.B. Goodenough, Enhanced cycling stability of hybrid Li-air batteries enabled by ordered Pd_3Fe intermetallic electrocatalyst, J. Am. Chem. Soc. 137 (2015) 7278–7281.

[70] Z. Li, W. Niu, Z. Yang, N. Zaman, W. Samarakoon, M. Wang, A. Kara, M. Lucero, M.V. Vyas, H. Cao, H. Zhou, G.E. Sterbinsky, Z. Feng, Y. Du, Y. Yang, Stabilizing atomic Pt with trapped interstitial F in alloyed PtCo nanosheets for high-performance zinc-air batteries, Energy Environ. Sci. 13 (2020) 884–895.

[71] G. Wang, J. Chang, S. Koul, A. Kushima, Y. Yang, CO_2 bubble-assisted Pt exposure in PtFeNi porous film for high-performance zinc-air battery, J. Am. Chem. Soc. 143 (2021) 11595–11601.

[72] M. Kim, H. Ju, J. Kim, Highly efficient bifunctional catalytic activity of bismuth rhodium oxide pyrochlore through tuning the covalent character for rechargeable aqueous Na-air batteries, J. Mater. Chem. A. 6 (2018) 8523–8530.

[73] M. Kim, H. Ju, J. Kim, Dihydrogen phosphate ion functionalized nanocrystalline thallium ruthenium oxide pyrochlore as a bifunctional electrocatalyst for aqueous Na-air batteries, Appl. Catal. B Environ. 245 (2019) 29–39.

11 1D Materials as Electrocatalysts for Metal-Air Batteries

Merve Gençtürk and Emre Biçer

CONTENTS

- 11.1 Introduction 151
- 11.2 Carbon-Based 1D Materials 151
  - 11.2.1 Carbon Nanotubes (CNT) 153
  - 11.2.2 Carbon Nanofibers (CNF) 153
  - 11.2.3 Carbon Nanorods 153
  - 11.2.4 N-Doped Carbon Materials 153
  - 11.2.5 Metal and Metal Oxides with 1D Carbon Composites 157
- 11.3 One-Dimensional Metal Electrocatalysts 159
  - 11.3.1 One-Dimensional Precious Materials 159
  - 11.3.2 One-Dimensional Non-Precious Materials 159
- 11.4 Other Materials 160
- 11.5 Conclusion 160
- References 160

11.1 INTRODUCTION

Energy is life, without it, there is nothing to do. Energy permeates our lives more with the increasing mobile applications, electric vehicles, etc. The ascending specific energy values with Li-ion batteries made these applications more popular. For three decades, Li-ions were capable of meeting our energy needs, but they no longer encounter the need at the desired level. Therefore, researchers investigate different kinds of batteries with more specific energy and power. Metal-air batteries are one of the focused battery technologies and they demonstrate approximately ten-fold more energy density compared with conventional Li-ion batteries. This redundancy can be made possible by the use of lithium metal in rechargeable metal-air batteries. Together with lithium metal, different types of electrocatalysts are used to provide charge/discharge reversibility. Oxygen reduction reaction (ORR) and oxygen evaluation reaction (OER) are the key reactions providing this ability with the appropriate electrocatalyst used in the cathode. Carbonaceous, non-carbonaceous together with precious and non-precious materials are used as electrocatalysts in metal-air batteries. Nanostructured materials including one-dimensional (1D) nanowires/tubes, two-dimensional (2D) nanosheets, and three-dimensional (3D) hollow spheres/cubes are widely employed in literature [1]. This chapter reviews the usage of 1D materials in metal-air batteries.

11.2 CARBON-BASED 1D MATERIALS

Among the nanomaterials used in metal-air batteries, carbon nanomaterials take great attention due to their unique properties of large surface area, high conductivity, and low cost. 1D nanostructured carbon materials are also recognized as a highly promising material for metal-air batteries with easy electrode accessibility, fast electron transport, and reduced ion diffusion [2]. To provide the electrochemical reactions in metal-air batteries, the following properties should be met: (i) high porosity to enable the transportation of O_2 and Li^+ ions, (ii) high catalytic activity to realize oxygen

DOI: 10.1201/9781003295761-11

reduction reaction (ORR) and oxygen evaluation reaction (OER), (iii) high conductivity for fast electron transfer, and (iv) stable interface to prevent the side reactions. The electronic structure of carbon has great importance in electrochemical activity involving adsorption of reactants and charge transfer and mass transport. Another significant issue is Li_2O_2 decomposition which reverses the discharge reaction [3].

Both ORR and OER involve multi-electron exchange steps and different oxygen intermediates according to electrolyte media. ORR can be two-electron, four-electron, or hybrid process:

Four-Electron Pathway of ORR in Alkaline Conditions

$$O_2 + H_2O + e^- \longrightarrow *OOH + OH^-$$
$$*OOH + e^- \longrightarrow *O + OH^-$$
$$*O + H_2O + e^- \longrightarrow *OH + OH^-$$
$$*OH + e^- \longrightarrow OH^-$$

Overall Reaction: $O_2 + 2H_2O + 4e^- \longrightarrow 4OH^-$

Two-Electron Pathway of ORR in Alkaline Conditions

$$O_2 + H_2O + e^- \longrightarrow *OOH + OH^-$$
$$*OOH + H_2O + e^- \longrightarrow H_2O_2 + OH^-$$

Overall Reaction: $O_2 + 2H_2O + 2e^- \longrightarrow 2OH^- + H_2O_2$

Four-Electron Pathway of ORR in Acidic Conditions

$$O_2 + H^+ + e^- \longrightarrow *OOH$$
$$*OOH + H^+ + e^- \longrightarrow *O + H_2O$$
$$*O + H^+ + e^- \longrightarrow *OH$$
$$*OH + H^+ + e^- \longrightarrow H_2O$$

Overall Reaction: $O_2 + 4H^+ + 4e^- \longrightarrow 2H_2O$

Two-Electron Pathway of ORR in Acidic Conditions

$$O_2 + H^+ + e^- \longrightarrow *OOH$$
$$*OOH + H^+ + e^- \longrightarrow H_2O_2$$

Overall Reaction: $O_2 + 2H^+ + 2e^- \longrightarrow H_2O_2$

*O, *OH, and *OOH species represent the reactive oxygen intermediates produced when oxygen is combined with the electrocatalyst active sites. The dissociation energy of O-O bond in H_2O_2 is calculated as 98.7 kJ/mol less than that of O_2 itself (498 kJ/mol), thus the reduction potential of the two-electron pathway (0.695 V) is lower than that of the four-electron pathway (1.23 V). Therefore, the electrochemical reaction goes through with the two-electron pathway if the catalytic activity of the electrocatalyst is not good. 1D-nanostructured materials appear as nanowires, nanotubes, nanoribbons, nanobelts, and nanorods. Also, the derivation of these 1D carbon nanostructured materials is studied in the literature. This chapter discusses 1D carbon nanostructured materials.

11.2.1 Carbon Nanotubes (CNT)

The good conductivity of carbon nanotubes, because of the interpenetration of each tube, made them a suitable cathode electrocatalyst in Li-air batteries, particularly in the charging process wherefore insulation feature of Li_2O_2 [4]. Studies with CNT and carbon powder (CP) in Li-air battery cathodes were reported to be a weight percentage rate vs. capacity (Figure 11.1A–C). Although CP and CNT showed different pore volume, surface area (CP has larger surface area than that of CNT), better capacities were observed as the CNT:CP ratio increased. This feature was elucidated by the increase in CNT rate in composite, but particle agglomeration size was decreased so that the oxygen transportation advanced and thus, the capacity increased. Besides, bare carbon powder could not achieve capacity retention [5]. A porous CNT manufactured by the colloidal template method was employed as a cathode catalyst in Li-air batteries. This binder-free cathode exhibited 4683 mAh/g capacity at 50 mA/g and capacity retention up to 40 cycles without fading [6]. A multiwalled CNT (MWCNT) exhibited an outstanding capacity value of 34.6 mAh/g at 500 mA/g. This highest capacity accounted for the existence of a high amount of void space in MWCNTs thus accommodating large quantities of Li_2O_2 particles. A continuous 50 charge/discharge cycles demonstrated a capacity of 1000 mAh/g at 250 mA/g current density (Figure 11.1D, E, F) [4]. Huang et al. published a mind-blowing article ascribing Li_2O_2 role in Li-air batteries. Artificial defects were created on the surface of CNT cathodes by using Ar plasma, eliciting nucleation sites for Li_2O_2 particles thus, increasing the density of the particles. As a result, a capacity raise was obtained in the defected cathodes when compared with the bare CNT. However, the formation of side products was observed, and this effectively reduced the cycle life of the Li-air battery [7]. Another study was reported by Li et al. explaining the partially cracked MWCNTs usage in Li-air batteries as an electrocatalyst on the cathode side. It was mentioned that two-fold capacity raises were recorded when compared with bare MWCNTs (Figure 11.1G). Partially cracked MWCNTs (CCNT) showed 1513 mAh/g capacity value at initial discharge at 0.01 mA/cm^2 current density. However, CCNT did not respond to an expected C-rate capability, and a capacity fading of 40% reduction was observed whilst the current density increased from 0.01 mA/cm^2 to 0.2 mA/cm^2 (Figure 11.1H) [8].

11.2.2 Carbon Nanofibers (CNF)

One-dimensional CNF materials are often used in metal-air batteries as electrocatalysts due to their high conductivity, large surface area, small size effect, quantum size effect, and thermal stability [9]. Mesoporous carbon coated with carbon nanofibers (mesoC/CNF) were fabricated by Song et al. indicating a good electrochemical activity due to the large surface area and conductivity in Li-air batteries (Figure 11.2A–2B). A PAN-based CNF was introduced by the electrospinning method. An initial capacity of 4000 mAh/g was achieved at 0.05 mA/cm^2 (Figure 11.2C). Additionally, the calculated conductivity was given 4.638 S/cm and a surface area of 2194 m^2/g [10].

11.2.3 Carbon Nanorods

Nitrogen and sulfur-doped 1D carbon microrod material was employed in the Al-air battery as cathode electrocatalyst with enhanced electrochemical performance. N-doped carbon dots were used as the precursor for these electrocatalysts. Together with 1D microrod electrocatalyst, 2D nanosheets and 3D framework were synthesized. However, a better result was obtained with N, S-doped 2D nanosheet electrocatalyst when compared with 1D and 3D materials, which was attributed to the higher amount of N and S doping [11].

11.2.4 N-Doped Carbon Materials

It is believed that N-doping might lead to a chemical activation on the passive surfaces of carbon materials [12] and also N atoms provide the distribution of charge to neighboring carbon atoms [13].

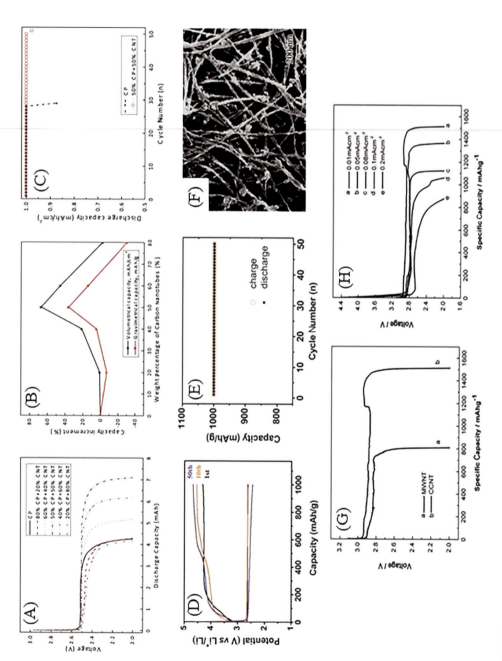

FIGURE 11.1 (A) C-Rate capability of CP/CNT material. (B) Capacity increment vs. weight of CNT. (C) Cycle life of bare CNT and 1:1 CP/CNT. (A, B, C: Adapted with permission [5]. Copyright (2014), Elsevier). (D) Charge and discharge capacity values vs. voltage. (E) Capacity vs. cycle number for MWCNTP material. (F) MWCNT fibers after first discharge. (D, E, F: Adapted with permission [4]. Copyright (2013), Royal Society of Chemistry). (G) Initial discharge capacity of partially cracked MWCNT (CCNT) and bare MWCNT. (H) Capacity fading at different current density values of CCNT. (G, H: Adapted with permission [8]. Copyright (2013), IOP Publishing, Ltd.).

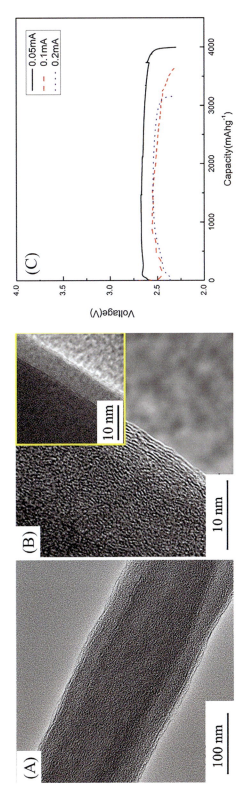

FIGURE 11.2 (A-B) TEM image of mesoC/CNF. (C) Plot of capacity vs. voltage of mesoC/CNF cathode material in Li-air battery. (A, B, C: Adapted with permission from Reference [10], Copyright (2014), Elsevier.)

Mi et al. reported employing CNT and N-doped CNT (N-CNT) in Li-air batteries as electrocatalysts on the cathode side. Nitrogen-doping increased the electrochemical performance of Li-air batteries both in ether and carbonate-based electrolytes. Namely, in the first discharge 3516 mAh/g and 4187 mAh/g capacity were observed in CNT and N-CNT-based electrocatalysts, respectively. However, the electrochemical stability of N-CNT was poor to demonstrate 50% capacity fading in the 5th cycle. Another N-CNT study was reported by Li et al. demonstrating a specific capacity of 866 mAh/g at 75 mA/g current density. It presented a 1.5 times higher capacity when compared with bare CNT. The capacity increase can be attributed to a better Li_2O_2 decomposition activity; hence it demonstrated developing reversibility in Li-air batteries [14]. Vertically aligned N-doped carbon fiber materials show good properties, particularly a low overpotential of 0.3 V, and also last up to 150 cycles with as high as 1000 mAh/g capacity in Li-air batteries (Figure 11.3A–C) [15]. Another vertically aligned N-doped CNT grown by plasma-enhanced chemical vapor deposition on carbon fiber paper gave an initial discharge capacity of 710 mAh/g at 0.5 mA/cm². Due to N-doping, a bamboo structure was formed, and this made rich surface defects causing the transportation of reactants to reaction sites (Figure 11.3D, 3E) [16].

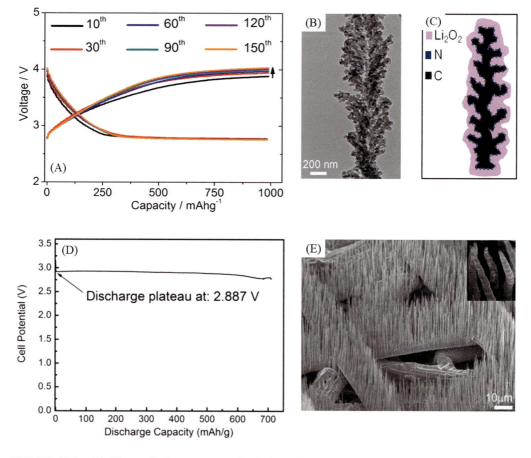

FIGURE 11.3 (A) Charge-discharge curves of VA-CNF from 10th to 150th cycles. (B) TEM image of VA-CNF. (C) A sketch of Li_2O_2 grown on VA-CNF. (A, B, C: Adapted with permission [15]. Copyright (2014), American Chemical Society.) (D) A plot of discharge capacity vs. voltage of VA-NCNT cathode material. € SEM image of carbon nanotube arrays (CNTA). (D, E: Adapted with permission [16]. Copyright (2013), Royal Society of Chemistry.)

11.2.5 METAL AND METAL OXIDES WITH 1D CARBON COMPOSITES

The metal and metal doping to carbon nanomaterials can change the surface area, electronic/ionic structure, pore size, charge distribution, conductivity, and also catalytic activity. Therefore, metal and metal oxide-modified carbon nanomaterials improve the charge distribution which contributes to electrochemical activity. Rod-like Fe_3O_4-carbon 1D composite (Figure 11.4A, 4B) structure demonstrated a discharge capacity of 1000 mAh but the capacity decreased in the first 30 cycles and then became stable (Figure 11.4C) [17]. γ-Fe_2O_3 embedded into porous carbon nanofibers was presented with 661 mAh/g initial capacity in a Zn-air battery. Also, an excellent ORR performance with 0.905 V half-wave potential with 5 mV drop after 5000 cycles [18]. Another iron-based material was used in Zn-air batteries. A metal-organic framework (MOF)-based Fe-MIL-88B precursor was used for having a Fe_2O_3 nanorod template together with a polypyrrole nanorod. After pyrolysis of this material Fe/N-doped carbon nanorod (N-CNR) was obtained. A specific capacity of 998 mAh/g was reported within this material. Active Fe-N_4 sites were observed in TEM images realizing the rich electron distribution explaining high ORR activity and capacity [19].

The use of CoO together with CNF (CoO/CNF) synthesized by one-step electrospinning as a composite cathode material in Li-air batteries was revealed by Huang et al. demonstrating 3882 mAh/g initial capacity at 0.2 mA/cm^2 current density (Figure 11.4D). However, after 8 cycles, the capacity dropped to 3302 mAh/g and after 25 cycles, the capacity was less than 500 mAh/g, therefore, a poor electrochemical cycle life was observed [20]. On the other hand, a ZIF-9 MOF-derived electrospun Co_3O_4/CNF was employed as a cathode material in a Li-air battery. It gave an initial capacity of 760 mAh/g at 500 mA/g current density, but the capacity remained at about 350 mAh/g after the 20th cycle [21]. Another Co-based material was reported by Chong et al. employing $Co_{0.68}Fe_{0.32}O$ with N-doped carbon nanostructure in a Zn-air battery. This material exhibited 673 mAh/g capacity at 1.5 mA/cm^3 [22]. A Co_3O_4/CNT composite material by polydopamine assistance was synthesized and electrochemical properties in Li-air battery were investigated by Yoon et al. An initial capacity of 5000 mAh/g at 100 mA/g and 3000 mAh/g at 200 mA/g current densities were reported (Figure 11.4E). Also, when the capacity was limited to 1000 mAh/g, the material remained stable and after 15 cycles, the capacity progressively decreased [23]. Needle-like Co_4N/CNF was reported to show active sites for ORR/OER and demonstrated 1000 mAh/g capacity with 100 cycles at 200 mA/g current density [24]. Besides, a mixed cobalt-manganese matrix together with N-doped carbon nanowires (Co/MnO/NC) exhibited good ORR and OER properties in the Zn-air battery. It exhibited a half-wave potential of 0.83 V for ORR performance which is comparable with commercial Pt/C (0.85 V). Moreover, Co/MnO/NC electrocatalyst gave 768, 730, and 692 mAh/g at 5, 25, and 50 mA/cm^2 current densities which are 93.6%, 89.0%, and 84.4% of theoretical capacities (Figure 11.4F) [25].

α-MnO_2 nanorods embedded on N-doped graphite nanofibers (N-GNF) showed a high capacity and low overpotential in Li-air batteries. It exhibited a superior C-Rate capability namely, 2907 mAh/g, 2943 mAh/g, and 3260 mAh/g at the current densities of 0.1, 0.2, and 0.3 mA/cm^2, respectively [26]. Additionally, electrospun carbon nanofibers decorated with Mn_3O_4 were employed and determined a discharge overpotential of 0.08 V, however, no capacity value was mentioned [27]. A very promising Li-air cathode electrocatalyst manganese (II) oxide coated onto carbon nanotubes (MnO/CNT) exhibited 6360 mAh/g at 0.1 mA/cm^2. Also, when the current densities were 0.2, 0.4, and 0.6 mA/cm^2, 5916, 4137, and 2527 mAh/g capacity values were observed which proved an excellent C-Rate capability. Besides, a good cyclability was determined. These better results were ascribed to a good conductive matrix occurring a fast and efficient electron transport [28].

Ruthenium mixed with 1D nanocarbon structures was investigated for metal-air batteries. A RuO_2-carbon nanotube composite in a Li-air battery exhibited superior charge and discharge overpotentials of 0.51 V and 0.21 V, respectively. The first discharge capacity was found to be 1130 mAh/g at 100 mA/g with a nearly 100% coulombic efficiency and also 1012 mAh/g and 790 mAh/g at 200 and 500 mA/g, respectively. The cycling stability was reported 100 cycles with no loss in

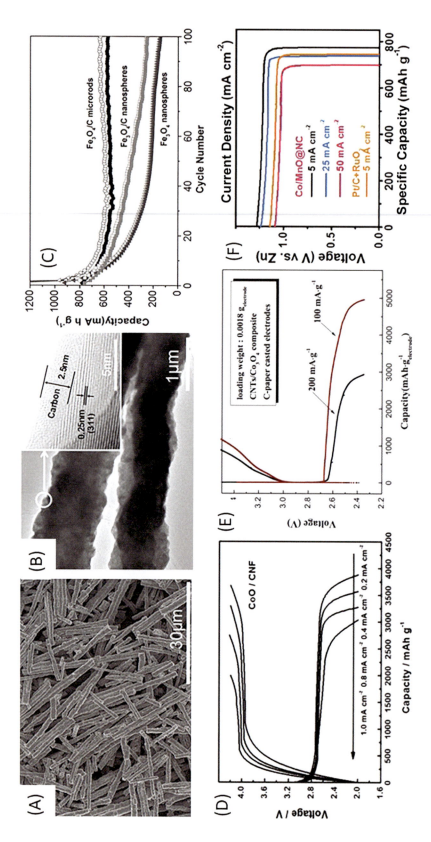

FIGURE 11.4 (A) SEM image of Fe_3O_4/C microrods. (B) TEM and HRTEM (inset) images of Fe_3O_4/C microrods. (C) Comparison of cycle number vs. capacity of Fe_3O_4/C microrods, nanosphere materials (A, B, C: Adapted with permission [17]. Copyright (2014) John Wiley and Sons). (D) C-Rate capability of CoO/CNF material at different current densities (D: Adapted with permission [20]. Copyright (2014) Elsevier.) (E) C-Rate dependence of Co_3O_4/CNF cathode material in Li-air battery (E: Adapted with permission [23]. Copyright (2014), Elsevier.) (F) C-Rate dependence of Co/MnO/NC material in Zn-air battery (F: Adapted with permission [25]. Copyright (2021), Elsevier.)

capacity with a cutoff capacity of 300 mAh/g at 500 mA/g [29]. Ruthenium oxide dispersed on multiwalled carbon nanotubes (RuO$_2$/MWCNT) also showed a good capacity and cyclic stability with an initial capacity of 1800 mAh/g and 20 cycles without capacity loss with a limited capacity of 1000 mA/g [30]. Another ruthenium decorated onto multiwalled carbon nanotube paper (Ru/MWCNTP) study for Li-air batteries was reported with a superior capacity value of 22600 mAh/g. To understand the cycling stability and prolong the cycles, a limiting capacity of 5000 mAh/g at 500 mA/g current density was applied to the Li-air battery and 50 stable cycles were observed. Also, a good C-rate capability was obtained within 1500 mA/g and 2000 mA/g [31].

Perovskite-based LaTi$_{0.65}$Fe$_{0.35}$O$_{3-\delta}$ with N-doped carbon nanorods was used as a Zn-air cathode electrocatalyst and exhibited a 440 mAh/g initial capacity with a discharge plateau of 1.16 V. A high electrocatalytic activity within bifunctional ORR and OER activities was reported [30]. Gold nanoparticles distributed onto vertically aligned carbon nanotubes (AuNP/VACNT) were studied for understanding the formation and decomposition of Li$_2$O$_2$ during charge/discharge cycles. The existence of Au nanoparticles assisted to occur nucleation sites for Li$_2$O$_2$ formation and also reduced the size of Li$_2$O$_2$ during discharge [32].

11.3 ONE-DIMENSIONAL METAL ELECTROCATALYSTS

11.3.1 One-Dimensional Precious Materials

1D precious materials demonstrate high electrocatalytic activity in metal-air batteries, however, they suffer higher costs when compared with other materials. Jung et al. proposed a 1D Ag nanowire structure with less than 1wt% chitin as a binder. This carbon-free cathode showed a good electrochemical stability of 160 cycles with 0.5 mAh/cm^2 capacity at 0.2 mA/cm^2 current density [33]. Another precious metal Ag-Pd-Pd tubular nanotubes gave a capacity of 1000 mAh/g after 100 cycles with 0.2 mA/cm^2 [34]. A vertically aligned 1D titanium nitride array decorated with platinum (Pt/TiN NTA) electrocatalyst material showed 2 times better mass activity and 1.6 times specific activity compared with commercial Pt/C. Also, a low overpotential of 0.09 V was observed [35]. A Pt-incorporated with CNT electrode for Li-air battery was reported by Lim et al. and achieved 1000 mAh/g capacity for more than 100 cycles at a high current density of 2 A/g. This capacity value along with cycle number is quite high among all metal-air batteries. This remarkable achievement is attributed to film-like discharge products and also the dual function of Pt in terms of ORR and OER [36].

11.3.2 One-Dimensional Non-Precious Materials

Non-precious materials show high ORR and OER properties in metal-air batteries, therefore 1D transitional metal oxides and also pure metals were employed. Natural abundance, relatively low cost, and high catalytic activity made transitional metal oxides popular in air cathodes. Debart et al. demonstrated a first discharge capacity of 3000 mAh/g by using α-MnO$_2$ nanowires as an electrocatalyst in an air cathode. However, cycle stability was to be poor [37]. On the other hand, Zhang reported δ-MnO$_2$ submicron tubes as an electrocatalyst in Li-air batteries and exhibited 6150 mAh/g at 25 mA/g and 2805 mAh/g at 600 mA/g at the first discharge [38]. Also, various nanoscale MnO$_2$ electrocatalyst was reported in the literature as nanowires [37, 39], nanotubes [40–42], and nanorods [43–44]. Co$_3$O$_4$ porous nanowire usage in Li-air batteries was reported by Liu et al. It showed an excellent cycling performance with a capacity of 1000 mAh/g at 100 mA/g current density [45].

Similarly, mesoporous CoO nanowire in Li-air battery was studied by Wu et al. It was found that the CoO nanowire achieved a capacity of 4800 mAh/g at 20 mA/g current density with a 50 cycles stability [46]. 1D Co$_3$O$_4$ nanofibers immobilized on graphene nanoflakes demonstrated an outstanding initial capacity value of 10.500 mAh/g and 80 stable cycles within a cutoff capacity of

1000 mAh/g. This high electrochemical activity was attributed to the large surface area, fast electron transport, and O_2 diffusion between the layers of composite [47]. Porous $NiCo_2O_4$ nanotube was synthesized by electrospinning method and exhibited superior electrocatalytic activity in Li-air battery [48].

The usage of perovskite oxides in metal-air batteries as an electrocatalyst was also reported. Due to their easy synthesis, high electrocatalytic activity for both ORR and OER, high electronic/ionic conductivity, and long durability, perovskite oxides are considered as promising materials [49] in metal-air batteries. 1D porous $La_{0.5}Sr_{0.5}CoO_{2.91}$ nanotubes in Li-air batteries were reported to present a high capacity of 7205 mAh/g with a plateau of 2.66 V at 100 mA/g between 2.2 and 4.4 V cutoff voltages. Furthermore, a good C-Rate capability was observed with 4507 mAh/g, 2542 mAh/g, and 593 mAh/g at 200, 500, and 1000 mA/g current densities, respectively. Successful cycling stability was achieved with 85 stable cycles without fading at a limiting discharge capacity of 1000 mAh/g [50]. A porous perovskite $La_{0.75}Sr_{0.25}MnO_3$ nanotubes together with Ketjen Black, on the other hand, achieved to reach 11000 mAh/g capacity value and maintained it after 5 cycles. A voltage plateau of 2.80 V was observed during discharge with a nearly 100% coulombic efficiency during 5 cycles [51].

11.4 OTHER MATERIALS

Metal-organic framework (MOF) derived materials were reported to be employed in metal-air batteries because of having flexible and template-based structures and distributed active sites. They are easy to obtain and have large specific surface area, high porosity, and pore size. However, MOF-derived metal-air batteries should be electrochemically active, particularly having low ORR voltage values and fast reaction kinetics [52].

11.5 CONCLUSION

1D materials are widely studied in the literature as an electrocatalyst in metal-air batteries. These materials are found to be highly promising materials due to fast-electron transport, reduced ion pathway, easy electrode accessibility, providing faster kinetics, better ion diffusion, and easy synthesis steps. Combined with the advantages of metallic lithium, 1D nanostructures present very superior properties compared with 2D and 3D nanomaterials. Besides, 1D nanomaterials display a broad surface area because of the interpretation of each tube. This unique feature causes these materials to give high capacity, and good C-rate capability when used as an electrocatalyst in metal-air batteries. Moreover, 1D nanomaterials particularly carbon-based nanomaterials exhibit a good conductivity assisting them with faster kinetics as well as longer higher capacity. However, although many encouraging steps are taken, a series of problematic issues such as side reactions, electrochemical degradation, low coulombic efficiency, high overvoltage, and rapid capacity loss has not been resolved.

In conclusion, 1D nanomaterials whether carbon-based or metal-based offer a promising design as a cathode electrocatalyst for metal-air batteries and make them a suitable and accessible material for commercial applications in the near future.

REFERENCES

[1] H.B. Wu, G. Zhang, L. Yu, X.W. Lou, One-dimensional metal oxide-carbon hybrid nanostructures for electrochemical energy storage, Nanoscale Horiz. 1 (2016) 27–40.

[2] C. Shi, K.A. Owusu, X.X.T. Zhu, G. Zhang, W. Yang, L. Mai, 1D carbon-based nanocomposites for electrochemical energy storage, Small. 15 (2019) 1902348.

[3] Q. Wei, F. Xiong, S. Tan, L. Huang, E.H. Lan, B. Dunn, L. Ma, Porous one-dimensional nanomaterials: Design, fabrication and applications in electrochemical energy storage, Adv. Mater. 29 (2017) 1602300.

[4] Y. Chen, F. Li, D.M. Tang, Z. Jian, C. Liu, D. Golberg, A. Yamada, H. Zhou, Multi-walled carbon nanotube papers as binder-free cathodes for large capacity and reversible non-aqueous Li-O_2 batteries, J. Mater. Chem. A. 1 (2013) 13076–13081.

[5] P. Tan, W. Shy, Z.H. Wei, L. An, T.S. Zhao, A carbon powder-nanotube composite cathode for non-aqueous lithium-air batteries, Electrochim. Acta. 147 (2014) 1–8.

[6] S. Liu, Z. Wang, C. Yu, Z. Zhao, X. Fan, Z. Linga, J. Qiu, Free-standing, hierarchically porous carbon nanotube film as a binder-free electrode for high-energy Li-O_2 batteries, J. Mater. Chem. 1 (2013) 12033–12037.

[7] S. Huang, W.W. Fan, X. Guo, F. Meng, X. Liu, Positive role of surface defects on carbon nanotube cathodes in overpotential and capacity retention of rechargeable lithium-oxygen batteries, ACS Appl. Mater. Interfaces. 6 (2014) 21567–21575.

[8] J. Li, B. Peng, G. Zhou, Z. Zhang, Y. Lai, M. Jia, Partially cracked carbon nanotubes as cathode materials for lithium-air batteries, ECS Electrochem. Lett. 2 (2013) A25–A27.

[9] X. Zhang, L. Wang, Research progress of carbon nanofiber-based precious-metal-free oxygen reaction catalysts synthesized by electrospinning for Zn-air batteries, J. Power Sources. 507 (2021) 230280.

[10] M.J. Song, M.W. Shin, Fabrication and characterization of carbon nanofiber@mesoporous carbon core-shell composite for the Li-air battery, Appl. Surface Sci. 320 (2014) 435–440.

[11] R. Cheng, M. Jiang, K. Li, M. Guo, J. Zhang, J. Ren, P. Meng, R. Li, C. Fu, Dimensional engineering of carbon dots derived sulfur and nitrogen co-doped carbon as efficient oxygen reduction reaction electrocatalysts for aluminum-air batteries, Chem. Eng. J. 425 (2021) 130603.

[12] R. Mi, H. Liu, H. Wang, K. Wong, J. Mei, Y. Chen, W.M. Lau, H. Yan, Effects of nitrogen-doped carbon nanotubes on the discharge performance of Li-air batteries, Carbon. 67 (2014) 744–752.

[13] X. Chen, J. Gao, S. Liu, Z. Yang, S. Wang, Z. Su, P. Zhu, X. Zhao, G. Wang, 1D bamboo-like N-doped carbon nanotubes with encapsulated iron-based nanoparticles as an advanced Zn-air battery cathode electrocatalyst, J. Alloys and Compds. 828 (2020) 154435.

[14] Y. Li, J. Wang, X. Li, J. Liu, D. Geng, J. Yang, R. Li, X. Sun, Nitrogen-doped carbon nanotubes as cathode for lithium-air batteries, Electrochem. Commun. 13 (2011) 668–672.

[15] J. Shui, F. Du, C. Xue, Q. Li, L. Dai, Vertically aligned N-doped coral-like carbon fiber arrays as efficient air electrodes for high-performance nonaqueous Li-O_2 batteries, ACS Nano. 8 (2014) 3015–3022.

[16] Y. Li, Z. Huang, K. Huang, D. Carnahan, Y. Xing, Hybrid Li-air battery cathode nanotube arrays directly grown on carbon fiber paper, Energy Environ. Sci. 6 (2013) 3339–3345.

[17] Y. Wang, L. Zhang, X. Gao, L. Mao, Y. Hu, X.W. Lou, One-pot magnetic field induced formation of Fe_3O_4/C composite microrods with enhanced lithium storage capability, Small. 10 (2014) 2815–2819.

[18] Z. Yao, Y. Li, D. Chen, Y. Zhang, X. Bao, J. Wang, Q. Zhong, γ-Fe_2O_3 clusters embedded in 1D porous N-doped carbon matrix as pH-universal electrocatalyst for enhanced oxygen reduction reaction, Chem. Eng. J. 145 (2021) 129033.

[19] X. Gong, J. Zhu, J. Li, R. Gao, Q. Zhou, Z. Zhang, H. Dou, L. Zhao, X. Sui, J. Cai, Y. Zhang, B. Liu, Y. Hu, A. Yu, S. Sun, Z. Wang, Z. Chen, Self-templated hierarchically porous carbon nanorods embedded with atomic Fe-N_4 active sites as efficient oxygen reduction electrocatalysts in Zn-air batteries, Adv. Funct. Mater. 31 (2021) 2008085.

[20] B.W. Huang, L. Li, Y.J. He, X.Z. Liao, Y.S. He, W. Zhang, Z.F. Ma, Enhanced electrochemical performance of nanofibrous CoO/CNF cathode catalyst for Li-O_2 batteries, Electrochim. Acta. 137 (2014) 183–189.

[21] M.J. Song, I.T. Kim, Y.B. Kim, M.W. Shin, Self-standing, binder-free electrospun Co_3O_4/carbon nanofiber composites for non-aqueous Li-air batteries, Electrochim. Acta. 182 (2015) 289–296.

[22] Y. Chong, Z. Pan, M. Su, X. Yang, D. Ye, Y. Qiu, 1D/2D hierarchical $Co_{1-x}Fe_xO$@N-doped carbon nanostructures for flexible zinc-air batteries, Electrochim. Acta. 323 (2020) 137264.

[23] T. Yoon, Y.J. Park, Polydopamine-assisted carbon nanotubes/Co_3O_4 composites for rechargeable Li-air batteries, J. Power Sources. 244 (2013) 344–353.

[24] K.R. Yoon, K. Shin, J. Park, S. Cho, C. Kim, J.W. Jung, J.Y. Cheong, H. Byon, H.M. Lee, I.D. Kim, Brush-like cobalt nitride anchored carbon nanofiber membrane: Current collector-catalyst integrated cathode for long cycle Li-O_2 batteries, ACS Nano. 12 (2018) 128–139.

[25] Y. Niu, X. Teng, S. Gong, X. Liu, M. Xu, Z. Chen, Boosting oxygen electrocatalysis for flexible zinc-air batteries by interfacing iron group metals and manganese oxide in porous carbon nanowires, Energy Storage Mater. 43 (2021) 42–52.

[26] A. Zahoor, M. Christy, H. Jang, K.S. Nahm, Y.S. Lee, Increasing the reversibility of Li-O_2 batteries with caterpillar structured a-MnO_2/N-GNF bifunctional electrocatalysts, Electrochim. Acta. 157 (2015) 299–306.

[27] K.-N. Jung, J.-I. Lee, S. Yoon, S.-H. Yeon, W. Chang, K.-H. Shin, J.-W. Lee, Manganese oxide/carbon composite nanofibers: Electrospinning preparation and application as a bi-functional cathode for rechargeable lithium-oxygen batteries, J. Mater. Chem. 22 (2012) 21845–21848.

[28] W.-B. Luo, S.-L. Chou, J.-Z. Wang, Y.-C. Zhai, H.-K. Liu, A facile approach to synthesize stable CNTs@MnO electrocatalyst for high energy lithium oxygen batteries, Scientific Rep. 5 (2015) 8012.

[29] Z. Jian, P. Liu, F. Li, P. He, X. Guo, M. Chen, H. Zhou, Core-shell structured CNT@RuO_2 composite as a high-performance cathode catalyst for rechargeable Li-O_2 batteries, Angew. Chem. 126 (2014), 452–456.

[30] M. Prabu, P. Ramakrishnan, P. Ganesan, A. Manthiram, S. Shanmugam, $LaTi_{0.65}Fe_{0.35}O_{3-d}$ nanoparticle-decorated nitrogen-doped carbon nanorods as an advanced hierarchical air electrode for rechargeable metal-air batteries, Nano Energy. 15 (2015) 92–103.

[31] F. Li, Y. Chen, D.-M. Tang, Z. Jian, C. Liu, D. Golberg, A. Yamada, H. Zhou, Performance-improved Li-O_2 battery with Ru nanoparticles supported on binder-free multi-walled carbon nanotube paper as cathode, Energy Environ. Sci. 7 (2014) 1648–1652.

[32] W. Fan, X. Guo, D. Xiao, L. Gu, Influence of gold nanoparticles anchored to carbon nanotubes on formation and decomposition of Li_2O_2 in nonaqueous Li-O_2 batteries, J. Phys. Chem. C. 118 (2014) 7344–7350.

[33] J.W. Jung, H.G. Im, D. Lee, S. Yu, J.H. Jang, K.R. Yoon, Y.H. Kim, J.B. Goodenough, J. Jin, I.-D. Kim, B.-S. Bae, Conducting nanopaper: A carbon-free cathode platform for Li-O_2 batteries, ACS Energy Lett. 2 (2017) 673–680.

[34] W.B. Luo, X.W. Gao, S.L. Chou, J.Z. Wang, H.K. Liu, Porous AgPd-Pd composite nanotubes as highly efficient electrocatalysts for lithium-oxygen batteries, Adv. Mater. 27 (2017) 6862–6869.

[35] A. Dong, X. Chen, S. Wang, L. Gu, L. Zhang, X. Wang, X. Zhou, Z. Liu, P. Han, Y. Duan, H. Xu, J. Yao, C. Zhang, K. Zhang, G. Cui, L. Chen, 1D coaxial platinum/titanium nitride nanotube arrays with enhanced electrocatalytic activity for the oxygen reduction reaction: Towards li-air batteries, ChemSusChem. 5 (2012) 1712–1715.

[36] H.-D. Lim, H. Song, H. Gwon, K.-Y. Pak, J. Kim, Y. Bae, H. Kim, S.-K. Jung, T. Kim, Y.H. Kim, X. Lepro, R. Ovalle-Robles, R.H. Baughman, K. Kang, A new catalyst-embedded hierarchical air electrode for high-performance Li-O_2 batteries, Energy Environ. Sci. 6 (2013) 3570–3575.

[37] A. Debart, A.J. Paterson, J. Bao, P.G. Bruce, α-MnO_2 nanowires: A catalyst for the O_2 electrode in rechargeable lithium batteries, Angew. Chem. 120 (2008) 4597–4600.

[38] P. Zhang, D. Sun, M. He, J. Lang, S. Xu, X. Yan, Synthesis of porous δ-MnO_2 submicron tubes as highly efficient electrocatalyst for rechargeable Li-O_2 batteries, ChemSusChem. 8 (2015) 1972–1979.

[39] O. Oloniyo, S. Kumar, K. Scott, Performance of MnO_2 crystallographic phases in rechargeable lithium-air oxygen cathode, J. Electron. Mater. 41 (2012) 921–927.

[40] T.T. Truong, Y.Z. Liu, Y. Ren, L. Trahey, Y.G. Sun, Morphological and crystalline evolution of nanostructured MnO_2 and its application in lithium-air batteries, ACS Nano. 6 (2012) 8067–8077.

[41] H.W. Park, D.U. Lee, L.F. Nazar, Z. Chen, Oxygen reduction reaction using MnO_2 nanotubes/nitrogen-doped exfoliated graphene hybrid catalyst for Li-O_2 battery applications, J. Electrochem. Soc. 160 (2013) A344–A350.

[42] A.K. Thapa, T. Ishihara, Mesoporous α-MnO_2/Pd catalyst air electrode for rechargeable lithium-air battery, J. Power Sources. 196 (2011) 7016–7020.

[43] X. Cai, L. Lai, J. Lin, Z. Shen, Recent advances in air electrodes for Zn-air batteries: Electrocatalysis and structural design, Mater. Horiz. 4 (2017) 945–976.

[44] P.-C. Li, C.-C. Hu, H. Noda, H. Habazaki, Synthesis and characterization of carbon black/manganese oxide air cathodes for zinc-air batteries: Effects of the crystalline structure of manganese oxides, J. Power Sources. 298 (2015) 102–113.

[45] Q. Liu, Y. Jiang, J. Xu, D. Xu, Z. Chang, Y. Yin, W. Liu, Z. Zhang, Hierarchical Co_3O_4 porous nanowires as an efficient bifunctional cathode catalyst for long life Li-O_2 batteries, Nano Res. 8 (2015) 576–583.

[46] B. Wu, H. Zhang, W. Zhou, M. Wang, X. Li, H. Zhang, Carbon-free CoO mesoporous nanowire array cathode for high-performance aprotic Li-O_2 batteries, ACS Appl. Mater. Interfaces, 7 (2015) 23182–23189.

[47] W.-H. Ryu, T.-H. Yoon, S.-H. Song, S. Jeon, Y.-J. Park, I.-D. Kim, Bifunctional composite catalysts using Co_3O_4 nanofibers immobilized on nonoxidized graphene nanoflakes for high-capacity and long-cycle Li-O_2 batteries, Nanolett. 13 (2013) 4190–4197.

[48] L. Li, L. Shen, P. Nie, G. Pang, J. Wang, H. Li, S. Dong, X. Zhang, Porous $NiCo_2O_4$ nanotubes as noble metal-free effective bifunctional catalysts for rechargeable $Li-O_2$ batteries, J. Mater. Chem. A. 3 (2015) 24309–24314.

[49] Y. Dai, J. Yu, C. Cheng, P. Tan, M. Ni, Mini-review of perovskite oxides as oxygen electrocatalysts for rechargeable zinc-air batteries, Chem. Eng. J. 397 (2020) 125516.

[50] P. Li, J. Zhang, Q. Yu, J. Qiao, Z. Wang, D. Rooney, W. Sun, K. Sun, One-dimensional porous $La_{0.5}Sr_{0.5}CoO_{2.91}$ nanotubes as a highly efficient electrocatalyst for rechargeable lithium-oxygen batteries. Electrochim. Acta, 165 (2015) 78–84.

[51] J.-J. Xu, D. Xu, Z.-L. Wang, H.-G. Wang, L.-L. Zhang, X.-B. Zhang, Synthesis of perovskite-based porous $La_{0.75}Sr_{0.25}MnO_3$ nanotubes as a highly efficient electrocatalyst for rechargeable lithium-oxygen batteries, Angew. Chem. Int. Ed. 52 (2013) 3887–3890.

[52] X. Wen, Q. Zhang, J. Guan, Applications of metal-organic framework-derived materials in fuel cells and metal-air batteries, Coor. Chem. Rev. 409 (2020) 213214.

12 2D Materials as Electrocatalysts for Metal-Air Batteries

Eren Kursun, Abdullah Uysal, and Solen Kinayyigit

CONTENTS

12.1 Introduction 165
12.2 Classification of 2D Materials as Electrocatalysts for Metal-Air Batteries 167
    12.2.1 Carbon-Based 2D Materials 167
        12.2.1.1 Graphene and Graphene-Like Structures 167
    12.2.2 2D Transition Metal-Based Derivatives 170
        12.2.2.1 MXenes 171
    12.2.3 Others 171
        12.2.3.1 Metal-Organic Frameworks 171
        12.2.3.2 Transition Metal Macrocycles 172
        12.2.3.3 Hexagonal Boron Nitride 172
        12.2.3.4 Covalent Organic Frameworks 173
        12.2.3.5 Bismuth Oxyhalides 173
12.3 Conclusion and Future Perspectives 174
References 174

12.1 INTRODUCTION

Metal-air batteries (MABs) promise very high theoretical energy densities (e.g. Zn-air batteries have ~1350 Wh/kg) compared to conventional lithium-ion batteries (LIBs). However, such high-density values are hardly met when it comes to applying them. The problem that lies here is mostly due to air electrode (cathode) interactions: intrinsic slow reaction kinetics, high overpotentials, and insufficient reversibility of oxygen chemistry [1]. The air cathode is made up of a catalyst, a current collector, and a gas diffusion layer [2]. It would be only fair to state that the most crucial part of the assembly is the electrocatalyst. Although there are plenty of MAB systems with a variety of anodes (Zn, Li, Mg, Al, Fe) and electrolytes (aqueous, non-aqueous), the importance of the electrocatalyst remains equally vital. MABs are governed by two fundamental reactions—oxygen reduction reaction (ORR) and oxygen evolution reaction (OER)—and appropriate electrocatalysts are sought after to hasten the kinetics of these reactions.

An ideal bifunctional electrocatalyst should effectively hasten both ORR and OER processes by lowering the overpotential [3, 4]. However, ORR and OER almost always have different requirements for active sites, therefore, it is a challenging task to systematically develop a bifunctional electrocatalyst [5]. Three important factors directly affect oxygen electrocatalytic reactions: mass transfer, electron transfer, and surface reaction. Any alterations that could affect these phenomena will have an impact on the electrocatalytic performance. For a good mass transfer, a thorough pore structure and pore distribution are desired while for fast and efficient electron transfer, high and uniform electrical conductivity is needed. In addition to these properties, adequate intrinsic activity and thermal, chemical, and mechanical durabilities are sought after [6]. It is agreed upon that there is a certain relation between electrical conductivity and catalytic activity; high electrical conductivity could promote faster reaction kinetics [5, 7–9]. During the discharge (ORR) process of a MAB,

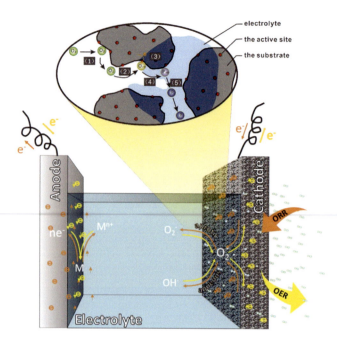

FIGURE 12.1 General structure and working principle of a MAB, which consists of five steps: 1) reactant diffusion, 2) reactant adsorption, 3) electron transfer, 4) product desorption, and 5) product diffusion. (Adapted with permission [6]. Copyright (2022), Elsevier.)

the phenomena that occur could be roughly described in five steps: 1) oxygen diffuses onto the catalyst surface from the atmosphere; (2) oxygen is adsorbed by the catalyst surface; 3) electrons emitted from the anode go for the oxygen molecules through the external circuit; 4) electrons react with the oxygen molecules, making the bond weaker; and 5) OH^- product gets removed from the catalyst surface and goes into the electrolyte (Figure 12.1) [2].

The Intrinsic activity is directly related to the strength of adsorption/desorption of some intermediates such as O^* or OH^* [6]. If the adsorption occurs through a strong interaction between the oxygen and the active site, the products are not so likely to easily detach. Also, weak interaction between the aforementioned species would most likely not be sufficient for adsorption [5]. From this logic, one could say that an efficient bifunctional electrocatalyst should have moderate adsorption energy. To find optimal points both for ORR and OER, volcano plots are generally constructed.

A wide variety of materials are being employed as electrocatalysts for MABs such as noble metals, metal alloys, carbon-derivatives, and transition metal-derivatives [2]. Noble metals (Pt, Au, Ir, Ru) are generally considered as state-of-the-art ORR electrocatalysts in terms of energy efficiency [1]. However, their scarcities, high costs, and low stabilities hinder them to be commercially and reasonably applied [10]. One of the prominent sub-classes of electrocatalysts for MABs is the 2D materials. Since the discovery of graphene, the 2D material research has sky-rocketed as they are evaluated as convenient and effective electrocatalysts due to their structural properties [11]. 2D structures offer more catalytic sites and higher intrinsic activity compared to their bulk counterparts and have shorter ion diffusion paths, all of which are attributes that could lower the overpotential [10, 12, 13]. In addition, having stacked layers of interstitially porous 2D materials could provide an easily accessible interface [14, 15]. These properties promote more effective electrocatalysis and more efficient catalysts may be systematically obtained by tuning the structural and experimental parameters [16]. It is important to note that a higher surface area does not always mean higher performance; pore size could be identified as a more important factor. For instance, when the pore sizes are considerably small, the entrance of the pores is easily blocked, which will in return cause the battery performance to deteriorate over time [17]. The durability of an electrocatalyst could be identified as a prerequisite for battery applications and methods such as chronoamperometry could be utilized to observe it [18]. The electrocatalytic performances of catalysts are often evaluated by linear sweep voltammetry (LSV) in an alkaline electrolyte at specific rotating speeds [9].

12.2 CLASSIFICATION OF 2D MATERIALS AS ELECTROCATALYSTS FOR METAL-AIR BATTERIES

To be able to systematically classify these materials, their precursor conditions and morphologies and the ultimate constituents have to be thoroughly investigated. Mainly, the performance effects caused by these alterations are due to synergistic effects, electronic conductivities, pore distributions, dopings, and defects [19]. The same strategies are viable for many of the material classes. In this part, light is shed upon three main branches of 2D materials, namely, carbon-based 2D materials, transition metal-based 2D materials, and others.

12.2.1 CARBON-BASED 2D MATERIALS

Carbon-based materials have been studied extensively, due to their good conductivities, low costs, large specific surface areas, and good chemical stabilities [20]. Carbon's unique chemistry allows it to have different structures and conformations such as graphite, graphene, fullerenes, carbon nanotubes, and more. Each of these structures could provide different pathways for both ORR and OER. Efficient electrocatalytic activities of carbon nanomaterials could be attributed to their highly exposed active sites and abundant porous defects [21, 22]. However, the electrocatalytic activities of pure carbon species are not so good; that is why they are being doped or defected. This would cause a rearrangement in the electronic structure and possibly enhance the electrocatalytic performance [20]. In addition, their electrocatalytic properties could also be altered by modifying the interior structure, termination, morphology, pore size, thickness, and crystallinity [1]. The practicality of 2D materials that was implemented in the introduction part is still valid for carbon-based materials. 2D carbon-based materials such as graphene, carbon and graphene nanosheets, or graphyne offer quite similar effects. It is also common to see carbonaceous materials as support substrates for some electrocatalysts, mostly the non-conductive ones since most of the carbonaceous materials are great electrical conductors. Doped carbon-based materials are not only electrocatalytically active but also can be used as support materials. This affects electron transfer for insulating or semiconducting catalysts where graphene, CNTs, and other hybrid carbon materials are being employed [6, 23].

12.2.1.1 Graphene and Graphene-Like Structures

Graphene, which is a near-planar honeycomb-like carbon allotrope that consists of sp^2 hybridized carbon atoms, is a very promising electrocatalyst material. It has long Π-Π conjugation with a specific surface area of ~2630 m^2/g, electrical conductivity of ~10^3–10^4 S/m, and thermal conductivity of ~5000 W/mK at room temperature [24, 25]. These features enable promising properties such as strong chemical stability, strong mechanical strength, and fast electron mobility. Due to their open and ultrathin structures and large surface areas, exposure to the highly abundant active sites is more favored, which makes them a good choice for efficient electrocatalysis [6]. It is only fair to state that perfect graphene is almost electrocatalytically inactive. To improve the behavior, a few strategies have been developed over the years. To diversify the utilization of these materials, doping or creating defects is a good way of increasing the electron transfer rate and synergistic effect, which would positively affect the electrocatalytical behavior [19]. Creating defects on graphene would break Π-conjugation which would cause the activation of Π electrons, which could increase the electrocatalytic performance. Doping graphene with N is a preferred strategy because of the similar atomic radii of C and N as well as the higher electronegativity of N which stimulates the delocalized p electrons. Ammonia (NH_3) immersion is a good way of doping carbon materials with N while being able to control the amount [21].

Zn is also being used as a precursor for controlling the pore size and the amount; it evaporates at 900 °C and enables the formation of pores of different sizes, mainly mesopores, during its evaporation, preventing the agglomeration of active sites [26]. After doping with N, three different moieties could be named; pyridinic-N, pyrrolic-N, and graphitic-N, all of which reside on different parts of the structure. It is worth knowing that most of the catalytic activity comes from graphitic- and pyridinic-N [19, 27]. Although the most commonly used doping type is N-, there are many options like B-, P-, S-, F-. For example, boron could be used as a dopant due to its lower electronegativity than carbon; it creates a reverse but still considerable positive effect on the system [28]. Either of these scenarios acts by reducing the band gap of the pristine structure, causing a better electrical conductivity which accelerates the charge transfer efficiency and hence hastening the kinetics. Currently, graphene nanosheets (GNSs) are considered great picks as electrocatalysts for MABs due to their large surface areas, high electronic conductivities, an abundance of active sites, and unique 3D diffusion pathways [1]. It is important to note that the latter could only be obtained by stacking layers. GNSs are generally exfoliated from graphite derivatives. They have plenty of edges and defects on the surface, which act as active sites in ORR and OER, making them the center of intensive research on the subject [24]. Metal-N co-doping is another way to endow the graphene structure with improved properties. In a metal-N-co-doped GNS, there are three plausible catalytically active sites which are being Me-N_x sites, N-C moieties on the basal plane, or N-C moieties on the edges [29].

Huang *et al.* have investigated the effect of N-doping amount on highly graphitized carbon catalysts: the pure carbon nanosheets (G-CNS) and N-doped carbon nanosheets (NG-CNS) (Figure 12.2) [2]. As expected, when applied in an Al-air battery, NG-CNS have performed a maximum power density of 130 mW/cm² (Pt/C = 138 mW/cm²) compared to G-CNS with 98 mW/cm². Also, the superior catalyst was applied in a Zn-air battery, which gave an open-circuit voltage (OCV) of 1.481 V and its maximum power density surpassed that of Pt/C [21].

Fan *et al.* have doped carbon nanosheets (CNSs) with N, P, and Fe, and with doping, the surface polarity got enhanced which provided more abundant active sites [3]. During synthesis, as the pyrolysis temperature increased, produced CNSs got thinner while their pore volumes got larger. The larger the sheet size and pore volumes, the easier the mass transport, aside from the increased number of exposed active sites [3]. The main variable was the pyrolysis temperature, and it was observed that up to some certain point (which is the point where agglomeration is promoted), the increase in pyrolysis temperature positively affected the electrochemical measurements such as onset potential and half-wave potential. A Tafel slope of 64 mV/dec was obtained for high-temperature pyrolyzed CNSs, considering the higher Tafel slope of Pt/C (75 mV/dec), it is fair to say that a fast ORR kinetics is achieved. In addition, overpotential was quite comparable to that of RuO_2, which is an indication of the fast OER kinetics. The effect of Fe_2P nanoparticles (NPs) on the electrocatalytic performance was also investigated where a higher Tafel slope was observed after the removal of Fe_2P NPs. This was attributed to the positive effect of Fe_2P NPs on overall ORR/OER performance [11].

Carbon nanoleaves (CNL) are another sub-branch of the nanosheet class which could also be doped, defected, or used as a support medium for other catalytically active species. In this regard, Huo *et al.* investigated the catalytic effects of Cu NPs and Co NPs on the CNL structure [27]. It was observed that the alloyed CuCo@CNL had a stronger cathodic peak than those of Cu@CNL and Co@CNL in their CV voltammograms, indicating a more efficient ORR performance [27]. This is a bare indication of the possible benefit that could be obtained by alloying transition metals and dispersing them in carbon matrices.

Other newly emerging carbon-based materials that are worth mentioning are graphyne and graphdiyne, which are also atomically thin and Π-conjugated carbon frameworks. However, unlike

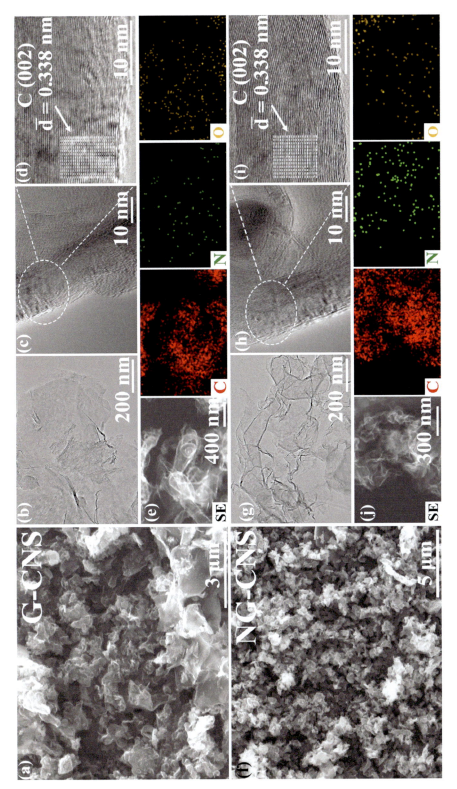

FIGURE 12.2 (a/f) Scanning electron microscopy (SEM) images; (b/g) transmission electron microscopy (TEM) images; (c/h) high-resolution TEM (HRTEM) images; (d/i) enlarged HRTEM images; and (e/j) high-angle annular dark-field scanning transmission electron microscopy (HAADF-STEM) images and energy dispersive spectrometry (EDS) mappings of obtained G-CNS and NG-CNS, respectively. (Adapted with permission [21]. Copyright (2022), Elsevier.

graphene, they are composed of a mixture of sp- and sp²-hybridized carbon atoms. If the number of connecting sp- carbon atoms between the hexagonal rings is two, it is identified as graphyne whereas if it is four, it is graphdiyne [28]. Kong *et al.* have investigated the effect of doping of graphyne with both B- and N and observed a similar trend to that of graphene and other 2D carbon structures [28].

12.2.2 2D Transition Metal-Based Derivatives

Transition metal derivatives could be separately identified as transition metal oxides, chalcogenides, carbides, sulfides, phosphates, and nitrides. What makes them a good choice of material as MAB electrocatalysts are their low costs, high abundance, and environmentally-friendly factors. Due to the possibility of existing in several valence states, even a single type of transition metal could offer materials with different structures and morphologies, providing a wide area of research [10]. Since most of the transition metal oxides are considered as semi-conductors, their electrical conductivities need improvement and this is generally achieved by carbon-based material supports [6, 19, 30, 31]. One huge drawback of transition metal compounds is their tendency to agglomerate, which decreases the overall stability [20]. This could also be prevented by support materials. Just like in carbon-based structures, surface engineering on the sheets has proven to be useful [31]. Wu *et al.* synthesized cobalt(II) oxide (CoO) nanosheets and investigated the effect of oxygen vacancies on the surface [32]. It was found that the vacancy content was directly proportional to catalytic activity. Investigation of the effect of temperature showed that aggregation of CoO particles above 800°C caused a significant drop in the oxygen vacancy content [32]. Additionally, Zhou *et al.* have synthesized 1 nm-thick CoO layers on top of Co-N-doped graphene substrate where the effect of thickness on ORR and OER performances was studied in detail [33]. Comparison of the rotating ring-disk electrode (RRDE) measurements on 4 nm and 1 nm structures showed higher ORR activity for the 1 nm species with a work function of 4.43 eV compared to 4.65 eV for that 4 nm sample. LSV analyses for the same species at a current density of 10 mA/cm² gave an overpotential of 370 mV for 1 nm thick structure which was very close to that of RuO_2, and much smaller than that of 4 nm structure, indicating a great OER performance [33]. This higher OER performance was attributed to the abundant active sites and the highly conductive substrate (which is N-doped graphene). Alloying transition metals together is another pathway to obtain more efficient electrocatalysts, caused mostly by the synergistic effect [27].

Transition metal dichalcogenides (TMDs) could be labeled under 2D transition metal derivatives; their structural composition is of the type MX_2 where M is a transition metal (Ni, Co, Fe) and X is a chalcogen (S, Se, Te). TMDs are abundant, cheap, and well suited for ORR [34]. Since the basal plane of TMDs is inert, they cannot be used for ORR, directly by themselves. However, by doping them with noble metals, the basal plane could be activated, hence a possible way to promote the ORR could be paved [10]. In a DFT-based novel work on 2D monolayer pristine $MoSe_2$, Upadhyay *et al.* created two structures: a pristine $MoSe_2$ and a Pt-doped $MoSe_2$ in which one of the Mo atoms was replaced by a single Pt atom [10]. According to their calculations, a semiconductor characteristic was observed for pristine $MoSe_2$ with a band gap of 2.21 eV. Surprisingly, the band gap decreased from 2.21 eV to 0 eV after doping the pristine structure with Pt, an indication of perfect metallic character. This was thought to occur due to the increase in carrier concentration after Pt-doping [10].

One fine example of 2D transition metal oxides is MnO_2 although its catalytical performance is usually hindered due to its low electrical conductivity. To overcome this issue, doping with transition metals such as Ni or Co was performed in recent studies. For instance, Mathur *et al.* synthesized 2D MnO_2 nanosheets in which the amount of Co as the dopant (5% and 10%) was controlled, and their ORR and OER behaviors were investigated in detail [35]. 5CMN (5% Co), 10CMN (10% Co) showed the best performance towards ORR and OER, respectively. This indicates that ORR and OER have different active sites [35]. In another work by Pargoletti *et al.*, the dopant type and its amount were investigated on MnO_2 where four different samples with 2% Fe, 5% Fe, 2% Co, and 5% Co were prepared and the best performance was observed from 2% Co sample with a decrease of 130 mV vs. GCE (glassy carbon electrode) overpotential [36]. It was also evident that higher doping percentages of Fe caused the formation of a segregated phase which possibly hindered the electrocatalytic activity.

Transition metal nitrides have also attracted a lot of attention due to their strong bondings, chemical stabilities, electrical conductivities, and catalytic activities [37]. However, especially M_4N-type 2D transition metal nitrides suffer from aggregation which causes the formation of thick plates due to low thermodynamic stability [31]. Transition metal sulfides, especially nickel sulfide (Ni_3S_2), are regarded as good candidates for OER electrocatalysis [38]. However, low electrical conductivity and tendency for agglomeration seriously restrict the use of these types of materials for ORR. Xu *et al.* have prepared Ni_3S_2 quantum dots and dispersed them in S, N- co-doped carbon nanosheets, which resulted in surpassing any of the reported catalytic activities of nickel sulfide-based catalysts [38].

12.2.2.1 MXenes

MXenes are promising electrocatalysts for ORR and OER with the general formula $M_{n+1}X_n$ where M is a transition metal and X is C or N, and are selectively etched by HF, HCl, etc. [39]. X atom occupies the inner octahedral site of these structures [40]. In addition, these materials could also be functionalized to take the form $M_{n+1}X_nT_x$ where T is O, OH, or F [40]. These functional groups make the basal plane of the structure more stable [41]. Large specific surface areas and the possibility of structure tuning via functional groups have made MXenes a good candidate for catalysis of ORR and OER for MABs [31]. Relatively weak van der Waals interactions between stacked layers of MXenes could easily be broken down with ultrasonic treatment or intercalation, which would help to obtain very thin, even single-layer, structures that are beneficial in terms of catalytic activity [31]. Additionally, it is known that 2D MXene platforms create excellent electronic pathways, promoting the electron transfer rate to the active sites of the electrocatalyst [42]. MXenes may also be used as self-sacrificial templates to synthesize 2D materials. For example, Wang *et al.* used $Ti_3C_2t_x$ MXene derivatives to produce highly graphited carbon-coated $FeTiO_3$ nanosheets at different annealing temperatures (400 °C, 450 °C, 500 °C, and 550 °C) [43]. The use of MXene templates prevented possible aggregation and increased the stability of the structure. The best bifunctional activity was achieved with the structure annealed at 500 °C, which exhibited a low overpotential of 323 mV at 10 mA/cm² and a small Tafel slope of 53 mV/dec. These values were comparable to those of the benchmark IrO_2/C catalyst [43]. Higher temperatures were found to lead to aggregation and stacking of layers whereas lower temperatures gave a lower degree of graphitization of carbon [43].

12.2.3 OTHERS

12.2.3.1 Metal-Organic Frameworks

Recently, metal-organic frameworks (MOFs) have attracted considerable attention due to their ability to self-assemble by the introduction of organic ligands and metal ions under appropriate conditions. This property of MOFs enables a uniform structure that prevents phenomena such as aggregation of metals and metal oxides and thus gives a positive impact on the electrocatalytic performance of the obtained material [44]. MOFs can be modified with specific surfactants to limit the vertical growth, allowing very thin (2D) structures [45]. Also, their structures and pore sizes are tunable by the introduction of different metals or organic ligands [46]. In addition, disordered arrangements or amorphous mixtures of precursors lead to uncontrollable active site regions, a major drawback when aiming for thorough catalysis [47]. Therefore, the tunable structure of MOFs makes them a good choice as precursors for electrocatalyst synthesis.

One prominent MOFs class is zeolitic imidazolate frameworks (ZIFs) which can be converted into porous N-doped carbon derivatives and metal nanoparticles by pyrolysis [47]. ZIFs are excellent templates for the controlled and easy synthesis of the catalyst with their large surface areas, high porosities, and N-self doping [14]. Although ZIFs are good candidates to be used to obtain 2D heteroatom-doped structures, ZIFs are not so commonly utilized as pure 2D structures since they tend to shrink and collapse during carbonation [15]. It is important to note that the pyrolysis temperature is an important parameter that affects the ultimate catalytic performance of ZIF-derived structures [48]. It is worth mentioning that the active sites are still ambiguous, and more research is required to be able to utilize these materials more effectively. Chen et al. recently developed a MOF-derived 2D carbon framework that is covered by N-doped CNTs with encapsulated, well-dispersed Co nanoparticles (NPs) on the tips [47]. This novel structure led to a 2D structure with a high surface in which Co NPs on tips of CNTs provided the active sites facilitating an improved mass transfer. The electrocatalytic activity of the obtained material was assessed by testing onset potential (0.93 V), half-wave potential (0.86 V), and limiting current density (5.61 mA/cm^2), an Tafel slope (79.3 mV/dec). These values were compared to those with no CNTs which gave an onset potential of 0.92), half-wave potential of 0.83 V, and a limiting current density of 4.93 mA/cm^2. The key point was the comparison of Tafel slopes, especially with the commercial Pt/C catalyst (87.7 mV/dec). The number of electrons transferred per oxygen molecule was obtained from Koutecky-Levich plots and at 0.40 V it came out to be 3.99 V, which indicates that an almost pure 4 electron route is present. This electrocatalyst was used in an assembled zinc-air battery and the specific capacity was measured as 789 mAh/g_{Zn} at a discharge current density of 5 mA/cm^2, a result superior to Pt/C-Ir/C performance with 698 mAh/g_{Zn} [47].

12.2.3.2 Transition Metal Macrocycles

Transition metal macrocycles are evaluated as a group of catalysts alternative to state-of-the-art precious metal catalysts. This group of materials is a sub-class of inorganic-organic composites composed of a macrocyclic structure (phthalocyanine, triallylamine, polypyrrole, etc.) and a chelated transition metal ion [8]. Due to the ease of tuning and thus controlling the configuration, they can be considered as future-promising materials. For these materials, the temperature of polymerization affects the electrical conductivity [8]. Iron and cobalt are the two most common transition metals residing within transition metal macrocycles. It is believed that the active sites in these systems are in the vicinity of M-N_x/C sites [1].

12.2.3.3 Hexagonal Boron Nitride

Boron nitride (BN), which could be utilized as a great supporting material, consists of 2D layers of sp^2-hybridized boron and nitrogen in a honeycomb-like structure which resembles the graphene structure. Hexagonal boron nitride (hBN) has excellent porosity, large surface area, low cost, and great thermal conductivity, which makes it a good support candidate [49]. However, due to the electrical insulator nature of the material, it cannot be used as an electrocatalyst by itself. It is generally considered a great supporting material for conducting electrocatalysts. The most common morphology of BN is the nanosheet structure that offers a high surface area for the catalysts and thus aids in the prevention of catalyst agglomeration [49]. In addition, hBN has very good corrosion and oxidation resistance, which is very much expected from support materials. This material is generally synthesized through three different methods: calcination, exfoliation, and chemical vapor deposition (CVD) [49]. It is agreed upon that hBN could enhance the performance of active materials for ORR and OER as a support material in MABs. TEM results display a hydrothermally synthesized bifunctional nanocomposite (CNT/BN) from carbon nanotubes (CNTs) and a hexagonal boron nitride (h-BN) (Figure 12.3a–b) [50]. In comparison with commercial Pt/C catalyst, the relatively higher ORR activity and improved onset potential (+0.86 V) versus RHE along with a current density of 5.78 mA/cm^2 was attributed to the large interface areas between CNT and h-BN (Figure 12.3 c-d) [50].

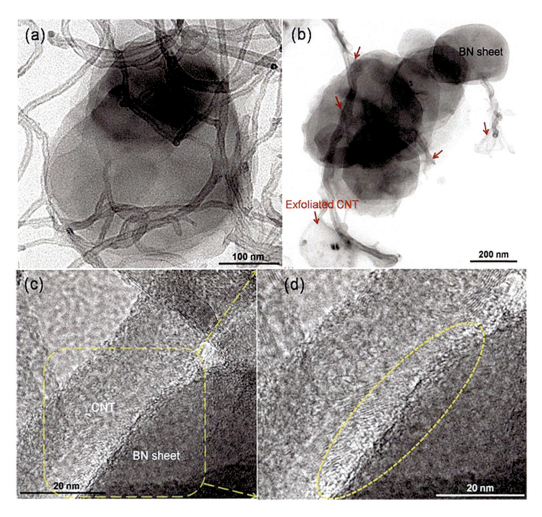

FIGURE 12.3 (a-b) Low-resolution TEM image of the CNT/BN composite catalyst; the CNTs are partially exfoliated to form a graphene-like sheet (as clearly seen in part (b); (c-d) HR-TEM images of CNT/BN catalyst, demonstrating an interface (designated by dotted area) between CNT and h-BN. (Adapted with permission [50]. Copyright (2017), John Wiley and Sons.)

12.2.3.4 Covalent Organic Frameworks

Covalent Organic Frameworks (COFs), an emerging class of porous and crystalline structures linked by covalent bonds between light elements, are promising in terms of ORR and OER activities with their tunable structures and active sites [51]. Liu *et al.* synthesized melem-cyanuric acid (MCAC) nanosheets that have a randomly stacked structure consisting of both crystalline and amorphous phases [51]. ORR performance of the produced material has surpassed that of N-doped GNSs and a comparison with the N-doped GNS gave lower overpotential (56 mV/dec vs. 78 mV/dec), higher half-wave potential (0.8 vs. 0.71 V), and better limiting current density (−6 vs. −4.2 mA/cm^2). High catalytic activity is attributed to the sponge-like and highly-conjugated structure that provides efficient paths for mass and charge transfer and hence accelerates the reaction [51].

12.2.3.5 Bismuth Oxyhalides

Bismuth oxyhalides, BiOX, (X = Cl, Br, I) are a newly emerging and promising class of materials that are expected to have good catalytic properties due to their intriguing chemical, electrical, and magnetic properties. Li *et al.* have synthesized BiOX nanosheets through a low-cost hydrothermal method and compared their Al-air battery performances by altering the halogen (Cl, Br, and I) in the structure [52]. Mesoporous structures were achieved with the abundance of the active sites. The best catalytic performance was obtained from BiOI, most likely due to the more favorable structure and morphology it offered [52].

12.3 CONCLUSION AND FUTURE PERSPECTIVES

A detailed summary of 2D materials used in electrocatalysis for metal-air batteries has been presented in this chapter with three main material types, namely, 2D carbon-based materials, 2D transition metal-based materials, and others (metal-organic frameworks, transition metal macrocycles, hexagonal boron nitride, coordination polymers, covalent organic frameworks, and bismuth oxyhalides). The strong chemical bonds in the planes of ordered layered atomic structure facilitate charge transfer. Besides, surface defects including atomic doping or deficiency enable more active sites by lowering the Gibbs free energy and optimizing the adsorption energy for the enhanced electrocatalytic activity. Despite the change in the materials, methodology and approach for improvement of electrocatalysis are quite similar for all cases, i.e., promoting defects/dopings or creating a porous structure with useful pore sizes. However, there is still a long way to go for the applications of doping strategy and defect engineering while reducing the thickness of 2D nanosheets. In situ characterization techniques should be employed to capture reaction intermediates for a better understanding of hetero-interface in 2D nanomaterial-based heterostructures during the electrocatalytic processes.

The electron transfer path and the adsorption behavior on active sites also have been focused on lately for a better prediction of the ORR/OER mechanism. Due to the possibility of reducing the overpotential of both ORR and OER for metal-air batteries, appropriate bifunctional electrocatalysts are sought after and it is highly believed that 2D materials, due to their superior characteristics, are very good candidates. Because of the bottleneck created by catalytic inefficiency, a significant difference is observed between the theoretical and practical energy densities of metal-air batteries. Other obstacles to the higher utilization of 2D electrocatalysis in MABs involve complex synthesis techniques, high-temperature treatment, precursors, and equipment cost constraints. Therefore, more efficient, economically feasible, and green synthesis strategies need to be developed.

REFERENCES

[1] F. Cheng, J. Chen, Metal-air batteries: From oxygen reduction electrochemistry to cathode catalysts, Chem. Soc. Rev. 41 (2012) 2172–2192.
[2] A.G. Olabi, E.T. Sayed, T. Wilberforce, A. Jamal, A.H. Alami, K. Elsaid, S.M.A. Rahman, S.K. Shah, M.A. Abdelkareem, Metal-air batteries: A review, Energies. 14 (2021) 7373.
[3] F. Lu, X. Cao, Y. Wang, C. Jin, M. Shen, R. Yang, A hierarchical $NiCo_2O_4$ spinel nanowire array as an electrocatalyst for rechargeable Li-air batteries, RSC Adv. 4 (2014) 40373–40376.
[4] Y. Li, J. Lu, Metal-air batteries: Will they be the future electrochemical energy storage device of choice?, ACS Energy Lett. 2 (6) (2017) 1370–1377.
[5] H. Wang, C. Tang, Q. Zhang, A review of precious-metal-free bifunctional oxygen electrocatalysts: Rational design and applications in Zn–Air batteries. Advanced Functional Materials, 28 (2018) 1803329.
[6] J. Deng, S. Fang, Y. Fang, Q. Hao, L. Wang, Y.H. Hu, Multiple roles of graphene in electrocatalysts for metal-air batteries. Catalysis Today. 409 (2022) 2–22.
[7] Y. Xiao, J. Wang, Y. Wang, W. Zhang, A new promising catalytic activity on blue phosphorene nitrogen-doped nanosheets for the ORR as cathode in nonaqueous Li-air batteries, Appl. Surf. Sci. 488 (2019) 620–628.
[8] W. Li, W. Ding, G. Wu, J. Liao, N. Yao, X. Qi, Cobalt modified two-dimensional polypyrrole synthesized in a flat nanoreactor for the catalysis of oxygen reduction, Chem. Eng. Sci. 135 (2015), 45–51.

[9] H.-P. Guo, X.-W. Gao, N.-F. Yu, Z. Zheng, W.-B Luo, C. Wu, H.-K. Liu, J.-Z. Wang, Metallic state two-dimensional holey-structured Co$_3$FeN nanosheets as stable and bifunctional electrocatalysts for zinc-air batteries, J. Mater. Chem. A. 7 (2019) 26549–26556.

[10] S.N. Upadhyay, S. Pakhira, Mechanism of electrochemical oxygen reduction reaction at two-dimensional Pt-doped MoSe$_2$ material: An efficient electrocatalyst, J. Mater. Chem. C. 9 (2021) 11331–11342.

[11] H. Fan, H. Liu, X. Hu, G. Lv, Y. Zheng, F. He, Fe$_2$P@mesoporous carbon nanosheets synthesized via an organic template method as a cathode electrocatalyst for Zn-air batteries, J. Mater. Chem. A. 7 (2019) 11321–11330.

[12] Z. Sun, L. Lin, M. Yuan, H. Li, G. Sun, C. Nan, Two-dimensional β-cobalt hydroxide phase transition exfoliated to atom layers as efficient catalyst for lithium-oxygen batteries, Electrochim. Acta 281 (2018) 420–428.

[13] B. Li, R. Zhu, H. Xue, Q. Xu, H. Pang, Ultrathin cobalt pyrophosphate nanosheets with different thicknesses for Zn-air batteries, J. Colloid Interface Sci. 563 (2020) 328–335.

[14] Y. Pan, K. Sun, S. Liu, X. Cao, K. Wu, W.C. Cheong, Z. Chen, Y. Wang, Y. Li, Y. Liu, D. Wang, Q. Peng, C. Chen, Y. Li, Core-shell ZIF-8@ZIF-67-derived CoP nanoparticle-embedded N-doped carbon nanotube hollow polyhedron for efficient overall water splitting, J. Am. Chem. Soc. 140 (2018) 2610–2618.

[15] L. Huang, L. Zuo, T. Yu, H. Wang, Z. He, H. Zhou, S. Su, T. Bian, Two-dimensional Co/Co$_9$S$_8$ nanoparticles decorated N, S dual-doped carbon composite as an efficient electrocatalyst for zinc-air battery, J. Alloys Compd. 897 (2022) 163108.

[16] H. Wan, F. Chen, W. Ma, X. Liu, R. Ma, Advanced electrocatalysts based on two-dimensional transition metal hydroxides and their composites for alkaline oxygen reduction reaction. Nanoscale. 42 (2020) 21479–21496.

[17] J. Lee, S. Tai Kim, R. Cao, N. Choi, M. Liu, K.T. Lee, J. Cho, Metal-air batteries with high energy density: Li-air versus Zn-air. Advanced Energy Materials. 1 (2011) 34–50.

[18] K. Liu, H. Huang, Y. Zhu, S. Wang, Z. Lyu, X. Han, Q. Kuang, S. Xie, Edge-segregated ternary Pd—Pt—Ni spiral nanosheets as high-performance bifunctional oxygen redox electrocatalysts for rechargeable zinc-air batteries, J. Mater. Chem. A. 10 (2022) 3808–3817.

[19] L. Zhang, Q. Shao, J. Zhang, An overview of non-noble metal electrocatalysts and their associated air cathodes for Mg-air batteries, Mater. Reports Energy. 1 (2021) 100002.

[20] W. Tang, B. Li, K. Teng, X. Wang, R. Liu, M. Wu, L. Zhang, P. Ren, J. Zhang, M. Feng, Advanced noble-metal-free bifunctional electrocatalysts for metal-air batteries, Journal of Materiomics. 8 (2022) 454–474.

[21] Q. Huang, Y. Xu, Y. Guo, L. Zhang, Y. Hu, J. Qian, S. Huang, Highly graphitized N-doped carbon nanosheets from 2-dimensional coordination polymers for efficient metal-air batteries, Carbon N.Y. 188 (2022) 135–145.

[22] H. Bin Yang, J. Miao, S.-F. Hung, J. Chen, H.B. Tao, X. Wang, L. Zhang, R. Chen, J. Gao, H.M. Chen, L. Dai, B. Liu, Identification of catalytic sites for oxygen reduction and oxygen evolution in N-doped graphene materials: Development of highly efficient metal-free bifunctional electrocatalyst, Sci. Adv. 2 (2016).

[23] Y. Li, H. Huang, S. Chen, X. Yu, C. Wang, T. Ma, 2D nanoplate assembled nitrogen doped hollow carbon sphere decorated with Fe$_3$O$_4$ as an efficient electrocatalyst for oxygen reduction reaction and Zn-air batteries, Nano Res. 12 (2019) 2774–2780.

[24] F. Li, T. Zhang, H. Zhou, Challenges of non-aqueous Li-O$_2$ batteries: Electrolytes, catalysts, and anodes, Energy & Environmental Science. 6 (2013) 1125–1141.

[25] Z. Sun, S. Fang, Y.H. Hu, 3D graphene materials: From understanding to design and synthesis control, Chem. Rev. 120 (2020) 10336–10453.

[26] H.S. Kim, J. Lee, J.-H. Jang, H. Jin, V.K. Paidi, S.-H. Lee, K.-S. Lee, P. Kim, S.J. Yoo, Waste pig blood-derived 2D Fe single-atom porous carbon as an efficient electrocatalyst for zinc—air batteries and AEMFCs, Appl. Surf. Sci. 563 (2021) 150208.

[27] M. Huo, B. Wang, C. Zhang, S. Ding, H. Yuan, Z. Liang, J. Qi, M. Chen, Y. Xu, W. Zhang, H. Zheng, R. Cao, 2D metal-organic framework derived CuCo alloy nanoparticles encapsulated by nitrogen-doped carbonaceous nanoleaves for efficient bifunctional oxygen electrocatalyst and zinc—air batteries, Chem. Eur. J. 25 (2019) 12780–12788.

[28] X. Kong, Y. Huang, Q. Liu, Two-dimensional boron-doped graphyne nanosheet: A new metal-free catalyst for oxygen evolution reaction, Carbon N.Y. 123 (2017) 558–564.

[29] C.H. Choi, H.-K. Lim, M.W. Chung, J.C. Park, H. Shin, H. Kim, S.I. Woo, Long-range electron transfer over graphene-based catalyst for high-performing oxygen reduction reactions: Importance of size, n-doping, and metallic impurities, J. Am. Chem. Soc. 136 (2014) 9070–9077.

[30] T. Yu, J. Fu, R. Cai, A. Yu, Z. Chen, Nonprecious electrocatalysts for Li-air and Zn-air batteries: Fundamentals and recent advances, IEEE Nanotechnol. Mag. 11 (2017) 29–55.

[31] J. Qiao, L. Kong, S. Xu, K. Lin, W. He, M. Ni, Q. Ruan, P. Zhang, Y. Liu, W. Zhang, L. Pan, Z. Sun, Research progress of Mxene-based catalysts for electrochemical water-splitting and metal-air batteries, Energy Storage Materials. 43 (2021) 509–530.

[32] M. Wu, G. Zhang, H. Tong, X. Liu, L. Du, N. Chen, J. Wang, T. Sun, T. Regier, S. Sun, Cobalt (II) oxide nanosheets with rich oxygen vacancies as highly efficient bifunctional catalysts for ultra-stable rechargeable Zn-air flow battery, Nano Energy. 79 (2021) 105409.

[33] T. Zhou, W. Xu, N. Zhang, Z. Du, C. Zhong, W. Yan, H. Ju, W. Chu, H. Jiang, C. Wu, Y. Xiw, Ultrathin cobalt oxide layers as electrocatalysts for high-performance flexible Zn-air batteries, Advanced Materials. 31 (2019) 1807468.

[34] Y. Wang, Y. Li, T. Heine, PtTe monolayer: Two-dimensional electrocatalyst with high basal plane activity toward oxygen reduction reaction, J. Am. Chem. Soc. 140 (2018) 12732–12735.

[35] A. Mathur, R. Kaushik, A. Halder, Photoenhanced performance of Cobalt-intercalated 2-D manganese oxide sheets for rechargeable zinc-air batteries, Mater. Today Energy. 19 (2021) 100612.

[36] E. Pargoletti, A. Salvi, A. Giordana, G. Cerrato, M. Longhi, A. Minguzzi, G. Cappelletti, A. Vertova, ORR in non-aqueous solvent for Li-air batteries: The influence of doped MnO_2-nanoelectrocatalyst, Nanomaterials. 10 (2020) 1735.

[37] F. Meng, H. Zhong, D. Bao, J. Yan, X. Zhang, In situ coupling of strung Co_4N and intertwined N-C fibers toward free-standing bifunctional cathode for robust, efficient, and flexible Zn-air batteries, J. Am. Chem. Soc. 138 (2016) 10226–10231.

[38] F. Xu, J. Wang, Y. Zhang, W. Wang, T. Guan, N. Wang, K. Li, Structure-engineered bifunctional oxygen electrocatalysts with Ni_3S_2 quantum dot embedded S/N-doped carbon nanosheets for rechargeable Zn-air batteries, Chem. Eng. J. 432 (2022) 134256.

[39] H. Wang, J.-M. Lee, Recent advances in structural engineering of MXene electrocatalysts, J. Mater. Chem. A. 8 (2020) 10604–10624.

[40] A. Ostadhossein, J. Guo, F. Simeski, M. Ihme, Functionalization of 2D materials for enhancing OER/ORR catalytic activity in Li-oxygen batteries, Commun. Chem. 2 (2019).

[41] B. Anasori, M.R. Lukatskaya, Y. Gogotsi, 2D metal carbides and nitrides (MXenes) for energy storage, Nature Reviews Materials. 2 (2017) 16098.

[42] M. Faraji, F. Parsaee, M. Kheirmand, Facile fabrication of N-doped graphene/$Ti_3C_2T_x$ (Mxene) aerogel with excellent electrocatalytic activity toward oxygen reduction reaction in fuel cells and metal-air batteries, J. Solid State Chem. 303 (2021) 122529.

[43] Y. Wang, J. Zhu, Y. Jiang, T. An, J. Huang, M. Jiang, M. Cao, Highly graphited carbon-coated $FeTiO_3$ nanosheets in situ derived from MXene: An efficient bifunctional catalyst for Zn-air batteries, Dalt. Trans. 51 (2022), 5706–5713.

[44] Z.-L. Wang, D. Xu, J.-J. Xu, X.-B. Zhang, Oxygen electrocatalysts in metal-air batteries: From aqueous to nonaqueous electrolytes. Chem. Soc. Rev. 43 (2014) 7746–7786.

[45] H. Li, M. Zhang, W. Zhou, J. Duan, W. Jin, Ultrathin 2D catalysts with N-coordinated single Co atom outside Co cluster for highly efficient Zn-air battery, Chem. Eng. J. 421 (2021) 129719.

[46] J. Tang, R.R. Salunkhe, J. Liu, N.L. Torad, M. Imura, S. Furukawa, Y. Yamauchi, Thermal conversion of core-shell metal-organic frameworks: A new method for selectively functionalized nanoporous hybrid carbon, J. Am. Chem. Soc. 137 (2015) 1572–1580.

[47] D. Chen, J. Yu, Z. Cui, Q. Zhang, X. Chen, J. Sui, H. Dong, L. Yu, L. Dong, Hierarchical architecture derived from two-dimensional zeolitic imidazolate frameworks as an efficient metal-based bifunctional oxygen electrocatalyst for rechargeable Zn-air batteries, Electrochim. Acta. 331 (2020) 135394.

[48] Y. Zhao, H.M. Zhang, Y. Zhang, M. Zhang, J. Yu, H. Liu, Z. Yu, B. Yang, C. Zhu, J. Xu, Leaf-like 2D nanosheet as efficient oxygen reduction reaction catalyst for Zn-air battery, J. Power Sources. 434 (2019) 226717.

[49] N. Ullah, R. Ullah, S. Khan, Y. Xu, Boron nitride-based electrocatalysts for HER, OER, and ORR: A mini-review. Frontiers of Materials Science. 15 (2021) 543–552.

[50] I.M. Patil, M. Lokanathan, B. Ganesan, A. Swami, B. Kakade, Carbon nanotube/boron nitride nanocomposite as a significant bifunctional electrocatalyst for oxygen reduction and oxygen evolution reactions, Chem. Eur. J. 23 (2017) 676–683.

[51] J. Liu, C. Wang, Y. Song, S. Zhang, Z. Zhang, L. He, M. Du, Two-dimensional triazine-based porous framework as a novel metal-free bifunctional electrocatalyst for zinc-air batty, J. Colloid Interface Sci. 591 (2021) 253–263.

[52] M. Li, J. Yuan, B. Nan, Y. Zhu, S. Yu, Y. Shi, M. Yang, Z. Wang, Y. Gu, Z. Lu, Ultrathin BiOX (X = Cl, Br, I) nanosheets as Al-air battery catalysts, Electrochim. Acta. 249 (2017) 413–420.

13 3D Materials as Electrocatalysts for Metal-Air Batteries

Demet OZER

CONTENTS

- 13.1 Introduction 179
- 13.2 Principles of Metal-Air Batteries... 181
- 13.3 3D Electrode Materials 184
  - 13.3.1 Carbon-Based Electrodes 184
  - 13.3.2 Oxide-Based Electrodes 186
  - 13.3.3 Chalcogenide-Based Electrodes 188
  - 13.3.4 Nickel Foam-Based Electrodes 189
  - 13.3.5 Metal-Organic Framework-Based Electrodes 189
- 13.4 Conclusions and Perspectives for Future Applications 191
- References 191

13.1 INTRODUCTION

In th^{-e} 21st century world, the rapid increase in energy needs with the increase in population and the gradual decrease in fossil fuel reserves have increased the search for renewable, environmentally friendly alternative energy sources. At the same time, harmful gases such as carbon dioxide, soot, nitrogen oxides, and sulfur oxides are released into the atmosphere with the combustion of fossil fuels. The increase in gas concentrations in the atmosphere may cause some health problems due to the increase in temperatures due to the greenhouse effect, as well as the ultraviolet rays falling to the world from the sun's rays with the effect of the atmosphere layers. For these reasons, alternative, clean, and renewable energy sources have been needed in recent years and different energy production systems have been studied intensively. Electricity is a highly efficient and environmentally friendly end-use energy. Electrochemical studies are of great importance in reducing costs in energy production, which is one of the most important problems of our age, and in finding and developing clean and renewable resources. Due to the instability of renewable energy sources, the need for energy storage systems for electricity generation has increased day by day. Batteries have great advantages in the conversion and storage of chemical energy. They are crucial for portable electronic devices due to their small size as compared to conventional energy storage techniques. Due to their high volumetric capacity, gravimetric capacity, and electrical and energy efficiency, Li-ion batteries have been among the most promising electrochemical energy technologies since the late 1990s. They now account for practically the entire market for rechargeable batteries. Large-scale applications of lithium-ion batteries are further constrained by their low energy density, poor safety, and high cost. Therefore, alternative energy conversion and storage systems are needed for future applications.

Recently, metal-air batteries have gained splendid concern for energy-storage applications because of their high theoretical density which is higher than lithium-ion batteries and free oxygen fuel from the atmosphere. A brief history of metal-air batteries is given in Figure 13.1a. Research

DOI: 10.1201/9781003295761-13

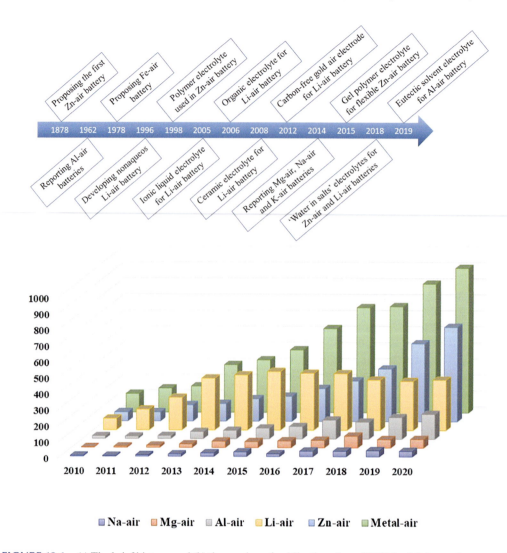

FIGURE 13.1 (a) The brief history, and (b) the number of publications from ISI Web of Science about metal-air batteries.

on metal-air batteries has advanced significantly since the discovery of the Zn-air battery. They are regarded as an environmentally benign, non-toxic, low-cost, and practical option because metal is abundant in nature. These metals are sodium, lithium, and potassium from the first group; magnesium and calcium from the second group; aluminum from the third group; and zinc and iron from transition metals. Among them, lithium-air batteries, sodium-air batteries, magnesium-air batteries, aluminum-air batteries, and zinc-air batteries are known as the most studied metal-air batteries according to Web of Science data (Figure 13.1b). It is seen that the studies on metal-air batteries have increased rapidly over the years, and metal-air batteries can be the most promising power storage devices with high density in the future. The design of metal-air batteries combines features from both fuel cells and conventional batteries. Despite their infancy, issues with metal anodes, air catalysts (cathodes), and electrolytes have slowed down the possible developments of metal-air batteries.

In this chapter, the components of metal-air batteries (metal electrode, air electrode, electrolyte, and separator) and especially 3D electrode materials are overviewed in detail. At the same time, some general difficulties encountered in the progress of metal-air batteries and their possible solutions are focused on.

13.2 PRINCIPLES OF METAL-AIR BATTERIES

Batteries are energy storage systems that can both store chemical energy and convert chemical energy into electrical energy with high efficiency without any gas output. With the introduction of commercial lithium-ion batteries (LIBs) by Sony in 1991, they have become the most popular energy storage system. Despite its superior features such as high energy densities, long cycle lives, high conversion stability, low self-discharge rates, and wide operating temperature range, the desired cost-profitability relationship in lithium-ion batteries has not yet been achieved and the production costs required for common use are quite high [1]. With lower cost and higher theoretical energy density than lithium-metal batteries, lithium-air batteries can have a great potential for future studies and metal-air batteries are a unique energy storage method because cathodic oxygen provides a limitless source of oxygen from the environment that does not need to be stored. Compared to lithium-ion batteries, lithium-air batteries, which have the same working principle, are systems that store more energy. Storage of 1500–2000 Wh/kg of the cathode is envisaged in lithium-air batteries (Li-air or Li-O_2), and thus it has drawn more and more attention as one of the upcoming energy storage techniques for automotive applications.

The electrochemical reaction of metal-air batteries is subject to oxidation in metal and reduction in the air (oxygen). The electrochemical cell consists of a metallic anode in a convenient electrolyte, separator, and air cathode (Figure 13.2) [2]. The air cathode does not store oxygen but obtains it from the atmosphere. It has an electrode overpotential-lowering electrocatalyst layer and an oxygen-diffusion layer that allow gaseous oxygen to diffuse between the environment and the catalyst surface. In lithium-air batteries, the cathode is the porous electrode where the discharge reaction products are stored. In non-aqueous electrolyte systems, due to the low oxygen solubility and diffusion coefficient, which are the kinetics of the discharge reaction, the amount of Li_2O_2 formed at a low discharge rate and thus the discharge capacity of the cell is determined by the porosity properties of the cathode material, such as surface area, pore volume, and pore size distribution.

It is often used as a component of porous cathode because catalysts are believed to improve oxygen reduction and evolution reaction. Electrocatalysts are also used because they reduce charge overvoltage. Metals such as manganese (Mn), cobalt (Co), ruthenium (Ru), platinum (Pt), iridium (Ir), and silver (Ag) are preferred as catalysts. By adding metal catalysts to the carbon electrode, it increases the oxygen reduction kinetics of the cathode, and an increase in the specific capacity is observed. Gibb's free-energy of the discharge reaction of the cathode active substance used in rechargeable lithium-air batteries must have a large negative value (high discharge potential). It should have a low molecular weight and be capable of a host-guest reaction with a large amount of lithium (high energy capacity), the chemical diffusion coefficient of lithium should be high (high power density), and the crystalline structure should change little or not at all during the reaction (high cycle number). It should be inexpensive, readily available, environmentally friendly, chemically stable, insoluble in electrolytes, and easily processed [3]. Detailed lithium-air battery studies to improve the cathode structure to increase efficient catalyst production, reaction kinetics, and discharge capacity are hampered by the weaknesses of the available electrolytes. Therefore, the occurrence of reversible Li_2O_2 conversion and oxidation is a difficult prerequisite for studying and developing the cathode structure. The development of a stable electrolyte will pave the way for accelerating future developments in cathode architecture.

The anode and cathode have been separated from each other using a separator to hinder mass transportation activities between the electrodes and prevent the short-circuit generated by metal dendrites. The important features affecting the choice of separator can be summarized as follows: they must be electrically insulating; electrolyte resistance (ionic) should be minimal; it should provide mechanical and dimensional stability; it should form an effective barrier against the displacement

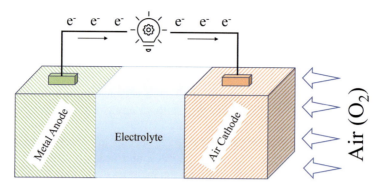

FIGURE 13.2 Schematic illustration of metal-air batteries.

TABLE 13.1
Theoretical Specific Capacities, and Energy Densities of Metal-Air Batteries

Batteries	Theoretical specific capacity (mAh g^{-1})	Theoretical energy density (Wh L^{-1})	Cell voltage (V)
Li-air	5928	7989	2.96
K-air	1187	1913	2.37
Na-air	1680	2466	2.30
Mg-air	5238	9619	3.09
Al-air	5779	10347	2.71
Zn-air	1218	6136	1.66
Fe-air	1080	3244	1.28

of particles and colloidal species between the two electrolytes; it should not enter into chemical reaction with the electrolyte, electrode materials, and other components in the cell; the separator should be easily wetted by the electrolyte; and thickness and other properties must be homogeneous.

As an anode material, alkali metals (Li, K, Na, etc.), alkaline earth metals (Ca, Mg, etc.), metalloids (Si, Al, etc.), and transition metal elements (Fe, Zn, etc.) can be used. The theoretical specific capacity and energy density of some metal-air batteries are summarized in Table 13.1 [4]. The theoretical energy densities are higher than the capacity of commercial Li-ion batteries. Among them, zinc and lithium-air batteries are the most studied metal-air batteries. Lithium-air batteries have superior theoretical performance. Although theoretically, lithium has a better energy density and specific capacity than zinc, its exorbitant price ($400–800/kWh) can be seen as a drawback when compared to zinc-air batteries ($10/kWh) [2]. Zinc-air batteries are the only cells in this category that have been successful and commercialized as primary cells, making them ideal for low-current applications like hearing aids. Zinc oxygen systems, despite their limited shelf life and charge capacity, may be the quickest and most dependable way to a practical secondary metal-air battery. Due to its exceptional energy density, the aluminum-air battery is a fantastic energy source for electric vehicles. Both aluminum- and magnesium-air batteries are also being developed for military applications in the aqueous salt solution system. The sodium-air battery has a high theoretical specific heat energy and is used in transportation because of the abundance of sodium, low cost, and environmentally beneficial nature. Table 13.1 shows theoretical specific capacities and energy densities of metal-air batteries.

The other crucial component of metal-air batteries is electrolytes. They are intimately related to the battery's effectiveness. The electrolytes must be consistent over a variety of environmental situations, non-toxic, low volatile, and high oxygen soluble. They have been chosen according to using metals and they can be both hydrous and anhydrous. Protic or aqueous electrolytes are used for metals that are not affected by water and humidity, while aprotic or non-aqueous electrolytes are used for others that can be affected by atmospheric moisture or water. When ionic liquid, solid, and polymer electrolytes have been used as non-aqueous electrolytes, alkaline, hybrid, room temperature ionic liquid, and quasi-solid flexible electrolytes have been applied as aqueous electrolytes for metal-air batteries [5]. The electrode reaction changes as to the type of electrolyte. For the metals such as iron, zinc, aluminum, magnesium, etc. generally aqueous electrolytes have been used and equations 1 and 2 were formed (M = metal, n = oxidation number of metal). The metal ions formed on the anode in the discharge step can react with the OH^- in an alkaline electrolyte. Conversions between oxygen and water take place at the cathode during the discharge/charge process [6].

$$\text{Metal electrode (anode): } M \leftrightarrow M^{n+} + ne^- \quad \text{(Equation 1)}$$
$$\text{Air electrode (cathode): } O_2 + 2H_2O + 4e^- \leftrightarrow 4OH^- \quad \text{(Equation 2)}$$

Nonaqueous and aprotic electrolytes are usually applied for lithium, sodium, and potassium metals which are sensitive to water and carbon dioxide in the air. Equations 3 and 4 are overtaken. On the cathode, oxygen combines with metal ions, and the discharge product on the air electrode can be metal superoxide or peroxide. According to the hard and soft acid-base theories, oxygen is converted to superoxide and coupled with metal ions. For example, if we describe it in terms of lithium, lithium metal is oxidized at the anode and forms Li^+ ions that migrate towards the cathode. Electrons formed in the oxidation stage at the anode pass through an external environment to do the work on the charge and return to the cathode to complete the electrochemical reaction of the Li^+ ions migrating from the anode. While oxygen is reduced by taking the existing electrons, it also causes the formation of any of the Li^+ ions, LiO_2, Li_2O_2, and Li_2O products. Larger cations such as sodium and potassium stabilize the superoxide anion more easily than the hard and soft acid/base theory [6].

$$\text{Metal electrode (anode): } M \leftrightarrow M^+ + e^- \quad \text{(Equation 3)}$$
$$\text{Air electrode (cathode): } xM^+ + O_2 + xe^- \leftrightarrow M_xO_2 \ (x = 1 \text{ or } 2) \quad \text{(Equation 4)}$$

The electrolyte is an important factor in describing the electrochemical properties of batteries, including charge and discharge capacity, current density, etc. Electrolytes should have features such as an admissible electrochemical window to minimize side reactions, high ionic conductivity, high mechanical stability, minimal reactivity in the environment and electrode material, cost-effectiveness, safety, and high thermal stability [4].

Metal-air battery performance is influenced by both rechargeability and discharge characteristics. Even though rechargeable metal-air batteries have a bright future, they now face challenges such as low power density, round-trip efficiency, and Coulombic efficiency. In lithium-air batteries, clogging of the pores of the porous carbon cathode leads to the formation of discharge by-products, a reduction in the required oxygen supply, and a decrease in the number of cycles, which is the most important parameter for commercialization. In addition, due to the dendritic growth of the lithium metal used in the anode, short circuits may occur in the cell. Lithium metal is sensitive to moisture and must be protected. The electrolytes used in lithium-air batteries directly affect battery performance. Aggressive oxygen decomposes oxygen-sensitive electrolytes and destabilizes the battery. In addition, the reactions that will occur during the discharge and the reaction products to be obtained differ according to the type of electrolyte used. As a result, the ability of lithium-air batteries to reach the theoretically estimated energy density is highly dependent on the electrolyte's high recyclability and ability to increase energy density. Aqueous, anhydrous, solid, and aqueous/

non-aqueous electrolytes are the types of electrolytes used in lithium-air batteries. Anhydrous electrolytes have a protective effect on lithium metal and reduce dendrite formation because they form SEI – Solid Electrolyte Interphase (solid electrolyte interface). Anhydrous electrolytes show promise in studies to develop lithium-air battery technology to ensure that the discharge products formed during the battery's cycle are highly recyclable and thus prevent pore clogging at the cathode [7].

In conclusion, it is important to first understand the basic oxygen chemistry to realize the full potential of metal-air cells. The development of efficient air electrodes and electrolytes for subsequent improvement of oxygen reduction (ORR) and/or oxygen oxidation (OER) reactions are key research goals. At the same time, the cycle life of rechargeable batteries is below the expected values due to the absence of stable bifunctional electrocatalysts and convertible metal anodes. When batteries are deeply cycled while being subjected to enormous currents, this flaw is compounded. This problem can be solved by designing and engineering both cathode and anode materials. In recent years, 3D electrodes have been applied to enhance power density, energy density, and rechargeability, and the studies about the design and fabrication of these electrodes increase day by day.

13.3 3D ELECTRODE MATERIALS

Low-rate capability, short lifetime, and a lack of effective, reliable bifunctional electrochemical reaction catalysts (ORR/OER) have all hindered the development of rechargeable air batteries for large-scale applications. Catalysts with precious noble metals (like Pt, Ag, Au, etc.), transition metals (like Ni, Co, etc.), and their oxides are typically employed for ORR and OER. Due to high activity and stability, the most used are cobalt-based materials up to now [8]. The commercialization of sustainable energy solutions has been constrained by transition metal catalysts' instability, unavailability, and inherently low electrical conductivity. Therefore, the creation of bifunctional electrocatalysts with high activity and extended durability for the oxygen reduction and evolution reaction has been required for the practical use of rechargeable metal-air batteries. Oxygen bubbles produced from the oxygen evolution reaction are intrinsically insulating, severely inhibiting the transfer of electrons/ions from the air electrode. As a result, creating high-throughput gas-breathing air electrodes is critical.

One of the biggest challenges for metal-air batteries is to develop effective electrocatalysts with high chemical stability in oxygen-containing electrolytes. For this purpose, three main strategies were developed to control catalyst chemical compositions. (a) The traditional or standard approach is to make powder catalysts using high-temperature pyrolysis, calcination, or hydrothermal methods, then fabricate electrodes by physically depositing an active material onto a carbon gas diffusion layer with the help of polymeric binders and extra conductive additives. (b) Hydrothermal process, electrodeposition, chemical vapor deposition, and atomic layer deposition technologies are used to generate electrocatalysts in situ on conductive substrates. (c) Freestanding catalytic films or wires using graphene sheets and analogs as active species. The latter two are also referred to as self-supported air electrodes and have the requisite advanced nanostructures and assembly without binders [9]. These methods show that the surface and structure modifications improved the electrochemical properties. One of the surface modification methods is the development of materials of different sizes. Considering the increase in the number of active centers, large surface area, enabled ion diffusion/charge transfer paths, and effective mass transfer, studies on 3D electrodes have been increasing recently. 3D electrodes can be classified under the main headings of carbon, oxide, chalcogenide nickel foam, and metal-organic framework-based electrodes.

13.3.1 CARBON-BASED ELECTRODES

The carbon nanomaterials such as porous carbons, carbon nitrides, graphene, carbon cloths, carbon nanotubes, and carbon quantum dots are applied as an effective electrocatalyst in both metal-free and composite forms because of high electronic conductivity, large specific surface area, and porous structure. The structural diversity of carbon materials, from 0D (fullerene), 1D (carbon nanotubes), and 2D (graphene) to 3D porous carbon nanostructures, makes them more adaptable for use with air electrodes [10]. Carbon materials provide ample space to house catalysts and prevent agglomeration. The surface properties increase the solid air contact efficiency, resulting in a large amount of oxygen adsorption. At the same time, it increases the electrical conductivity and electron transfer rate and provides more active centers to stimulate electrocatalytic activity [11]. To improve the performance of air batteries, doping metals (especially transition metals) and nonmetals into 3D graphene has been successfully applied to modify the electronic properties and surface polarities of electrodes. Lin and coworkers prepared 3D N-doped graphene using polypyrrole as a nitrogen source via the facile synthesis method. The obtained electrocatalyst demonstrates high catalytic activity and long-term stability for oxygen reduction reaction [12]. Nitrogen doping enhances the conductivity, provides more nucleation sites around nitrogen and reduces aggregation of discharge products [13].

He and coworkers produced 3D nitrogen-doped graphene by hydrothermal method and annealing process using melamine as a nitrogen source. The 3D N-doped graphene foam showed a higher specific capacity of 7300 mAh/g and longer cycle life (21 cycles) than 3D graphene (2250 mAh/g, 8 cycles) [14]. Xue et al. synthesized B, N-doped graphene foam by a chemical vapor deposition method. They investigated the effects of doping of boron (B), nitrogen (N), and both of them (BN). Boron acts as an electron acceptor, favors oxygen adsorption, and increases active sites. According to onset potentials and current densities, the BN-doped graphene shows higher electrocatalytic activity than B- and N-doped graphene for oxygen reduction reaction due to synergic effects of doping elements [15]. Chen and coworkers doped nitrogen (N) and nickel (Ni) to nanoporous graphene. The pre-addition of nitrogen by chemical vapor deposition followed by chemical etching significantly increases the nickel doping amount and stability, and the resulting N, Ni-doped graphene is an efficient electrocatalyst for both ORR and OER and electrodes with high cyclic stability and flexibility for rechargeable zinc-air batteries [16]. The schematic representation of the zinc battery was given in Figure 13.3a. Ni- and N-doped 3D graphene are used as air cathode, polyvinyl alcohol (PVA) is used as an electrolyte, and zinc foil is used as an anode. The discharge polarization curves (Figure 13.3b) of the co-doped np-graphene show a maximum power density of 83.8 mW/cm^2 with the longest testing period (43 hours), which is higher than the Pt/C + IrO$_2$ catalyst (74.5 mW/cm^2). After 258 cycles, the battery can be steadily charged and drained with only minimal performance loss, with an initial voltage gap of 0.77 V at 2 mA/cm^2 (Figure 13.3c) while Pt/C + IrO$_2$ shows a larger initial voltage gap of 0.95 V and obvious performance decay after 120 cycles. Additionally, the battery's performance is unaffected by varying degrees of bending (Figure 13.3d).

Chen and coworkers synthesized N, O-doped graphene-carbon nanotube hydrogel as a 3D flexible film with high surface area, mechanical durability, and superior conductivity. 100 nm carbon nanotube homogeneously disperse on 10-micrometer graphene and the obtained electrode showed high catalytic activity, high current density, excellent stability, and strong durability (Young's modulus, 27.3 MPa) [17]. Dai and coworkers prepared 3D-N, P doped mesoporous nanocarbon foams via pyrolysis of polyaniline aerogels in the presence of phytic acid. The obtained catalyst with a high surface area (1663 m^2/g) has high catalytic efficiency for both OER and ORR in rechargeable zinc-air battery [18]. Liu and coworkers developed 3D-S, N-doped C foams as an efficient electrocatalyst using the SBA-15 template. The surface area of foam is 1127 m^2/g and it has superior catalytic activity, strong stability, and high methanol tolerance in comparison to the commercial Pt/C electrocatalyst [19]. Carbon cloths are generally applied as a flexible surface for different active materials in electrochemical applications. The carbon nitride nanofibers co-doped with phosphorus and sulfur on carbon cloth were prepared as a flexible electrode material for rechargeable zinc-air batteries which had a high surface area (1649 m^2/g), more active sites, superior electrical conductivity, remarkable tensile modulus, and strength [20]. The obtained electrode with high flexibility, long-term cyclic durability, and stability demonstrated high power density (231 mW/cm^2), high energy density (785 Wh/kg for 5 mA/cm^2), and extraordinary lifespan (over 240 hours).

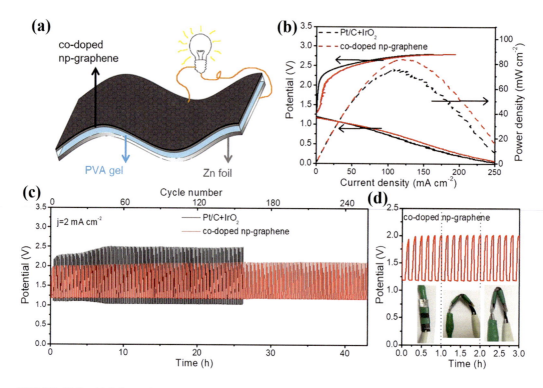

FIGURE 13.3 (a) Schematic representation of the 3D N, Ni-graphene-based Zn-air battery, (b) polarization and power density curves, (c) discharge/charge cycling curves, and (d) discharge/charge curves under different bending states. (Adapted with permission [16]. Copyright (2019), John Wiley and Sons.)

13.3.2 Oxide-Based Electrodes

Transition metal oxides, due to their polyvalent and abundant crystallographic structures resulting in rich redox electrochemistry and materials chemistry, provide ample opportunity for bifunctional applications because of their extraordinary activity and stability [21]. The mesoporous 3D cobalt oxide was prepared through a simple template-derived process using a polystyrene bead template. To facilitate oxygen passage into and out of the catalyst structure during the oxygen reduction and evolution events, respectively, a 3D honeycomb-like structure with microsized pores was created. As a result, the overpotential for the onsets of ORR and OER may be lowered because of the oxygen being easily transported into and out of the structure, increasing the catalyst's activity. Compared with other electrocatalysts, the 3D honeycomb-like mesoporous structure led to enhanced electrochemical durability, rechargeability, and cycle life for the zinc-air battery. Gao and coworkers used the cobalt oxide electrocatalyst for lithium-air batteries and to improve the electrocatalytic performance of oxides, they investigate the effects of the surface, interface, and doping metals [8]. The silver nanoparticles doping on the Co_3O_4@Co_3O_4 surface ensures 12000 mAh/g while Co_3O_4 and Co_3O_4/Ag have 4600 and 7500 mAh/g initial capacity, respectively at 200 mA/g current density. The composite also shows high-rate capability (4700 mAh/g @800 mA/g), low overpotential, long cycle life, and stability due to the synergic effect of both metal and surface environment, improvement of electronic conductivity, and increment of active sites.

In another study, the cobalt oxide nanowire on the stainless-steel surface which has high conductivity and permeability was prepared through a direct growth method which is a time-saving and practical process. The schematic representation of a 3D rechargeable Co_3O_4 nanowire on stainless steel was given in Figure 13.4a. The SEM image of stainless-steel current collector before the

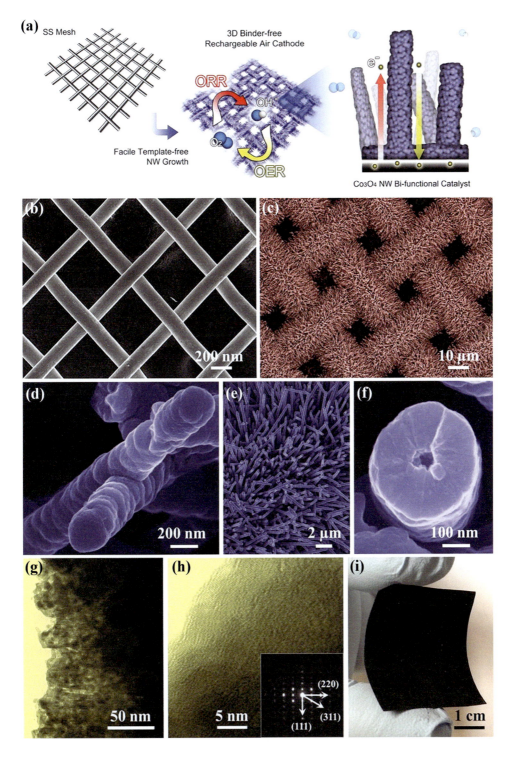

FIGURE 13.4 (a) Schematic representation of 3D rechargeable Co_3O_4 nanowire as an air cathode; SEM image of (b) stainless steel current collector before the growth, (c) densely coated Co_3O_4 nanowire array, (d) Co_3O_4 nanowires, (e) self-standing Co_3O_4, and (f) cross-section of Co_3O_4 (g) TEM image of a mesoporous Co_3O_4 nanowire wall. (h) HR-TEM image of the Co_3O_4 nanowire wall and FFT pattern of Co_3O_4 nanowires exhibiting polycrystallinity. (i) Optical image of a flexible air electrode. (Adapted with permission [22]. Copyright (2014), John Wiley and Sons.)

growth (Figure 13.4b), densely coated Co_3O_4 nanowire array (Figure 13.4c), Co_3O_4 nanowires approximately 300 nm (Figure 13.4d), self-standing Co_3O_4 with high surface area (Figure 13.4e), a cross-section of tubular Co_3O_4 with a circular hollow center of diameter 50 nm (Figure 13.4f), TEM and HR-TEM image of a mesoporous Co_3O_4 nanowire (Figure 13.4g–h) were shown, and as to the optical image of the electrode, it can be used for flexible device applications (Figure 13.4i). The non-conductive co-binder material removed from the electrode prevents degradation of the binder, improving the electrical property and improving electrochemical stability. Stainless steel acts as a support for the growth of Co_3O_4 nanowires and a current collector, simplifying battery design and significantly reducing internal resistance. The prepared highly active bifunctional electrode showed exceptional rechargeability with each pulse cycle lasting 10 min (5 min each for discharge/charge) at a fixed current of 50 mA and long-term durability using the extended cycling test (3 h discharge followed by 3 h charge) with charge and discharge potential retentions of 97% and 94%, respectively as a practical zinc-air battery, using natural air instead of pure oxygen [22].

The 3D hollow Co_3O_4/MnO_2-carbon nanotube hybrid material was prepared by coprecipitation method in conjunction with post-heat treatment and showed high specific capacity (775 mAh/g at 10 mA/cm^2), power density (340 mW/cm^2), and long life cycle for both primary and rechargeable zinc-air battery [23]. Due to the abundance of metal and cost-effectiveness, it can be an effective candidate as an electrocatalyst for industrial applications. The 3D network CoP/Co_2P/Co_3O_4 composite shows high efficiency and durability with high open-circuit and power density as an air cathode in zinc-air battery and has a specific capacity of 710 mAh/g, outperforming that of Pt/Ru (690 mAh/g) [24]. Although the electrochemical studies of cobalt-containing oxides are mostly successful, results have been obtained in metal-air batteries among other oxides. As an example, the 3D macroporous $LaFeO_3$ prepared as an electrocatalyst for lithium-air battery demonstrate low overpotential, high specific capacity, good rate capability, and cycle stability up to 124 cycles [25].

The 3D electrodes have been applied as an anode material for zinc-air batteries. The 3D porous structure provides long-range electronic conductivity throughout the charge-discharge, makes the current distribution more uniform throughout the electrode structure, and limits the shape change of the elements in the interior of the porous electrode in alkaline electrolytes by accelerating the saturation/dehydration of zincate to ZnO within the limited void volume [26]. As an example, the 3D zinc@zinc oxide sponge anode has 728 mAh/g specific capacity and 907 Wh/kg specific capacity for zinc-air battery.

13.3.3 Chalcogenide-Based Electrodes

Metal chalcogenides have shown high electrocatalytic efficiency due to their electronic properties and structural diversity. 3D flower-shaped SnS_2 nano pedals which had 56 m^2/g surface area and 2–20 nm pore size were prepared by the solvothermal method as a cathode material for sodium-air battery and show low overpotential (0.52 V), high power density (300 mW/g at 240 mA/g), and excellent life cycle (40 cycles) [27]. 3D nickel-cobalt dichalcogenides ($NiCo_2S_4$) nanosheet on carbon cloth prepared by the hydrothermal method used as an air cathode for a rechargeable zinc-air battery. They outperform commercial precious metal-based electrodes in terms of high discharge capacity (722 mAh/g), a small potential difference between discharge and charge, and higher cycle durability with reversibility due to their abundant redox chemistry and plentiful exposed edge sites [28]. The $CoSe_2$ is embedded in a nitrogen-doped carbon nanosheet array and nanotube ($CoSe_2$-NCNT NSA) to form a 3D electrocatalyst for a zinc-air battery. $CoSe_2$ is derived from a cobalt metal-organic framework through pyrolysis and annealing with selenide powder in an inert atmosphere (H_2/Ar). The $CoSe_2$-NCNT NSA delivered an open circuit voltage of 1.37 V, a lower voltage gap (1.03 V), and a higher peak power density (51.1 mW/cm^2), as compared to the $CoSe_2$-NCNT NP (1.59 V, and 20.1 mW/cm^2) and $CoSe_2$ NSA (1.76 V, and 12.4 mW/cm^2). The obtained electrocatalyst shows high mechanical flexibility, durability, structural stability, and electrical conductivity [29].

13.3.4 NICKEL FOAM-BASED ELECTRODES

To prevent carbon corrosion which can be seen at high charge potentials, carbon-free conductive materials are an attractive alternative to metal-air batteries. The microporosity of nickel foam increases gas transportation. Xu and coworkers prepared nanoporous nickel and AuNi alloy doped nickel foam as a cathode material for the lithium-air battery. Ni foam enhances the mechanical stability and electrical conductivity, while nanoporous Ni increases mass transfer and AuNi alloy improves catalytic activity with regulate the morphology. The obtained electrode shows ultrahigh specific capacity (22551 mAh/g at 1.0 A/g) and a long-life cycle. The most seen problem in lithium-air batteries is the formation of lithium peroxide and lithium superoxide caused by the poor electrochemical/chemical stability of the $Li-O_2$ system because they attack the cathode and electrolyte and cause the formation of unwanted by-products like lithium carbonate and cause battery death. Due to high surface area, porosity, conductivity, and stability, nickel in the composite prevent the reduction of lithium superoxide and supports the diffusion, and expedites the reaction kinetics [30]. At the same time, nickel foam as a current collector solves the instability of the carbon-based materials up against the lithium peroxide and superoxide and offers quick mass transfer, rapid electron transport, and enough void space to contain the discharge products. The nickel foam-supported carbon nanosheets decorated with ultrafine cobalt phosphide nanoparticle was prepared for OER, ORR, overall water splitting, and zinc-air batteries [31]. While the commercial Pt/C/NF has 60 mW cm^{-2} power density, the obtained CoP_x@CNS has 110 mW/cm^2 power density with a high lifetime (130 hours) and cyclic stability (40 cycles).

13.3.5 METAL-ORGANIC FRAMEWORK-BASED ELECTRODES

Metal-organic frameworks are attractive materials because of their adaptable and designable topologies, ordered crystal structures, tunable porosity for mass transfers, large specific surface areas for sufficient interactions, high-density active sites, and favorable stability [32]. 3D iron and nitrogen-doped MIL 53@CNT/graphene composite are highly active and stable in both alkaline and acidic electrolytes in catalyzing the ORR and has a higher open voltage (1.414 V) and specific capacity (637.4 mAh/g) than the commercial Pt/C catalyst in Zn-air battery [33]. In recent years, to promote the surface properties of electrodes, MOFs have been used as a template. 3D zeolitic-imidazolate framework prepared using Zn^{2+} cluster and methylimidazole ligand has a high surface area, porosity, and flexibility. ZIF-derived cobalt-nitrogen doped carbon (Co-N_x/C) nanorod has superior electrocatalytic activity and high stability for oxygen evolution and reduction reactions and shows the high performance of Zn-air batteries with high cycling stability even at a high current density (853.12 Wh/kg which is about 78.6% of the theoretical energy density (1086 Wh/kg)) [34]. In another study, 3D Co_4N/CNW/CC electrode was prepared via pyrolyze of ZIF-67 on polypyrrole (PPy) nanofibers network rooted on carbon cloths as shown in Figure 13.5a [35]. Both the 3D shape of polypyrrole nanofiber and ZIF-67 nanoparticles were found in SEM images (Figure 13.5b–d). It shows high OER activity (Figure 13.5e), OER kinetic as to TAFEL plot (Figure 13.5f), and ORR activity and stability (Figure 13.5g). The flexibility of the 3D structure increases the surface area and active sites and enhances the electrocatalytic performance. From galvanostatic discharge and charge cyclic curves at different current densities (Figure 13.5h), it can be said that the obtained electrode shows excellent rechargeable performance and at higher densities, it has still stable cycles for flexible zinc-air battery.

Zou and coworkers prepared the first single crystal 1D MOF 74 nanotube with cobalt acetate and 2,5-dihydroxy terephthalic acid using a recrystallization method with an 848.8 m^2/g surface area and used it as a precursor to obtaining 3D nitrogen and cobalt doped carbon nanotube on carbon nanofibers (179.3 m^2/g surface area). When the cobalt and nitrogen enhance the electrocatalytic oxidation-reduction activity, carbon nanotube and nanofiber increase the electrical conductivity [36]. The modifications on the electrocatalyst surface improve the active sites and when compared to the

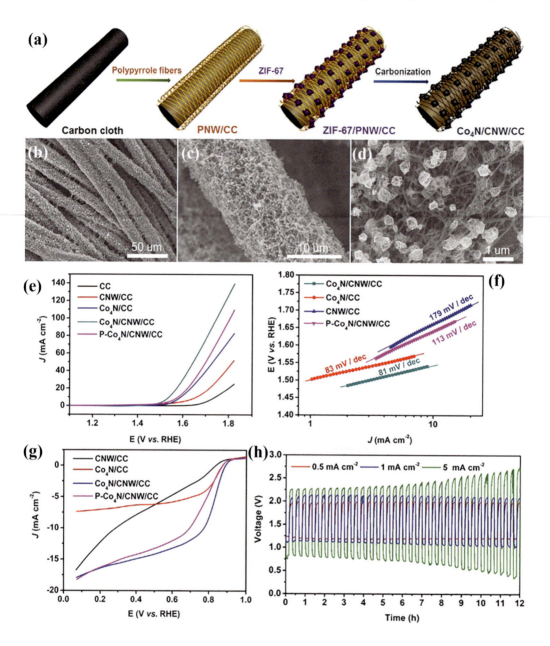

FIGURE 13.5 (a) Schematic representation of the formation of $Co_4N/CNW/CC$ electrode, (b–d) SEM images of the electrode in different magnitude, (e) OER polarization curves, (f) TAFEL plots, (g) ORR polarization curves, (h) Galvanostatic discharge-charge cycling curves at 0.5, 1, and 5 mA/cm². (Adapted with permission [35]. copyright (2016), American Chemical Society.)

other carbon templates like graphene, NCo@CNT-NF shows higher electrocatalytic performance in ORR reaction with superior durability. When used as a cathode for zinc-air battery, show high power density (220 mW/cm²) with excellent stability. Yi and coworkers synthesized cobalt-based metal-organic framework using 3,5-dimethyl-4-carboxypyrazol as ligand via the solvothermal method, mixed with carbon black and derived to form $Co-CoO-Co_3O_4$/N-doped carbon electrocatalyst. MOF is used as an excellent platform and MOF-derived material has a 173.7 m²/g surface area. The obtained material is highly efficient for OER and ORR. When compared 20 wt. % the platinum/carbon, it shows higher power and current density and is devised as a high-performance cathode electrocatalyst for zinc-air battery [37].

13.4 CONCLUSIONS AND PERSPECTIVES FOR FUTURE APPLICATIONS

Increasing energy needs because of the increase in the world population, the rapid depletion of fossil fuel resources, and the search for alternatives to sustainable energy sources is a process that is constantly increasing day by day. CO_2 emissions and the resulting air pollution can be reduced by using zero-emission electric vehicles (EV) instead of an internal combustion engine. In the energy network of the future, power storage systems are one of the indispensable devices for buffering unregulated energy production and renewable energy sources. Low-cost batteries with high power density, long shelf life, and environmental friendliness are urgently needed. Metal-air batteries, whose cathode reaction is by reduction of oxygen, have high energy density since the cathode active material is not stored in the battery and is supplied from the air atmosphere. Metal-air batteries with ultra-high energy density, such as Zn and Li-air batteries, have shown considerable promise for future large-scale applications as a next-generation electrochemical storage device. In the 2030s, it is foreseen that lithium-air batteries will be used as an energy storage system in electric vehicles. Li-air batteries can theoretically store 10,000 W/hr more energy per kilogram. The theoretical energy capacity that can be stored in lithium-air batteries is 10 times greater than that of commercially used lithium-ion batteries. Although lithium-air batteries are promising systems, they still contain many aspects that need to be optimized. These problems are the inability of the battery cell to maintain its stability at high cycles, the degradation of lithium by the penetration of moisture in the air into the cell, and the costs of consumables and equipment. Among the problems encountered, the main problem that prevents the commercialization of Li-air batteries is the cyclic instability of the battery, which occurs due to the aggressive character of oxygen. This is caused by using an oxygen-sensitive and often very low-stability electrolyte. Li-air batteries benefit from developments in relatively older battery systems, as they are a current technology and need a great deal of improvement. Although this situation mostly creates positive results, differences have occurred in terms of electrolyte performances due to differences in mechanisms. The efficiency of air batteries depends on factors that need to be controlled, such as catalyst selection, electrode microstructure, electrolyte components, and oxygen permeability. Solutions are developed for all these factors, and efforts have been made to extend the cycle life of air batteries. Current primary Zn-air batteries have an energy density approaching 500 Wh/kg. With the recent development of better ORR catalysts and optimization of the cathode architecture, the energy and power density is expected to increase in the near future. If all of the current technical difficulties can be successfully overcome, anhydrous metal-air batteries with high theoretical energy density may provide a long-term solution for large-scale electrochemical energy storage. The production of metal-air batteries has been crucial for both practical and theoretical reasons. Despite their current low performance, their enormous potential and promising nature are undeniable facts.

REFERENCES

[1] N. Chawla, Recent advances in air-battery chemistries, Materials Today Chemistry. 12 (2019) 324–331.
[2] H.-F. Wang, Q. Xu, Materials design for rechargeable metal-air batteries, Matter. 1(3) (2019) 565–595.
[3] M.M. Thackeray, Manganese oxides for lithium batteries, Progress in Solid State Chemistry. 25(1–2) (1997) 1–71.
[4] Q. Liu, Z. Chang, Z. Li, X. Zhang, Flexible metal-air batteries: Progress, challenges, and perspectives, Small Methods. 2(2) (2018) 1700231.
[5] X. Zhang, X.-G. Wang, Z. Xie, Z. Zhou, Recent progress in rechargeable alkali metal-air batteries, Green Energy & Environment. 1(1) (2016) 4–17.

[6] Y. Li, J. Lu, Metal-air batteries: Will they be the future electrochemical energy storage device of choice?, ACS Energy Letters. 2(6) (2017) 1370–1377.

[7] A.G. Olabi, E.T. Sayed, T. Wilberforce, A. Jamal, A.H. Alami, K. Elsaid, S.M.A. Rahman, S.K. Shah, M.A. Abdelkareem, Metal-air batteries—a review, Energies. 14(21) (2021) 7373.

[8] R. Gao, Z. Yang, L. Zheng, L. Gu, L. Liu, Y. Lee, Z. Hu, X. Liu, Enhancing the catalytic activity of Co_3O_4 for $Li-O_2$ batteries through the synergy of surface/interface/doping engineering, ACS Catalysis. 8(3) (2018) 1955–1963.

[9] M. Wu, G. Zhang, M. Wu, J. Prakash, S. Sun, Rational design of multifunctional air electrodes for rechargeable Zn-Air batteries: Recent progress and future perspectives, Energy Storage Materials. 21 (2019) 253–286.

[10] Y. Huang, Y. Wang, C. Tang, J. Wang, Q. Zhang, Y. Wang, J. Zhang, Atomic modulation and structure design of carbons for bifunctional electrocatalysis in metal-air batteries, Advanced Materials. 31(13) (2019) 1803800.

[11] Y. Liu, Q. Sun, W. Li, K.R. Adair, J. Li, X. Sun, A comprehensive review on recent progress in aluminum-air batteries, Green Energy & Environment. 2(3) (2017) 246–277.

[12] Z. Lin, G.H. Waller, Y. Liu, M. Liu, C.-P. Wong, 3D nitrogen-doped graphene prepared by pyrolysis of graphene oxide with polypyrrole for electrocatalysis of oxygen reduction reaction, Nano Energy. 2(2) (2013) 241–248.

[13] Y.-J. Wang, B. Fang, D. Zhang, A. Li, D.P. Wilkinson, A. Ignaszak, L. Zhang, J. Zhang, A review of carbon-composited materials as air-electrode bifunctional electrocatalysts for metal—air batteries, Electrochemical Energy Reviews. 1(1) (2018) 1–34.

[14] M. He, P. Zhang, L. Liu, B. Liu, S. Xu, Hierarchical porous nitrogen doped three-dimensional graphene as a free-standing cathode for rechargeable lithium-oxygen batteries, Electrochimica Acta. 191 (2016) 90–97.

[15] Y. Xue, D. Yu, L. Dai, R. Wang, D. Li, A. Roy, F. Lu, H. Chen, Y. Liu, J. Qu, Three-dimensional B, N-doped graphene foam as a metal-free catalyst for oxygen reduction reaction, Physical Chemistry Chemical Physics. 15(29) (2013) 12220–12226.

[16] H.J. Qiu, P. Du, K. Hu, J. Gao, H. Li, P. Liu, T. Ina, K. Ohara, Y. Ito, M. Chen, Metal and nonmetal codoped 3D nanoporous graphene for efficient bifunctional electrocatalysis and rechargeable Zn-air batteries, Advanced Materials. 31(19) (2019) 1900843.

[17] S. Chen, J. Duan, M. Jaroniec, S.Z. Qiao, Nitrogen and oxygen dual-doped carbon hydrogel film as a substrate-free electrode for highly efficient oxygen evolution reaction, Advanced Materials. 26(18) (2014) 2925–2930.

[18] J. Zhang, Z. Zhao, Z. Xia, L. Dai, A metal-free bifunctional electrocatalyst for oxygen reduction and oxygen evolution reactions, Nature Nanotechnology. 10(5) (2015) 444–452.

[19] Z. Liu, H. Nie, Z. Yang, J. Zhang, Z. Jin, Y. Lu, Z. Xiao, S. Huang, Sulfur-nitrogen co-doped three-dimensional carbon foams with hierarchical pore structures as efficient metal-free electrocatalysts for oxygen reduction reactions, Nanoscale. 5(8) (2013) 3283–3288.

[20] S.S. Shinde, J.-Y. Yu, J.-W. Song, Y.-H. Nam, D.-H. Kim, J.-H. Lee, Highly active and durable carbon nitride fibers as metal-free bifunctional oxygen electrodes for flexible Zn-air batteries, Nanoscale Horizons. 2(6) (2017) 333–341.

[21] M.G. Park, D.U. Lee, M.H. Seo, Z.P. Cano, Z. Chen, 3D ordered mesoporous bifunctional oxygen catalyst for electrically rechargeable zinc-air batteries, Small. 12(20) (2016) 2707–2714.

[22] D.U. Lee, J.Y. Choi, K. Feng, H.W. Park, Z. Chen, Advanced extremely durable 3D bifunctional air electrodes for rechargeable zinc-air batteries, Advanced Energy Materials. 4(6) (2014) 1301389.

[23] X. Li, N. Xu, H. Li, M. Wang, L. Zhang, J. Qiao, 3D hollow sphere Co_3O_4/MnO_2-CNTs: Its high-performance bi-functional cathode catalysis and application in rechargeable zinc-air battery, Green Energy & Environment. 2(3) (2017) 316–328.

[24] W. Zou, J. Xiang, H. Tang, Three-dimensional nano-framework $CoP/Co_2P/Co_3O_4$ heterojunction as a trifunctional electrocatalyst for metal—air battery and water splitting, New Journal of Chemistry. 46(18) (2022) 8786–8793.

[25] J.-J. Xu, Z.-L. Wang, D. Xu, F.-Z. Meng, X.-B. Zhang, 3D ordered macroporous $LaFeO_3$ as efficient electrocatalyst for $Li-O_2$ batteries with enhanced rate capability and cyclic performance, Energy & Environmental Science. 7(7) (2014) 2213–2219.

[26] J.F. Parker, C.N. Chervin, E.S. Nelson, D.R. Rolison, J.W. Long, Wiring zinc in three dimensions rewrites battery performance—dendrite-free cycling, Energy & Environmental Science. 7(3) (2014) 1117–1124.

[27] Z. Khan, N. Parveen, S.A. Ansari, S. Senthilkumar, S. Park, Y. Kim, M.H. Cho, H. Ko, Three-dimensional SnS2 nanopetals for hybrid sodium-air batteries, Electrochimica Acta. 257 (2017) 328–334.

[28] S. Hyun, S. Shanmugam, Hierarchical nickel—cobalt dichalcogenide nanostructure as an efficient electrocatalyst for oxygen evolution reaction and a zn—air battery, ACS Omega. 3(8) (2018) 8621–8630.

[29] W. Liu, D. Zheng, L. Zhang, R. Yin, X. Xu, W. Shi, F. Wu, X. Cao, X. Lu, Bioinspired interfacial engineering of a CoSe 2 decorated carbon framework cathode towards temperature-tolerant and flexible Zn-air batteries, Nanoscale. 13(5) (2021) 3019–3026.

[30] J.-J. Xu, Z.-W. Chang, Y.-B. Yin, X.-B. Zhang, Nanoengineered ultralight and robust all-metal cathode for high-capacity, stable lithium-oxygen batteries, ACS central science. 3(6) (2017) 598–604.

[31] C.C. Hou, L. Zou, Y. Wang, Q. Xu, MOF-mediated fabrication of a porous 3D superstructure of carbon nanosheets decorated with ultrafine cobalt phosphide nanoparticles for efficient electrocatalysis and zinc-air batteries, Angewandte Chemie International Edition. 59(48) (2020) 21360–21366.

[32] D. Ozer, E.T. Tunca, Design and construction of MOF nanomaterials, in Metal-organic framework-based nanomaterials for energy conversion and storage (pp. 35–65), Elsevier, 2022.

[33] W. Yang, Y. Zhang, X. Liu, L. Chen, J. Jia, In situ formed Fe-N doped metal organic framework@ carbon nanotubes/graphene hybrids for a rechargeable Zn-air battery, Chemical Communications. 53(96) (2017) 12934–12937.

[34] I.S. Amiinu, X. Liu, Z. Pu, W. Li, Q. Li, J. Zhang, H. Tang, H. Zhang, S. Mu, From 3D ZIF nanocrystals to Co—Nx/C nanorod array electrocatalysts for ORR, OER, and Zn—air batteries, Advanced Functional Materials. 28(5) (2018) 1704638.

[35] F. Meng, H. Zhong, D. Bao, J. Yan, X. Zhang, In situ coupling of strung Co4N and intertwined N—C fibers toward free-standing bifunctional cathode for robust, efficient, and flexible Zn—air batteries, Journal of the American Chemical Society. 138(32) (2016) 10226–10231.

[36] L. Zou, C.-C. Hou, Z. Liu, H. Pang, Q. Xu, Superlong single-crystal metal—organic framework nanotubes, Journal of the American Chemical Society. 140(45) (2018) 15393–15401.

[37] X. Yi, X. He, F. Yin, B. Chen, G. Li, H. Yin, Co-CoO-Co3O4/N-doped carbon derived from metal-organic framework: The addition of carbon black for boosting oxygen electrocatalysis and Zn-Air battery, Electrochimica Acta. 295 (2019) 966–977.

14 Carbon-Based Electrocatalysts for Metal-Air Batteries

Zongge Li and Guoxin Zhang

CONTENTS

14.1 Brief Introduction of Aqueous MABs 195
14.2 Mechanism of Oxygen Catalysis Involving MABs Cathode 196
14.3 Metal-Free Heteroatom-Doped Carbon Materials as MABs Cathode 198
14.4 Atomically Dispersed Metal-Nitrogen-Carbon Materials as MABs Cathodes 200
14.5 Flexible Integrated Carbon Fiber Films as MABs Cathodes 203
14.6 Conclusions and Perspectives 204
References 205

14.1 BRIEF INTRODUCTION OF AQUEOUS MABs

As a promising energy device in the new generation of electronic products, electrical energy storage, and electric transportation, metal-air batteries (MABs) are one of the representatives of the new generation of green secondary batteries, with low costs, no toxicity, no pollution, high specific energy, etc. [1]. The negative electrode of MABs is widely selected, such as zinc, magnesium, aluminum, etc., and the use of an aqueous electrolyte as a medium can maintain the safety of battery operation [2]. As summarized in Table 14.1, the theoretical specific energy of zinc-air batteries (ZnABs) is as high as 1086 Wh/kg, and the current practical utilization can reach 200–300 Wh/kg, and there is still great potential for achieving improved energy density [3]. In addition, it also has the advantages of stable discharge voltage, high safety, and light weight. The magnesium-air batteries (MgABs) have high output voltage and energy density (6800 Wh/kg), especially the abundant reserves of magnesium in the crust and seawater render great advantages [4]. The theoretical specific energy of aluminum-air batteries (AlABs) can reach 8100 Wh/kg, and it also has an extremely high volumetric energy density, second only to lithium-air batteries (LiABs) (11429 Wh/kg) among common metals [1]. Comparatively, the specific energy of MABs is 5–8 times higher than that of lead-acid batteries and 3–5 times higher than that of nickel-hydrogen batteries, while the unit cost is about the same as that of lead-a¼ batteries, which is 1/4 of the cost of nickel-hydrogen batteries. In addition, Al is inexpensive and widely obtainable as a negative electrode material for MABs, which is of great attractiveness for the new energy vehicle industry [5].

The cathode electrode of MABs is the key to regulating the charge-discharge reaction processes, mainly involving electrochemical oxygen reduction (ORR) and oxygen evolution reactions (OER) (if rechargeable) [6]. The oxygen electrocatalysis is mainly occurred at a three-phase interface (TPI) constructed by the solid catalyst, liquid electrolyte, and air channel; therefore, to achieve an efficient oxygen electrocatalytic process, it requires high electron conductivity and high electrocatalytic properties of electrocatalyst, and smooth transportation of ions between bulk solution and TPI, and oxygen gas between bulk air and TPI [7].

Generally speaking, a MABs cathode is mainly composed of conductive support, electrocatalyst, gas-diffusion layer (GDL), and waterproof breathable layer. Therefore, it is crucial to use electrocatalysts with high catalytic activity to improve battery performance [8]. At present, the electrocatalysts used for the assembly of MABs cathode can be mainly divided into carbon materials, noble

TABLE 14.1
The Basic Information of Al/Mg/Zn-Air Batteries

Type of MABs	Theoretical Voltage (V)	Practical Voltage (V)	Theoretical Energy (kWh/kg)	Practical Energy (kWh/kg)
Al	2.71	1.2–1.6	8.1	0.3–0.4
Mg	3.09	1.2–1.4	6.8	0.3–0.4
Zn	1.66	0.9–1.2	1.3	0.2–0.3

metals and their compounds, transition metal oxy/nitrides, and other electrocatalysts. Although noble metal (Pt/Ru/Pd, *etc.*) materials possess unprecedented ORR and OER activity, their high costs severely limit the development of their large-scale applications [9]. There have been numerous reports demonstrating that carbon-based electrocatalysts, of low price, wide accessibility, and high electrical conductivity, can be gifted with rich electrochemical properties by decorating with alien components such as metal-free heteroatoms, metal atoms, or crystalline metal compounds. Their electrochemical behaviors can be further improved by regulating structures and porosities [10].

In terms of efficient functionalization of carbon-based electrocatalysts, three main categories can be sorted: metal-free heteroatom doping, atomic metal doping, and the compositing of nanocrystals [11].

First, introducing heteroatoms (N, O, F, S, *etc.*) of different atomic sizes and electronegativity into the carbon substrate induces beneficial charge transfer between carbon atoms and heteroatoms, enabling the local charge re-arrangement [12]. This leads to apparent changes in physiochemical properties such as electronic states, geometry, and wettability, which in turn improves the electrocatalytic performance of doped carbon materials [13].

Second, atomic metal stabilized by nitrogen-doped carbon materials represents another type of hot electrocatalysts for promoting ORR and OER in MABs; the active sites in this type of material can be identified as M-Nx, and it has been confirmed that the interaction between the coordinated metal atom and ligands is very strong, leading to intensive charge transfer and geometric reconstruction [14]. At present, atomic non-noble transition metal-doped carbon electrocatalysts have achieved fruitful advancements in electrode material synthesis and theoretical research studies for MABs. Usually, they have active sites of the same atomic structure, including the coordination structure and geometric configurations, forming strong links to the molecular catalysts in heterogeneous catalysis [15]. This ensures high activity and selectivity simultaneously, suppressing harmful side reactions that may occur in the nanocrystal cases.

Third, supporting non-precious metal nanocrystals on a carbon substrate is also hot due to their controllable sizes, high site density, and well-defined crystalline faces that enable direct investigation of structure-activity relationship and exploration of high-performance applications [16].

In the following section, we will discuss the design principle and synthesis of metal-free doped carbon and atomically dispersed metal-nitrogen-carbon materials for the assembly of aqueous MABs.

14.2 MECHANISM OF OXYGEN CATALYSIS INVOLVING MABs CATHODE

MABs have broad application prospects in the storage and conversion of renewable electrical energy. On the air side of a MAB, the ORR occurs during discharge, and the OER occurs during charge. However, the kinetic processes of the ORR and OER reactions are slow, thus limiting the practical application of MABs. Therefore, the development of high-performance ORR and OER electrocatalysts is particularly critical for the development of MABs, and disclosing their reaction mechanisms at the same time poses important guiding significance for the implantation of MABs [17]. Due to the high complexity of ORR reaction kinetics, the exploration of its atomic-scale mechanism is still developing. There are currently two recognized ORR reaction pathways: the 4e⁻ pathway and the 2e⁻ pathway [18].

The 4e⁻ path: In an acidic medium, direct reduction generates H_2O; in an alkaline medium, OH^- is generated. The specific reaction equation is as follows:

$$\text{Acidic: } O_2 + 4H^+ + 4e^- \rightarrow 2H_2O \text{ (1.229 V vs. NHE)} \tag{1}$$

$$\text{Alkaline: } O_2 + 2H_2O + 4e^- \rightarrow 4OH^- \text{ (0.401 V vs. NHE)} \tag{2}$$

The reaction equation for the 2e⁻ path:

$$\text{Acidic: } O_2 + 2H^+ + 2e^- \rightarrow H_2O_2 \text{ (0.670 V vs. NHE)} \tag{3}$$

$$H_2O_2 + 2H^+ + 2e^- \rightarrow 2H_2O \text{ (1.770 V vs. NHE)} \tag{4}$$

$$\text{Alkaline: } O_2 + H_2O + 2e^- \rightarrow HO_2^- + OH^- \text{ (−0.065 V vs. NHE)} \tag{5}$$

$$HO_2^- + H_2O + 2e^- \rightarrow 3OH^- \text{ (0.867 V vs. NHE)} \tag{6}$$

To understand the reaction mechanism more intuitively, the preceding process can be transformed into a flow chart. The 2e⁻ pathway of ORR: O_2 molecules are adsorbed on the active site (*O_2), which occurs when two proton-coupled electron transfer reaction generates intermediates *OOH and *H_2O_2, respectively, which are adsorbed on the active site, and then *H_2O_2 is desorbed from the active site to obtain H_2O_2. On the other hand, the 4e⁻ pathway undergoes four proton-coupled electron transfer reactions to generate intermediates *OOH, respectively, and *O and *OH are converted at the active site. Therefore, to obtain a highly active 4e⁻ ORR catalyst, it is necessary to suppress the occurrence of the H_2O_2 reaction pathway. In addition, due to thermodynamic instability, the resulting adsorbed form of *H_2O_2 may undergo electrochemical dissociation. This indicates the dissociation of *H_2O_2 into two *OH radicals, which are converted to H_2O through proton coupling, and this dissociation is believed to occur simultaneously. The *H_2O_2 dissociation involves two consecutive proton-coupling reactions. At the same time, the generated H_2O_2 can be decomposed by disproportionation reaction $H_2O_2 \rightarrow 2H_2O + O_2$, or can be re-adsorbed by active sites for electrochemical decomposition [19].

From the reaction mechanism equation, it can be seen that the catalytic activity of ORR is independently completed on the atomic-level active sites, and the higher the site density, the stronger the catalytic capability [20]. Existing studies have shown that when nanoparticles or clusters undergo electrocatalytic reactions, only surface atoms act as active sites, while the embedded metal atoms are not available, resulting in low atom utilization efficiency. Atomically dispersed non-noble metal catalysts disperse metal atoms on conductive substrates (mainly carbon materials) and achieve excellent electrocatalytic performance by using heteroatoms to adjust the coordination configurations, and the loading amount, or the types of conductive substrates [21].

The mechanism of OER in rechargeable MABs is generally considered to be the inverse process of ORR, and this reaction typically involves four-electron transfer. The overall reaction equation is $2H_2O \rightarrow 2H_2 + O_2$. The slow reaction kinetics of OER results in a high overpotential, which is the main limiting factor for charging a MABs cell [22]. The reaction mechanism is as follows:

$$\text{Acidic: } M + H_2O_{(l)} \rightarrow M-OH + H^+ + e^- \tag{7}$$

$$M-OH + OH^- \rightarrow M-O + H_2O_{(l)} + e^- \tag{8}$$

$$M-O + H_2O_{(l)} \rightarrow M-OOH + H^+ + e^- \quad (9)$$

$$M-OOH + H_2O_{(l)} \rightarrow M + O_{2(g)} + H^+ + e^- \quad (10)$$

$$\text{Alkaline: } M + OH^- \rightarrow M-OH \quad (11)$$

$$M-OH + OH^- \rightarrow M-O + H_2O_{(l)} \quad (12)$$

$$M-O + OH^- \rightarrow M-OOH + e^- \quad (13)$$

$$M-OOH + OH^- \rightarrow M + O_2 + H_2O_{(l)} \quad (14)$$

In an acidic or alkaline medium, the intermediate products are all in the form of OH*, O*, and OOH*. The difference is that the catalytic result of each step on the surface of the catalyst in an acidic medium interacts with H_2O, while in an alkaline medium, it is combined with OH^-. Therefore, the O element in acid comes from H_2O in solution, while O in basic comes from OH^- ionized from the solution. In the DFT calculation, the overpotentials of the intermediate-state OH* and OOH* are one of the important credentials to measure the catalyst activity [23]. OER does not require high electrical conductivity of catalyst materials, and the current precious metals RuO_2 and IrO_2 have excellent catalytic performance. Similar to ORR, the atomically dispersed sites can complete catalytic reactions with high efficiency and high throughput [24]. Based on the aforementioned reaction mechanism, bifunctional electrocatalysts have been widely explored and studied. Such catalysts are generally used as cathode materials for rechargeable MABs.

14.3 METAL-FREE HETEROATOM-DOPED CARBON MATERIALS AS MABs CATHODE

Conventional carbon materials, such as conductive carbon black, carbon nanotubes, carbon nanosheets, and graphene, generally exhibit inert ORR and OER due to the lack of active sites for activating O_2 and its derived intermediates. Recently, enormous efforts including heteroatom doping, surface modification, and defect engineering have been devoted to the functionalization of carbon materials. Among them, the most common and effective strategy is heteroatom doping [25]. For instance, in the process of materials synthesis, the dopant is introduced into the framework of the carbon material precursor in situ, and the unit heteroatom or multi-element heteroatom-doped carbon can be obtained by post-processing.

So far, there have been many reports that non-metallic atoms (N, B, F, P, and S, *etc.*) have been successfully doped into carbon-based frameworks to improve their ORR/OER performance [12]. Given the differences in atomic size, binding state, and electronegativity between carbon atoms and heteroatoms, the charge and spin distribution can be optimized by tuning the π-electron delocalization of nearby carbon atoms by heteroatom dopants, thereby endowing carbon materials with highly enhanced ORR/OER catalytic activity for MABs. Among the examples of heteroatom doping, the most classic one belongs to N-doped carbon [26]. Since it has an atomic size similar to carbon atoms, this allows it to easily replace carbon atoms without significant lattice distortion. The N-doped carbon materials can be bestowed with electronic sp^2 charge distribution similar to that of aromatic sp^2 carbon, making them the most widely studied non-metal-doped carbon electrocatalysts.

In 2009, Dai's group for the first time reported doped carbon nanomaterials for ORR, uncovering that the nitrogen-doped carbon nanotube arrays can be developed into very good electrocatalysts comparable to Pt [27]. It is confirmed that the C atoms around the pyrrolic-like N species have much higher positive charges than the C atoms around the pyridinic-like N species. It is known that the N dopants in carbon materials mainly exist in four different forms, namely graphitic N, pyridine N, pyrrolic N, and oxidized N. These N dopants are usually co-present in carbon materials, and their content largely depends on the chemical properties of the precursors and the pyrolysis temperature used for the preparation of doped carbon materials. Subsequent efforts have been made to unravel the origin of the catalytic activity of N-doped carbon materials for use in oxygen catalysis before further generalization of N-doped carbon for application in MABs. For instance, Yang and co-workers increased the doping amount of N to 9.59 wt% and optimized the ratio of graphitic N and pyridinic N in 3D N-doped reduced graphene, obtaining very good ORR performance compared with these of same loading of commercial Pt/C (20 wt%).

A hierarchical porous N-doped carbon material was prepared by Zhang et al. [28]. As shown in Figure 14.1(A-B), the as-made carbon material annealed at 900 °C exhibited better performance compared to 20 wt% Pt/C: 10 mV higher onset potential and 19.0 mV higher half-wave potential. The reaction kinetics were investigated by Tafel plots, and the slopes of Pt and PDC-900 rapidly increased to 108 mV/dec at high current densities, indicating that the $O_2 \rightarrow OOH^*$ process was a rate-determining step (Figure 14.1C). According to the EIS in Figure 14.4D, PDC-900 exhibits the same first semicircle (range I) and the same short 45° phase-shift segment (range II) as Pt in the high-frequency region, implying a fast charge-exchange rate for O_2 conversion. Huang et al. developed a simple method to realize N doping in carbon materials in which one carbon atom in each benzene ring of graphene (GDY) was substituted by pyridinic N (PyN-GDY) [26]. The PyN-GDY-assembled ZnABs exhibited superior performance relative to Pt/C. Density functional theory calculations unraveled the active sites in PyN-GDY for ORR to be the acetylenic carbon atom close to the pyridinic N.

Compared with ORR-active N-doped carbon catalysts, fewer mechanistic studies have been reported on the OER capability of N-doped carbon. Nakanishi and co-workers are the first to report

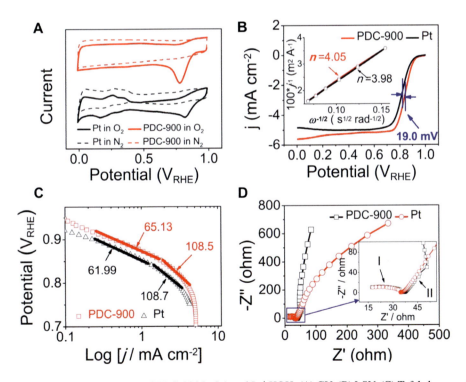

FIGURE 14.1 ORR performance of PDC-900 in 0.1 mol L^{-1} KOH. (A) CV, (B) LSV, (C) Tafel slope, and (D) EIS of PDC-900 relative to 20 wt% Pt/C. LSV curves in (B) were collected at the rotation rate of 1600 rpm. The inset of (B) shows the K-L plot of PDC-900 and 20 wt% Pt/C (0.45 V RHE). (Adapted with permission [29]. Copyright (2016), Elsevier.)

an N-doped carbon material capable of efficiently catalyzing OER in alkaline media [29]. At an OER current of 10 mA/cm^2 ($E_{j=10}$), the N-rich polymer-derived catalyst exhibits a low overpotential of 0.38 V, which is comparable to the Ir-based catalysts. Most rechargeable MABs using N-doped carbon suffer from large overpotentials during the charging process (OER), requiring voltages over 2.0 V to fully charge the battery. Therefore, the high overpotential of OER is considered to be one of the main challenges for the ZnABs application of N-doped carbon cathodes. In principle, all the alien atoms that are more electronegative than C can act as efficient dopants to activate carbon materials into metal-free catalysts. The aforementioned basic research further promotes the application of N/X dual-doped carbon in bifunctional oxygen electrocatalysis and rechargeable MABs. Dai and his colleagues developed an efficient method to fabricate bifunctional N, P-co-doped carbon foam through the pyrolysis of polyaniline/phytic acid aerogel [30]. The N, P-co-doped carbon foam-assembled ZnABs exhibited an open-circuit potential of 1.48 V, a specific capacity of 735 mAh/g_{Zn} (corresponding to an energy density of 835 Wh/kg_{Zn}), a peak power density of 55 mW/cm^2, and can be stable operated for over 240 h after mechanical recharging. Its assembled two-electrode rechargeable ZnABs could be stably cycled for 180 cycles at 2 mA/cm^2. Density functional theory calculations suggested that the N, P co-doping, and graphene edge effects are contributing to the excellent bifunctional electrocatalytic activity.

Furthermore, defective structures including edge defects have recently been shown to be active sites for oxygen electrocatalysis, which can also tune the electronic and chemical properties of carbon materials. Edge defects in carbon atoms provide a small fraction of active unpaired π electrons, effectively facilitating electron transfer. Yu's group prepared an edge-rich N-doped carbon material (termed as NKCNPs-900) by alkali activation and N-doping, which exhibited bifunctional oxygen electrocatalytic activity in alkaline media [31]. The N doping near the edge sites further contributes to the bifunctional electrocatalytic capability, thereby enhancing the ORR and OER activities and providing a relatively low ΔE, *i.e.*, $E_{j=10}$, and the half-wave potential ($E_{1/2}$) of the ORR is 0.92 V. As a result, a good rechargeable MAB performance was achieved with a peak power density of 131.4 mW/cm^2, an energy density of 889.0 Wh/kg, and a stable charge-discharge cycle of 575 times.

14.4 ATOMICALLY DISPERSED METAL-NITROGEN-CARBON MATERIALS AS MABs CATHODES

With the improvement of characterization instruments and the deepening of the understanding of nanomaterials, the atomic structures and properties of active components in nanomaterials can be identified and visualized at atomic scales such as by AC-TEM HAADF-STEM measurement and XAS analysis. It has been confirmed that when the size of active components is further reduced to the atomic level, their physical and chemical properties can be dramatically changed, leading to largely enhanced catalytic activity and selectivity. Also, the strong interaction between a single metal atom and the surrounding ligands renders a catalytically active site [32]. Single-atom catalysts can be dated back to research conducted by Zhang and his co-workers in 2011; they reported the synthesis of a single platinum atom (Pt_1/FeO_x) supported by iron oxide, and the activity of CO oxidation and selective oxidation at low temperature was sharply improved by at least 1–2 orders of magnitude compared to particulate Pt [33].

The empowered active metal by atomic dispersion also benefits the downward costs of electrocatalysts for both noble metal and non-precious metal catalysts. So far, noble metal catalysts are still the most efficient catalysts in various electrocatalytic processes, such as Pt, Pd, Ir, and Ru, but their low reserves and high prices prevent them from wide implantation. The development of single-atom catalysts reduces the loading amount and improves the utilization efficiency of noble metal components on the one hand, and on the other hand, dramatically improves the catalytic activity of non-precious metal components to a new horizon that is comparable with noble metals. The much higher unsaturation of atomically dispersed metal on carbon supports allows the coordinated metal atoms to be finely tuned by adjusting its embedded coordination microenvironment, thus regulating the electronic structure and the adsorption/desorption properties of active sites for reactants and intermediate products, and eventually improving the energy conversion efficiency of various catalytic reactions such as ORR and OER, as well as the energy storage efficiency of MABs.

Graphitic carbon materials with rich N anchors have commonly used substrate for the stabilization of atomic metals, which is normally known in a M-Nx structure, where M = Co, Fe, Ni, Mn, Cu, Zn, *etc.*, and x = 2, 3, 4, or 5. Therefore, enhancing the catalytic activity of atomic metal-nitrogen-carbon (M-N-C) materials can be achieved by regulating the coordination number of N atoms (including axial coordination of Cl, N, and O), the introduction of metallic environment, the adjacent heteroatom doping (such as S, B, and P), etc. [34]. For instance, Zang *et al.* used a single Co atom supported on a porous N-doped carbon material to simultaneously enhance the ORR activity and reduce the OER overpotential [35]. Benefiting from the mechanical flexibility of this composite and its bifunctional electrocatalytic activity, this single-atom Co catalyst exhibits promising performance in solid-state ZnABs, exhibiting a high open circuit potential of 1.411 V and excellent cycle stability of 125 cycles.

Xu *et al.* reported the design and synthesis of dual-carbon-supported single-atom Co with a precise $Co-N_3-C$ structure. Density function theoretical calculations suggest that the $Co-N_3-C$ is beneficial to lowering the energy barrier of the reaction intermediates, thereby accelerating the ORR kinetics. The prepared CoN_3-C catalyst exhibits unprecedented ORR activity with a half-wave potential of 0.891 V relative to a reversible hydrogen electrode (RHE) and achieves excellent stability under alkaline conditions [36]. A π-electron conjugated structure was constructed between a single Fe site and graphene using intermolecular synergy by Zhang *et al.* The as-prepared pyrolysis-free Fe-NC catalyst (named pfSAC-Fe-0.2) exhibited better performance than the commercial Pt/C catalyst in driving ORR and its assembled ZnABs. The pfSAC-Fe-0.2-assembled ZnABs exhibit better discharge behavior, while Pt/C exhibits higher power density (123.43 mW/cm^2, compared with 113.81 mW/cm^2) and larger specific capacity with a large zinc utilization of 89.3% (732 mAh/g$_{Zn}$ at 100 mA/cm^2) [37].

A feasible method was developed to construct atomic binary Ni-Fe bonds stabilized by multiple N ligands, achieving the optimal synthesis of atomically dispersed NiFe-NC with excellent bifunctional ORR/OER performance [38]. The as-synthesized NiFe-NC contain abundant NiN_3-FeN_3 functional moieties, and the dual-site mode allows more negative charges to reside in local Ni-Fe sites. ORR measurements revealed that the atomic NiFe-NC achieved excellent bifunctional oxygen electrocatalytic performance under alkaline conditions, delivering a high ORR half-wave potential of 0.85 V and a low OER overpotential of 467 mV at 10 mA/cm^2 (Figure 14.2A). Furthermore, its assembled ZnAB offers large power density, robust rate capability, and cycling stability (Figure 14.2B–D). DFT calculations indicated that Fe sites in the structure of $Ni(N_3)(OH)-Fe(N_3)$ were responsible for ORR and OER (Figure 14.2E). Adjacent Ni sites coordinated to OH serve as good mediators for tuning the properties of Fe sites, resulting in efficient formation, association, and dissociation of OOH* [38]. The calculated overpotentials of Fe sites in $Ni(N_3)(OH)-Fe(N_3)$ can be significantly reduced to 0.436 V and 0.420 V for ORR and OER, respectively (Figure 14.2F). The preceding work shows that the activity of atomic metal sites can be effectively regulated by constructing the metal environment.

Defects in carbon structure also play a crucial role in the expression of electrocatalytic activity of M-N-C materials. It has been reported that more defects in carbon benefit the landing of M atoms in the first place, leading to higher MN_x content. This suggests that nitrogen species are preferentially trapped in the defective carbon, improving the anchoring efficiency of atomic metals. In addition, the ORR activity of a central metal atom can be largely tuned by the presence of defective carbon nearby, but the nitrogen element favors the integration of metal atoms into the carbon matrix. For instance, DFT calculations reveal that the Fe atoms trapped at the edges of N-doped divacancies are more active than the basally trapped Fe sites in a common structure of $Fe-N_4$ [39].

The activity of highly dispersed M-Nx sites can be further enhanced by adjusting their microenvironment such as by neighboring adjacent heteroatom doping, metal cluster/particles, etc. For

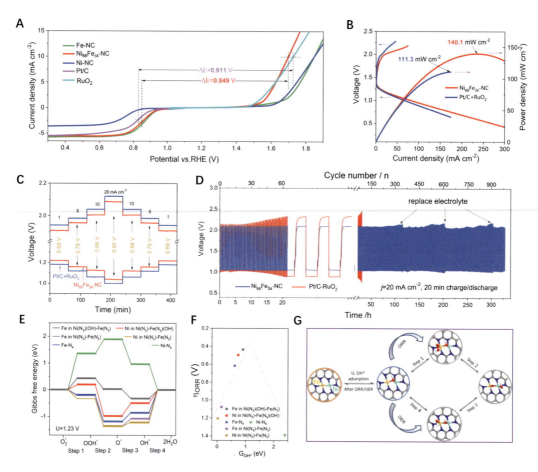

FIGURE 14.2 (A) Combined ORR and OER polarization curves of Ni-NC, $Ni_{66}Fe_{34}$-NC, Fe-NC, and Pt/C-RuO_2 in O_2-saturated 0.1 mol/L KOH solution. Evaluations of $Ni_{66}Fe_{34}$-NC- and Pt/C-RuO_2-assembled ZnABs: (B) charge/discharge polarization curves and power densities; (C) galvanostatic charge and discharge polarization plots at different current densities; (D) long-term cycling performance with a current density of 20.0 mA/cm^2; (E) reaction-free energy diagrams of possible responsible sites for ORR and OER (U = 1.23), including single metal Fe or Ni-N_4-C_n, Ni or Fe on bimetal Ni(N_3)-Fe(N_3)-C_n, and Ni or Fe on Ni(N_3)-Fe(N_3)-C_n with the other metal atom saturated by OH. (F) volcano plots for (e) ORR and (f) OER with a descriptor of ΔG_{OH^*}; (G) proposed ORR and OER mechanism on the Fe site in Ni(N_3)(OH)-Fe(N_3)-C_n. (Adapted with permission [38]. Copyright (2020), Elsevier.)

instance, Jia *et al.* developed a facile method to synthesize atomically dispersed Fe-NC material with sulfur at precise locations on the periphery of Fe (termed Fe-NSC) [40]. X-ray absorption near-edge structural analysis confirms that the central Fe atom is stabilized in a specific configuration of Fe(N_3)(N-C-S). By enabling precisely localized S doping, the electronic structure of the Fe-N_4 moiety can be largely tuned to beneficially tune the absorption/desorption properties of O-containing reactants/intermediates on the active site. Electrochemical measurements show that the Fe-NSC samples exhibit significantly enhanced ORR performance at 0.1 KOH compared to the S-free Fe-NC material (referred to as Fe-NC). In fact, in typical MNC electrocatalysts, the synergistic effect between metal species and N dopants can be largely contributed to the enhancement of ORR/OER activity. However, there has been speculation about how the metal center affects catalytic performance. Metal nanoparticles not only induce the catalytic growth of graphitic carbon but also stimulate the surrounding nitrogen-doped carbon layer, which is thought to be related to the activity of ORR or OER. For instance, Yan *et al.* synthesized a fluffy peony-like FeCo-NC structure based on the controllable thermal transition of atomically dispersed FeCo-NC materials [41]. By precisely controlling the metal dose in the precursor, appropriately sized FeCo alloy nanoparticles are embedded under the atomically dispersed FeCo-NC layer during pyrolysis. The resulting composite, composed of abundant FeCo-Nx sites and beneficial FeCo alloy nanoparticles, provides excellent bifunctional catalytic performance for ORR and OER. Notably, its assembled air cathodes show significantly enhanced performance of ZnABs in terms of specific power, charge/discharge voltage, rate capability, and cycling stability.

14.5 FLEXIBLE INTEGRATED CARBON FIBER FILMS AS MABs CATHODES

Normally, binders such as PTFE, and PVDF are inevitably used for gluing powder electrocatalysts and adhering the catalytic layer to the substrate, which has been verified to potentially promote the synthesis of highly reactive superoxide radicals that may result in severe degradation of electrode performance. To avoid the use of binders, it is necessary to develop free-standing air electrodes by directly growing catalysts on flexible substrates. To optimize the structure and improve the conductivity of the cathode, active materials have been directly grown on flexible substrates by chemical or electrochemical methods, ultimately forming a free-standing cathode. Compared with the common techniques for fabricating air cathodes, such as spraying and printing, the free-standing design has its advantages:

- Easy design of the porous structure that facilitates mass transfer and wettability;
- Absence of polymeric binders enhances the stability of cathode structure;
- Direct contact between the active material and current collector benefits improved electron transport and reaction kinetics.

Wang *et al.* reported the synthesis of free-standing metal-free carbon nanofiber film (named NPCNF-O) by regulating the oxygen species in N-doped porous carbon nanofibers using the electrostatic spinning method [42]. The NPCNF-O film contained a beneficial porous structure inside the carbon nanofibers and rich oxygen-containing functional groups. Density functional theory calculations confirmed that the carboxyl groups regulated the local charge density of the N-doped carbon matrix, optimizing the adsorption energy of the O-containing intermediates, thus improving the ORR activity. The NPCNF-O film showed excellent ORR ($E_{1/2}$ = 0.85 V vs. RHE) and OER ($E_{j=10}$ = 1.556 V vs. RHE) activities; both are superior to the noble metal electrocatalysts. Furthermore, its assembled ZnABs realized a maximum power density of 125.1 mW/cm², a specific capacity of 726 mAh/g, and long-term durability. The NPCNF-O film with excellent flexibility can be further used as a self-supporting air electrode for bendable and durable quasi-solid-state ZnABs.

Free-standing MABs electrodes decorated with atomically dispersed M-N-C active sites can be obtained through electrospinning technology. Wang *et al.* developed a cost-effective and scalable preparation strategy for large-area flexible CNFs films with uniformly distributed diatomic FeN_3-CoN_3 sites [43], as shown in Figure 14.3A–C. Due to the abundant distribution of atomically dispersed highly active FeN_3-CoN_3 sites, the Fe_1Co_1-CNF thin films show excellent bifunctionality for the paired ORR and OER (Figure 14.3D–F). When assembled into liquid ZnAB devices, the Fe_1Co_1-CNF achieved a high specific power of 201.7 mW/cm² and excellent cycling stability. Owing to the good mechanical strength and flexibility of Fe_1Co_1-CNF, a portable ZnAB with excellent shape deformability and stability can be demonstrated, as shown in Figure 14.3G–I, in which the Fe_1Co_1-CNF was used as an integrated self-supporting membrane electrode. These findings provide a facile strategy for fabricating high-yield flexible multifunctional catalytic electrodes. Considering that the electrospinning process can continuously fabricate flexible carbon substrates on a large scale, it is expected to facilitate the development of flexible MABs for future applications.

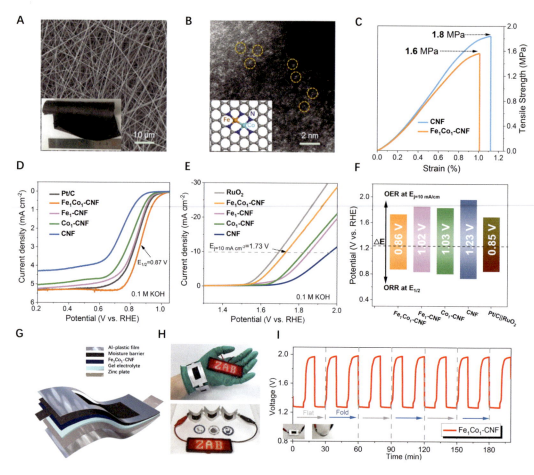

FIGURE 14.3 (A) SEM images of Fe_1Co_1-CNF, inset shows the digital image of integrated Fe_1Co_1-CNF film. (B) Magnified HAADF-STEM image of Fe_1Co_1-CNF, inset shows the proposed structure of $FeCoN_6$ site in Fe_1Co_1-CNF. (C) Mechanical properties of Fe_1Co_1-CNF film. (D) ORR and (E) OER polarization curves of Fe_1Co_1-CNF film with comparison to control samples and Pt/C. (F) Potential gaps between the ORR $E_{1/2}$ (half-wave potential) and the OER $E_{j=10\ mA/cm^2}$. (E) Scheme of rechargeable solid-state flexible ZnABs using Fe_1Co_1-CNF as cathode catalyst layer. (G) Digital photos show the application of wearable and flexible ZABs. (H) The charge-discharge curves of the Fe_1Co_1-CNF-based solid-state ZnABs at 2.0 mA/cm² under alternately folding and releasing conditions. (Adapted with permission [43]. Copyright (2021), Elsevier.)

14.6 CONCLUSIONS AND PERSPECTIVES

Aqueous metal-air batteries (MABs) are promising electrochemical energy storage systems, as they have many fascinating merits including high theoretical voltage, high energy density, and high operation safety. However, the cathodic ORR is sluggish in kinetics, which needs efficient electrocatalysts to facilitate. Currently, the best electrocatalysts for ORR are noble metal Pt/Ir/Ru-based materials, yet their reserves in the earth's crust are very limited and prices are very high. Carbon-based electrocatalysts are representative sort of a cost-effective alternative for the implantations of MABs. In this chapter, we mainly focus on the introduction of two types of carbon-based electrocatalysts for the assembly of MABs, including the metal-free heteroatom-doped carbon materials and the atomically dispersed M-N-C materials. In addition, the optimization of free-standing cathode materials with the decoration of heteroatom doping and atomic metals for flexible MABs is also covered.

Despite their important progress in the advancements in the design and synthesis of cathode materials and their assembled MABs, the challenge of sustainable synthetic routes for large-scale production remains formidable. The customizable chemical composition of the dopant elements and the low site density of M-Nx are the most critical impediments to the implantation of MABs. The rational design and practical synthesis of advanced carbon-based electrocatalysts that are capable of alleviating the slow rate of ORR and OER will surely facilitate the development and application of MABs. Therefore, future research should focus on the following aspects to advance the practical MABs applications.

Increasing the loading of heteroatom and defect species is important to mediate the properties of carbon substrate, help the stabilization of atomic metals, and prevent the aggregation of active metal atoms. Therefore, developing new synthetic routes for the obtainment of doped carbon materials with heavy heterodoping is highly desirable. To further pin the structure-activity relationship of atomic metals on carbon substrates, it is highly recommended to design and realize the heteroatom ligands with a precise layout. As previously mentioned, nanoparticles can act as electron reservoirs for the encapsulating M-N-C shells; however, the exact role of metal nanoparticles in the expression of electrocatalytic activity remains elusive, and large nanoparticles are usually inactive and unstable under practical conditions. Therefore, the design and fabrication of atomically dispersed metal species and the fine introduction of heteroatoms on carbon substrate will help to improve the higher site density of $M-N_x$ species, thereby increasing the catalytic activity of M-N-Cs and optimizing MABs cell efficiency.

Many current characterization techniques can identify the morphology, chemical composition, and structural properties of carbon-based electrocatalysts at the macroscopic and microscopic levels. However, the geometric and electronic structures of the catalyst are more related to the clear understanding of reactions occurring at the MABs cathode, while still being hard to unravel. Therefore, the development of various advanced characterization techniques, especially aberration-corrected electron microscopy, X-ray absorption, photoelectron spectroscopy, and advanced scanning transmission electron microscopy, will provide helps to gain deep insight into the configuration and bonding of ligands and heterodopant nearby at the atomic level. *In situ* characterization techniques will further contribute to identifying the structural and electronic behavior of oxygen electrocatalysis at the interface of electrocatalysts in actual working conditions.

REFERENCES

[1] H.-F. Wang, Q. Xu, Materials design for rechargeable metal-air batteries, Matter. 1(3) (2019) 565–595.
[2] Y. Sun, X. Liu, Y. Jiang, J. Li, J. Ding, W. Hu, C. Zhong, Recent advances and challenges in divalent and multivalent metal electrodes for metal—air batteries, J Mater Chem A. 7(31) (2019) 18183–18208.
[3] X. Zhu, C. Hu, R. Amal, L. Dai, X. Lu, Heteroatom-doped carbon catalysts for zinc—air batteries: Progress, mechanism, and opportunities, Energy Environ Sci. 13(12) (2020) 4536–4563.
[4] Z. Guo, S. Zhao, T. Li, D. Su, S. Guo, G. Wang, Recent advances in rechargeable magnesium-based batteries for high-efficiency energy storage, Adv Energy Mater. 10(21) (2020) 1903591.
[5] M. Jiang, C. Fu, P. Meng, J. Ren, J. Wang, J. Bu, A. Dong, J. Zhang, W. Xiao, B. Sun, Challenges and strategies of low-cost aluminum anodes for high-performance al-based batteries, Adv Mater. 34(2) (2022) 2102026.
[6] Y. Liu, Z. Chen, Z. Li, N. Zhao, Y. Xie, Y. Du, J. Xuan, D. Xiong, J. Zhou, L. Cai, Y. Yang, CoNi nanoalloy-Co-N4 composite active sites embedded in hierarchical porous carbon as bi-functional catalysts for flexible Zn-air battery, Nano Energy. 99 (2022) 107325.
[7] L. Wan, Z. Xu, Q. Cao, Y. Liao, B. Wang, K. Liu. Nanoemulsion-coated Ni-Fe hydroxide self-supported electrode as an air-breathing cathode for high-performance zinc-air batteries, Nano Lett. 22(11) (2022) 4535–4543.
[8] J. Wu, B. Liu, X. Fan, J. Ding, X. Han, Y. Deng, W. Hu, C. Zhong, Carbon-based cathode materials for rechargeable zinc-air batteries: From current collectors to bifunctional integrated air electrodes, Carbon Energy. 2(3) (2020) 370–386.

[9] C.X. Zhao, J.N. Liu, J. Wang, D. Ren, B.Q. Li, Q. Zhang, Recent advances of noble-metal-free bifunctional oxygen reduction and evolution electrocatalysts, Chem Soc Rev. 50(13) (2021) 7745–7778.
[10] Y. Huang, Y. Wang, C. Tang, J. Wang, Q. Zhang, Y. Wang, J. Zhang, Atomic modulation and structure design of carbons for bifunctional electrocatalysis in metal-air batteries, Adv Mater. 31(13) (2019) 1803800.
[11] C. Hu, R. Paul, Q. Dai, L. Dai, Carbon-based metal-free electrocatalysts: From oxygen reduction to multifunctional electrocatalysis, Chem Soc Rev. 50(21) (2021) 11785–11843.
[12] M. Fan, J. Cui, J. Wu, R. Vajtai, D. Sun, P.M. Ajayan, Improving the catalytic activity of carbon-supported single atom catalysts by polynary metal or heteroatom doping, Small. 16(22) (2020) 1906782.
[13] R. Paul, Q. Dai, C. Hu, L. Dai, Ten years of carbon-based metal-free electrocatalysts, Carbon Energy. 1(1) (2019) 19–31.
[14] Y. Mu, T. Wang, J. Zhang, C. Meng, Y. Zhang, Z. Kou, Single-atom catalysts: Advances and challenges in metal-support interactions for enhanced electrocatalysis, Electrochem Energy Rev. 5(1) (2021) 145–186.
[15] C.-X. Zhao, J.-N. Liu, J. Wang, C. Wang, X. Guo, X.-Y. Li, X. Chen, L. Song, B.-Q. Li, Q. Zhang, A clicking confinement strategy to fabricate transition metal single-atom sites for bifunctional oxygen electrocatalysis, Sci Adv. 8(11) (2022) eabn5091.
[16] X. Zheng, X. Cao, Z. Sun, K. Zeng, J. Yan, P. Strasser, X. Chen, S. Sun, R. Yang, Indiscrete metal/metal-N-C synergic active sites for efficient and durable oxygen electrocatalysis toward advanced Zn-air batteries, Appl Catal, B. 272 (2020) 118967.
[17] S. Wang, E. Zhu, Y. Huang, H. Heinz, Direct correlation of oxygen adsorption on platinum-electrolyte interfaces with the activity in the oxygen reduction reaction, Sci Adv. 7(24) (2021) eabb1435.
[18] Y. Ji, M. Yang, H. Dong, L. Wang, T. Hou, Y. Li, Monolayer group IVA monochalcogenides as potential and efficient catalysts for the oxygen reduction reaction from first-principles calculations, J Mater Chem A. 5(4) (2017) 1734–1741.
[19] H. Yin, Y. Dou, S. Chen, Z. Zhu, P. Liu, H. Zhao, 2D electrocatalysts for converting earth-abundant simple molecules into value-added commodity chemicals: Recent progress and perspectives, Adv Mater. 32(18) (2020) 1904870.
[20] F. Liu, L. Shi, X. Lin, D. Yu, C. Zhang, R. Xu, D. Liu, J. Qiu, L. Dai, Site-density engineering of single-atomic iron catalysts for high-performance proton exchange membrane fuel cells, Appl Catal, B. 302 (2022) 120860.
[21] J. Zhang, Y. Zhao, C. Chen, Y.C. Huang, C.L. Dong, C.J. Chen, R.S. Liu, C. Wang, K. Yan, Y. Li, G. Wang, Tuning the coordination environment in single-atom catalysts to achieve highly efficient oxygen reduction reactions, J Am Chem Soc. 141(51) (2019) 20118–20126.
[22] J. Song, C. Wei, Z.F. Huang, C. Liu, L. Zeng, X. Wang, Z.J. Xu, A review on fundamentals for designing oxygen evolution electrocatalysts, Chem Soc Rev. 49(7) (2020) 2196–2214.
[23] F. Shi, X. Zhu, W. Yang, Micro-nanostructural designs of bifunctional electrocatalysts for metal-air batteries, Chin J Catal. 41(3) (2020) 390–403.
[24] J. Liu, J. Xiao, B. Luo, E. Tian, G.I.N. Waterhouse, Central metal and ligand effects on oxygen electrocatalysis over 3d transition metal single-atom catalysts: A theoretical investigation, Chem Eng J. 427 (2022) 132038.
[25] H. Wang, Y. Shao, S. Mei, Y. Lu, M. Zhang, J.K. Sun, K. Matyjaszewski, M. Antonietti, J. Yuan, Polymer-derived heteroatom-doped porous carbon materials, Chem Rev. 120(17) (2020) 9363–419.
[26] Q. Lv, N. Wang, W. Si, Z. Hou, X. Li, X. Wang, F. Zhao, Z. Yang, Y. Zhang, C. Huang, Pyridinic nitrogen exclusively doped carbon materials as efficient oxygen reduction electrocatalysts for Zn-air batteries, Appl Catal, B. 261 (2020) 118234.
[27] K. Gong, F. Du, Z. Xia, M. Durstock, L. Dai, Nitrogen-doped carbon nanotube arrays with high electrocatalytic activity for oxygen reduction, Science. 323(5915) (2009) 760–764.
[28] G. Zhang, H. Luo, H. Li, L. Wang, B. Han, H. Zhang, Y. Li, Z. Chang, Y. Kuang, X. Sun, ZnO-promoted dechlorination for hierarchically nanoporous carbon as superior oxygen reduction electrocatalyst, Nano Energy. 26 (2016) 241–247.
[29] Y. Zhao, R. Nakamura, K. Kamiya, S. Nakanishi, K. Hashimoto, Nitrogen-doped carbon nanomaterials as non-metal electrocatalysts for water oxidation, Nat Commun. 4 (2013) 2390.
[30] J. Zhang, Z. Zhao, Z. Xia, L. Dai, A metal-free bifunctional electrocatalyst for oxygen reduction and oxygen evolution reactions, Nat Nanotechnol. 10(5) (2015) 444–452.

[31] Q. Wang, Y. Lei, Y. Zhu, H. Wang, J. Feng, G. Ma, Y. Wang, Y. Li, B. Nan, Q. Feng, Z. Lu, H. Yu, Edge defect engineering of nitrogen-doped carbon for oxygen electrocatalysts in Zn-air batteries, ACS Appl Mater Interfaces. 10(35) (2018) 29448–29456.

[32] X. Wang, Z. Li, Y. Qu, T. Yuan, W. Wang, Y. Wu, Y. Li, Review of metal catalysts for oxygen reduction reaction: From nanoscale engineering to atomic design, Chem. 5(6) (2019) 1486–1511.

[33] B. Qiao, A. Wang, X. Yang, L.F. Allard, Z. Jiang, Y. Cui, J. Liu, J. Li, T. Zhang, Single-atom catalysis of CO oxidation using Pt1/FeOx, Nat Chem. 3(8) (2011) 634–641.

[34] Y. Wang, Y. Liang, T. Bo, S. Meng, M. Liu, Orbital dependence in single-atom electrocatalytic reactions, J Phys Chem Lett. 13 (2022) 5969–5976.

[35] W. Zang, A. Sumboja, Y. Ma, H. Zhang, Y. Wu, S. Wu, H. Wu, Z. Liu, C. Guan, J. Wang, S.J. Pennycook, Single co atoms anchored in porous N-doped carbon for efficient zinc–air battery cathodes, ACS Catal. 8(10) (2018) 8961–8969.

[36] H. Xu, H. Jia, H. Li, J. Liu, X. Gao, J. Zhang, M. Liu, D. Sun, S. Chou, F. Fang, R. Wu, Dual carbon-hosted co-N3 enabling unusual reaction pathway for efficient oxygen reduction reaction, Appl Catal, B. 297 (2021) 120390.

[37] P. Peng, L. Shi, F. Huo, C. Mi, X. Wu, S. Zhang, Z. Xiang, A pyrolysis-free path toward superiorly catalytic nitrogen-coordinated single atom, Sci Adv. 5(8) (2019) eaaw2322.

[38] M. Ma, A. Kumar, D. Wang, Y. Wang, Y. Jia, Y. Zhang, G. Zhang, Z. Yan, X. Sun, Boosting the bifunctional oxygen electrocatalytic performance of atomically dispersed Fe site via atomic Ni neighboring, Appl Catal, B. 274 (2020) 119091.

[39] J. Li, H. Zhang, W. Samarakoon, W. Shan, D.A. Cullen, S. Karakalos, M. Chen, D. Gu, K.L. More, G. Wang, Z. Feng, Z. Wang, G. Wu, Thermally driven structure and performance evolution of atomically dispersed FeN4 sites for oxygen reduction, Angew Chem, Int Ed. 58(52) (2019) 18971–18980.

[40] Y. Jia, X. Xiong, D. Wang, X. Duan, K. Sun, Y. Li, L. Zheng, W. Lin, M. Dong, G. Zhang, W. Liu, X. Sun, Atomically dispersed Fe-N4 modified with precisely located S for highly efficient oxygen reduction, Nano-Micro Lett. 12(1) (2020) 116.

[41] Y. Wang, G. Zhang, M. Ma, Y. Ma, J. Huang, C. Chen, Y. Zhang, X. Sun, Z. Yan, Ultrasmall NiFe layered double hydroxide strongly coupled on atomically dispersed FeCo-NC nanoflowers as efficient bifunctional catalyst for rechargeable Zn-air battery, Sci China Mater. 63(7) (2020) 1182–1195.

[42] F. Qiang, J. Feng, H. Wang, J. Yu, J. Shi, M. Huang, Z. Shi, S. Liu, P. Li, L. Dong, Oxygen engineering enables N-doped porous carbon nanofibers as oxygen reduction/evolution reaction electrocatalysts for flexible zinc-air batteries, ACS Catal. 12(7) (2022) 4002–4015.

[43] Y. Wang, Z. Li, P. Zhang, Y. Pan, Y. Zhang, Q. Cai, S.R.P. Silva, J. Liu, G. Zhang, X. Sun, Z. Yan, Flexible carbon nanofiber film with diatomic Fe-Co sites for efficient oxygen reduction and evolution reactions in wearable zinc-air batteries, Nano Energy, 87 (2021) 106147.

15 Metal Oxide-Based Electrocatalysts for Metal-Air Batteries

Bhugendra Chutia, Chiranjita Goswami, and Pankaj Bharali

CONTENTS

15.1 Introduction 209
    15.1.1 Metal-Air Batteries (MABs): Importance and Applicability 209
    15.1.2 Components and the Working Principles of MAB 211
    15.1.3 Kinetics of ORR/OER in MABs..212
    15.1.4 Need for Electrocatalyst in MABs...213
15.2 Roadblocks in MABs 214
    15.2.1 Development of Advanced Electrocatalysts for MABs..216
15.3 Metal Oxides as Bifunctional Electrocatalysts for MABs 217
    15.3.1 Noble Metal Oxides...217
    15.3.2 Non-Noble Metal Oxides...217
        15.3.2.1 Single Metal Oxides 217
        15.3.2.2 Spinels 218
        15.3.2.3 Perovskites..218
        15.3.2.4 Other Mixed Metal Oxides 219
15.4 Tailoring the Structure of Metal Oxide Electrocatalysts 219
    15.4.1 Morphology..219
    15.4.2 Doping..221
    15.4.3 Oxygen Vacancy and Defects..221
    15.4.4 Interface 221
15.5 Conclusions and Future Scopes 222
Acknowledgements..222
References..223

15.1 INTRODUCTION

15.1.1 METAL-AIR BATTERIES (MABS): IMPORTANCE AND APPLICABILITY

The 21st century largely relies on the development of alternative energy technologies to ensure a clean and sustainable future. This is mainly due to the rapid increase in population growth, degradation of fossil fuels, fluctuation in oil costs, and the subsequent global issues. The energy storage technologies are mainly classified into four different types, *viz.*, electrical, mechanical, chemical, and electrochemical. The electrochemical approach is quite popular as it possesses both conversion and storage abilities. Owing to its fascinating features such as low maintenance, negligible emission, superior round trip efficiency, long life cycle, and energy characteristics, batteries are recognized as outstanding and reliable energy storage devices [1]. Various types of battery technologies include Li-ion batteries, Li-air batteries, lead-acid batteries, Zn-air batteries, redox flow batteries, Na-ion

DOI: 10.1201/9781003295761-15

batteries, etc. [2]. The commercialization of electric vehicles in the modern-day world is generally powered by conventional Li-ion batteries. However, the low specific energy densities of the Li-ion batteries propelled scientists toward the development of various metal-air batteries (MABs). MABs are persuasive due to their high energy density and safer working mode than the existing alternative devices [3]. Additionally, the use of oxygen from the air as the cathode helps to minimize the cost as well as the weight of the entire system. The characteristics of both fuel cells and conventional batteries are found in a MAB. The theoretical energy density of MABs is even higher in comparison to that of rechargeable lead-acid batteries, primary $Zn-MnO_2$ batteries, and nickel-metal hydride (Ni-MH) batteries, since the oxygen is mostly derived from the air but not stored in the cell [4]. Hence, they find wide application in electrical vehicles, stationary as well as portable electronic devices [5]. The primary Zn-air battery is used in hearing-aid devices, whereas Mg-air and Al-air batteries have been used in underwater propulsion [6]. To date, Zn-air batteries and Li-air batteries have significantly attracted the attention of the various MABs [7]. The performance of a MAB is evaluated based on theoretical energy density, apart from other factors. A comparison of the theoretical energy densities for different types of MABs is shown in Figure 15.1. It can be perceived that the theoretical energy density and battery voltage of Li-air battery is the highest among all other MABs. Yet, Zn-, Al-, and Fe-air batteries are also of significance due to their safety and cost-effective nature.

MABs can be classified based on different metal species involved and their variable reaction mechanisms, which demand the design of unique cell components of a particular type. Typically, they are divided into two major types depending on the electrolyte used in the cell system, namely, aqueous and non-aqueous electrolytes. The former is not affected by moisture, while the latter is moisture-sensitive and uses an aprotic solvent [8]. Similar to the other battery systems, MABs also have certain technological and scientific challenges that are required to be resolved. The sluggish reaction kinetics of the cathode and low utilization efficiency of the anode are the major shortfalls that limit practical applications of MABs. During the reaction process, species like hydroxide, metal oxides (MOs), etc. accumulate at the anode's surface from the electrolytes which leads to corrosion. This impedes further accessibility of the electrolyte thereby preventing the discharge of

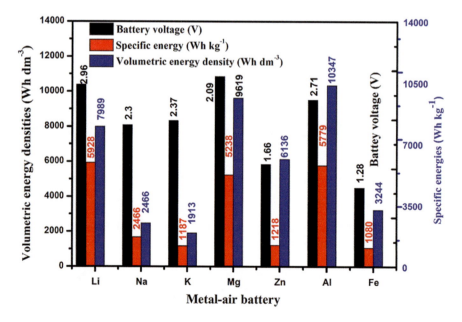

FIGURE 15.1 Theoretical specific energies, volumetric energy densities, and nominal battery voltages of various metal-air batteries (MABs). (Adapted with permission [9]. Copyright The Authors, some rights reserved; exclusive licensee MDPI. Distributed under a Creative Commons Attribution License 4.0 (CC BY).)

remaining active material. The corrosion and self-discharge of the anode lower its coulombic efficiency and curb the battery lifetime. The main disadvantages of cathode include high overpotential, intrinsic slow reaction rate, and poor reversibility of oxygen chemistry [6].

The two fundamental electrochemical reactions on which the MABs operate are the oxygen reduction reaction (ORR) and oxygen evolution reaction (OER), which relate to the discharge and charge processes, respectively [10]. This chapter primarily summarizes the development and importance of MABs in the field of energy and its mechanistic pathways. The role of different types of MO-based electrocatalysts will be discussed, especially underlining the morphology, effect of doping, presence of defects, and interface engineering. The challenges encountered during the development of a MAB and the electrocatalysts and also the future perspectives for improving the electro-kinetics of MOs for ORR/OER are taken into consideration.

15.1.2 Components and the Working Principles of MAB

The basic components of MAB include a metal electrode (anode), an air electrode (cathode), and an aqueous or non-aqueous based electrolyte. The metal electrodes *viz.*, Zn, Al, Li, Na, and Mg are normally investigated. Li, Na, and K are highly active metals but are unstable in aqueous systems due to which non-aqueous aprotic electrolytes are widely employed. However, Al, Mg, Fe, and Zn are relatively inactive and hence they require alkaline aqueous electrolytes during the reaction. Another component of MAB is the separator, which is used to separate two different electrolytes. It also inhibits certain mass transport processes that occur between the electrodes and prevent the short-circuit caused mainly by the metal dendrites [7]. A characteristic open cell structure in MAB draws oxygen—the cathode active material which is continuously supplied from an external source i.e., air [4, 6]. The metal anode comprises either alkali metals (commonly Li, Na, and K) or alkaline earth metals or the first-row transition metals like Fe and Zn having high electrochemical equivalence. The working principle of a MAB is based on the transformation of metals or alloys into metallic ions at the anode, while, the conversion of oxygen to hydroxide ions at the cathode [9].

The choice of the electrolyte, whether aqueous or non-aqueous, depends on the nature of the anode used. A redox reaction between the metal and oxygen from the air is responsible for the generation of electricity in MABs [11]. Figure 15.2a,b schematically represents the working principle

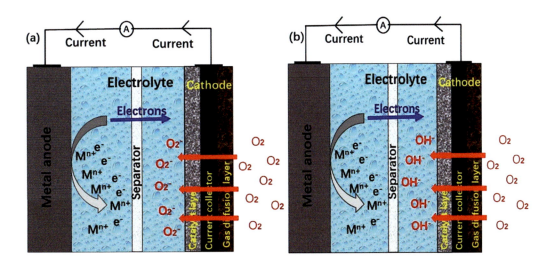

FIGURE 15.2 Schematic diagrams of MABs working principles for (a) non-aqueous electrolyte and (b) aqueous electrolyte. (Adapted with permission [9]. Copyright The Authors, some rights reserved; exclusive licensee MDPI. Distributed under a Creative Commons Attribution License 4.0 (CC BY).)

of an aqueous and a non-aqueous MAB. When an aqueous electrolyte is used, oxygen diffuses through the gas diffusion layer towards the battery, thereafter forming oxygen anions after combining with the electrons. In the case of a non-aqueous electrolyte, oxygen molecules are simply converted to their corresponding ions after receiving the electrons [9].

15.1.3 Kinetics of ORR/OER in MABs

The chemistry involved in MABs is the conversion between molecular O_2 and its reduced forms reversibly via a one-, two-, or four-electron process. The preferred pathway depends on the relative stability of the intermediates and the products formed [12]. In MABs, both metal and oxygen participate in electrochemical reactions. The fundamental reactions involved are as follows [9]:

$$\text{At anode: } M \leftrightarrow M^{n+} + ne^- \tag{1}$$

$$\text{At cathode: } O_2 + 2H_2O + 4e^- \leftrightarrow 4OH^- \tag{2}$$

The electrode reactions are well-dependent on the metal electrodes, electrocatalysts as well electrolytes used [7]. An ideal electrolyte should fulfill certain properties like non-toxicity, low moisture absorption, low volatility, high boiling point, high oxygen solubility, high stability against metal electrodes used, and reduced oxygen species. ORR proceeds via two different mechanisms, namely, a four-electron pathway and a two-electron pathway based on the oxygen adsorption species. When two O atoms bind to the catalyst surface, the O_2 adsorption is supposed to follow the four-electron pathway. The half-cell reactions can be represented as follows [13]:

$$\text{Overall: } O_2 + 2H_2O + 4e^- \leftrightarrow 4OH^- \tag{3}$$

$$O_2 + 2H_2O + 2e^- \rightarrow 2OH_{ads} + 2OH^- \tag{4}$$

$$2OH_{ads} + 2e^- \rightarrow 2OH^- \tag{5}$$

When only one O atom is adsorbed on the surface of the catalyst, the favored pathway is the two-electron pathway.

$$\text{Overall: } O_2 + H_2O + 2e^- \rightarrow OH^- + O_2H^- \tag{6}$$

$$O_2 + H_2O + e^- \rightarrow OH^- + O_2H_{ads} \tag{7}$$

$$O_2H_{ads} + e^- \rightarrow O_2H^- \tag{8}$$

ORR in the air electrode, being a slower reaction, is the rate-determining step. Hence, effective electrocatalysts are required to accelerate its rate and lower its overpotential. Higher is the overpotential, lower will be output performance and energy storage efficacy. Other important factors that affect the reaction rate are the high mass transport, low-rate capability, poor round-trip efficiency, low Coulombic efficiency, and inadequate cycle life [14]. All these difficulties can be resolved by rational designing materials with high catalytic activity.

Like ORR, OER is the limiting factor that obstructs metal-air batteries from practical use because a total of four electrons are required for the generation of O_2. The kinetics involve the transfer of one electron at a time via multiple steps. In an alkaline medium, OH^- groups are oxidized to O_2 and H_2O, while O_2 and $4H^+$ are formed from two molecules of water in an acidic medium, during OER [15]. The anodic half-reactions at 25 °C and 1 atm are shown next:

$$4OH^- \leftrightarrow 2H_2O\,(l) + O_2\,(g) + 4e^-; \; E_a^\circ = 0.404 \text{ V } vs. \text{ RHE (alkaline medium)} \tag{9}$$

$$2H_2O\,(l) \leftrightarrow O_2\,(g) + 4H^+ + 4e^-; \; E_a^\circ = 1.23 \text{ V } vs. \text{ RHE (acidic medium)} \tag{10}$$

15.1.4 Need for Electrocatalyst in MABs

The two fundamental reactions involved in MABs, i.e., ORR and OER suffer from their sluggish kinetics, and high voltage hysteresis and produce low energy density. Hence, scientists nowadays are chiefly concerned with overcoming those loopholes. Catalysts are thereby required to promote the reactions via an alternative low-energy pathway, accelerate the redox kinetics, and improve the overall electrochemical performance of MABs [16]. Previous literature revealed that Pt and IrO_2/RuO_2 are used as benchmark catalysts to catalyze ORR and OER, respectively [15–18]. The noble metal catalysts are recognized as the best material for ORR in terms of electrocatalytic performance. They are also used to catalyze OER in Li–O_2 batteries. Lee's group studied the catalytic activity of reduced graphene oxide supported Pt, Pd, and Ru nanomaterials as a cathode catalyst in Li–O_2 cell, using a $LiCF_3SO_3$–tetra(ethylene glycol) dimethyl ether (TEGDME) salt-electrolyte system [19]. These noble metal catalysts which are expensive and rare also have poor stability and durability during long-run electrochemical measurements [20]. Therefore, alternative electrocatalysts are highly demanding to overcome such drawbacks and for the productive utilization of MABs. Due to its low price and good ORR activity, non-noble or transition MOs such as MnO_x and CoO_x find widespread applications in MABs. For example, α-MnO_2 has been extensively used as an electrocatalyst in non-aqueous Li-air batteries. The enhanced activity can be attributed to its crystal structure which enables Li-ion coordination and easy oxygen decomposition [3]. However, non-noble metal-based electrocatalysts generally have mono-catalytic ORR/OER reactive sites. It has been observed that ORR catalysts of the type M-N-C (M = Mn, Fe, Co, etc.) have higher catalytic activity in alkaline solution as compared to that of Pt/C. Similarly, the OER activity of certain transition MOs and sulfides also outperforms Ir/C and RuO_2 [21]. Yet, these materials can't meet the requirements of a bifunctional catalyst, which can simultaneously catalyze both ORR and OER processes. The materials having good ORR performance may not be equally active towards OER and vice-versa, which is a challenge for the scientific community. The development of bifunctional catalysts is highly desirable because it can substantially influence the commercialization of MABs and also reduces the cost of catalysts [21]. The basic criterion for an electrocatalyst to have bifunctional property is to have a low variance index (ΔE) value [22, 23]. Recent years show the marked development of bifunctional catalysts that include noble metals, noble metal-based alloys or composites, carbon materials, transition MOs or hydroxides, chalcogenides, etc. [22, 24, 25].

Single-atom catalysts (SACs) are another class of crucial materials that have received great attention in this field. SACs effectively reduce the dependence on noble metals and enhance the intrinsic activity of metal atoms. The theoretical atomic utilization efficiency of SACs is almost 100%, which is quite appreciable [16]. For instance, atomically dispersed Fe-Ni single atoms embedded in nitrogen-doped carbon matrix (FeNi SAs/NC) are developed by Peng and co-workers for electrocatalytic ORR/OER in MABs. The as-synthesized catalyst was found to exhibit a high onset potential (E_{onset}) of 0.98 V vs. RHE and half-wave potential ($E_{1/2}$) of 0.84 V towards ORR. Also, OER displayed a low overpotential of 270 mV at 10 mA cm^{-2} [26]. Similarly, Han et al. [27] and Wang et al. [28] have developed noble metal-free Fe-N_x-C and Co-N-C SAC electrocatalysts for Zn-air batteries with outstanding results. The obtained Fe-N_x-C electrocatalyst act as an excellent cathode catalyst for primary Zn-air batteries, resulting in a high open-circuit voltage (OCV) and a high-power density of 1.51 V (vs. RHE) and 96.4 mW/cm^2, respectively [27]. Further, the Co-N-C SAC electrocatalyst exhibited bifunctional activity with a ΔE of 0.81 V (vs. RHE) in an alkaline medium. The performance of Co-N-C SAC as an air cathode catalyst was also tested in self-made primary and rechargeable Zn-air batteries besides the conventional electrochemical measurements. The data obtained are depicted in Figure 15.3(a-d) [28].

Although the carbonaceous materials are efficient bifunctional electrocatalysts, yet, suffer from oxidation or corrosion in presence of O_2 at high electrode potentials. This can be mitigated by doping carbon composites with heteroatom(s). In doing so, the ORR/OER actives sites, electronic conduction, stability, and durability of the resultant catalyst increase [23]. It is to be noted that undoped

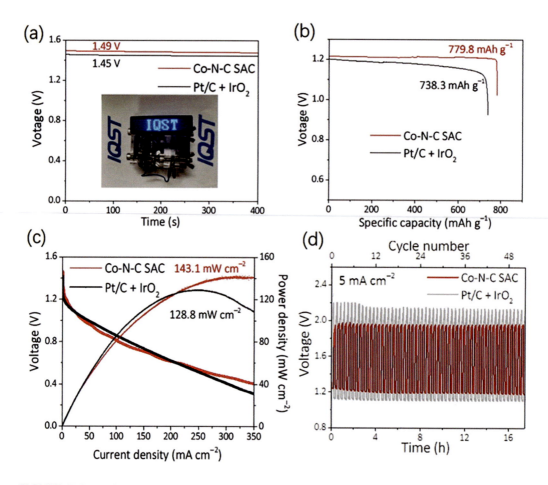

FIGURE 15.3 (a) Open-circuit plots of Co-N-C SAC and 20% Pt/C + IrO$_2$ (inset: Optical image of an LED light array powered by two ZABs in series using Co-N-C SAC as the air cathode). (b) Specific capacities of the ZABs using Co-N-C SAC and 20% Pt/C + IrO$_2$ as catalysts, which are normalized to the mass of the completely consumed Zn. (c) Polarization and power density curves for primary ZABs. (d) Charge-discharge cycling performance of Co-N-C SAC-based ZABs and 20% Pt/C+IrO$_2$-based ZABs with a duration of 20 min per cycle at 5 mA/cm^2. (Adapted with permission [28]. Copyright The Authors, some rights reserved; exclusive licensee MDPI. Distributed under a Creative Commons Attribution License 4.0 (CC BY).)

pristine carbon materials in an aqueous medium exhibit better electrocatalytic activity for ORR and OER as compared to that of carbon-based materials in non-aqueous electrolytes. It can be perceived that the catalyst's activity varies with the electrolyte used. Perovskites or perovskite oxides also act as effective bifunctional electrocatalysts owing to their crystal structure, abundant stoichiometry, and many metal-active centers [22]. Thus, the types of catalysts for MABs come in a broad spectrum. In this chapter, we will confine our study exclusively to the MO-based electrocatalysts for MABs. The role of morphology, doping, defects, and interface engineering in the MOs will also be discussed in the succeeding sections.

15.2 ROADBLOCKS IN MABs

Despite the tremendous research for the advancement of MAB technology over the last decades, several technical hurdles must be resolved before the commercial execution of the concept. The activity of rechargeable MAB is fundamentally assessed from the recharging and discharging efficiency of the device. The up-gradation of the MAB operation dealing with both efficient discharge and rechargeability remains a great challenge. The key issues that counter MABs commercialization are associated with the inadequate functionality of metal anodes, air cathodes, and electrolytes. Consequently, it is vital for endorsing further advancement in the MABs to resolve the key challenges from the approach of materials science, exploring the material design of metal electrodes, air electrodes, electrolyte, and separator materials. Figure 15.4 presents the schematic scenario for current issues and material design strategies for the advancement of MAB performances.

The critical challenges associated with the metal anode are corrosion, passivation, and dendritic formation over the electrode surface. The side reactions between the electrolyte and the metal electrode are accountable for the undesirable physicochemical phenomenon over the anode surface. The corrosion reactions at the metal anode can be depicted as

$$M + (2 + x) H_2O \rightleftharpoons 2 M(OH) + H_2 \tag{11}$$

$$M + H_2O \rightleftharpoons 2 MO_x + H_2 \tag{12}$$

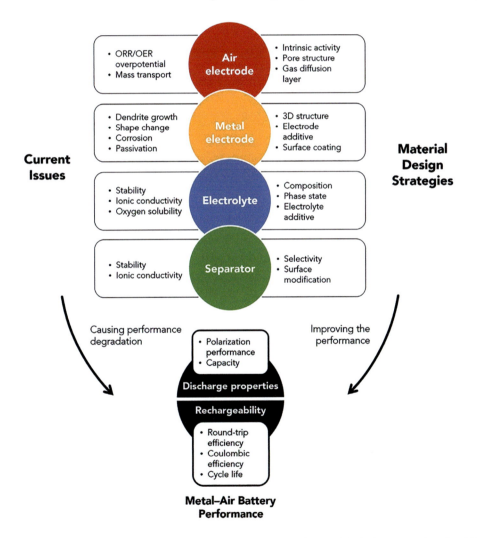

FIGURE 15.4 Evaluation parameters, material challenges, and material design strategies for MABs. (Adapted with permission [7]. Copyright (2022), Elsevier.)

The hydrogen evolution reaction (HER) is a common reaction phenomenon that appears nearly in all kinds of MABs because of the more feasible HER redox potential than that of M/MO redox potential. The reaction is also responsible for the corrosion of the anode surface which in turn reduces the coulombic efficiency of the anode. In another process, a significant roadblock appears during reaction due to the growth of insulating layers over the electrode surfaces which is termed as passivation. The insulating layers prevent the discharged ions to roam around the metal anode, which eventually goes to the cell in a state of high internal electrical resistance. The dendritic formation can make a gradual change in the shape of the metal anode making uneven roughness. This occurs due to the accumulation of metal ions during the electrode cycling in alkaline electrolytes. This can produce an unstable MAB system.

15.2.1 Development of Advanced Electrocatalysts for MABs

The catalyst layer is a major cathode component that has a significant impact on the efficiency of a characteristic MAB. The design of bifunctional ORR/OER electrocatalysts can promote both the discharge and charge processes of MABs. A competent bifunctional electrocatalyst is needed to enhance the ORR/OER rate and to reduce the overpotential as the normal kinetics of the oxygen reaction are inadequately vigorous. The ORR and OER are taking place at the air electrode (cathode), where ORR takes place throughout the discharge cycle while OER occurs during the charge cycle. Numerous electrocatalysts systems, including noble metal-based materials (such as Pt, Pd, IrO_2, RuO_2, etc.), metallic alloys, transition metal oxides (TMOs), mixed metal oxides (MMOs), perovskites, metal-organic frameworks (MOFs), etc. anchored on different carbonaceous materials have been extensively studied as cathode catalysts [29, 30].

Figure 15.5 shows the different development strategies for metal-based catalysts to further enhance the active sites and the intrinsic activities. To date, Pt-group metal-based materials

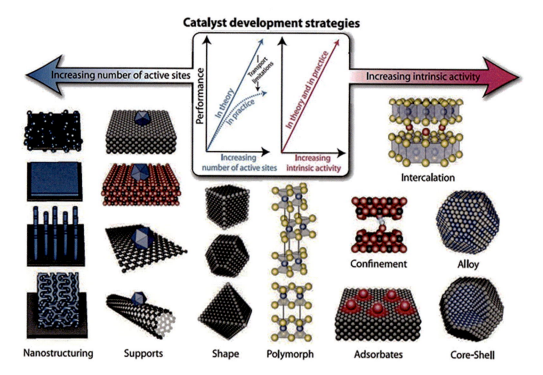

FIGURE 15.5 Development strategies for metal catalysts. (Adapted with permission [29]. Copyright The Authors, some rights reserved; exclusive licensee MDPI. Distributed under a Creative Commons Attribution License 4.0 (CC BY).)

such as Pt and Ir-/Ru-based alloys or oxides have been established to be highly active in oxygen electrocatalysis [15–18]. Nevertheless, noble metal catalysts are mostly incompetent to bifunctional activity, thus obstructing their application in rechargeable MABs. Consequently, the development of high-performance non-precious metal-based bifunctional electrocatalysts is crucial for practical application. In recent times, non-noble MO-based composites have attracted massive attention as the bifunctional oxygen electrocatalyst owing to its inexpensive, abundant, and high activity toward ORR and OER. Because of the high O_2-diffusion capability, synergistic effect, anionic defects, and significant hetero-structure interfacial mechanism, these materials demonstrate excellent electrocatalytic activities. To facilitate the maximized electrocatalytic efficiency, the research community is further concentrating their work on the suitable morphological and structural engineering of the catalyst materials which will be summarized in this section.

15.3 METAL OXIDES AS BIFUNCTIONAL ELECTROCATALYSTS FOR MABs

MO electrocatalysts include the oxides of noble metals, transition metals, and inner transition metals. MOs with different stoichiometric or even non-stoichiometric compositions have recently attracted growing research interest worldwide. Benefiting from their adjustable chemical constitution, structural diversity, easy preparation, and astonishing synergistic and electrochemical properties, these materials play significant roles in environmentally friendly energy storage/conversion technologies [17]. Additionally, offering greater scope for cultivation of the intrinsic activity with structural engineering for high electrochemically exposed surface area, MOs can be a potential candidate as the electrocatalysts for MABs.

15.3.1 Noble Metal Oxides

Noble metals such as Au, Pt, Ag, etc. usually have little tendency to unite with O_2 in the free state and under red heat conditions, they may react with and do not alter its composition. Noble MOs are readily decomposed by heat as a consequence of the weak affinity between the metal and oxygen. It is noteworthy that RuO_2 and IrO_2 are well recognized to be highly active toward OER and regarded as the state-of-the-art OER electrocatalyst [15–18]. Nevertheless, RuO_2 and IrO_2 are less active toward ORR and hence not efficient as a bifunctional electrocatalyst for ORR and OER.

15.3.2 Non-Noble Metal Oxides

15.3.2.1 Single Metal Oxides

Single MOs i.e., oxides with identical metal centers, can be classified into this category. These materials show significant activity and stability in the alkaline medium for ORR and OER [31]. For example, TMOs of the type M_xO_y (M = Mn, Fe, Co, Ni, etc.) supported on carbon are observed to be highly active toward oxygen electrocatalysis. Manganese oxides (Mn_xO_y) have attracted the researcher because of their flexible valences and plentiful structures, giving rise to rich redox couples. As an illustration, Cheng et al. reported the electrocatalytic activities of MnO_2 correlating with the crystallographic structure, following the order of α- > β- > γ-MnO_2 [32]. Gorlin et al. designed a thin-film analog consisting of a nanostructured Mn(III) oxide by using an electrodeposition technique that exhibits an excellent electrocatalytic activity for the ORR and OER in alkaline solution [33]. Its bifunctional ORR/OER performance is comparable to that of noble metal catalysts. Furthermore, apart from Mn_xO_y, other single MOs such as Fe_xO_y, NiO, CuO, and Co_xO_y were also demonstrated to be efficiently active toward the electrochemical oxygen reaction [25, 31, 34].

15.3.2.2 Spinels

Spinel-type oxide is one of the most significant bifunctional O_2 electrocatalysts among TMOs. Typical spinel oxides can generally be designated as group oxides with the formula AB_2O_4, where A and B stand for divalent and trivalent metal ions, respectively and their structure is assembled upon a cubic close-packed array of O^{2-} ions, where A(II) cations usually occupy the parts or all tetrahedral sites and B(III) cations occupy octahedral sites. Based on the spatial distribution of A(II) and B(III) cations in the unit cells, spinel oxides can be classified as normal, inverse, and complex spinel oxides. Owing to its exceptional crystal structure it can provide effective catalytic active sites for both ORR and OER. Hence, the structure of spinel oxides has a straight connection to the ORR and OER performance. Many researchers have been working on the fine-tuning and structural design of spinel oxides directing from different aspects. Rational doping in A/B sites, designing oxygen vacancy defects, adjusting the crystal plane and structure, morphology-controlled synthesis, etc. are some fascinating methodologies that can be used to modulate the spinel oxides to enhance the ORR/OER performance. Recently, Han et al. synthesized morphology-controlled Co_3O_4 (cubic, octahedron, and polyhedron) nanocrystals anchored on N-doped reduced graphene by a simple template-free hydrothermal method [35]. Nanostructured Co_3O_4 polyhedrons with the major exposed facets of (112) displayed greater ORR/OER performance relative to that of cubic and octahedron Co_3O_4 nanocrystals due to the different surface abundance of Co^{3+} and Co^{2+} active sites. It was suggested that the greater population of octahedrally coordinated Co^{3+} on (112) planes boosted the adsorption, desorption, and dissociation properties of O_2 species, which enhances the activity. Similarly, Li et al. synthesized a three-dimensional flower-like heterostructure of $NiCo_2O_4$ with N-doped carbon derived from $g-C_3N_4$, under a hydrothermal approach [36]. The $NiCo_2O_4$-CN displays improved ORR activity with a $E_{1/2}$ of 0.81 V and high OER activity with a smaller overpotential (383 mV). In a test with a Zn-air battery, the $NiCo_2O_4$-CN shows similar performance to the commercial Pt/C and IrO_2. The enhanced bifunctional ORR/OER activity is attributed to the combined effects of porous 3D structure, high electrochemically active surface area (ECSA), and abundant oxygen vacancies incorporated in the catalyst. Mixed valence spinel oxides have also been observed to be bifunctionally active toward oxygen electrocatalysis. In the late '90s, Rios et al. reported the ORR/OER activity of the $Mn_xCo_{3-x}O_4$ [37]. Similarly, mixed-valence spinel oxides such as $Cu_xCo_{3-x}O_4$, $Ni_xCo_{3-x}O_4$, $CoMn_2O_4$, etc. have also been well demonstrated as efficient bifunctional ORR/OER activity.

15.3.2.3 Perovskites

Perovskites are a class of unique MOs that are bifunctionally active for ORR and OER. They can be typically represented as ABO_3, where A usually represents a positively charged rare earth metal or alkaline earth metal cation and B is a TM ion with a charge that is more positive relative to A. The perovskite unit cell comprises BO_6 octahedrons where A introduced in the void among the BO_6 structures. Perovskites have extensive flexibility for engineering the crystal structure relative to spinel oxides as almost all kinds of metallic elements (more than 90% metallic element) can be doped into perovskite structures and a large range of oxygen non-stoichiometry is allowed and because of these characteristic features, perovskites have been largely studied for the advancement of their bifunctional (ORR/OER) performances. Modification of the cation in A/B sites has been an effective technique to tune the electronic structure, which generates varieties of redox couple and lattice defects in perovskites. For instance, Lee et al. designed Ni-doped Co-based double perovskite ($PrBa_{0.5}Sr_{0.5}Co_{1.9}Ni_{0.1}O_{5+\delta}$, PBSCN1) employing sol-gel and combustion technique [38]. The doping of Ni triggered the coexistence of Co cations with various oxidation states with enlarged surface sites for O_2 adsorption/desorption thereby exhibiting enhanced OER activity. Analogously, Fe-doped $PrBa_{0.5}Sr_{0.5}Co_{1.5}Fe_{0.5}O_{5+\delta}$ nanofibers were formulated via electrospinning [39]. The mesoporous $PrBa_{0.5}Sr_{0.5}Co_{1.5}Fe_{0.5}O_{5+\delta}$ nanofiber displayed high ORR and OER activity with enhanced stability in the Zn-air battery.

15.3.2.4 Other Mixed Metal Oxides

Heterostructures containing one oxide on another oxide are kind of unique MOs that have several implications in catalytic industries including oxygen electrocatalysis. Taking into account catalytic chemists, MMOs are the oxygen-containing unification of two or more metallic ions in proportions that may either vary or be defined by strict stoichiometry. They are usually attained in the form of powder or single crystals. MMOs have characteristics of rich oxygen defects, multivalent redox, and oxide-oxide interfacial phenomenon. Owing to these properties, non-noble MMOs exhibit excellent bifunctional ORR/OER performance and stand as a promising alternative to expensive Pt-based electrocatalysts. Recently, we reported a hybrid heteronanostructure of α-MnO_2/Mn_3O_4/CeO_2 with the atomic-level coupled nanointerface supported on carbon synthesized via a facile solvothermal approach as an advance bifunctional electrocatalyst [17]. Figure 15.6 displays the oxide-oxide interfaces between α-MnO_2, Mn_3O_4, and CeO_2 nanostructures over the carbon matrix, particle size distribution, ORR/OER activities, and the schematic ORR/OER mechanism on α-MnO_2/Mn_3O_4/CeO_2/C hybrid nanostructure. CeO_2 plays an essential role as an O_2 buffer and in increasing the surface $Mn^{2+/3+/4+}$ in α-MnO_2/Mn_3O_4/CeO_2/C for the enhanced ORR/OER processes.

Similarly, Liu et al. synthesized Co_3O_4-CeO_2/KB using a simple two-step hydrothermal method as a high-performance ORR catalyst for Al-air batteries [40]. The activity and stability of Co_3O_4/KB toward ORR were notably amplified when it was mixed with CeO_2 NPs. Goswami et al. reported the synthesis of CuO_x-CeO_2/C with rich oxide-oxide and oxide-carbon interfaces that exhibited high bifunctional ORR/OER activities in an alkaline medium [41]. They demonstrated the lattice spacing incorporated with characteristics of oxide-oxide and oxide-carbon interface structure in CuO_x-CeO_2/C along with the ORR/OER performance exhibited by the catalyst. The higher ORR/OER performance of CuO_x-CeO_2/C was attributed to the synergistic effects, enriched oxygen vacancies owing to the presence of Ce^{4+}/Ce^{3+}, abundant oxide-oxide, and oxide-carbon heterointerfaces, and regular distribution of oxides over the carbon matrix that facilitates faster electronic conduction. Wang and coworkers synthesized another type of MMO, $CoNiO_2$/SNC nanocomposites with the unique waxberry-like hollow architecture as trifunctional electrocatalysts for OER, ORR, and HER [42]. The electrocatalyst demonstrates remarkable activity for OER, ORR, and HER in an alkaline medium.

15.4 TAILORING THE STRUCTURE OF METAL OXIDE ELECTROCATALYSTS

15.4.1 Morphology

It is well-recognized from the crystallographic correlations, that the shape of a nanoparticle strongly influences the pattern of the atoms at its surface, i.e., its structure. The catalytic behavior of a material is inherently related to the structural shape and size of the catalyst. Hence, the shape and size control of MOs are considered as an influential command for maneuvering the properties of MOs towards oxygen electrocatalysis [43]. However, the correlation between morphology and electrocatalysis has not been well explored or understood yet. Zhang et al. investigated Cu_2O with three different morphologies (sphere, octahedron, and truncated octahedron) using the potentiostatic electrodeposition technique [44]. The ORR performance exhibited by the Cu_2O with truncated octahedron shape was found to be superior and based on the comprehensive study of electrocatalysis experiments and DFT calculations, it was observed that truncated Cu_2O octahedron preferentially exposes the (100) facet, which is favorable to effective adsorption and activation of O_2 on the surface of Cu_2O. Similarly, Yang and the group reported single-crystal $(Mn,Co)_3O_4$ octahedra, synthesized via a precipitation-aging process [45]. Based on the Mn/Co ratio the $(Mn,Co)_3O_4$ octahedra expose (111) and (011) facets of cubic and tetragonal phases, respectively. The single-crystal octahedra of Mn_2CoO_4 and $Mn_{2.5}Co_{0.5}O_4$ with (011) facets exhibited efficient selectivity towards ORR in an alkaline medium.

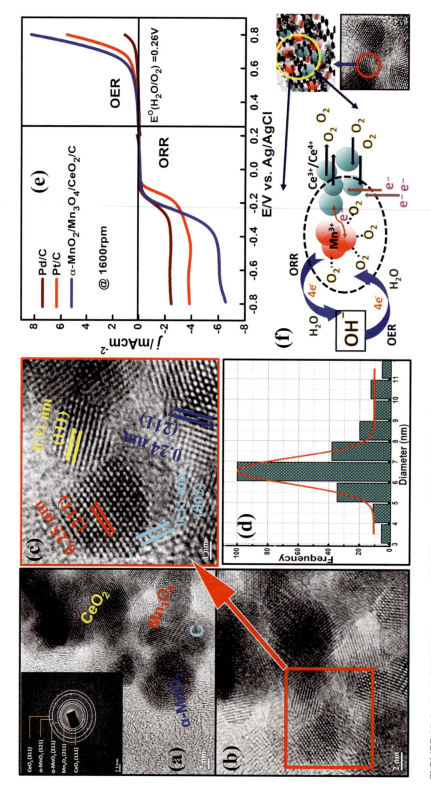

FIGURE 15.6 (a–c) HRTEM images and SAED pattern of α-MnO$_2$/Mn$_3$O$_4$/CeO$_2$/C hybrid nanostructure, (d) particle size distribution, (e) ORR and OER activities of α-MnO$_2$/Mn$_3$O$_4$/CeO$_2$/C compared with Pd/C and Pt/C, and (f) schematic illustration of α-MnO$_2$/Mn$_3$O$_4$/CeO$_2$/C catalyzed ORR and OER. (Adapted with permission [17]. Copyright (2021), American Chemical Society.)

15.4.2 Doping

The methodical doping of species into host structures technique has long been applied to modify the electronic, magnetic, and physical structure of semiconductors to increase their working efficiency [46]. Likewise, in catalytic systems, the selective metal doping into a MO matrix can result in the establishment of a large number of structural defects such as oxygen vacancies, interstitial defects, etc. [47] which is of great importance in industrial catalytic processes including oxygen electrocatalysis. Owing to this, tuning MOs with the doping method has become a potential technique for the development of advanced oxygen electrocatalysts. For instance, Zhang et al. reported Ru doped NiO/Co_3O_4 heterostructure as an advance trifunctional electrocatalysts for ORR/OER/HER [48]. The catalyst produced a current density of 100 mA cm^{-2} in alkaline solution at only 269 mV (OER) overpotential and a high $E_{1/2}$ of 0.88 V (ORR). In a test for the water-splitting device, the catalyst can function steadily for a longer duration (> 40 h), which is much better than Pt/C and RuO_2 at high current densities.

15.4.3 Oxygen Vacancy and Defects

Defect engineering is an inspiring strategy for fabricating highly efficient and stable electrocatalysts that have recently attracted widespread research attention. It can either modulate the adjacent electronic structure of defect sites for controlling activation energy of various adsorbed intermediates or serve as "docking" sites to stabilize the atomic moieties and form an additional unique synergetic coordination assembly as the active sites [49]. Defects in nanomaterials are usually recognized as the active sites for electrocatalytic processes because of the exceptional electronic and surface properties in the local vicinity. Defects in MOs are synonymous with cationic or anionic defects, broadly, which is characterized as oxygen vacancy and metal vacancy. In ORR/OER electrocatalysis, oxygen vacancy plays a crucial role in providing plentiful adsorption sites for a competent catalysis process. Consequently, defect engineering has been broadly implemented for fabricating MOs to enhance their electrocatalytic reactions.

15.4.4 Interface

In heterogeneous catalyst systems, interfaces are the edges between adjacent substances of the catalyst material which possess unavoidable defects that make it unique relative to its parent materials. Recently, interface engineering of MO-based heterostructures has received massive attention as it has been recognized as an important strategy for the design and advancement of oxygen electrocatalysts. A schematic illustration of the recent research attempts for the surface and interface engineering of metal nanocrystals in electrocatalysis is depicted in Figure 15.7 [50].

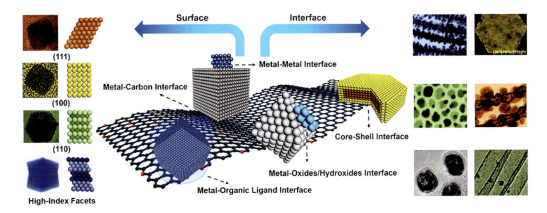

FIGURE 15.7 Schematic illustration of the research efforts for surface and interface engineering of metal nanocrystals in electrocatalysis. (Adapted with permission [50]. Copyright (2018), Elsevier.)

The interface in heterostructures provides a broader scope for modulating the reaction environment such as optimization of chemisorption for reaction intermediates, controlling the electron/mass transportation, and preventing active components from aggregating [51]. Interface engineering is accomplished by the building of heterojunction. The process involves the integration of different components such as metal-metal, metal-oxide, oxide-oxide, and metal-nonoxide (e.g., carbon, phosphate, chalcogenide, nitride, and carbide). A typical interfacial catalyst system of CuO_x and CeO_2 anchored on a carbon bed (CuO_x-CeO_2/C) was synthesized by Goswami et al. using a facile hydrothermal approach [41]. They established the oxide-oxide interfaces in the electrocatalyst which is not only between C (002) and CeO_2 (111) facets but also the oxide-oxide interface within CeO_2 (200) and CeO_2 (111) facets exposing abundant active sites for a favorable electrocatalytic process for ORR/OER.

15.5 CONCLUSIONS AND FUTURE SCOPES

Although, a huge development in the field of energy technologies can be witnessed in today's world, yet, the rational design of highly stable and active bifunctional ORR/OER catalysts for MABs is still challenging. This chapter summarized the current trends and advancements of various MO-based electrocatalysts for MABs. A brief discussion on the study of single MOs, perovskites, spinel oxides, and MMO along with their structure-activity relationship is included in this chapter. In recent years, the development of electrocatalysts is focused on tailoring interfaces and defective sites present in TMOs. Irrespective of the advancements in battery technology, many challenges are yet to be addressed. Aqueous MABs such as Al-air, Zn-air, and Mg-air batteries still undergo anode degradation and poor rechargeability. This needs to be overcome by developing various ORR and OER electrocatalysts to improve the anode quality, rechargeability along with reaction kinetics. The main challenges faced during the development of non-aqueous MABs include (a) synthesis of highly efficient bifunctional electrocatalysts that can reduce the overpotential of both ORR and OER processes; (b) lack of basic understanding of the reaction mechanism based on the type of electrocatalysts used; (c) instability of carbon-based cathodes and electrolytes which causes unwanted reactions to occur simultaneously, thereby lowers the cycling performances; (d) rational designing of the air cathode so that its open cell structure can be well utilized to achieve high specific capacity; and (e) the highly reactive nature of Li dendrites (mainly in Li-air batteries) formed during the reaction also needed to be rectified before commercialization. A correlation between the experimental and theoretical findings will help to explain the reaction mechanism of ORR and OER, for which advanced computing technologies such as theoretical calculations, model fitting, etc. need to be performed. Owing to the high surface area and high electronic conductivity, carbon-based nanomaterials can be greatly utilized for both ORR and OER in the future. The unstable nature of carbon-based materials can be overcome by doping with a suitable element. Moreover, it is quite necessary to understand the positive and negative impacts of different catalysts so that the gap between the research on functional materials and their applications in electrochemical energy storage is closed. Another important factor is to introduce an acceptable standard testing condition for the catalysts. It is really difficult to compare the catalytic performances of the catalysts that are already reported in the literature. This is because the choice of the cathode, choice of electrolyte and its volume, choice of salt and its concentration, loading of catalysts, etc. widely varies for different studies. Thus, a standard testing condition will greatly influence the kinetics and the thermodynamics of the system. Apart from the development of various catalysts, technical up-gradation of the components of MABs is very important. Thus, the design of cost-effective and new MAB technologies that can make the use of bifunctional catalysts is extremely demanding, without compromising their performance.

ACKNOWLEDGEMENTS

B.C. acknowledges CSIR, New Delhi for a senior research fellowship and Tezpur University, India for Research and Innovation grant.

REFERENCES

[1] M.A. Rahman, X. Wang, C. Wen, High energy density metal-air batteries: A review, J. Electrochem. Soc. 160 (2013) A1759–A1771.

[2] J. Yi, P. Liang, X. Liu, K. Wu, Y. Liu, Y. Wang, Y. Xia, J. Zhang, Challenges, mitigation strategies and perspectives in development of zinc-electrode materials and fabrication for rechargeable zinc-air batteries, Energy Environ. Sci. 11 (2018) 3075–3095.

[3] R. Cao, J.S. Lee, M. Liu, J. Cho, Recent progress in non-precious catalysts for metal-air batteries, Adv. Energy Mater. 2 (2012) 816–829.

[4] Z.L. Wang, D. Xu, J.J. Xu, X.B. Zhang, Oxygen electrocatalysts in metal-air batteries: From aqueous to nonaqueous electrolytes, Chem. Soc. Rev. 43 (2014) 7746–7786.

[5] A.G. Olabi, E.T. Sayed, T. Wilberforce, A. Jamal, A.H. Alami, K. Elsaid, S. Mohammod, A. Rahman, S.K. Shah, A.G. Olabi, E.T. Sayed, T. Wilberforce, A. Jamal, A.H. Alami, K. Elsaid, S. Mohammod, A. Rahman, S.K. Shah, Metal-air batteries: A review. Energies. 14 (2021) 7373.

[6] F. Cheng, J. Chen, Metal-air batteries: From oxygen reduction electrochemistry to cathode catalysts, Chem. Soc. Rev. 41 (2012) 2172–2192.

[7] H.F. Wang, Q. Xu, Materials design for rechargeable metal-air batteries, Matter. 1 (2019) 565–595.

[8] J.S. Lee, S.T. Kim, R. Cao, N.S. Choi, M. Liu, K.T. Lee, J. Cho, Metal-air batteries with high energy density: Li-air versus Zn-air, Adv. Energy Mater. 1 (2011) 34–50.

[9] C. Wang, Y. Yu, J. Niu, Y. Liu, D. Bridges, X. Liu, J. Pooran, Y. Zhang, A. Hu, Recent progress of metal–air batteries–A mini review, Applied Sciences. 9 (2019) 1–22.

[10] D.U. Lee, P. Xu, Z.P. Cano, A.G. Kashkooli, M.G. Park, Z. Chen, Recent progress and perspectives on bi-functional oxygen electrocatalysts for advanced rechargeable metal-air batteries, J. Mater. Chem. A 4 (2016) 7107–7134.

[11] Y. Li, J. Lu, Metal-air batteries: Will they be the future electrochemical energy storage device of choice?, ACS Energy Lett. 2 (2017) 1370–1377.

[12] Q. Dong, D. Wang, Catalysts in metal-air batteries, MRS Commun. 8 (2018) 372–386.

[13] L. Li, Z. wen Chang, X.B. Zhang, Recent progress on the development of metal-air batteries, Adv. Sustain. Syst. 1 (2017) 1–51.

[14] J. Pan, Y.Y. Xu, H. Yang, Z. Dong, H. Liu, B.Y. Xia, Advanced architectures and relatives of air electrodes in Zn-air batteries, Adv. Sci. 5 (2018) 1700691.

[15] M. Tahir, L. Pan, F. Idrees, X. Zhang, L. Wang, J.J. Zou, Z.L. Wang, Electrocatalytic oxygen evolution reaction for energy conversion and storage: A comprehensive review, Nano Energy. 37 (2017) 136–157.

[16] Y. Wang, F. Chu, J. Zeng, Q. Wang, T. Naren, Y. Li, Y. Cheng, Y. Lei, F. Wu, Single atom catalysts for fuel cells and rechargeable batteries: Principles, advances, and opportunities, ACS Nano. 15 (2021) 210–239.

[17] B. Chutia, N. Hussain, P. Puzari, D. Jampaiah, S.K. Bhargava, E.V. Matus, I.Z. Ismagilov, M. Kerzhentsev, P. Bharali, Unraveling the role of CeO_2 in stabilization of multivalent mn species on α-MnO_2/Mn_3O_4/CeO_2/C surface for enhanced electrocatalysis, Energy & Fuels. 35 (2021) 10756–10769.

[18] C. Goswami, K.K. Hazarika, Y. Yamada, P. Bharali, Nonprecious hybrid metal oxide for bifunctional oxygen electrodes: Endorsing the role of interfaces in electrocatalytic enhancement, Energy & Fuels. 35 (2021) 13370–13381.

[19] Y.S. Jeong, J.B. Park, H.G. Jung, J. Kim, X. Luo, J. Lu, L. Curtiss, K. Amine, Y.K. Sun, B. Scrosati, Y.J. Lee, Study on the catalytic activity of noble metal nanoparticles on reduced graphene oxide for oxygen evolution reactions in lithium-air batteries, Nano Lett. 15 (2015) 4261–4268.

[20] K.K. Hazarika, Y. Yamada, E.V. Matus, M. Kerzhentsev, P. Bharali, Enhancing the electrocatalytic activity via hybridization of Cu(I/II) oxides with Co_3O_4 towards oxygen electrode reactions, J. Power Sources. 490 (2021) 229511.

[21] Y.L. Zhang, K. Goh, L. Zhao, X.L. Sui, X.F. Gong, J.J. Cai, Q.Y. Zhou, H. Da Zhang, L. Li, F.R. Kong, D.M. Gu, Z.B. Wang, Advanced non-noble materials in bifunctional catalysts for ORR and OER toward aqueous metal-air batteries, Nanoscale. 12 (2020) 21534–21559.

[22] X. Wu, C. Tang, Y. Cheng, X. Min, S.P. Jiang, S. Wang, Bifunctional catalysts for reversible oxygen evolution reaction and oxygen reduction reaction, Chem.—A Eur. J. 26 (2020) 3906–3929.

[23] Y.J. Wang, H. Fan, A. Ignaszak, L. Zhang, S. Shao, D.P. Wilkinson, J. Zhang, Compositing doped-carbon with metals, non-metals, metal oxides, metal nitrides and other materials to form bifunctional electrocatalysts to enhance metal-air battery oxygen reduction and evolution reactions, Chem. Eng. J. 348 (2018) 416–437.

[24] A.A. Ambalkar, U.V. Kawade, Y.A. Sethi, S.C. Kanade, M.V. Kulkarni, P.V. Adhyapak, B.B. Kale, A nanostructured SnO_2/Ni/CNT composite as an anode for Li ion batteries, RSC Adv. 11 (2021) 19531–19540.

[25] T. Zhou, W. Xu, N. Zhang, Z. Du, C. Zhong, W. Yan, H. Ju, W. Chu, H. Jiang, C. Wu, Y. Xie, Ultrathin cobalt oxide layers as electrocatalysts for high-performance flexible Zn-air batteries, Adv. Mater. 31 (2019) 1–8.

[26] D. Yu, Y. Ma, F. Hu, C.C. Lin, L. Li, H.Y. Chen, X. Han, S. Peng, Dual-sites coordination engineering of single atom catalysts for flexible metal—air batteries, Adv. Energy Mater. 11 (2021) 1–9.

[27] J. Han, X. Meng, L. Lu, J. Bian, Z. Li, C. Sun, Single-atom $Fe-N_x-C$ as an efficient electrocatalyst for zinc—air batteries, Adv. Funct. Mater. 29 (2019) 1808872.

[28] L. Wang, Z. Xu, T. Peng, M. Liu, L. Zhang, J. Zhang, Bifunctional single-atom cobalt electrocatalysts with dense active sites prepared via a silica xerogel strategy for rechargeable zinc—air batteries, Nanomaterials. 12 (2022) 381.

[29] M.V. Ramos-Garcés, J.L. Colón, Preparation of zirconium phosphate nanomaterials and their applications as inorganic supports for the oxygen evolution reaction, Nanomaterials. 10 (2020) 822.

[30] S. Patowary, R. Chetry, C. Goswami, B. Chutia, P. Bharali, Oxygen reduction reaction catalysed by supported nanoparticles: Advancements and challenges, ChemCatChem. 14 (2022) e202101472.

[31] Y.-M. Zhao, F.-F. Wang, P.-J. Wei, G.-Q. Yu, S.-C. Cui, J.-G. Liu, Cobalt and iron oxides co-supported on carbon nanotubes as an efficient bifunctional catalyst for enhanced electrocatalytic activity in oxygen reduction and oxygen evolution reactions, ChemistrySelect. 3 (2018) 207–213.

[32] F. Cheng, Y. Su, J. Liang, Z. Tao, J. Chen, MnO_2-based nanostructures as catalysts for electrochemical oxygen reduction in alkaline media, Chem. Mater. 22 (2010) 898–905.

[33] Y. Gorlin, T.F. Jaramillo, A bifunctional nonprecious metal catalyst for oxygen reduction and water oxidation, J. Am. Chem. Soc. 132 (2010) 13612–13614.

[34] Y. Chen, K. Rui, J. Zhu, S.X. Dou, W. Sun, Recent progress on nickel-based oxide/(oxy)hydroxide electrocatalysts for the oxygen evolution reaction, Chem.—A Eur. J. 25 (2019) 703–713.

[35] X. Han, G. He, Y. He, J. Zhang, X. Zheng, L. Li, C. Zhong, W. Hu, Y. Deng, T.Y. Ma, Engineering catalytic active sites on cobalt oxide surface for enhanced oxygen electrocatalysis, Adv. Energy Mater. 8 (2018) 1702222.

[36] Y. Li, Z. Zhou, G. Cheng, S. Han, J. Zhou, J. Yuan, M. Sun, L. Yu, Flower-like $NiCo_2O_4$—CN as efficient bifunctional electrocatalyst for Zn-Air battery, Electrochim. Acta. 341 (2020) 135997.

[37] E. Rios, J.L. Gautier, G. Poillerat, P. Chartier, Mixed valency spinel oxides of transition metals and electrocatalysis: Case of the $Mn_xCo_{3-x}O_4$ system, Electrochim. Acta. 44 (1998) 1491–1497.

[38] H. Lee, O. Gwon, C. Lim, J. Kim, O. Galindev, G. Kim, Advanced electrochemical properties of $PrBa_{0.5}Sr_{0.5}Co_{1.9}Ni_{0.1}O_{5+\delta}$ as a bifunctional catalyst for rechargeable zinc-air batteries, ChemElectroChem. 6 (2019) 3154–3159.

[39] Y. Bu, O. Gwon, G. Nam, H. Jang, S. Kim, Q. Zhong, J. Cho, G. Kim, A highly efficient and robust cation ordered perovskite oxide as a bifunctional catalyst for rechargeable zinc-air batteries, ACS Nano. 11 (2017) 11594–11601.

[40] K. Liu, X. Huang, H. Wang, F. Li, Y. Tang, J. Li, M. Shao, Co_3O_4-CeO_2/C as a highly active electrocatalyst for oxygen reduction reaction in Al-air batteries, ACS Appl. Mater. Interfaces. 8 (2016) 34422–34430.

[41] C. Goswami, Y. Yamada, E.V. Matus, I.Z. Ismagilov, M. Kerzhentsev, P. Bharali, Elucidating the role of oxide-oxide/carbon interfaces of CuO_x-CeO_2/C in boosting electrocatalytic performance, Langmuir. 36 (2020) 15141–15152.

[42] Q. Zhang, W. Han, Z. Xu, Y. Li, L. Chen, Z. Bai, L. Yang, X. Wang, Hollow waxberry-like cobalt-nickel oxide/S,N-codoped carbon nanospheres as a trifunctional electrocatalyst for OER, ORR, and HER, RSC Adv. 10 (2020) 27788–27793.

[43] Y.X. Chen, S.P. Chen, Z.Y. Zhou, N. Tian, Y.X. Jiang, S.G. Sun, Y. Ding, L.W. Zhong, Tuning the shape and catalytic activity of Fe nanocrystals from rhombic dodecahedra and tetragonal bipyramids to cubes by electrochemistry, J. Am. Chem. Soc. 131 (2009) 10860–10862.

[44] X. Zhang, Y. Zhang, H. Huang, J. Cai, Electrochemical fabrication of shape-controlled Cu_2O with spheres, octahedrons and truncated octahedrons and their electrocatalysis for ORR, New J. Chem. 42 (2018) 458–464.

[45] H. Liu, X. Zhu, M. Li, Q. Tang, G. Sun, W. Yang, Single crystal $(Mn,Co)_3O_4$ octahedra for highly efficient oxygen reduction reactions, Electrochim. Acta. 144 (2014) 31–41.

[46] H. Ohno, Making nonmagnetic semiconductors ferromagnetic, Science. 281 (1998) 951–956.

[47] D.M. Smyth, Effects of dopants on the properties of metal oxides, Solid State Ionics. 129 (2000) 5–12.
[48] J. Zhang, J. Lian, Q. Jiang, G. Wang, Boosting the OER/ORR/HER activity of Ru-doped Ni/Co oxides heterostructure, Chem. Eng. J. 439 (2022) 135634.
[49] Y. Jia, K. Jiang, H. Wang, X. Yao, The role of defect sites in nanomaterials for electrocatalytic energy conversion, Chem. 5 (2019) 1371–1397.
[50] Y. Yang, M. Luo, W. Zhang, Y. Sun, X. Chen, S. Guo, Metal surface and interface energy electrocatalysis: Fundamentals, performance engineering, and opportunities, Chem. 4 (2018) 2054–2083.
[51] R. Zhao, Q. Li, X. Jiang, S. Huang, Interface engineering in transition metal-based heterostructures for oxygen electrocatalysis, Mater. Chem. Front. 5 (2021) 1033–1059.

16 Enhanced Performance of Lithium-Air Batteries by Improved Cathode Materials

B. Jeevanantham and M. K. Shobana

CONTENTS

16.1 Introduction 227
16.2 Electrolyte-Based Electrode Material..229
16.3 Problems Faced by Lithium-Air Batteries (LABs).......................................230
16.4 Rectification of Problems 231
    16.4.1 Carbonaceous Cathode..231
        16.4.1.1 Carbon Nanotubes (CNTs)...232
        16.4.1.2 Carbon Nanofibers (CNFs)...232
        16.4.1.3 Graphene/Graphene Oxide 233
        16.4.1.4 Mesoporous Carbon 235
    16.4.2 Metal/Metal Oxide-Based Cathodes 235
        16.4.2.1 Precious Metal/Metal Oxides 236
        16.4.2.2 Non-Precious Metals/Metal Oxides 236
    16.4.3 Other Cathode-Based Materials..241
16.5 Conclusion and Perspective..243
References..244

16.1 INTRODUCTION

Lithium-based electrode materials resolve the environmental impacts by the influence of excellent electrochemical material on automobile industries. Depending on the battery type, theoretical energy densities vary from 250 Wh/kg for lithium-ion batteries (LIBs) to nearly 2500 Wh/kg for lithium-sulfur batteries (LSB) and about 11000 Wh/kg for lithium-air batteries (LABs). This review focuses primarily on LABs due to their high practical energy density (in comparison to gasoline density (Figure 16.1). In the 1970s, the idea of LABs was proposed for electric and hybrid vehicles, but their scientific interest waned in the late 2000s due to the advancement of material sciences. Further, the LABs contain four major components, namely: the anode, cathode, electrolyte, and separator. Lithium anode is the suitable electrode material for oxygen reduction reaction (ORR) and oxygen evolution reaction (OER) processes. Carbon is the highly favored cathode material. The electrolyte is of four different types, namely aqueous, non-aqueous, hybrid, and solid-state, and the separator, which separates the electrode materials. Here, the cathode material with a porous structural framework consists of a conductive agent, a binder, and a catalyst. In which the oxygen (O_2) evolution and insertion take place, where the oxygen reacts with lithium ions and gets converted to lithium oxide. This lithium-air system provides energy density through lithium-oxide conversion. Likewise, lithium, aluminum, zinc, etc., are the different elements used as anode materials. Here, lithium metal is the best choice for anode material because it helps provide higher energy densities

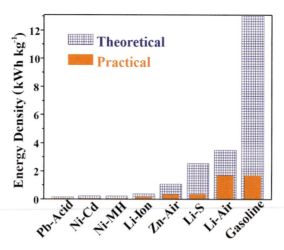

FIGURE 16.1 Specific energy density of gasoline compared with different rechargeable batteries. (Adapted with permission [10]. Copyright (2015), Royal Society of Chemistry.)

compared to the other interaction electrodes. In this case, the cathode (i.e., air electrode) contains a carbon matrix with a catalyst (which increases the reaction efficiency). Moreover, lithium metals have higher electro positivity (−3.04V vs. standard hydrogen electrode (SHE)) [1]. Similarly, the non-aqueous electrolyte systems were given more interest due to their higher energy density and excellent rechargeability than other electrolyte systems. In 1976, LABs with aqueous electrolyte systems were first proposed by Littauer and Tsai [2]. Then, after two decades of metal-air studies, Abraham et al. utilized organic polymer electrolytes for LABs [3]. Additionally, the separator in a LAB prevents intermediate reactants, oxygen from the cathode, and electrons from passing through the electrolyte.

In LABs, the ORR and OER are the two different reactions that are responsible for the discharge and charging process, respectively. In the discharging process, the oxygen atoms from the cathode get converted to a superoxide anion and react well with lithium ions to form lithium peroxide (LiO_2). This LiO_2 reacts with additional oxygen ions and gets transformed into Li_2O_2. Likewise, the mechanism of charging, or the OER process, takes place with the oxidation of Li_2O_2 to Li^+ and O_{2-} ions. In this case, the oxygen ions combine with electrons and get evolved from the system [4]. The electrochemical reaction of ORR and the OER process is as follows.

$$2Li^+ + O_2 + 2e^- \rightarrow Li_2O_2 \text{ (ORR)} \quad (1)$$

$$Li_2O_2 \rightarrow 2Li^+ + 2e^- + O_2 \text{ (OER)} \quad (2)$$

In ORR, LiO_2 is the intermediate product before Li_2O_2 and, further, the direct decomposition takes place at Li_2O_2 with the evolution of O_2. This reaction depends strongly on the kinetics of ORR [1].

LABs are capable to deliver high round-trip efficiency, good specific capacity, excellent rate capability, and outstanding cyclic performance. But there are several challenges associated with the design of a practical LAB. There are numerous problems including low energy efficiency, poor rate capability, higher charge overpotential, and shorter cycle life span, which are primarily the result of slow OER and ORR kinetics and sluggish cathode mass transport [5]. Also, these cells are sensitive to the anode passivation with lower operating current density due to the contamination of air and moisture [6]. Likewise, the dry air contains different constituents like CO_2 (0.03%), O_2 (21%), N_2 (78%), and various gases, whereas the open system undergoes different reactions and affects the battery system effectively. Further, the complicated reactions and the half-open system mechanism of LABs require high catalytic activity, excellent electronic conductivity, and high porosity to design the cathode material. Here, the high catalytic activity of the LAB not only facilitates the increase in the OER/ORR process but also reduces the overpotential. Likewise, the electronic conductivity of the air battery improves the fast electrochemical reactions, and its high porosity generates and stores the discharge products [5]. Many challenges remain unsolved in the LAB system. Thus, the ultimate aim of LABs is to achieve good round-trip efficiency, high practical capacity, and long cyclability. In this review, the lithium-air cathodes are mainly focused on their performance challenges and advancements.

16.2 ELECTROLYTE-BASED ELECTRODE MATERIAL

High energy density LABs consist of four different types of electrolytes, where each electrolyte classifies various electrode systems. In Figure 16.2, electrolyte systems contain aqueous, aprotic (non-aqueous), mixed hybrid, and solid-state systems. An aqueous system contains a carbon cathode, a lithium anode, and lithium salts that are dissolved in water as the electrolyte. In this case, the cathode material clogging is highly rectified due to the water-soluble reaction products. But the lithium anode reacts with water and shows serious effects. Therefore, an artificial secondary electrolyte interface (SEI) is required for the anode material. Also, at the air cathode, water molecules undergo the redox reaction and are shown in the following equation.

$$2Li + \frac{1}{2}O_2 + H_2O \rightarrow 2LiOH \tag{3}$$

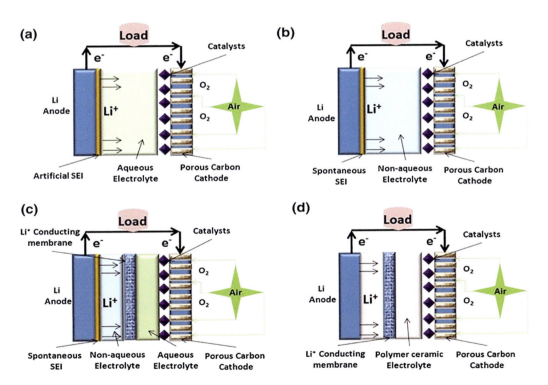

FIGURE 16.2 Electrolyte-based cathode material (a) aqueous, (b) aprotic, (c) hybrid, and (d) solid-state type. (Adapted with permission [1]. Copyright (2013), Springer Nature.)

Similarly, aprotic electrolytes are highly favored electrolytes due to their large energy density and charge reversibility. Lithium salts such as $LiPF_6$, $LiSO_3CF_3$, $LiAsF_6$, etc., esters, ethers, and organic carbonates have been used as non-aqueous electrolytes. The discharge product is Li_2O_2, which is insoluble in organic solvents and accumulates at the cathode/electrolyte interface, causing clogging and volume expansion. Further, the transport of O_2 gets affected due to the solubility of O_2 within the electrolyte. Thus, dual-pore material inhibits these drawbacks and offers an excellent capacity. Here, one pore serves as an oxidation product store, and the other one is for O_2 transport [7]. Likewise, a mixed hybrid electrolyte system combines the advantages of aqueous and aprotic electrolyte systems. In which the aprotic (non-aqueous) electrolyte is on the side of the lithium anode and the aqueous solution is in contact with the lithium cathode, drawbacks from each electrode material are highly rectified. Similarly, a solid-state electrolyte system contains polymer (glass)-ceramic composites, which reduce the battery's overall impedance. Further, the lithium-ion charge transfer is enhanced at the anode and couples the cathode electrochemically with the electrolyte. Also, the solid-state electrolyte eliminates the chance of ignition from rupture [8]. Moreover, the lower conductivity of lithium ions compared with liquid electrolytes is the main drawback in the case of solid-state [9].

16.3 PROBLEMS FACED BY LITHIUM-AIR BATTERIES (LABs)

LABs cause different effects, like larger overpotential, oxygen singlet, reactive oxygen species (ROS), and impurity gas. Here, the large overpotential causes severe decomposition of the carbon, electrolyte, and binders with weak electrochemical stability and also provides insoluble side products and impurity gases. The second is singlet oxygen, where the degradation of carbon, redox mediators, and electrolyte takes place due to the effect of the newly found reactive intermediate charge/discharge process. Third, during charging/discharging, ROS with low chemical stability (O_2^-/Li_2O_2 and O_2^-/LiO_2) attack the functional groups of carbon, binder, and electrolyte. Finally, impurity gases such as CO_2, N_2, and H_2O produced by electrolyte or carbon dioxide degradation alter the final discharge products and reaction pathways. In addition, it results in the continuous consumption of lithium ions and an unstable SEI, which hinders the li-ion intercalation and leads to cell death [4].

Further, in an aprotic electrolyte system, the decomposition between the electrolyte and the oxygen forms an irreversible organic and inorganic carbonate rather than Li_2O_2 [10]. Further, in LABs, the open cathode system undergoes reactions with ambient air (i.e., CO_2 and moisture) and causes a serious effect on oxygen transport. The following reactions show the effects of open system lithium-air with ambient air.

$$2Li + O_2 + H_2O \rightarrow Li_2O_2 \quad (4)$$

$$4Li + O_2 + 2H_2O \rightarrow 4LiOH \quad (5)$$

$$4Li + O_2 + 2CO_2 \rightarrow 2Li_2CO_3 \quad (6)$$

$$2Li_2O_2 + 2CO_2 \rightarrow 2Li_2CO_3 + O_2 \quad (7)$$

$$2Li_2O_2 + 2H_2O \rightarrow 4LiOH + O_2 \quad (8)$$

In ambient air, carbon dioxide (CO_2) is highly soluble in organic solvents and shows high reactivity with Li, LiOH, and Li_2O_2 and gets transformed into Li_2CO_3 (destroyer of LABs). In addition, the product formed is much more stable than the Li_2O_2, which mitigates the electrochemical charging behavior. In addition, the obtained Li_2CO_3 increases the electrode polarization in the discharge process and reduces the voltage-cyclic efficiency in the charging process [11]. Furthermore, the lithium ions that combine with O_2 (50%) and CO_2 (50%) show good discharge capacity, but the Li_2CO_3 obtained during this process acts as an electrical insulator, and it is difficult to decompose [12].

Further, the decomposition of Li_2CO_3 takes place at higher voltages (above 4V) with reduced round-trip efficiency, so the voltage should be maintained at 4.38–4.61V [11]. Likewise, during discharge, lithium-ions react with O_2 and form LiO_2 and Li_2O_2. Further, the obtained products fill in the air cathode pores, preventing ORR and affecting both the specific and volumetric energy densities. Also, LiO_2 is not oxidizable during charging and its formation is favored at high discharge rates [1]. Besides O_2, lithium ions also react with nitrogen (78% of the earth's atmosphere) in the air. Zhang et al. show this reversible reaction in the potential range of 0.8–1.2V by using the reaction,

$$6Li + N_2 \rightarrow 2Li_3N \qquad (9)$$

But the effect of N_2 on LABs is much smaller when compared with CO_2 and moisture [13].

Furthermore, it is problematic in the case of lithium anode when lithium anode reacts with electrolyte and forms an SEI over the anode surface. This SEI leads to uneven current distribution, which paves the way for dendrite growth and short circuits [1]. By increasing the anode's surface area, the dendrite growth consumes a large amount of electrolyte and produces a new passivation layer. Further, due to the electrolyte consumption, the performance of the battery deteriorated after a few cycles. In addition, the dendrite growth damages the separator and causes a short circuit by reaching the cathode layer [14].

16.4 RECTIFICATION OF PROBLEMS

Rechargeable LABs with a non-aqueous electrolyte system must possess a few characteristics like high specific energy, good round-trip efficiency, low overpotential, and high-rate capability. The uniform distribution of highly active catalyst and optimum catalyst over the electrode material enhances the cathode structure by improving the specific capacity. Electrocatalyst reduces the charge/discharge overpotential and improves the round-trip efficiency [1]. Moreover, due to slow kinetics, there will be a deviation in the charge and discharge voltages, which leads to 100% coulombic efficiency with lower round-trip efficiency. Furthermore, the use of bi-functional catalysts accelerates the OER and ORR kinetic processes, which increases the discharge voltage and round-trip efficiency, as well as decreases the charging voltage. In addition, the rate capability and cyclic performance are improved by the better active catalyst and help to enhance the Li_2O_2 reduction rate during charging [10]. Electrocatalysts used in various works, specifically carbon and graphene, metal oxide and their components, and precious metal alloys aid to form an efficient cathode. Likewise, in an aqueous electrolyte system, the artificial SEI should be placed over the lithium anode to avoid corrosion. Further, there are numerous attempts made to overcome the drawbacks. First, choose a less reactive electrolyte (liquid or polymer) with a lithium anode. Second, place an ultrathin polymer over the lithium metal surface. Third, HF addition in the electrolyte solution gives a smooth surface by transforming into LiF on the metal surface. During the charge and discharge process, the catalyst should be integrated with the cathode to reduce overpotentials. Moreover, the embedded catalyst not only provides good cyclability but also excellent energy efficiency [15]. This chapter discusses the role of the non-aqueous cathode system, which includes the cathode reaction mechanism and its properties [1].

16.4.1 CARBONACEOUS CATHODE

Several scientific hurdles have to be overcome to design a rechargeable LAB, including: lack of discharge due to the block of discharge products, alkyl carbonate formation, unstable electrode in moist air, higher charge overpotential compared to discharge, inadequate knowledge of catalytic effect, and decomposition of carbonate during discharge. In addition, the synthesis and design of a highly conductive novel porous carbon must provide sufficient O_2 diffusion channels, a three-dimensional (3D) porous structure (for mass transport and Li_2O_2 deposition), high stability (for ROS), and high catalytic activity (to boost round-trip efficiency). Furthermore, they must be electrolyte-wettable and maintain a three-phase interface during charging and discharging. Additionally, it should prevent catalyst corrosion and the ingression of CO_2 and H_2O into the cell.

Moreover, carbon materials aid in various fields, namely electrode materials in fuel cells, supercapacitors, batteries, conductive agents, and catalytic supporters. It is the most highly favored cathode material for LABs, because of its low cost, light weight, large surface area, tunable pore structure, and high conductivity [5]. Subsequently, due to their high reaction activities, large pore volume, and economic merits, they have been employed as supporters or catalysts. Further, there is no difference in ORR potential with or without a catalyst, because the porous carbon reveals enough catalytic activity in LABs without the other catalyst [15]. Moreover, the decomposition of carbon materials takes place due to the highly oxidizing environment and discharge products of LABs. In addition, the incompatibility of carbon materials and the electrolyte continually brings about side reactions that degrade the energy efficiency and the cycle life.

For the cathode, carbon material modification, advanced material engineering, and nanocarbon materials utilization should be adopted to mitigate the parasitic reaction and enhance the catalytic activity. Here, porous carbon materials such as carbon nanotubes, carbon nanofiber, and graphene are used as cathode materials. But, as it is difficult to develop the 3D network, they are combined with current collectors and hardly contribute to the charge/discharge reaction with insufficient surface area.

16.4.1.1 Carbon Nanotubes (CNTs)

Several researchers have investigated carbon nanotubes because of their high conductivity, regular pore structure, and excellent thermal and chemical stability. CNTs have a nanofibrous structure with good tolerance to mechanical stress. During the deposition or decomposition of discharge solid, it retains the conductive network in the electrode. CNTs that are prepared through simple vacuum filtration of suspension have been used as lithium-air cathodes (not as additional current collectors) with moderate surface area. However, the meso- and macro-pore voids formed naturally and contribute to the high discharge capacity [16]. Ushijima et al. synthesized CNTs through the liquid pulse technique (LPI) in a high aspect ratio (that is twice larger than the carbon black and carbon paper) and made them mix with thicker fibrous carbon. As a result, the gas permeation resistance decreases with an increase in discharge capacity at high current densities [17]. Similarly, partially cracked carbon nanotubes were used as cathodes by Li et al., which showed two times higher discharge capacity (1513 mAh/g) than the raw multi-walled carbon nanotubes (MWCNT) [18]. Further, to improve the conductivity of CNT, nitrogen (N) doping is highly favored. Because N-doping improves the chemical activation on the CNT passive surface as well as assures good corrosion resistance and high electrical conductivity. Using the floating catalyst chemical vapor deposition method, Mi et al. synthesized N-CNT, which provides more nucleation sites and promotes higher dispersion of discharge products [19]. Furthermore, in 2019, Li et al. prepared N-CNT on carbon paper with a conductive carbon-black layer as an air cathode. It achieved a higher reversible capacity of 2203 mAh/g at 2.81V than active carbon (497 mAh/g) [20]. A similar composite (RGO/CNT) with a high surface area (564.48 m^2/g) was developed by Salehi et al. which has a discharge capacity of 9000 mAh/g at 100 A/g [21]. The mesoporous Co_3O_4-rods-entangled carbonized polyaniline nanotubes (Co_3O_4-e-cPANI) electrode was developed by Li et al. via a hydrothermal method (Figure 16.3). Here, these electrodes have a high discharge capacity of 32478 mAh/g at 100 mA/g of current density with a cyclability of 430 cycles at 500 mA/g [22].

16.4.1.2 Carbon Nanofibers (CNFs)

CNFs are less expensive when compared with CNTs, but the aspect ratio is too low to be formed into sheets without a binder. The high aspect ratio of CNF sheets using the LPI technique aids in increasing the discharge capacity by decreasing the gas permeation resistance [18]. In addition, the

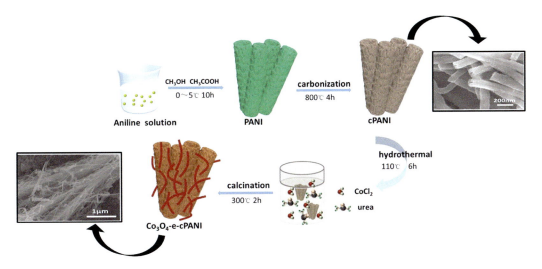

FIGURE 16.3 Preparation of Co_3O_4-e-cPANI via hydrothermal technique. (Adapted with permission [22]. Copyright (2018), American Chemical Society.)

electrospinning technique helps to obtain high aspect ratio CNF sheets with an increased diameter. Han et al. prepared a CNF web with different graphitization degrees, such as 1000, 1200, and 1400 °C. Here, the 3D CNF web enables the uniform pore structure and provides electron transport along one-dimensional (1D) geometry. Further, pyrolyzing the CNF web at 1400°C improves cyclic retention and lowers polarization [23]. Additionally, Crespiera et al. developed cobalt (Co) doped CNF that displays a large reversible capacity of 3700 mAh/g. Compared with cobalt-free cathodes, Co-containing cathodes have lower polarization and better active sites [24]. Likewise, Song et al. synthesized a binder-free Co_3O_4/carbon nanofiber composite via electrospinning and postthermal treatment of zeolite imidazolate framework-9/polyacrylonitrile (ZIF-9/PAN) nanofiber. The cathode material possesses higher discharge-charge capacities, enhanced cyclic performance, and lower charging overpotentials due to the presence of Co_3O_4 within the carbon nanofiber [25]. In addition, Zhang et al. fabricated CoO_x nanoparticles on N-doped carbon nanofiber via electrospinning, which significantly improves the electrocatalytic activity by modulating the Co^{2+}/Co^{3+} ratio. Further, this catalyst demonstrates an outstanding discharge capacity of 7763.7 mAh/g, good performance towards the ORR and OER, and high cyclic stability [26]. Likewise, through coaxial electrospinning, Bui et al. distributed platinum (Pt) metal nanoparticles over the CNF surface. Here, the Li_2O_2 film formed over the cathode reduces the charge/discharge overpotentials and enables a prolonged cycle life of about 163 cycles. Further, at 500 mA/g, it shows a high specific discharge value of 1000 mAh/g [27].

16.4.1.3 Graphene/Graphene Oxide

Graphene/graphene oxide act as cathode materials for LABs with the benefit of high conductivity, large surface area, good electron transfer, and excellent round trip efficiency. Electron distribution in homogenous sp^2 and sp^3 hybridization allows graphene to act as an excellent two-dimensional (2D) and 3D supporter for ORR and OER electrocatalysts [28]. Li et al. first employed graphene nanosheets as a LAB cathode. In this, the graphene sheets provide ideal porosity; after discharge, the products get deposited over the edges of the sheets. Thus, a thicker or darker color was observed (Figure 16.4 (a-d)), which shows the discharge capacity of 8705.9 mAh/g (Figure 16.4e) [29]. Salehi et al. used ultrasonic irradiation to prepare the GO/CNT composite and converted it to rGO/CNT by heating (thermal and reducing gent procedure). It shows the capacity of 9000 mAh/g at 100 mA/g density and a voltage plateau at 2.78V with a 564.48 m^2/g surface area [21]. Likewise, Sun et al. synthesized graphene with different pore structure diameters (60 nm, 250 nm, and 400 nm), where the 250 nm pore-sized graphene achieves the highest capacity of the other pores. Further, they discovered ruthenium (Ru) nanocrystals with graphene (Figure 16.4f), which promote the OER with excellent reversible capacity (17700 mAh/g), outstanding specific capacity at large current density, and reduced charge/discharge overpotential (about 0.355V) (Figure 16.4g) [30]. Further, to facilitate the ORR, OER, and many other oxidation reactions, noble metal nanoparticles are better known for their catalytic activity [31]. Furthermore, because of the stronger interaction over the functionalized graphene surface compared to conventional carbon, 2D graphene (rGO) is highly favored to interact with noble metals [32]. Thus, these catalysts show less aggregation and better dispersion. Further, the catalytic properties are enhanced by the inherent porous structure and large surface area. In addition, the noble metal nanoparticles that interact with rGO reduce the charge overpotentials and make them insensitive to discharge potentials. Using the one-pot polyol synthesis method, noble metal nanoparticles such as Ru, Pt, and Pd loaded on rGO were assessed as OER electrocatalysts. In which, the Ru-rGO composite shows stable cycling performance and lower charge potential. Here, the discharge products are decomposed at lower potentials due to the Ru nano catalysts. Additionally, in Pt- and Pd-rGO composites, the electrolyte decomposition causes fluctuations in the voltage profiles during cycling. Although electrolyte decomposition occurs after electrochemical cycling for Pt- and Pd-rGO catalysts, no electrolyte stability is observed for Ru-rGO [33].

Moreover, to enhance the catalytic activity, Ren et al. synthesized PdSnCo nanoparticles supported by nitrogen-doped graphene oxide (PdSnCo/NG) via a solvothermal method. With the 6% Pd

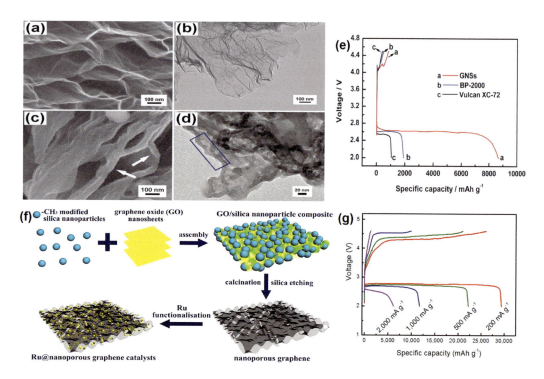

FIGURE 16.4 Scanning electron microscope (SEM) and transmission electron microscope (TEM) images of graphene nanosheets (a, b) before discharge, (c, d) after discharge, (e) graphene nanosheet discharge capacity comparison with other carbon materials. (Adapted with permission [29], Copyright (2011), Royal Society of Chemistry.) (f) Synthesis of Ru-functionalized nanoporous graphene catalysts, (g) charge/discharge profile of Ru-functionalized nanoporous graphene at different current densities. (Adapted with permission [30]. Copyright (2014), American Chemical Society.)

content, the cathode delivers an excellent reversible capacity value of 6750 mAh/g, lower charge voltage, and good cyclability [34]. Further, the highly interconnected hydrophobic 3D graphene sheets aid in a highly porous structure for oxygen and electrolyte ion diffusion and act as moisture-resistive and efficient charge transport cathode materials. Zhong et al. synthesized a densely packed, binder-free, hydrophobic 3D graphene cathode membrane that operates under ambient conditions. Further, they selectively generate O_2/H_2O and retard the diffusion of moisture in the air and provide an excellent high capacity of 5700 mAh/g for 20 cycles and high cycling performance of greater than 2000 cycles at 140 mAh/g and above 100 cycles at 1400 mAh/g [35].

16.4.1.4 Mesoporous Carbon

A carbon membrane with a 2 to 50 nm pore size is known as mesoporous carbon. Due to its remarkable porosity for oxygen diffusion, mesoporous carbon is the most attractive cathode material. Moreover, they are classified into two types, namely, ordered and disordered mesoporous carbon. Because of outstanding electrical conductivity, faster mass kinetics, and higher specific surface area, ordered mesoporous carbon (OMC) is highly preferred for aprotic electrolyte systems. Sun et al. synthesized a mesoporous carbon (CMK-3), an OMC prepared by rod-shaped silica SBA-15 with hollow rod-shaped pores with a 5nm pore size that has comparable pore volume and specific surface area to Ketjenblack (KB). In addition, the CMK-3 loaded with CoO nanoparticles via the wet impregnation method acts as a cathode material for LABs. It exhibits higher capacity retention than the Super-P and pristine CMK-3 carbons. Furthermore, up to 15 cycles, this cell exhibits excellent cycling stability of 1021 mAh/g capacity than the Super P carbon. During discharging, the CMK-3 carbon facilitates oxygen diffusion and releases products like CO_2 and H_2O. Here, the CoO nanoparticles increase the round-trip efficiency by reducing the charge overpotential [36]. Likewise, carbon replica (CR), which is synthesized with an 11 nm pore size and a 919 m^2/g surface area, has also been used in LABs and shows similar properties as that of KB [37]. Yui et al. incorporated $Pt_{10}Ru_{90}$ electrocatalysts with CMK-3 and CR. Further, they are examined in a dry air atmosphere under a 0.1 mA/cm^2 density within the voltage range between 2.0 and 4.2V. Here, the $Pt_{10}Ru_{90}$/CMK-3 cell delivers a less discharge capacity of about 103 mAh/g, while the $Pt_{10}Ru_{90}$/CR composite delivers a high discharge capacity value of 1000 mAh/g, and up to 9 cycles, it achieves a cyclic strength of 828 mAh/g. Meanwhile, the incorporated CMK-3 and CR with or without electrocatalysts show an increased coulombic efficiency of 39 to 78% and 62 to 100% for the first cycle [38].

Similarly, Song et al. coated mesoporous carbon with carbon nanofibers via an electrospinning and nano-casting strategy without any binder or catalysts. By linking nanofibers with one another, the mesoporous coated carbon nanofibers formed a 3D cross-linked structure. Further, the coating results in an enlarged surface area (i.e., from 708 m^2/g to 2194 m^2/g) and achieves higher electrically conductive kinetics of 4.638 S/cm than the nanofiber (which shows only about 3.0759 S/cm). In addition, with electrochemical testing, the coated nanofiber exhibits a discharge capacity of nearly 4000 mAh/g, whereas the pristine nanofiber achieves only 2750 mAh/g. Also, it has the greatest potential to be employed as a lightweight and efficient electrode material for LABs [39]. Further, to improve the degree of graphitization of the mesoporous carbon nanofiber (MCN), they have doped it with cobalt (Co) nanoparticles. In MCN, the homogeneously distributed Co nanoparticles show a specific surface area in the range of 40–300 m^2/g. Further, the doped MCN achieves a pore size of 10–50 nm with 3700 mAh/g as specific capacity and excellent cycling performance. Here, the Co embedded in MCN reduces the discharge polarization and charge overpotential and also enhances the cycling durability at curtailed capacity [40].

16.4.2 Metal/Metal Oxide-Based Cathodes

To improve the performance and charge-discharge process of the metal-air battery, metal/metal oxide catalyst-based cathodes have been used in LABs. Here, metal ions in metal oxides are partially substituted in other composite structures, which reduces the electrode overpotential, improves the OER and ORR activity, and enhances electron transportation. It further regulates the position of metal atoms and the electron structure and aids in the outstanding performance of the LABs [5]. In addition, metal and metal oxide have been categorized into different types, namely, precious metal/metal oxide and non-precious metal/metal oxide.

16.4.2.1 Precious Metal/Metal Oxides

For ORR and OER chemical reactions, precious metals are considered the most important catalysts. During the charging-discharging process, employing precious metals in LABs not only enhances the overall performance but also reduces the electrodes' overpotential. However, the precious metal/metal oxide catalysts are rarely reported because of their scarce reserves and high price. In addition, the most investigated precious metal cathode catalysts include palladium (Pd) [41], ruthenium (Ru) [42], platinum (Pt) [43], and gold (Au) [44]. Shen et al. reported a porous Pd-CNT sponge cathode prepared via a chemical vapor deposition (CVD) method followed by an electrochemical deposition technique. Here, the Pd nanoparticle existence achieves the higher ORR catalytic activity. Further, in this case, the ceramic electrolyte protects the positive electrode as well as non-volatile ionic liquids in the cathode. By controlling the ionic liquid amount, the sponge becomes wet and allows for the continuous passage of oxygen, lithium ions, and electrons. Furthermore, this battery withstands high humid air levels and attains a high specific discharge capacity of about 9092 mAh/g [41]. However, due to this catalytic activity, there will be a decomposition of the weak electrolyte, so a stable electrolyte is highly required for LABs. Jung et al. developed rGO with Ru-based nanomaterials as a cathode and TEGDME-LiCF$_3$SO$_3$ as an electrolyte. Here, the rGO composite with Ru and RuO$_2$.0.64H$_2$O shows a pore size of less than 2.5 nm (Figure 16.5(a-f)), which functions as electrocatalysts and maintains excellent cycling stability without electrolyte decomposition up to 30 cycles. In addition, both Ru-rGO and RuO$_2$.0.64H$_2$O composites show excellent catalytic activity in the OER reaction, but the latter has a larger reversible capacity of 5000 mAh/g than Ru-rGO and reduces the average charge potential to ~3.7V at a current density of 500 mA/g [42].

Further, the bi-functional catalysts (combining two precious metals) demonstrate high round-trip efficiency but do not provide a good specific capacity. Also, this combination makes a high-cost air cathode, so new catalysts with reduced costs are in higher demand. In precious metals, Pt increases the OER process, and Au enhances the kinetics of the ORR process. Similarly, the Au-Pd nanoparticles prepared with mesoporous β-MnO$_2$ composite via a hydrothermal method act as an excellent cathode catalyst with better morphology (Figure 16.5 (g-i)). Silica KIT-6 aids in synthesizing this composite. Further, this cathode catalyst improves the oxygen evolution and reduction reactions; it enhances the energy and coulombic efficiencies of lithium-air cells. Moreover, it attains a high specific discharge value of 775 mAh/g at a 0.13 mA/cm^2 current density (Figure 16.5j). Here, the observed capacity value is due to the oxidation of Li$^+$ ions to Li$_2$O$_2$ at 2V discharge [43]. Further, the deposition of Pt$_x$Co$_y$ (x:y = 4, 2, 1, and 0.5) composite over the Vulcan XC-72 (Vxc-72/Pt$_x$Co$_y$) carbon via the chemical reduction method shows the increased specific capacity. The increase of Co content in Pt$_x$Co$_y$ significantly improves the catalytic activity of the LABs. Here, the PtCo$_2$ catalyst achieves a high specific capacity of 3040 mAh/g, which is three times larger than the bare carbon, and the relationship between the cycle number and specific capacity is presented in Figure 16.5k [44].

16.4.2.2 Non-Precious Metals/Metal Oxides

The high cost of precious metals creates an unavoidable drawback to the LABs that drastically affect their practical applications. Further, to replace these metals, several researchers chose cheap metals (transition metals) for LABs. Moreover, this section demonstrates the different transition metal-based cathode materials, including metals, metal oxides, sulfides, carbides, and nitrides.

Enhanced Lithium-Air Batteries by Improved Cathode Materials

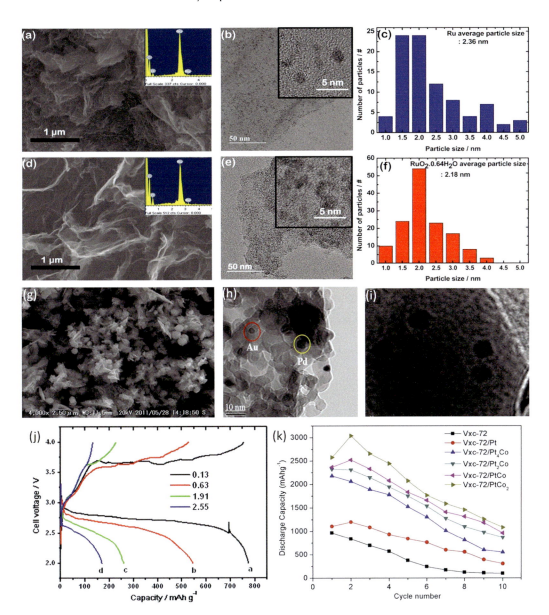

FIGURE 16.5 (a, d) SEM images of Porous Ru-rGO and $RuO_2 \cdot 0.64H_2O$-rGO samples; (inset) SEM-EDX images of Ru-rGO and $RuO_2 \cdot 0.64H_2O$-rGO composites, (b, e) TEM images of Ru-rGO and $RuO_2 \cdot 0.64H_2O$-rGO composites; (inset: high-resolution transmission electron microscope (HRTEM)) images), (c, f) Ru-rGO and $RuO_2 \cdot 0.64H_2O$-rGO hybrids particle size distribution. (Adapted with permission [42]. Copyright (2013), American Chemical Society.) (g) SEM, (h) TEM, and (i) HRTEM images of Au-Pd nanoparticles with mesoporous β-MnO_2, (j) Charge/discharge profile of Au-Pd supported β-MnO_2. (Adapted with permission [43]. Copyright (2012), Elsevier.) (k) Capacity vs. cycle number curves of Pt_xCo_y (x:y = 4, 2, 1, and 0.5)/Vulcan XC-72 carbon electrodes. (Adapted with permission [44]. Copyright (2013), Elsevier.)

16.4.2.2.1 Transition Metals/Metal Oxides

LABs with excellent ORR and OER activity have utilized highly catalytic transition metal oxides as cathode catalysts. They also act as promising catalytic substitutions for precious metals, and they are categorized into three types, namely, transition metal chalcogenides, oxides, and hydroxides. In which, the transition metal oxides and metal hydroxides are investigated for mono- and bi-functional ORR/OER to enhance the LABs efficiency, because the transition metals provide efficient catalytic activity, controllable structure, and low cost. Significantly, 2D oxides and hydroxides provide excellent ion transport by providing a large surface area and also provide a large platform to break the chemical bonds, which improves the electroactivity reaction in LABs. Similarly, transition metal chalcogenides have higher catalytic activity and excellent conductivity than oxides [28]. Also, the remarkable activity of transition metals provides durability, abundance, and environmental friendliness. Catalysts like nickel (Ni) and transition metal oxides show excellent OER activity, while Co, Pt, and Pd demonstrate OER activity. Ji et al. designed a bifunctional catalyst from a 3D hierarchical structure of carbon with transition metals like Co, Fe, and Ni for LABs. This composite exhibits a high catalytic activity with a lower overpotential of 430 mV at 10 mA cm^{-2} density for OER and 0.87V as a positive half-wave potential for ORR, and shows a good cyclic test life of over 120 h even in open air [45]. Similarly, Chang et al. encapsulated N-doped graphene sheets with a bifunctional catalyst of Co-Ni alloy via a hydrothermal method. Whereas Co and Ni exhibit excellent ORR and OER activity. In addition, the nanoalloy with N-doped graphene thin layer shows high structural stability, good catalytic activity, and a lower voltage potential of about 0.55V. Further, at 1.4 mA/cm^2 current density, it demonstrates a charge and discharges voltage gap of 0.55V. Moreover, it shows the high power and energy densities of 134.2 W/m^2 and 3158 Wh/kg at 7 mA/cm^2 [46].

Furthermore, the most studied metal oxides as catalysts, namely, TiO_2, Co_3O_4, SnO_2, Al_2O_3, etc. Here, TiO_2 provides excellent catalytic properties, is environmentally friendly, and shows high stability. TiO_2 exhibits outstanding performance during battery operation and protects the cathode layer from discharge products. Thus, Yang et al. coated TiO_2-X as a thin layer over the CNTs by the atomic layer deposition method. Here, TiO_2-X not only provides high catalytic activity toward the electrode reactions but also protects the cathode layer by inhibiting the reaction between carbon and electrolyte (excellent corrosion resistance). Further, this improvement in catalytic activity is due to the presence of Ti^{3+} and $Ti^{3.5+}$ ions in TiO_2. In addition, they provide a high reversible specific capacity of 11000 mAh/g at 300 mA/g density with nearly 1.8 times improved cyclability than pristine CNT [47]. Likewise, Co_3O_4 acts as a promising cathode catalyst due to its excellent catalytic properties in various applications. Yoon et al. synthesized a Co_3O_4/CNT (where Co_3O_4 is attached to the sidewall of a CNT) via a hydrothermal method. Based on electrode mass, the Co_3O_4/CNT composite demonstrates a reversible capacity value of 600 to 800 mAh/g and shows lower overvoltage during the charge/discharge process [48]. Peng et al. developed a LAB with a different hybrid electrolyte system. Figure 16.6 (a-d) depicts the morphologies of Co_3O_4/graphene cathode catalysts prepared by the hydrothermal method. Further, these composites have been investigated to find the optimum load of Co_3O_4 in graphene sheets, where Co_3O_4 is taken in different weight amounts like 33.7 wt%, 48.2 wt%, and 62.5 wt%. Moreover, the 48.2 wt% Co_3O_4 with a 14.2 m^2/g surface area and an 87.5 Å pore size shows a balanced weight for charge-transfer mediated graphene and good dispersibility of the catalyst on the graphene sheets. The observed result shows good catalytic activity with reduced polarization and the ambient air of 70–100% humidity [49].

Furthermore, at 500 °C, Lee et al. studied the solvothermally prepared SnO_2 cathodes in the size ranging between 100 and 200 nm with electrochemical properties. Where the amorphous-structured SnO_2 exhibits a discharge capacity value of 350 mAh/g at 10 mA/g [50]. Likewise, Li et al. used the bifunctional concept and synthesized graphene-like m-SnS_2-SnO_2/C (m—glucose amount) material via the facile hydrothermal method and then followed by high-temperature treatment with argon (Ar) atmosphere. In this case, the glucose content is given different weights to make the different SnS_2-SnO_2/C compositions. In which the increase in glucose content improves the performance of LABs. Further, in comparison with specific surface area and conductivity, the 225-SnS_2-SnO_2/C showed excellent performance [51]. Similarly, vanadium pentoxide (V_2O_5) has been employed in industrial applications. VO_4 (active sites) species present over the V_2O_5 surface aid in catalytic activity. Further, after 12 cycles, it provides an initial discharge capacity of 715 mAh g^{-1} and attains a maximum specific discharge capacity of 2260 mAh/g. In addition, they retain a higher capacity of

1.24% during 2 to 8 cycles [52]. Additionally, Lim et al. synthesized V_2O_5 supported by Al_2O_3 via a wet impregnation method. Here, Al_2O_3 that is coated over the V_2O_5 acts as a solid electrolyte interface with high lithium-ion conductivity (10^{-7} S/cm) and low electronic conductivity. Subsequently, it reaches the reversible capacity of 2875 mAh/g and attains the maximum reversible capacity of 3250 mAh/g. Furthermore, the capacity decreases with cycling [53]. Moreover, manganese oxide is one of the most suitable cathode catalysts for LABs due to its excellent round-trip efficiency.

16.4.2.2.2 Manganese Oxide

Until now, one of the most studied metal oxides is manganese oxide, which is widely employed as a cathode catalyst because of its increased specific capacity and good round-trip efficiency. MnO_2 in bulk form shows only a small number of active sites on the surface, which limits the catalytic activity and resists ionic and electronic diffusion. Moreover, nanoscale MnO_2 is highly preferred, which provides diffusion paths of ions and catalytic active sites. In addition, due to its highly active electrical properties, template/surfactant-based MnO_2 has a very complicated synthesis procedure. Further, different inter-linked octahedral MnO_6 subunits have various crystallographic forms such as α, β, λ, etc. Thapa et al. synthesized a cathode material (i.e., mesoporous polythiophene birnessite (b-MnO_2)) by a modified interface method (aqueous/organic medium). Here, the b-MnO_2 has a high surface area, which gets exposed to electrochemical measurements, and it shows a discharge capacity of 162, 128, and 95 mAh/g at different current densities of 200, 500, and 600 mA/g (69% of its theoretical capacity), respectively. Further, the b-MnO_2 cathode demonstrates a high surface area, which leads to high current density, and a thin pore wall, which reduces the lithium-ion diffusion path [54]. In addition, with MnO_x, CeO_2 also acts as an attractive candidate for cathode catalyst because of its high oxygen ion conductivity and excellent mechanical resistance. Thus, Zhu et al. decorated MnO_2 with CeO_2 nanorods via in-situ redox reactions. In this case, CeO_2 acts as an oxygen buffer to relieve oxygen insufficiency. Further, this catalyst shows high electrochemical performance, i.e., a high discharge capacity of 2617 mAh/g at 100 mA/g, lower overpotential, excellent rate capability, and good cyclic stability, where, at 200 mA/g current density, only about 1.1% voltage is obtained after 30 cycles [55]. Likewise, Thapa et al. synthesized mesoporous β-MnO_2 bifunctionally supported on Au-Pd using the KIT-6 template. It enhances the coulombic and energy efficiency of the LABs and improves the ORR and OER reaction kinetics. In addition, they attain a specific discharge value of 775 mAh g^{-1} at a 0.13 mA/cm^2 [43]. Similarly, Lu et al. used MnO_x with bifunctional Au-Pd (AMP) as a cathode catalyst. Here, the Au and Pd reduce the overvoltage of OER and ORR. Furthermore, the electrochemical measurements of AMP have been compared with Super-P and bare MnO_x. In Figure 16.6 (e, f), the AMP has a 3200 mAh/g specific capacity at 500 mA/g, which is 3.2 and 9.8 times higher than the bare MnO_x and Super-P [56].

Likewise, Maenetja et al. employed density functional theory (DFT) to examine the surfaces of Li-MO_2 (where M = Mn, Ti, and V). Here, the oxygen adsorption on Li-MO_2 aids in the charging and discharging of LABs. Further, the oxygen adsorption has stabilized the presence of lithium on the surface, where the oxygen atoms bind with lithium atoms as well as coordinate with Mn surface atoms. Figure 16.6 (g-i) shows the graphs between surface free energy and oxygen chemical potential, indicating that MnO_2 is preferred as a cathode catalyst over TiO_2 and VO_2 with much more stable surface free energies than TiO_2 and VO_2 [57]. As discussed in section 16.4.1.3, graphene is proven as an efficient host for nanomaterials with a large surface area. Thus, Zahoor et al. prepared a δ-MnO_2/N-rGO (MNGC) composite via a simple non-template hydrothermal technique, which provides numerous oxygen and electrolyte pathways to facilitate lithium-ion transportation. In this case, the δ-MnO_2 is grown on the surface of N-rGO, which enhances the ORR process when compared to MnO_2 and rGO. Further, the MNGC electrode tested with a LAB demonstrates a reversible discharge value of 5250 mAh/g at 0.2 mA/cm^2 with decreased overpotential [58]. Wang et al. deposited manganese oxide on the hierarchical carbon nanocage to develop a highly efficient Li-O_2 battery (LOB). At 0.1–1.0 A/g, this composite exhibits a lower overpotential of 0.73–0.99V and excellent electrocatalytic activity. In this case, the MnO_x boosts the ORR/OER reversibility, the

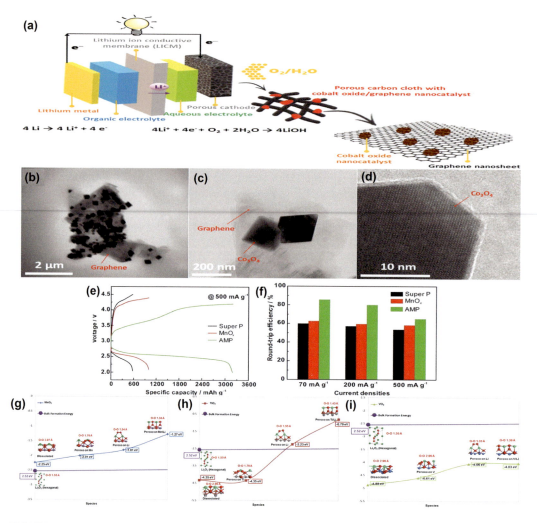

FIGURE 16.6 (a) Hybrid electrolyte lithium battery working at discharge state, (b-d) different TEM magnification images of CO_3O_4/graphene composite. (Adapted with permission [49]. Copyright (2020), Elsevier.) (e, f) Au-Pd (AMP) specific capacity comparison at 500 mA/g with MnO_x and Super-P. (Adapted with permission [56]. Copyright (2016), Royal Society of Chemistry.) Surface (110) adsorption and lithium peroxide bulk energetics of (g) MnO_2, (h) TiO_2, (i) VO_2. (Adapted from [57], Copyright (2021), The Electrochemical Society.)

interconnecting composite pore structure increases the Li_2O_2 storage and facilitates the mass/ion transfer, and carbon nanocages reduce the side reactions. Further, the transfer of composite electrodes in LOBs results in a high reversible discharge capacity of 27135 mAh/g at 0.1 A/g, (i.e., much higher compared with pristine nanocage and MnO_x/XC-72 (carbon)) [59].

16.4.2.2.3 Perovskite Oxides

Perovskites with an ABO_3 structure are widely employed in different applications. Here, the face-centered cubic (FCC) lattice is formed by A cations and oxygen, while the octahedral sites in the FCC lattice are occupied by B cations and are fully surrounded by oxygen neighbors. At high temperatures, it provides the good catalytic activity. Lu et al. fabricated $La_{0.8}Sr_{0.2}MnO_3$ nanorods via the soft template method, which shows a large surface area of 20.6 m²/g, which is more beneficial for OER and ORR processes. Further, in LAB, the microporous composite achieves a lower overpotential, a high discharge capacity of 6890 mAh/g at 200 mA/g, and excellent rate capability up to 400 mA/g [60]. Similarly, Cheng et al. synthesized a cathode catalyst $La_{0.6}Sr_{0.4}Co_{0.2}Fe_{0.8}O_3$ composite for LABs via the sol-gel method. Here, these composites readily decrease the overpotential and aid in beneficial OER and ORR catalytic processes. Further, in LABs, the $La_{0.6}Sr_{0.4}Co_{0.2}Fe_{0.8}O_3$ composite with a discharge capacity value of 9678 mAh/g increases to the reversible discharge capacity of 13979 mAh/g. In addition, it also enhances the cyclability from 10 to 20 cycles at a current density of 200 mA/g [61]. Further, in ambient air, Cheng et al. tested the $La_{0.6}Sr_{0.4}Co_{0.2}Fe_{0.8}O_3$ cathode catalyst with LAB, which showed an enhanced specific capacity of 6027 mAh/g and an enhanced cycle number of 156 cycles at 400 mA/g. Also, during the discharge and charge process, it provides more oxygen pathways and active sites for product nucleation and decomposition [62]. By microemulsion method, Lim et al. synthesized a nano-size $Nd_{0.67}Sr_{0.33}CoO_{3-\delta}$ with a particle size of about 20–50 nm. Further, the perovskite, with a surface area of 12.759 m^2/g, exhibits a 33.68 mA/cm^2 current density at 0.9V vs. (Hg/HgO). Here, the presence of Co ions leads to a larger surface area and excellent catalytic activity. Further, the Co-ions that provide the catalytic active sites and the surface area increase the active sites for reactants and catalyst particles. Further, it achieves excellent cyclic performance and stability for 60 hours with a potential difference of less than 1.1V [63].

16.4.3 OTHER CATHODE-BASED MATERIALS

However, other than carbonaceous and metal/metal oxide-based materials, there are some other cathode materials that help to improve the efficiency of the LABs. Some of them are listed, namely, transition metal carbides, sulfides, nitrides, and carbon-free cathodes. Here, metal carbides have superior electrochemical performance for OER and ORR. Thotiyl et al. reported the TiC as a cathode-based material to overcome the disadvantages of nano-porous gold and carbon. Also, compared with carbon, TiC reduces the side reaction at the cathode/electrolyte interface and is more stable than gold. Furthermore, TiC is very light, easy to handle, and less expensive compared to gold. In addition, it exhibits better reversibility in the formation and decomposition of Li_2O_2 compared to nano-porous gold and delivers greater than 98% retention after 100 cycles, whereas gold shows 95% [64]. Likewise, Christy et al. comprised a carbide composite with molybdenum and tungsten with a nanowire structure. In this case, the bimetallic catalysts do not participate in any side reactions. Here, molybdenum carbide is a suitable candidate for catalytic activity, and tungsten carbide shows a large surface area and excellent catalytic activity. Further, it demonstrates an outstanding reversible capacity value of 5000 mAh/g at 0.1 mA/g with a lower overpotential of 0.9–1.2V [65]. Thakur et al. in 2022 decorated carbon nanospheres with iron carbide (FeN_xC_y) connected with graphene sheets. Further, it has a higher surface area (mesoporous) with active storage sites for Li_2O_2. Additionally, during charging, the nitrogen-doped and iron-doped carbon leads to Li_2O_2 facile dissociation, which leads to lower overpotential. Furthermore, this composite electrode in LABs exhibits a high specific capacity of about 9000 mAh/g with 2920 Wh/kg as energy density and 18 mW/cm^2 as power density [66].

Due to the similar structure of oxygen and sulfur, transition metal sulfides also play the same role in electrochemical devices. Figure 16.7(a-d) shows the SEM and TEM images of nickel sulfides synthesized by Ma et al. with flower-like (f-NiS) and rod-like (r-NiS) morphologies. Here, both the nickel sulfides provide good catalytic activity compared to other carbons. In particular, the f-NiS is highly favored for air battery applications. At 75 mA/g, the f-NiS offers the lowest charge voltage of 4.24V and the highest capacity value of 6733 mAh/g (Figure 16.7e), which is significantly higher than the r-NiS and super P (Figure 16.7f) [67]. Furthermore, Zhan et al. in 2021 synthesized a porous yolk-shell structure CoS_2@NC (where nitrogen is coordinated with ZIF-67) with a carbon layer. Here, transition metal sulfides with large surface area and volume help in high intrinsic activities and show excellent catalytic activity for OER and ORR processes. A metal-organic framework (MOF) made up of organic linkers and metal clusters provides high O_2 accessibility, tunable chemical composition, and open metal sites. Based on these efforts, the cathode CoS_2@NC-400/AB could

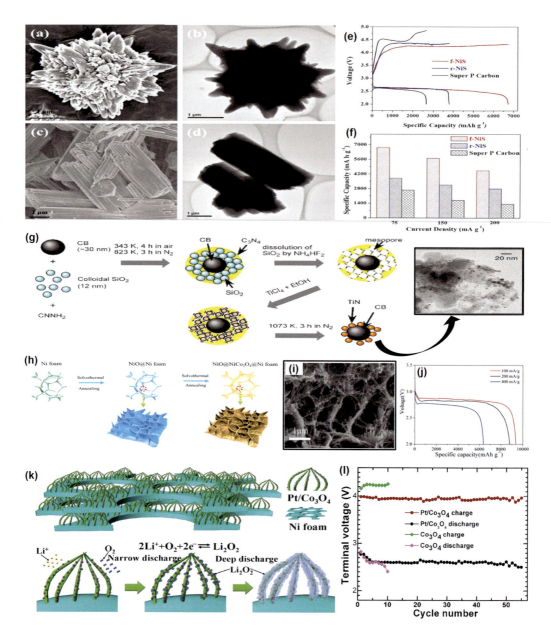

FIGURE 16.7 (a, c) SEM and (b, d) TEM images of f-NiS and r-NiS, (e) charge-charge curves of NiS and Super-P catalysts at 75 mA/g, (f) specific capacities of f- NiS, r-NiS and Super-P carbon at different current densities. (Adapted with permission [67]. Copyright (2013), Springer Nature.) (g) schematic illustration of mpg-C_3N_4 on carbon black, followed by TiN formation. (Adapted with permission [69]. Copyright (2010), Royal Society of Chemistry.) (h) Design and preparation of NiO@$NiCo_2O_4$ on Ni foam, (i) field emission scanning electron microscope (FESEM) images of NiO@$NiCo_2O_4$ composite, (j) NiO@$NiCo_2O_4$ electrodes discharge curves at different current densities. (Adapted with permission [74]. Copyright (2021), Royal Society of Chemistry.) (k) Schematic architecture and working mechanism of Ni foam-Pt/Co_3O_4, (l) terminal voltages of Pt/Co_3O_4 and Co_3O_4 cathodes at 100 mA/g. (Adapted with permission [76]. Copyright (2015), American Chemical Society.)

deliver a specific discharge value of about 3529.8 mAh/g at 0.1 mA/cm^2 [68]. Similarly, transition metal nitrides have the capability of providing high thermal stability, high corrosion resistance, and robust hardness. Chen et al. prepared TiN nanoparticles over carbon black by using the mpg-C_3N_4/CB composite template (Figure 16.7g). Here, this composite shows an excellent improvement in ORR current with enhanced performance [69]. Likewise, Sun et al. coated vanadium nitride over the nitrogen-doped carbon nano-ribbon that was grown on the carbon paper (VN@C) through a synchronous nitridation-pyrolysis process. Here, the N-doped carbon enhances the stability of vanadium nitride as well as improves the catalytic activity. The synthesized sample has been made through different heat treatments like 750, 850, and 950 °C, respectively. The VN@C-850 cathode exhibits a charge and discharge voltage gap of about 0.88V with a reversible discharge capacity of 8269 mAh/g and a longer cyclability of 183 cycles at 1000 mAh/g cut-off capacity [70].

LABs work efficiently with carbon electrodes, but some chronic problems need to be rectified. Thus, an air cathode without carbon would be the better idea. Further, before considering the carbon-free electrode, the non-carbon material must satisfy the following criteria: high electrochemical stability and outstanding electronic conductivity. Ma et al. proposed a new idea to develop a mixed ionic-electronic conductor by substituting a 3D transition metal into the perovskite structure, $Li_xLa_yMO_3$ (M = Ti, Mn, Cr, Co, and Fe). In addition, it has a very low activation barrier (below 0.411 eV) for lithium-ion diffusion. Further, the $Li_{0.34}La_{0.55}MnO_{3-\delta}$ perovskite has a high electronic conductivity of 2.04×10^{-3} S cm^{-1}, a lithium conductivity of about 8.53×10^{-5} S/cm, and excellent reversibility at 0.21 mAh/cm^2 density after 100 cycles [71]. Likewise, Beyer et al. prepared antimony-doped tin oxide cathodes via a hydrothermal strategy. Here, the cathode material with 190 m^2/g surface area and a 4.5 nm particle size enables the Li_2O formation and prevents the formation of carbonate on the surface. Further, gas evolution during the charge-discharge process is measured, which resists the CO_2 evolution beyond 5V [72]. Similarly, to avoid harm from the decomposition of carbon materials and organic binders, Zhao et al. deposited RuO_2 catalyst over the TiO_2 directly via the electrodeposition method without the use of a binder and made it grow on the Ti foams. RuO_2 was selected as a cathode catalyst due to its high electron conductivity and catalytic activity, while TiO_2 was chosen because it is stable, biocompatible, and environmentally friendly. Thus, RuO_2 aids in maintaining the charge voltage of 3.7V throughout the battery cycle and exhibits 130 cycle discharge/charge [73]. Wang et al. in 2021 decorated Ni foam with 3D $NiCo_2O_4$ nanowires covered in NiO nanosheets by using a two-step solvothermal method (Figure 16.7h), and the morphologies are shown in Figure 16.7i. Here, 3D $NiCo_2O_4$ improves the ORR/OER process, while NiO promotes the deposition of carbonate and carboxylate species. Further, these carbon-free cathodes deliver a highly reversible discharge capacity of 9361.4 mAh/g at 100 mA/g and at 200 mA/g, it shows 176 cycles of stability with a discharge capacity of 500 mAh/g (Figure 16.7j) [74]. Additionally, Huang et al. prepared Co_3O_4 nanowires with Ru nanoparticles, which show improved cyclic stability (when capacity is limited to 500 mAh/g) and sustain the cycling of 29 times, while the bare Co_3O_4 catalyst sustains only 9 cycles. Furthermore, at 300 mAh/g limited capacity, it shows stability up to 122 cycles [75]. Likewise, an electrode design containing Pt/CO_3O_4 nanowires incorporated into Ni foam has been proposed by Cao et al. (Figure 16.7k). Here, Pt promotes the deposition of Li_2O_2 on the nanowires (where this deposition aids in Li_2O_2 decomposition during recharge that reduces the side reactions). Further, this catalytic cathode sustains 50 cycles at a discharge capacity value of about 500 mAh/g and its terminal voltage graph is shown in Figure 16.7l [76].

16.5 CONCLUSION AND PERSPECTIVE

In this featured review, we summarize the progress made in research on different cathode materials used in LABs. Each cathode and the cathode catalysts deliver high specific capacity and excellent energy density. Even though they show higher energy density than other metal-air batteries and resolve the energy-related issues by being considered as a future technology, there are some difficulties to be overcome, such as humidity, carbon microstructure at the nano and mesoscale, catalysts, additives, overall cell design, and depth of discharge. For instance, it needs to have better round-trip efficiency, practical capacity, and cycling life to be commercially viable.

First, 3D porous networks were created by assembling 2D materials to provide numerous pathways for electrons, oxygen, and Li+ as well as storage space for discharge products. Dopants, defects, and other active phases are introduced into cathode materials to improve ORR/OER catalytic activity. Graphene-based materials are viable cathode materials, in particular, which may demonstrate significant capacity when coupled with a different active element. Further, the electrolyte solution makes an unwanted reaction with the cathode and causes the deposition of lithium byproducts, which causes 5 to 10% capacity loss per cycle. Also, in non-aqueous systems, the more complicated reactions take place at the anode and the electrolyte, thus passivation film is more important to protect the electrode. SEI, which blocks and eliminates air contaminants and lithium deposited products, helps obtain an efficient LAB. Furthermore, the low ionic conductivity severely limits battery development; hence, ionic conductive materials are added to improve conductivity and limit dendrite growth. Likewise, the efficient catalysts in the cathode material may wisely avoid the side products. Significantly, developing a highly stable and active bifunctional catalyst may reduce the charge/discharge overpotentials.

Further, to develop a lightweight cathode material, combine the advantages of carbon and carbon-free cathodes instead of using heavy-weight carbon-free cathodes. However, a membrane that reduces the ionic liquid electrolyte evaporation is also required. Also, the different synthesizing methods lead to different catalytic properties. Thus, preparing a cathode material with a highly homogenous morphology and pure phase would be an efficient task. Meanwhile, understanding the Mn-based catalytic mechanism is an attractive aspect that helps design a suitable catalyst. In addition, matching an electrolyte with suitable catalysts will produce the most efficient cathode materials. Further, from the summary, it is concluded that N-doped carbon and metal are highly suitable for chemical activation, and also, they assure good corrosion, high electrical conductivity, and excellent resistance. Usage of MOF as cathode catalysts possesses high charge/discharge conductivity and also enhances the cyclability and lower overpotential. Additionally, Pt and Pd help in the ORR and OER processes more than the non-precious metals. Transition metal nitrides, carbides, sulfides, and perovskite oxides have high catalytic activity as well as help to develop electrochemical devices.

Furthermore, for the design of a realistic rechargeable $Li-O_2$ cell, major device components such as the cathode, electrolyte, separator, catalyst, and anode, and their interfacial matching should be fully considered. Finally, the mechanical properties of the LABs should be given more attention, particularly for flexible LABs. Though there are numerous drawbacks, LABs will emerge as viable in future commercial applications.

REFERENCES

[1] Md Rahman, Xiaojian Wang, Cuie Wen, A review of high energy density lithium-air battery technology, J. Appl. Electrochem. 44 (2014) 5–22.
[2] Ernest L. Littauer, Keh Chi Tsai, Anodic behavior of lithium in aqueous electrolytes: ii. Mechanical passivation, J. Electrochem. Soc. 123 (1976) 964.
[3] K.M. Abraham, Z. Jiang, A polymer electrolyte-based rechargeable lithium/oxygen battery, J. Electrochem. Soc. 143 (1996) 1.
[4] Peng Zhang, Mingjie Ding, Xiaoxuan Li, Caixia Li, Zhaoqiang Li, Longwei Yin, Challenges and strategy on parasitic reaction for high-performance nonaqueous lithium—oxygen batteries, Adv. Energy Mater. 10 (2020) 2001789.
[5] Chengyi Wang, Zhaojun Xie, Zhen Zhou, Lithium-air batteries: Challenges coexist with opportunities, APL Mater. 7 (2019) 040701.

[6] B. Jeevanantham, Youngseok Song, Heeman Choe, M.K. Shobana, Structural and optical characteristics of cobalt ferrite nanoparticles, Mater. Lett. X. 12 (2021) 100105.
[7] Ralph E. Williford, Ji-Guang Zhang, Air electrode design for sustained high power operation of Li/air batteries, J. Power Sources. 194 (2009) 1164–1170.
[8] Binod Kumar, Jitendra Kumar, Robert Leese, Joseph P. Fellner, Stanley J. Rodrigues, K.M. Abraham, A solid-state, rechargeable, long cycle life lithium-air battery, J. Electrochem. Soc. 157 (2009) A50.
[9] M.K. Shobana, B. Jeevanantham, Metal—organic framework based cathode materials in lithium—sulfur batteries, in Lithium-sulfur batteries (pp. 333–360). Elsevier, 2022.
[10] Zhong Ma, Xianxia Yuan, Lin Li, Zi-Feng Ma, David P. Wilkinson, Lei Zhang, Jiujun Zhang, A review of cathode materials and structures for rechargeable lithium-air batteries, Energy Environ. Sci. 8 (2015) 2144–2198.
[11] Dongsheng Geng, Ning Ding, T.S. Andy Hor, Sheau Wei Chien, Zhaolin Liu, Delvin Wuu, Xueliang Sun, Yun Zong, From lithium-oxygen to lithium-air batteries: Challenges and opportunities, Adv. Energy Mater. 6 (2016) 1502164.
[12] Xueping Zhang, Xiaowei Mu, Sixie Yang, Pengfei Wang, Shaohua Guo, Min Han, Ping He, Haoshen Zhou, Research progress for the development of Li-air batteries: Addressing parasitic reactions arising from air composition, Energy Environ. Mater. 1 (2018) 61–74.
[13] Tao Zhang, Jun Yang, Jinhui Zhu, Jingjing Zhou, Zhixin Xu, Jiulin Wang, Feilong Qiu, Ping He, A lithium-ion oxygen battery with a Si anode lithiated in situ by a Li 3 N-containing cathode, Chem. Comm. 54 (2018) 1069–1072.
[14] Peng Tan, H.R. Jiang, X.B. Zhu, Liang An, C.Y. Jung, M.C. Wu, Le Shi, Wei Shyy, T.S. Zhao, Advances and challenges in lithium-air batteries, Appl. Energy. 204 (2017) 780–806.
[15] Hyungsub Woo, Joonhyeon Kang, Jaewook Kim, Chunjoong Kim, Seunghoon Nam, Byungwoo Park, Development of carbon-based cathodes for Li-air batteries: Present and future, Electron. Mater. Lett. 12 (2016) 551–567.
[16] Akihiro Nomura, Kimihiko Ito, Yoshimi Kubo, CNT sheet air electrode for the development of ultra-high cell capacity in lithium-air batteries, Sci. Rep. 7 (2017) 1–8.
[17] Keita Ushijima, Shinichiroh Iwamura, Shin R. Mukai, Simple and cost-effective method to increase the capacity of carbon nanotube sheet cathodes for lithium—air batteries, ACS Appl. Energy Mater. 3 (2020) 6915–6921.
[18] Jie Li, Bin Peng, Geng Zhou, Zhian Zhang, Yanqing Lai, Ming Jia, Partially cracked carbon nanotubes as cathode materials for lithium-air batteries. ECS Electrochem. Lett. 2 (2012) A25.
[19] Rui Mi, Hao Liu, Hao Wang, Ka-Wai Wong, Jun Mei, Yungui Chen, Woon-Ming Lau, Hui Yan, Effects of nitrogen-doped carbon nanotubes on the discharge performance of Li-air batteries, Carbon. 67 (2014) 744–752.
[20] Yu Li, Zhonglin Zhang, Donghong Duan, Yunxia Han, Kunlei Wang, Xiaogang Hao, Junwen Wang, Shibin Liu, Fanhua Wu, An integrated structural air electrode based on parallel porous nitrogen-doped carbon nanotube arrays for rechargeable Li-air batteries, Nanomaterials. 9 (2019) 1412.
[21] Masoumeh Salehi, Zahra Shariatinia, Abbas Sadeghi, Application of RGO/CNT nanocomposite as cathode material in lithium-air battery, J. Electroanal. Chem. 832 (2019) 165–173.
[22] Chengxing Li, Daobin Liu, Yukun Xiao, Zixuan Liu, Li Song, Zhipan Zhang, Mesoporous Co3O4-rods-entangled carbonized polyaniline nanotubes as an efficient cathode material toward stable lithium—air batteries, ACS Appl. Energy Mater. 2 (2019) 2939–2947.
[23] Hyungkyu Han, Yeryung Jeon, Zhiming Liu, Taeseup Song, Highly graphitic carbon nanofibers web as a cathode material for lithium oxygen batteries, Appl. Sci. 8 (2018) 209.
[24] Martinez Crespiera Sandra, David Amantia, Etienne Knipping, Christophe Aucher, Laurent Aubouy, Julia Amici, Juqin Zeng, Usman Zubair, Carlotta Francia, Silvia Bodoardo, Cobalt-doped mesoporous carbon nanofibres as free-standing cathodes for lithium—oxygen batteries, J. Appl. Electrochem. 47 (2017) 497–506.
[25] Myeong Jun Song, Il To Kim, Young Bok Kim, Moo Whan Shin, Self-standing, binder-free electrospun Co3O4/carbon nanofiber composites for non-aqueous Li-air batteries, Electrochim. Acta. 182 (2015) 289–296.
[26] Xiuling Zhang, Wei Fan, Shuyu Zhao, Ran Cao, Congju Li, An efficient, bifunctional catalyst for lithium—oxygen batteries obtained through tuning the exterior Co 2+/Co 3+ ratio of CoO x on N-doped carbon nanofibers, Catal. Sci. Technol. 9 (2019) 1998–2007.

[27] Hieu Trung Bui, Dong Wook Kim, Jungdon Suk, Yongku Kang, Carbon nanofiber@ platinum by a coaxial electrospinning and their improved electrochemical performance as a Li– O2 battery cathode, Carbon. 130 (2018) 94–104.

[28] Yajun Ding, Yuejiao Li, Min Wu, Hong Zhao, Qi Li, Zhong-Shuai Wu, Recent advances and future perspectives of two-dimensional materials for rechargeable Li-O2 batteries, Energy Storage Mater. 31 (2020) 470–491.

[29] Yongliang Li, Jiajun Wang, Xifei Li, Dongsheng Geng, Ruying Li, Xueling Sun, Superior energy capacity of graphene nanosheets for a nonaqueous lithium-oxygen battery, Chem. Comm. 47 (2011) 9438–9440.

[30] Bing Sun, Xiaodan Huang, Shuangqiang Chen, Paul Munroe, Guoxiu Wang, Porous graphene nanoarchitectures: An efficient catalyst for low charge-overpotential, long life, and high capacity lithium-oxygen batteries, Nano Lett. 14 (2014) 3145–3152.

[31] Yan-Juan Gu, Wing-Tak Wong, Electro-oxidation of methanol on Pt particles dispersed on RuO2 nanorods, J. Electrochem. Soc. 153 (2006) A1714.

[32] Rong Kou, Yuyan Shao, Donghai Wang, Mark H. Engelhard, Ja Hun Kwak, Jun Wang, Vilayanur V. Viswanathan et al., Enhanced activity and stability of Pt catalysts on functionalized graphene sheets for electrocatalytic oxygen reduction, Electrochem. Commun. 11 (2009) 954–957.

[33] Yo Sub Jeong, Jin-Bum Park, Hun-Gi Jung, Jooho Kim, Xiangyi Luo, Jun Lu, Larry Curtiss, et al., Study on the catalytic activity of noble metal nanoparticles on reduced graphene oxide for oxygen evolution reactions in lithium-air batteries, Nano Lett. 15 (2015) 4261–4268.

[34] Xiangzhong Ren, Biyan Liao, Yongliang Li, Peixin Zhang, Libo Deng, Yuan Gao, Facile synthesis of PdSnCo/nitrogen-doped reduced graphene as a highly active catalyst for lithium-air batteries, Electrochim. Acta. 228 (2017) 36–44.

[35] Xing Zhong, Benjamin Papandrea, Yuxi Xu, Zhaoyang Lin, Hua Zhang, Yuan Liu, Yu Huang, Xiangfeng Duan, Three-dimensional graphene membrane cathode for high energy density rechargeable lithium-air batteries in ambient conditions, Nano Res. 10 (2017) 472–482.

[36] Bing Sun, Hao Liu, Paul Munroe, Hyojun Ahn, Guoxiu Wang, Nanocomposites of CoO and a mesoporous carbon (CMK-3) as a high performance cathode catalyst for lithium-oxygen batteries, Nano Res. 5 (2012) 460–469.

[37] Toshiyuki Yokoi, Yasuhiro Sakamoto, Osamu Terasaki, Yoshihiro Kubota, Tatsuya Okubo, Takashi Tatsumi, Periodic arrangement of silica nanospheres assisted by amino acids, J. Am. Chem. Soc. 128 (2006) 13664–13665.

[38] Yuhki Yui, Shuhei Sakamoto, Masaya Nohara, Masahiko Hayashi, Jiro Nakamura, Kota Suzuki, Masaaki Hirayama, Ryoji Kanno, Takeshi Komatsu, Highly ordered mesoporous carbon support materials for air electrode of lithium-air secondary batteries, Electrochem. 85 (2017) 128–132.

[39] Myeong Jun Song, Moo Whan Shin, Fabrication and characterization of carbon nanofiber@ mesoporous carbon core-shell composite for the Li-air battery, Appl. Surf. Sci. 320 (2014) 435–440.

[40] Martinez Crespiera Sandra, David Amantia, Etienne Knipping, Christophe Aucher, Laurent Aubouy, Julia Amici, Juqin Zeng, Usman Zubair, Carlotta Francia, Silvia Bodoardo, Cobalt-doped mesoporous carbon nanofibres as free-standing cathodes for lithium—oxygen batteries, J. Appl. Electrochem. 47 (2017) 497–506.

[41] Yue Shen, Dan Sun, Ling Yu, Wang Zhang, Yuanyuan Shang, Huiru Tang, Junfang Wu, Anyuan Cao, Yunhui Huang, A high-capacity lithium—air battery with Pd modified carbon nanotube sponge cathode working in regular air, Carbon. 62 (2013) 288–295.

[42] Hun-Gi Jung, Yo Sub Jeong, Jin-Bum Park, Yang-Kook Sun, Bruno Scrosati, Yun Jung Lee, Ruthenium-based electrocatalysts supported on reduced graphene oxide for lithium-air batteries, ACS Nano. 7 (2013) 3532–3539.

[43] Arjun Kumar Thapa, Tae Ho Shin, Shintaro Ida, Gamini U. Sumanasekera, Mahendra K. Sunkara, Tatsumi Ishihara, Gold—palladium nanoparticles supported by mesoporous β-MnO2 air electrode for rechargeable Li-air battery, J. Power Sources. 220 (2012) 211–216.

[44] Dawei Su, Hyun-Soo Kim, Woo-Seong Kim, Guoxiu Wang, A study of PtxCoy alloy nanoparticles as cathode catalysts for lithium-air batteries with improved catalytic activity, J. Power Sources. 244 (2013) 488–493.

[45] Dongxiao Ji, Shengjie Peng, Dorsasadat Safanama, Haonan Yu, Linlin Li, Guorui Yang, Xiaohong Qin, Madhavi Srinivasan, Stefan Adams, Seeram Ramakrishna, Design of 3-dimensional hierarchical architectures of carbon and highly active transition metals (Fe, Co, Ni) as bifunctional oxygen catalysts for hybrid lithium—air batteries, Chem. Mater. 29 (2017) 1665–1675.

[46] Zheng Chang, Feng Yu, Zaichun Liu, Shou Peng, Min Guan, Xiaoxiao Shen, Shulin Zhao, Nian Liu, Yuping Wu, Yuhui Chen, Co-Ni alloy encapsulated by N-doped graphene as a cathode catalyst for rechargeable hybrid Li-air batteries, ACS Appl. Mater. Interfaces. 12 (2019) 4366–4372.
[47] Jingbo Yang, Dingtao Ma, Yongliang Li, Peixin Zhang, Hongwei Mi, Libo Deng, Lingna Sun, Xiangzhong Ren, Atomic layer deposition of amorphous oxygen-deficient TiO2-x on carbon nanotubes as cathode materials for lithium-air batteries, J. Power Sources. 360 (2017) 215–220.
[48] Tack Han Yoon, Yong Joon Park, Carbon nanotube/Co3O4 composite for air electrode of lithium-air battery, Nanoscale Res. Lett. 7 (2012) 1–4.
[49] Si-Han Peng, Tse-Hsi Chen, Chih-Hsun Lee, Hsin-Chun Lu, Shingjiang Jessie Lue, Optimal cobalt oxide (Co3O4): Graphene (GR) ratio in Co3O4/GR as air cathode catalyst for air-breathing hybrid electrolyte lithium-air battery, J. Power Sources. 471 (2020) 228373.
[50] Yoon-Ho Lee, Heai-Ku Park, The electrochemical properties of SnO 2 as cathodes for lithium-air batteries, J. Korean Electrochem. Soc. 22 (2019) 164–171.
[51] Jingjuan Li, Xiaoyan Hou, Ya Mao, Chunyan Lai, Xianxia Yuan, Enhanced performance of aprotic electrolyte Li-O2 batteries with SnS2—SnO2/C heterostructure as efficient cathode catalyst, Energy Fuels. 34 (2020) 14995–15003.
[52] Sung Hoon Lim, Bok Ki Kim, Woo Young Yoon, Catalytic behavior of V2O5 in rechargeable Li-O2 batteries, J. Appl. Electrochem. 42 (2012) 1045–1048.
[53] Sung Hoon Lim, Do Hyung Kim, Ji Young Byun, Bok Ki Kim, Woo Young Yoon, Electrochemical and catalytic properties of V2O5/Al2O3 in rechargeable Li-O2 batteries, Electrochim. Acta. 107 (2013) 681–685.
[54] Arjun Kumar Thapa, Bill Pandit, Rajesh Thapa, Tulashi Luitel, Hem Sharma Paudel, Gamini Sumanasekera, Mahendra K. Sunkara, Nanda Gunawardhana, Tatsumi Ishihara, Masaki Yoshio, Synthesis of mesoporous birnessite-MnO2 composite as a cathode electrode for lithium battery, Electrochim. Acta. 116 (2014) 188–193.
[55] Yongqiang Zhu, Shanhu Liu, Chao Jin, Shiyu Bie, Ruizhi Yang, Jiao Wu, MnO x decorated CeO 2 nanorods as cathode catalyst for rechargeable lithium-air batteries, J. Mater. Chem. A. 3 (2015) 13563–13567.
[56] Xueyi Lu, Long Zhang, Xiaolei Sun, Wenping Si, Chenglin Yan, Oliver G. Schmidt, Bifunctional Au—Pd decorated MnO x nanomembranes as cathode materials for Li-O$_2$ batteries, J. Mater. Chem. A 4 (2016), 4155–4160.
[57] Khomotso P. Maenetja, Phuti E. Ngoepe, First principles study of oxygen adsorption on Li-MO2 (M= Mn, Ti and V)(110) surface, J. Electrochem. Soc. 168 (2021) 070556.
[58] Awan Zahoor, Raza Faizan, Khaled Elsaid, Saud Hashmi, Faaz Ahmed Butt, Zafar Khan Ghouri, Synthesis and experimental investigation of δ-MnO2/N-rGO nanocomposite for Li-O2 batteries applications, Chem. Eng. J. Adv. 7 (2021) 100115.
[59] Baoxing Wang, Chenxia Liu, Lijun Yang, Qiang Wu, Xizhang Wang, Zheng Hu, Defect-induced deposition of manganese oxides on hierarchical carbon nanocages for high-performance lithium-oxygen batteries, Nano Res. (2022) 1–5.
[60] Fanliang Lu, Yarong Wang, Chao Jin, Fan Li, Ruizhi Yang, Fanglin Chen, Microporous La0. 8Sr0. 2MnO3 perovskite nanorods as efficient electrocatalysts for lithium—air battery, J. Power Sources. 293 (2015) 726–733.
[61] Junfang Cheng, Ming Zhang, Yuexing Jiang, Lu Zou, Yingpeng Gong, Bo Chi, Jian Pu, Li Jian, Perovskite La0. 6Sr0. 4Co0. 2Fe0. 8O3 as an effective electrocatalyst for non-aqueous lithium-air batteries, Electrochim. Acta. 191 (2016) 106–115.
[62] Junfang Cheng, Yuexing Jiang, Ming Zhang, Yu Sun, Lu Zou, Bo Chi, Jian Pu, Li Jian, Aprotic lithium—air batteries tested in ambient air with a high-performance and low-cost bifunctional perovskite catalyst, ChemCatChem. 10 (2018) 1635–1642.
[63] Chaehyun Lim, Changmin Kim, Ohhun Gwon, Hu Young Jeong, Hyun-Kon Song, Young-Wan Ju, Jeeyoung Shin, Guntae Kim, Nano-perovskite oxide prepared via inverse microemulsion mediated synthesis for catalyst of lithium-air batteries, Electrochim. Acta. 275 (2018) 248–255.
[64] Ottakam Thotiyl, Muhammed M., Stefan A. Freunberger, Zhangquan Peng, Yuhui Chen, Zheng Liu, Peter G. Bruce, A stable cathode for the aprotic Li-O2 battery, Nat. Mater. 12 (2013) 1050–1056.
[65] Maria Christy, Anupriya Arul, Young-Beom Kim, Carbide composite nanowire as bifunctional electrocatalyst for lithium oxygen batteries, Electrochim. Acta. 300 (2019) 186–192.
[66] Pallavi Thakur, Anand B. Puthirath, Pulickel M. Ajayan, Tharangattu N. Narayanan, Iron carbide decorated carbon nanosphere-sheet hybrid based rechargeable high-capacity non-aqueous Li-O2 batteries, Carbon. 196 (2022) 320–326.

[67] Zhong Ma, Xianxia Yuan, Zhenlin Zhang, Delong Mei, Lin Li, Zi-Feng Ma, Lei Zhang, Jun Yang, Jiujun Zhang, Novel flower-like nickel sulfide as an efficient electrocatalyst for non-aqueous lithium-air batteries, Sci. Rep. 5 (2015) 1–9.

[68] Yang Zhan, Shun-zhi Yu, Shao-hua Luo, Jian Feng, Qing Wang, Nitrogen-coordinated CoS2@ NC yolk-shell polyhedrons catalysts derived from a metal-organic framework for a highly reversible Li-O2 battery, ACS Appl. Mater. Interfaces. 13 (2021) 17658–17667.

[69] Jia Chen, Kazuhiro Takanabe, Ryohji Ohnishi, Daling Lu, Saori Okada, Haruna Hatasawa, Hiroyuki Morioka, Markus Antonietti, Jun Kubota, Kazunari Domen, Nano-sized TiN on carbon black as an efficient electrocatalyst for the oxygen reduction reaction prepared using an mpg-C 3 N 4 template, Chem. Comm. 46 (2010) 7492–7494.

[70] Kailing Sun, Mingrui Liu, Siyan Yu, Huiyu Song, Jianhuang Zeng, Xiuhua Li, Shijun Liao, In-situ grown vanadium nitride coated with thin layer of nitrogen-doped carbon as a highly durable binder-free cathode for Li-O2 batteries, J. Power Sources. 460 (2020) 228109.

[71] Sang Bok Ma, Hyuk Jae Kwon, Mokwon Kim, Seong-Min Bak, Hyunpyo Lee, Steven N. Ehrlich, Jeong-Ju Cho, Dongmin Im, Dong-Hwa Seo, Mixed ionic—electronic conductor of perovskite LixLayMO3– δ toward carbon-free cathode for reversible lithium-air batteries, Adv. Energy Mater. 10 (2020) 2001767.

[72] H. Beyer, M. Metzger, J. Sicklinger, X. Wu, K.U. Schwenke, H.A. Gasteiger, Antimony doped tin oxide-synthesis, characterization and application as cathode material in Li-O2 cells: implications on the prospect of carbon-free cathodes for rechargeable lithium-air batteries, J. Electrochem. Soc. 164 (2017) A1026.

[73] Guangyu Zhao, Yanning Niu, Li Zhang, Kening Sun, Ruthenium oxide modified titanium dioxide nanotube arrays as carbon and binder free lithium—air battery cathode catalyst, J. Power Sources. 270 (2014) 386–390.

[74] Hongjiao Wang, Bo Fan, Zhongkuan Luo, Qixing Wu, Xuelong Zhou, Fang Wang, A unique hierarchical structure: NiCo 2 O 4 nanowire decorated NiO nanosheets as a carbon-free cathode for Li—O 2 battery, Catal. Sci. Technol. 11 (2021) 7632–7639.

[75] Liliang Huang, Yangjun Mao, Guoqing Wang, Xueke Xia, Jian Xie, Shichao Zhang, Gaohui Du, Gaoshao Cao, Xinbing Zhao, Ru-decorated knitted Co 3 O 4 nanowires as a robust carbon/binder-free catalytic cathode for lithium—oxygen batteries, New J. Chem. 40 (2016) 6812–6818.

[76] Jingyi Cao, Shuangyu Liu, Jian Xie, Shichao Zhang, Gaoshao Cao, Xinbing Zhao, Tips-bundled Pt/Co3O4 nanowires with directed peripheral growth of Li2O2 as efficient binder/carbon-free catalytic cathode for lithium-oxygen battery, ACS Catal. 5 (2015) 241–245.

17 Aqueous Electrolytes

Rijith S, Sarika S, Akhila M, and Sumi V S

CONTENTS

17.1 Introduction 249
17.2 Electrolytes in Batteries..250
    17.2.1 Liquid Electrolytes...250
    17.2.2 Solid-State Electrolytes..251
17.3 Aqueous Electrolytes..253
    17.3.1 Alkaline Electrolytes..253
    17.3.2 Neutral Electrolytes 254
    17.3.3 Acidic Solutions...254
17.4 Factors Influencing the Performance of Aqueous Metal-Air Batteries 255
    17.4.1 Band Structure 255
    17.4.2 Transport Property 255
    17.4.3 Viscosity...256
17.5 Principle behind Metal-Air Batteries for Aqueous Electrolytes..................................257
17.6 General Aqueous Electrolytes for Metal-Air Batteries and Their Drawbacks 258
17.7 Strategies for Designing a Better Aqueous Electrolyte System 259
    17.7.1 Effect of Additives and pH Regulation 260
    17.7.2 Salt Concentration..260
    17.7.3 Gelling of Electrolytes 260
    17.7.4 Synthesis of Hybrid-Solvent Electrolytes 261
    17.7.5 Beyond Concentrated Electrolytes...261
    17.7.6 Interface Tuning 261
17.8 Future Perspective and Challenges 261
17.9 Conclusion..262
Acknowledgments..262
References..263

17.1 INTRODUCTION

The need for energy on a worldwide scale has been increasing day by day. Energy generation from non-renewable resources is unsustainable and causes massive emissions of greenhouse gases [1]. Thus, the use of renewable energy sources has been explored as a reasonable option. However, because of the range of weather conditions, the natural sources, such as solar or wind, are intuitively varying and erratic, necessitating the development of energy storage systems as the future energy conception is unpredictable [2]. The major goals of energy storage devices are a greener environment, well-managed energy conservation, reduced consumption of fuel, and buffering of the intermittent renewable sources. In this regard, the role of batteries is promising in terms of their energy storage and conversion reactions. Rechargeable lithium-ion batteries (LiBs) are the pioneers in this area and have excellent energy efficiency and a long life span [3]. The energy density of LiBs still needs to be improved due to the conventional intercalation-based technology, though [4]. Moreover, the expensive and noticeable safety concerns pull it back from wide applications. Thus, storage devices with high energy density have received great research interest. The metal-air batteries break all the conventional mechanisms in that it generates electricity by the use of oxygen from the atmosphere through a redox reaction with the metal, unlike the conventional batteries which store all the reagents in the cathode making them bulkier. Hence, it has a wide range of applications in the electronics field, electric vehicles, power plants, etc. [5]. The metal-air batteries (MAB) such as Zn-air batteries and Li-air batteries have high energy density due to the metal-air bonding which is thirty times more than the traditional rechargeable lead-acid, nickel-metal hydride (Ni-MH), and LiBs [6]. The metal anode consists of suitable electrolytes whereas oxidation-reduction reaction (ORR) as well as oxygen evolution reaction (OER) occur at the cathode and are different from the typical intercalation mechanisms.

17.2 ELECTROLYTES IN BATTERIES

The choice of electrolyte materials has a vital role in battery electrochemistry since they are the media via which the electrodes can interact with each other during cell operation. An ideal electrolyte material should have some beneficial characteristics such as non-toxic, environmentally benign, and cost-efficient along with high thermal stability, proper safety, low chemical side reaction, wide electrochemical stability window (ESW), and enhanced interfacial kinetics which can be industrially manufactured by simple strategies [7]. A group of electrolytes has been exploited in metal-air batteries and it can be classified based on their physical appearance such as solid and liquid electrolytes as depicted in Figure 17.1.

17.2.1 Liquid Electrolytes

Liquid electrolytes are classified as aqueous or non-aqueous electrolytes based on their operating systems. The aqueous alkaline solutions such as KOH, NaOH, and LiOH have been utilized as an electrolyte in MAB owing to their better ionic conductivity. Among these, the high relative stability of KOH makes them a more ideal choice to be employed in the fabrication of MAB than the latter ones. Even though aqueous electrolytes have high ionic conductivity, comparable overpotential

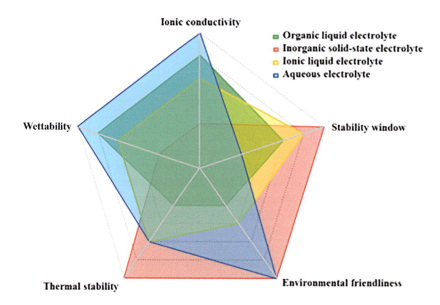

FIGURE 17.1 Various parameters of different electrolytes and their semi-quantitative comparison for the overall performance of energy storage system.

Aqueous Electrolytes

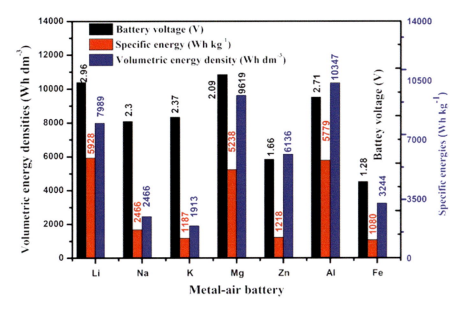

FIGURE 17.2 Electrolytes based on liquid systems for metal-air batteries.

(1.09 V), and better cycle life, they fail to achieve a wide ESW range [8] and thus cannot meet the demand for high energy density. Thus, new strategies should be adapted to overcome these issues for the fabrication of efficient aqueous electrolyte-based energy storage systems. Recent works authenticate that metals like Al, Fe, and Zn are suitable for aqueous systems whereas Li, Na, and K are innately suitable for non-aqueous electrolytes.

The non-aqueous MABs such as Li-air batteries, Na-air batteries, and K-air batteries are gaining attention now and the studies on non-aqueous MABs have primarily concentrated on lithium-air batteries more than the later ones due to the complexity of molding procedure during the processing of potassium and sodium. Electrolytes based on sulfones, ethers, and amides are extensively used in non-aqueous liquid systems for MABs and they can handle a broader electrochemical window than aqueous systems, as illustrated in Figure 17.2. For instance, lithium triflate in tetraethylene glycol dimethyl ether (TEGDME) has the best electrochemical performance and relative stability among the various electrolyte systems for manufacturing MABs [9]. The non-flammability, excellent thermal stability, strong oxidizing potential, and broad potential window of ionic liquids make them employed as better electrolytes for MABs. However, the depletion in kinetics and hike in cost create a barrier to limiting the application in MABs.

Even though non-aqueous electrolytes have certain advantages since the batteries operate in atmospheric conditions, bending, twisting, and folding are unavoidable which can cause a leakage issue during a long time running the battery. This leads to environmental problems and cannot be considered a promising option for future developments in battery operations.

17.2.2 Solid-State Electrolytes

Solid-state electrolytes are categorized into solid inorganic and solid organic electrolytes based on their chemical composition. In Li-ion batteries, solid inorganic electrolytes such as perovskite [10], NASICON [11], LISICON [12], LiPON [13], Li_3PO_4-Li_4SiO_4 [14], and others are widely studied. The aforementioned inorganic solid electrolytes, on the other hand, are inflexible and unsuitable for building flexible MABs; to make matters worse, their ionic conductivity is also abysmal. However, the materialization of solid organic electrolytes acts as a good option for this challenge. The addition of polymers and metal salts to organic electrolytes increases their mechanical characteristics although these results in reduced ionic conductivity than liquid electrolytes. The hybrid system of PVA and PEO is commonly reported as an electrolyte in MABs due to their water solubility and high stability with a wide range of compounds with high conductivity of 10^{-2} S/cm [15].

The ionic conduction mechanisms of solid and liquid electrolytes are different from each other. In liquid electrolytes due to the comparatively quick exchange between solvated and conductive ions, the potential energy diagram of mobile ion is stated as flat. On the contrary, a solid-state crystalline structure requires periodic transitions for the diffusion of mobile species where an energy barrier known as bottleneck sites divides two local minima, which are generally crystallographic locations for ions along the minimum energy path. So, in liquid electrolytes, the ion conduction mechanism is due to concentration diffusion, whereas ions in solids jump between the stable ground state and transition metastable state of the crystalline lattice, especially in crystal solid electrolytes. In the crystalline case, interstitials or vacancies of cations are seen as the mobile charged species, and their migration mechanisms can be categorized as ion migration to a nearby vacant site. A revised Arrhenius relationship can be used to describe how the ionic conductivity of a solid crystal electrolyte is thermally activated, and the Nernst-Einstein relationships can be used to connect the ionic mobility, μ, to the ion diffusion coefficient, D. Since polymer and glass are two solid electrolytes that lack long-range structural order, their mechanisms differ following this. For the use of solid-state batteries glass is also used as a promising electrolyte. The mechanism proceeds via the same Arrhenius theory and the Nernst-Einstein relationship. The "paddlewheel" mechanism is frequently seen in high-temperature crystalline polymorphs in which cation migration happens in electrolytes through a process that the cation combines with a large number of anions, and its orientation changes always [16]. The orientation of molecules happens frequently due to the weak molecular interaction and it changes according to the phase changes that are in the rotor phase and ordered phase the σ orientation is different. The potential energy diagram of migration in solid as well as liquid electrolytes are depicted in Figure 17.3. The relationship curve between σ and $\frac{1}{T}$ in polymer electrolyte is curved and best fits the behavior of VTF (after Vogel, Tammann, and Fulcher). The T_0 (equilibrium glass-transition temperature), is connected to the

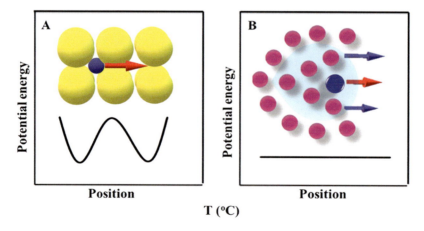

FIGURE 17.3 Potential energy of (A) migration of mobile ion in interstitial mobile ion in solid electrolyte and (B) charge species with solvation shell of liquid electrolytes. (Adapted with permission [17]. Copyright (2016), American Chemical Society.)

measured glass transition temperature. Numerous aqueous electrolytes were designed to fulfill the performance of metal-air batteries.

17.3 AQUEOUS ELECTROLYTES

Aqueous electrolytes outperform organic liquids, polymers, inorganic solid-state, and ionic liquid electrolytes in terms of ionic conductivity, interfacial wettability, safety, and environmental friendliness in battery systems. Aqueous rechargeable batteries (ARBs) based on aqueous electrolytes have gained importance, to prevent the possible flammability and explosion of organic liquid electrolyte-based batteries, as well as complex (atmosphere controlled) manufacturing procedures [18]. Furthermore, the magnitude of ionic conductivity of aqueous electrolytes is significantly better than that of organic liquid electrolytes. The use of flame retardant agents, redox shuttle additives for the prevention of overcharging, or less volatile electrolytes like ionic liquids are just a few of the solutions that have been put forth so far to address the safety concerns with organic electrolytes. The solid electrolyte interface (SEI) can play a crucial role in providing organic liquid electrolytes with outstanding power conversion efficiency, as well as electrochemical and thermodynamic stability [19]. The efficacy of room-temperature polymer electrolytes is limited by their weak ionic conductivity. Despite their promising characteristics, electrolytes are still a long way from being commercialized since they need complicated manufacturing procedures and are carried out in extremely dry conditions even if equipped with technology. Furthermore, the magnitude of ionic conductivity of aqueous electrolytes is significantly better than that of organic liquid electrolytes. The nitrate and sulfate salts dissolved aqueous electrolytic system is less expensive and ecofriendly in nature. Overcharge causes oxygen to be generated at the positive electrode to permeate through the separators and at the negative electrode is reduced to water. This "oxygen-cycle" makes the battery more resistant to overcharging. ARBs have excellent cyclic stability and high energy density because of the high permittivity and low viscosity of water. As a result of the aforementioned advantages of aqueous electrolytes, ARBs can be used to create ultrafast charge-discharge energy storage systems on a grid-scale [20, 21]. Magnesium, Aluminum, Zinc, and Iron anodes are all compatible with aqueous electrolytes, in contrast to nonaqueous systems. They have been extensively researched, and have energy densities similar to lithium-air batteries. Since the electrolyte plays a major role in the battery electrochemistry and directly influences the battery's rechargeability and cell voltage, thus a proper electrolyte is generally essential for the battery system to work satisfactorily.

17.3.1 ALKALINE ELECTROLYTES

Alkaline electrolytes are used more frequently in aqueous-based MABs because they have faster kinetics and less overpotential so they have better ORR kinetics than acidic electrolytes. But alkaline electrolytes have a chance of forming carbonate formation due to the reaction of atmospheric CO_2 with the electrolyte which can obstruct the positive electrode material's porous structure and reduce cathode effectiveness. The superior ionic conductivity, high oxygen diffusion coefficient, low viscosity, and good activity of KOH leads it to use for both the Zn anode and the air cathode and has predominantly been employed as the electrolyte for Zn-air batteries [22]. Many studies have authenticated KOH as an excellent electrolyte. KOH with a 30 wt. % possesses high conductivity of 640 mS/cm at 25 °C whereas a 37 wt. % solution possesses low corrosion gassing along with high conductivity [23]. Corrosion is a major issue faced by most energy storage systems. In Al-air batteries, the usage of NaOH solution will result in the self-corrosion of the aluminum anode as well as parasitic generation of hydrogen during battery discharge that lowers the performance of the

TABLE 17.1
Effect of Additive on Alkaline Electrolyte (NaOH) for Al-Based Anode

Additives	Anode	Results	References
Ethanol	Al	HER rate diminishes H+ in ethanol was much less reactive than H$_2$O.	[24]
L-cysteine and cerium nitrate	Al	Prevents self-corrosion of Al electrode by the formation a complex film on the surface of electrode by synergistic effect of L-cysteine and cerium nitrate.	[25]
CMC and ZnO	AA5052 (Al alloy)	Prevents self-corrosion of the anode by the formation of a complex film on the surface of alloy by CMC and ZnO.	[26]
1-Allyl-3-methylimidazolium bis(trifluoromethylsulfonyl) imide (IL)	Al	Prevents self-corrosion of the anode as well as retards HER. The complex formed by the reaction of IL with anode serves as the reaction surface.	[27]

battery. Many research works were done to solve these issues by adding a suitable additive to NaOH electrolytes and were summarized in Table 17.1.

Therefore, to enhance the performance of metal-air batteries and simplify their applications, the traditional alkaline electrolyte must be modified and replaced.

17.3.2 Neutral Electrolytes

Neutral electrolytes mainly used in Al air batteries have higher activity and reaction kinetics than alkaline electrolytes. Due to the lack of support material, the substrates undergo severe corrosion which results in significant loss of catalyst, which inversely affects the long-term stability of the battery system. To address the corrosion issue associated with the use of alkaline electrolytes, recently neutral electrolytes that are ecologically benign have gained attention.

According to Sumboja et al. [28] for zinc-air batteries, the reduced carbon corrosion of the air cathode and decreased carbonation in the neutral electrolyte resulted in a satisfactory voltage profile and excellent cycling stability for up to 90 days. For Zn-air batteries, a concentrated Zn-ion electrolyte with (1 m Zn(TFSI)$_2$ + 20 m LiTFSI) which was capable of holding water in an open environment, and since TFSI is present around the Zn^{2+} rather than H$_2$O, the H$_2$ evolution was successfully stopped, leading to dendrite-free Zn plating stripping with reversibility and a high coulombic efficiency of with an enhanced energy density. Some of the main issues faced by the Zn-air batteries are low conductivity, solubility, and reversibility attributed to the formation of ZnO from Zn(OH)$_4^{2-}$. Recent works demonstrate that the alcohols form a zincate-based compound Zn(OH)$_4$(OR)$_n^{2-}$ with Zn(OH)$_4^{2-}$, which does not readily give ZnO as Zn(OH)$_4^{2-}$. Thus, the addition of alcohols to the electrolytes retards the formation of ZnO to a great extent in Zn-air batteries [29]. Additives such as water-soluble graphene (WSG), phosphates, and decyl glucoside can serve as electrolyte additives, which provide a better anti-corrosion nature. The recent additives for neutral electrolytes and the enhanced electrochemical battery parameters in various battery systems were depicted in Table 17.2.

Even though neutral chloride electrolytes are a more durable substitute for alkaline electrolytes, it is still necessary to fabricate catalysts with better catalytic activity to resolve the slow electrochemical kinetics for both the oxygen evolution and oxygen reduction reactions. It is indeed critical to significantly improve the conductivities of neutral electrolytes, which have lower conductivities.

17.3.3 Acidic Solutions

The H+ ions in the solutions of an acidic electrolyte can reduce the battery performance by reacting with the metal. Thus, acidic electrolyte is not a suitable choice for metal-air batteries.

TABLE 17.2
Effect of Additive on Neutral Electrolyte for Various Battery systems

Batteries	Neutral electrolyte	Additives	Improvements	Ref
Zn-air batteries	Cl- based	Alcohols	The rate of formation of ZnO reduced, thus the retention capacity, as well as reversibility, increased	[29]
Zn-air batteries	Cl- based	MnO_2	Discharge and charge voltage of 1 and 2 V at 1 mA/cm^2, respectively, and excellent cycling stability up to 90 days of continuous cycle test	[28]
Zn-air batteries	Zn-ion	$Zn(TFSI)_2$ and LiTFSI	Retards HER and leading to dendrite-free Zn plating with reversibility and a high coulombic efficiency of almost 100% along with an enhanced energy density of 300 Wh/kg	[30]
Mg-air batteries	NaCl	WSG	Corrosion of Mg anode decreased, along with an increase in discharge current density from 13.24 to 19.33 mA/cm^2	[31]
Mg-air batteries	NaCl	DG	Inhibit Mg anode corrosion	[32]
Mg-air batteries	NaCl	Phosphate	Inhibit anode corrosion and the anodic efficiency increased to 72.5%	[33]

17.4 FACTORS INFLUENCING THE PERFORMANCE OF AQUEOUS METAL-AIR BATTERIES

17.4.1 BAND STRUCTURE

The electrochemical stability window (ESW) of the electrolytes directly relies on the HOMO and LUMO band gap which originates from its structure and energy levels [34]. The band gap energy (E_g) of the lowest unoccupied molecular orbital (LUMO) and highest occupied molecular orbital (HOMO) gives rise to the electrolyte's ESW. For aqueous electrolytes, a reasonable design is required to ensure adequate electrolyte/electrode interface stability. Solvents can take part in the oxidation and reduction process even if they seem to be stable in terms of HOMO and LUMO energies. The critical factors for the accurate identification of the aqueous electrolytes ESW are the reduction and oxidation potentials. The design of an electrolyte depends on the limiting electrochemical potentials for the anode (μ_A) and cathode (μ_C), i.e, their Fermi energies as well as the physical states of the reactants on both sides (solid, liquid, or gaseous; processing consideration) [35]. μ_A and μ_C should be situated in the ESW of the electrolyte to achieve thermodynamic stability in a battery. The open circuit potential (V_{OC}) is defined as follows.

$$eV_{OC} = \mu_A - \mu_C < E_g \tag{1}$$

where "e" is the magnitude of the electronic charge. A schematic representation of the energy level diagram of electrodes and aqueous electrolytes is shown in Figure 17.4.

17.4.2 TRANSPORT PROPERTY

Electrolytes require conducting ions, which affects how rapidly the energy stored in the electrodes can be supplied. The ion transport in liquid electrolytes depends on the dissociation of ionic compounds and migration of the solvated ions. Ionic conductivity is classified as follows on grounds of solvation and migration [37].

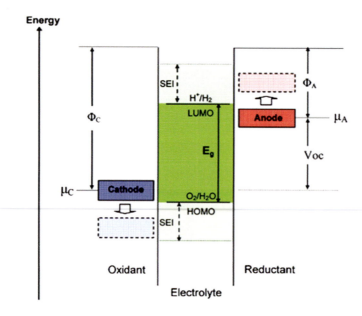

FIGURE 17.4 Schematic open-circuit energy diagram of an aqueous electrolyte. (Adapted with permission [36]. Copyright (2010), American Chemical Society.)

$$\sigma = \sum n_i \mu_i Z_i e \qquad (2)$$

The total mobility of both anions and cations determines the transport property. However, in the case of batteries, the current carried by the cations is taken into account and the principle of transference number in terms of cations (t_+) is given as [38]

$$t_+ = \frac{\mu_+}{\sum_i \mu_i} \qquad (3)$$

As dilution increases, the mobility of cations is found to be less than that of anion which was authenticated by the t_+ value in the range of 0.2 to 0.4. The $t_+ < 1$ is probably due to the excessive solvation shield around the cation. By the ensuing dominant anion movement and saturation towards the surface of the electrode resulted the transference number becomes less than one and which significantly produces concentration polarization when the battery is in use, which necessitates a relatively bare cation to increase the cation mobility.

17.4.3 Viscosity

According to the Stokes-Einstein relationship, solvents with reduced viscosity (η) can improve ion mobility and reduce concentration polarization [38]

$$\mu_i = \frac{1}{6\pi\eta r} \qquad (4)$$

where "r" is the solvation radius. A higher μi is associated with a lower η while a lesser cation solvation degree (solvation radius) leads to a greater μ_i. Furthermore, the dielectric constant (ε) is also important, because it determines the charge carrier number (n). Ions have an increased chance of

Aqueous Electrolytes

remaining free with a higher ε [39]. As a result, an ideal ionic conductivity solvent should have a high ε and a low η, which is challenging to achieve in organic solvent systems, however, water can be an optimum choice, providing desirable ionic conductivity to high power density ARBs.

17.5 PRINCIPLE BEHIND METAL-AIR BATTERIES FOR AQUEOUS ELECTROLYTES

The principle of MABs differs from aqueous electrolyte to nonaqueous electrolyte. At the anode during the discharge process metal is oxidized and the metal ion combines with hydroxides ions to produce metal hydroxides.

$$M + n\,OH^- \rightarrow M(OH)_n + ne^- \quad (5)$$

where M and n represent the metal and valence of metal ions respectively.

At the cathode, the released electrons combine with the external oxygen whereas the hydroxide ion is reduced by the reduction of water (ORR)

$$\frac{n}{4}O_2 + ne^- + \frac{n}{2}H_2O \rightarrow nOH^- \quad (6)$$

From the cathode to the anode the produced hydroxide ions are transported through the separator to complete the circuit.

$$M + \frac{n}{4}O_2 + \frac{n}{2}H_2O \rightarrow M(OH)_n \quad (7)$$

The change in an electronic structure and the catalyst affects the ORR reaction mechanisms. In an aqueous electrolyte, the pace of the ORR or OER process can be increased by using an oxygen electrocatalyst. However, without any catalyst, the kinetics of the reaction is somewhat slow. Recent discoveries showed that the electronic structure plays an important role in determining the oxide catalytic performance. In ORR, the OH^- regeneration and competition between OH^- and O_2^{2-} on catalyst surface influence the rate of reaction. The schematic representation of MABs is depicted in Figure 17.5.

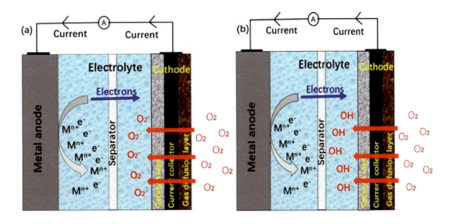

FIGURE 17.5 Schematic diagrams of MABs working principles for (a) non-aqueous electrolyte and (b) aqueous electrolyte.

The reaction mechanism of Fe-air battery in an aqueous alkaline medium.

$$\text{Anode: } 3Fe(OH)_2 + 2OH^- \rightarrow Fe_3O_4 + 4H_2O + 2e^- \tag{8}$$

$$\text{Cathode: } O_2 + 2H_2O + 4e^- \leftrightarrow 4OH^- \tag{9}$$

$$\text{Overall reaction: } 2O_2 + 3Fe \leftrightarrow Fe_3O_4 \tag{10}$$

The reaction kinetics of Al-air

$$\text{Anode: } Al + 4OH^- \leftrightarrow AlOH_4^- + 3e^- \tag{11}$$

$$\text{Cathode: } O_2 + 2H_2O + 4e^- \leftrightarrow 4OH^- \tag{12}$$

$$\text{Overall reaction: } 3O_2 + 2Al \leftrightarrow Al_2O_3 \tag{13}$$

In Zn-air batteries, the oxygen is reduced at the cathode to form hydroxyl ions and zincate ions ($Zn(OH)_4$) formed by combining with hydroxyl ions and zinc ions from the anode.

In Zn-air battery, the overall reactions in an aqueous alkaline electrolyte are shown in the following.

$$\text{Cathode: } O_2 + 2H_2O + 4e^- \leftrightarrow 4OH^- \tag{14}$$

$$\text{Anode: } Zn + 4OH^- - 2e^- \leftrightarrow 2ZnOH_4 \tag{15}$$

$$\text{Overall reaction: } 2Zn + O_2 + 2H_2O + 4OH^- \leftrightarrow 2Zn(OH)_4^{+2} \tag{16}$$

At present aqueous electrolyte Zn-air batteries may be recharged electrically, whereas Al-air and Mg-air batteries must be refilled mechanically by replacing the metal anodes, which in an aqueous electrolyte cannot be converted directly from ions to metals [40].

17.6 GENERAL AQUEOUS ELECTROLYTES FOR METAL-AIR BATTERIES AND THEIR DRAWBACKS

By using earth-abundant and non-toxic precursors and opting for low-cost manufacturing processes AQUION Inc. developed patented technology of Aqueous Hybrid Ion (AHI™) chemistry, employing a Na-based aqueous electrolyte, carbon-titanium phosphate composite anode, manganese oxide cathode, and cotton as a separator. In 1994, Li et al. published an ARMB that used $LiMn_2O_4$ as the cathode, VO_2 as the anode, and 5 mol/L aqueous $LiNO_3$ solution as the electrolyte. In terms of gravimetric energy performance, they demonstrated that this cell chemistry is a safe and cost-effective technology that can compete with nickel-cadmium and lead-acid batteries [41]. See et al. successfully used a KOH-based electrolyte in a flexible zinc-air battery and achieved a high activity [22].

However, the main issue related to the ESW of the aqueous electrolyte as it is in a narrow range, which affects the cycle life and increased the cost of running the battery systems. The long-term charge-discharge and recharge may be hindered due to the electrode dissolution. In addition to that, the side reaction occurred as a result of the decomposition of water and the corresponding decomposition voltage is 1.23 V, which has an immediate impact on the energy and power densities of the cell. Additionally, the application of ARBs at high or low temperatures may be restricted by the phase transition of water with temperature variation, constraining the operating temperature range of aqueous electrolytes.

17.7 STRATEGIES FOR DESIGNING A BETTER AQUEOUS ELECTROLYTE SYSTEM

The common conventional concepts for better aqueous electrolytes are a wide range of working temperatures with appreciable thermal stability, enhanced ionic conductivity, large ESW, low side reactions, and electrode dissolution. The recent studies on aqueous electrolytes depict that its increase in efficiency mainly relies on some parameters such as the additive, salt concentration, pH regulation, gelling, solvent-hybrid, and tuning of electrode-electrolyte interface and beyond salt concentration as shown in Figure 17.6.

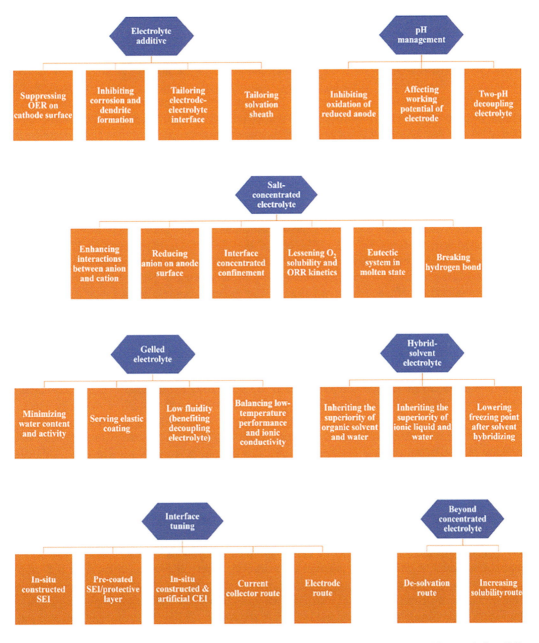

FIGURE 17.6 Various strategies for a better aqueous electrolyte system. (Adapted with permission [36]. Copyright (2010), American Chemical Society.)

17.7.1 Effect of Additives and pH Regulation

The incorporation of additives in the forms of metals, ions, and organic or inorganic constituents into aqueous electrolytes can be beneficial in tailoring the ESW as well as the electrode-electrolyte interfaces and results in enhancing the ion kinetics. Chao et al. [42] show that the addition of Mn ion suppresses the oxygen evolution reaction (OER) and widens the ESW. The addition of $PbSO_4$ results in a corrosion resistance nature and ensures dendrites-free Zn anodes [43]. The pH of the relevant aqueous electrolytes can theoretically be used as a benchmark to assess if the electrodes are stable against them. In reality, altering pH can be another way to control the stability of the battery system.

17.7.2 Salt Concentration

The salt content is a key factor that affects the electrode/electrolyte interfaces, ionic conductivity, liquid electrolyte, ESWs via viscosity, solvation, mobility of ions, and morphology of electrolyte structure [44]. Based on this, tailoring of liquid electrolytes may be done simply and effectively by adjusting the salt levels in the electrolytes. Liquid electrolytes with higher salt content will exhibit unique physicochemical and electrochemical properties that are significantly different from those of traditional dilute electrolytes due to improved contacts between ions/solvents and a decrease in the number of free-state solvent molecules [45].

The idea of "water-in-salt" electrolytes (WiSEs) brought a new era to salt-concentrated electrolytes and was one notable breakthrough achievement by Wang and Xu's group in 2015 [46]. WiSEs are described as the salt in a binary system that exceeds the solvent in terms of volume as well as weight, causing the formation of thick interphase on the surface of the anode. Hydrate melt electrolytes (HMEs), which employ a eutectic system, were another important electrolyte among the salt-concentrated systems. Early, LiTFSI and $LiN(SO_2C_2F_5)_2$ (LiBETI) were preferred for HMEs due to their weak basic nature by weakly interacting with Li^+ ions, favoring their solvation by H_2O rather than the profuse generation of ion pairs. Additionally, the anions had a "plasticizing" action that prevented the salts and hydrates from crystallizing. The eutectic composition of $Li(TFSI)_{0.7}(BETI)_{0.3}$ was found to exhibit the highest miscibility, forming a stable, transparent liquid with an extremely low water content ($H_2O:Li^+$ = 2.0 molar ratio). For aqueous electrolytes with a high salt content, the cation/water ratio of WiSEs and HMEs often rises over 2.0, but those that do not fit the criteria or reach the ratio are classified under regular super concentration electrolytes (RSCEs). Various MD simulations and spectrum analyses explored the thermodynamic and kinetic aspects of the ESW of a $LiNO_3$-based RSCE ($LiNO_3:H_2O$ = 1: 2.5). By generating $(Li^+(H_2O)_2)_n$ polymer-like chains instead of the common H bonding between H_2O molecules, the structure of the Li^+-water interaction was identified at a super-concentration, and thereby the ESW was widened to 2.55 V. In a nutshell, salt-concentrated electrolytes can be a universal method to increase the energy density and cycle stability of ARBs, which have been used to broaden operating temperature windows. Although these systems have achieved desirable results, their high cost likely prevents them from being used on a broad scale.

17.7.3 Gelling of Electrolytes

The gelled aqueous electrolytes can be called hydrogel electrolytes because they are made of a matrix composed of cross-linked polymer chains having interstitial spaces filled with H_2O molecules, providing them a quasi-solid nature and flexible characteristic. In addition to their physical characteristics, hydrogel electrolytes also have extremely accessible and adaptable polymer chemistries that enable their fabrication and use in flexible and elastic ARBs. The creation of soft electronics, implantable strain sensors, and wearable medical devices is made possible by hydrogel-based ARBs with shape memory, self-healing, and stretchability properties. Recent works demonstrate that higher stability can be achieved by minimizing the water content using gelled electrolytes

Aqueous Electrolytes

[47]. Also, the poor fluidity arises from the quasi-solid nature of gelled electrolytes that offer a high energy density. Thus, gelled aqueous electrolytes are a solvent technique to solidify and hence inactivate H_2O solvent molecules and thereby widen ESWs, enhance the stability, and broaden the operating temperature range, which is practical and affordable for superior performance. The following part will examine hybrid solvent, another technique for manipulating solvents.

17.7.4 Synthesis of Hybrid-Solvent Electrolytes

Hybrid solvent electrolytes are a new class of electrolytes formed by the combination of aqueous and non-aqueous solvents, effectively addressing the conflicts regarding the economic viability, ambient reaction control, environmental concerns, and performance, while inherent benefits of each system. Zhang et al. reported Na^+ supported solvent-hybrid electrolyte $NaTi_2(PO_4)_3/Na_3V_2(PO_4)_3$, which shows an ESW of up to 2.8 V [48]. Ionic liquids have special qualities including the capacity to withstand flames, nonvolatility, and great thermal stability, yet the high viscosity limits their potential uses. Therefore, a "water in salt/ionic liquid" electrolyte made of 1-ethyl-3-methylimidazolium bis(trifluoromethane sulfonyl) imide (EMIM-TFSI) salt and LiTFSI salt was combined with a very tiny amount of water (1:1:2). The ESW in this electrolyte was greatly expanded without reducing ionic conductivity, enabling access to a commercial Nb_2O_5 full potential at a low potential window (1.6 V vs. Ag/AgCl) [49]. Hybridizing aqueous electrolytes with organic solvents of low freezing points improves its low-temperature performance. Acetonitrile (AN) has a freezing point of 48 ^0C, with a high dielectric constant, high oxidation stability, and high water miscibility, all of which allow it to form a redox pair to demonstrate high capacity at low and high temperatures [49]. If a balance can be established between wide ESW, strong ionic conductivity, low freezing point, and affordability, it may be said that the hybridizing technique for solvents is promising.

17.7.5 Beyond Concentrated Electrolytes

The advantages of salt-concentrated electrolytes for ARBs have been extensively discussed here, however since the electrolyte requires salts with high solubility, the practical applicability is sometimes constrained. In this regard, introducing new economically viable salts with low solubility, and expanding ESW become viable. The methods opted for this purpose mainly consist of de-solvation and solubility augmentation [50].

17.7.6 Interface Tuning

The concept of interface tuning technique is a key factor in improving the performance of aqueous electrolytes. The main objectives of tuning may be broadly divided into thermodynamics (thermal and chemical stability) and kinetics (charge and mass transportation). Since the aqueous electrolytes have excellent interfacial kinetics and wettability, it is more crucial to address the thermodynamics. Electrolytes can be kinetically stabilized by the SEI at potentials far above their thermodynamic stability thresholds, enabling reversible cell processes. Before the discovery of WiSE, it did not take place in aqueous electrolytes. By integrating several spectroscopic, electrochemical, and computational methodologies, it has been suggested that two routes may be involved in the formation of an SEI in WiSE: (1) reduction of anion complexes or clusters, and (2) reduction of O_2 and CO_2 dissolved in the electrolyte.

17.8 FUTURE PERSPECTIVE AND CHALLENGES

Several strategies, including electrolyte additive, pH management, salt concentration, gelling, hybrid solvent, interface tailoring, and resolving salt solubility limitation, reported extensively over the past five years help us to minimize the difficulties of limited ESWs, thermal instability, and side reactions of aqueous electrolytes. The related accomplishments, particularly those in salt-concentrated electrolytes, are impressive and are quickly boosting the characteristics of aqueous electrolytes in the overall performance of the battery system. For high-energy and high-stability, the salt-concentrated approach is currently the most effective and efficient; nevertheless, the cost of salt presents a problem for large-scale industrial applications.

There are still several possible obstacles to the ongoing commercialization of these despite the substantial advancements. First, aqueous electrolytes require better characterization and modeling. The variables of DFT analysis as well as MD simulation for tuning and solvation of sheath respectively should be pre-set consciously rather than purposely to support the as-obtained experimental results. Raman, FTIR, and NMR spectroscopy are typically utilized as indirect methods in the experimental assessment of the ion solvation framework, and just a few reports have used direct methods. The WiSEs were explored using X-ray nano-imaging, which provided conclusive proof of their stability when compared to traditional aqueous electrolytes. Even though concentrated electrolytes offer more efficiency, elevated salt concentrations were always not necessary. In certain contexts, excessive concentration can be quite harmful as too diluted, which demonstrates that appropriate concentration and proper interfacial design seem to be more important than attempting to synthesize more concentrated electrolytes. In salt-concentrated systems, the slow and sluggish reaction kinetics and mass transfer attributes from their high viscosity are often unfavorable and this can be rectified to some extent by opting for nanosized electrodes which can balance the broad ESWs of the electrolyte and the high capacity of the electrode. In practice, reporting battery performance based on limited parameters does not accurately reflect the performance needed for real-world applications. The development of innovative electrolyte ideas has allowed for a thorough investigation of aqueous energy storage devices, but there are still issues to be resolved before they can be used on a wide basis.

17.9 CONCLUSION

The innovations on the electrolyte side have a significant impact on aqueous battery advancements. Choosing a suitable aqueous electrolyte for metal-air batteries is crucial, particularly when lithium, sodium, or potassium metal is employed as the anode. But often, batteries run the risk of short-circuiting and catastrophic accidents. In addition to the difficulties mentioned earlier, it is crucial to have a thorough grasp of how metal-air batteries behave when they cycle. Both experimental and theoretical models are required for a deeper understanding of the electrochemical processes. Technical research is just as crucial for the development of flexible metal-air batteries as basic research because the assembly process has a significant impact on the battery's energy density.

Due to the inherent benefits, including the high-power output associated with rising ionic conductivity, high safety for eschewing the flammable organic electrolytes, cost-effectiveness resulting from the use of salt abundance, and environmental friendliness of aqueous electrolyte is of great significance for large-scale energy storage. Due to innate issues including anode corrosion, side reactions, and dendrite development, it is required to modify the traditional alkaline electrolytes by combining them with innovative electrolyte possibilities. In conclusion, as this chapter has demonstrated, the information obtained regarding the chemistry of aqueous electrolytes for use in high-energy metal-air batteries can serve as the foundation for the fabrication of a wide range of future aqueous energy storage systems.

ACKNOWLEDGMENTS

I would like to express our deep appreciation and indebtedness particularly to the following: Dr. Sarika S, Dr. Sumi V S, and Akhila Muhammed for their endless support and understanding for the compilation of the chapter. I also express sincere thanks to thank the Principal, Sree Narayana College, Kollam, Kerala, India, and Professor and Head, Department of Chemistry, Sree Narayana College, Kollam, Kerala, India for providing the library facilities.

REFERENCES

[1] M. Armand, J.M. Tarascon, Building better batteries, Nature. 451(7179) (2008) 652–657.
[2] S. Park, Y. Shao, J. Liu, Y. Wang, Oxygen electrocatalysts for water electrolyzers and reversible fuel cells: Status and perspective, Energy & Environmental Science. 5(11) (2012) 9331–9344.
[3] P. Gu, M. Zheng, Q. Zhao, X. Xiao, H. Xue, H. Pang, Rechargeable zinc—air batteries: A promising way to green energy, Journal of Materials Chemistry A. 5(17) (2017) 7651–7666.
[4] B. Dunn, H. Kamath, J.M. Tarascon, Electrical energy storage for the grid: A battery of choices, Science. 334 (6058) (2011) 928–9235.
[5] P.G. Bruce, S.A. Freunberger, L.J. Hardwick, J.M. Tarascon, Li-O_2 and Li—S batteries with high energy storage, Nature Materials. 11(1) (2012) 19–29.
[6] Y. Li, J. Lu, Metal-air batteries: Will they be the future electrochemical energy storage device of choice?, ACS Energy Letters. 2(6) (2017) 1370–1377.
[7] J. Janek, W.G. Zeier, A solid future for battery development, Nature Energy. 1(9) (2016) 1–4.
[8] F. Meng, H. Zhong, D. Bao, J. Yan, X. Zhang, In situ coupling of strung Co_4N and intertwined N-C fibers toward free-standing bifunctional cathode for robust, efficient, and flexible Zn-air batteries, Journal of the American Chemical Society. 138(32) (2016) 10226–10231.
[9] Y. Chen, S.A. Freunberger, Z. Peng, F. Bardé, P.G. Bruce, Li-O_2 battery with a dimethylformamide electrolyte, Journal of the American Chemical Society. 134(18) (2012) 7952–7957.
[10] P. Birke, S. Scharner, R.A. Huggins, W. Weppner, Electrolytic stability limit and rapid lithium insertion in the fast-ion-conducting $Li_{0.29}La_{0.57}TiO_3$ perovskite-type compound, Journal of the Electrochemical Society. 144(6) (1997) 167.
[11] A. Morata-Orrantia, S. García-Martín, E. Morán, M.Á. Alario-Franco, A new $La_{2/3}Li_xTi_{1-x}Al_xO_3$ solid solution: Structure, microstructure, and Li+ conductivity, Chemistry of Materials. 14(7) (2002) 2871–2875.
[12] M. Murayama, R. Kanno, M. Irie, T. Hata, N. Sonoyama, Y. Kawamoto, Synthesis of new lithium ionic conductor thio-LISICON-lithium silicon sulfides system, Journal of Solid State Chemistry. 168(1) (2002) 140–148.
[13] S.J. Lee, J.H. Bae, H.W. Lee, H.K. Baik, S.M. Lee, Electrical conductivity in Li—Si—P—O—N oxynitride thin-films, Journal of Power Sources. 123(1) (2003) 61–64.
[14] J.F. Whitacre, W.C. West, Crystalline Li_3PO_4/Li_4SiO_4 solid solutions as an electrolyte for film batteries using sputtered cathode layers, Solid State Ionics. 175(1–4) (2004) 251–255.
[15] T. Osaka, T. Momma, Y. Matsumoto, Y. Uchida, Effect of carbon dioxide on lithium anode cyclability with various substrates, Journal of Power Sources. 68(2) (1997) 497–500.
[16] J.G. Smith, D.J. Siegel, Low-temperature paddlewheel effect in glassy solid electrolytes, Nature Communications. 11(1) (2020) 1–1.
[17] J.C. Bachman, S. Muy, A. Grimaud, H.H. Chang, N. Pour, S.F. Lux, O. Paschos, F. Maglia, S. Lupart, P. Lamp, L. Giordano, Inorganic solid-state electrolytes for lithium batteries: Mechanisms and properties governing ion conduction, Chemical Reviews, 116(1) (2016) 140–162.
[18] X. Yuan, F. Ma, L. Zuo, J. Wang, N. Yu, Y. Chen, Y. Zhu, Q. Huang, R. Holze, Y. Wu, T. van Ree, Latest advances in high-voltage and high-energy-density aqueous rechargeable batteries, Electrochemical Energy Reviews. 4(1) (2021) 1–34.
[19] Z. Liu, Y. Huang, Y. Huang, Q. Yang, X. Li, Z. Huang, C. Zhi, Voltage issue of aqueous rechargeable metal-ion batteries, Chemical Society Reviews. 49(1) (2020) 180–232.
[20] N. Alias, A.A. Mohamad, Advances of aqueous rechargeable lithium-ion battery: A review, Journal of Power Sources. 274 (2015) 237–251.
[21] W. Tang, Y. Zhu, Y. Hou, L. Liu, Y. Wu, K.P. Loh, H. Zhang, K. Zhu, Aqueous rechargeable lithium batteries as an energy storage system of superfast charging, Energy & Environmental Science. 6(7) (2013) 2093–2104.
[22] D.M. See, R.E. White, Temperature and concentration dependence of the specific conductivity of concentrated solutions of potassium hydroxide, Journal of Chemical & Engineering Data. 42(6) (1997) 1266–1268.
[23] M. Xu, D.G. Ivey, Z. Xie, W. Qu, Rechargeable Zn-air batteries: Progress in electrolyte development and cell configuration advancement, Journal of Power Sources. 283 (2015) 358–371.
[24] M. Jingling, W. Jiuba, Z. Hongxi, L. Quanan, Electrochemical performances of Al0.5Mg0.1Sn0.02In alloy in different solutions for Al-air battery, Journal of Power Sources. 293 (2015) 592–598.

[25] D. Wang, H. Li, J. Liu, D. Zhang, L. Gao, L. Tong, Evaluation of AA5052 alloy anode in alkaline electrolyte with organic rare-earth complex additives for aluminium-air batteries, Journal of Power Sources. 293 (2015) 484–491.

[26] J. Liu, D. Wang, D. Zhang, L. Gao, T. Lin, Synergistic effects of carboxymethyl cellulose and ZnO as alkaline electrolyte additives for aluminium anodes with a view towards Al-air batteries, Journal of Power Sources. 335 (2016) 1–1.

[27] M.A. Deyab, 1-Allyl-3-methylimidazolium bis (trifluoromethylsulfonyl) imide as an effective organic additive in aluminum-air battery, Electrochimica Acta. 244 (2017) 178–183.

[28] A. Sumboja, X. Ge, G. Zheng, F.T. Goh, T.A. Hor, Y. Zong, Z. Liu, Durable rechargeable zinc-air batteries with neutral electrolyte and manganese oxide catalys, Journal of Power Sources. 332 (2016) 330–336.

[29] J. Lee, B. Hwang, M.S. Park, K. Kim, Improved reversibility of Zn anodes for rechargeable Zn-air batteries by using alkoxide and acetate ions, Electrochimica Acta. 199 (2016) 164–171.

[30] F. Wang, O. Borodin, T. Gao, X. Fan, W. Sun, F. Han, A. Faraone, J.A. Dura, K. Xu, C. Wang, Highly reversible zinc metal anode for aqueous batteries, Nature Materials. 17(6) (2018) 543–549.

[31] M.M. Dinesh, K. Saminathan, M. Selvam, S.R. Srither, V. Rajendran, K. V. Kaler, Water soluble graphene as electrolyte additive in magnesium-air battery system, Journal of Power Sources. 276 (2015) 32–38.

[32] M.A. Deyab, Decyl glucoside as a corrosion inhibitor for Magnesium-air battery, Journal of Power Sources. 325 (2016) 98–103.

[33] Y. Zhao, G. Huang, C. Zhang, C. Peng, F. Pan, Effect of phosphate and vanadate as electrolyte additives on the performance of Mg-air batteries, Materials Chemistry and Physics. 218 (2018) 256–261.

[34] N. Mahmood, T. Tang, Y. Hou, Nanostructured anode materials for lithium ion batteries, progress, challenge and perspective, Advanced Energy Materials. 6(17) (2016) 1600374.

[35] J.B. Goodenough, Design considerations, Solid State Ionics. 69(3–4) (1994) 184–198.

[36] J.B. Goodenough, Y. Kim, Challenges for rechargeable Li batteries, Chemistry of Materials. 22 (2010) 587–603.

[37] S. Chen, M. Zhang, P. Zou, B. Sun, S. Tao, Historical development and novel concepts on electrolytes for aqueous rechargeable batteries, Energy & Environmental Science. 15(5) (2022) 1805–1839.

[38] W.B. Kirby, Linden's handbook of batteries (5th edn). McGraw-Hill Education, New York, 2019.

[39] A.K. Kontturi, K. Kontturi, L. Murtomäki, D.J. Schiffrin, Effect of preferential solvation on Gibbs energies of ionic transfer, Journal of the Chemical Society, Faraday Transactions. 90(14) (1994) 2037–2041.

[40] T. Zhang, Z. Tao, J. Chen, Magnesium-air batteries: from principle to application, Materials Horizons. 1(2) (2014) 196–206.

[41] W. Li, J.R. Dahn, D.S. Wainwright, Rechargeable lithium batteries with aqueous electrolytes, Science. 264(5162) (1994) 1115–1117.

[42] D. Chao, W. Zhou, C. Ye, Q. Zhang, Y. Chen, L. Gu, K. Davey, S.Z. Qiao, An electrolytic Zn-MnO$_2$ battery for high-voltage and scalable energy storage, Angewandte Chemie. 131(23) (2019) 7905–7910.

[43] T.K. Hoang, M. Acton, H.T. Chen, Y. Huang, T.N. Doan, P. Chen, Sustainable gel electrolyte containing Pb^{2+} as corrosion inhibitor and dendrite suppressor for the zinc anode in the rechargeable hybrid aqueous battery, Materials Today Energy. 4 (2017) 34–40.

[44] K Xu, Nonaqueous liquid electrolytes for lithium-based rechargeable batteries, Chemical Reviews. 104(10) (2004) 4303–4418.

[45] Y. Yamada, J. Wang, S. Ko, E. Watanabe, A. Yamada, Advances and issues in developing salt-concentrated battery electrolytes, Nature Energy. 4(4) (2019) 269–280.

[46] L. Suo, O. Borodin, T. Gao, M. Olguin, J. Ho, X. Fan, C. Luo, C. Wang, K. Xu, "Water-in-salt" electrolyte enables high-voltage aqueous lithium-ion chemistries, Science. 350(6263) (2015) 938–943.

[47] G. Wang, X. Lu, Y. Ling, T. Zhai, H. Wang, Y. Tong, Y. Li, LiCl/PVA gel electrolyte stabilizes vanadium oxide nanowire electrodes for pseudocapacitors, ACS Nano. 6(11) (2012) 10296–10302.

[48] H. Zhang, B. Qin, J. Han, S. Passerini, Aqueous/nonaqueous hybrid electrolyte for sodium-ion batteries, ACS Energy Letters. 3(7). (2018) 1769–1770.

[49] Q. Dou, Y. Wang, A. Wang, M. Ye, R. Hou, Y. Lu, L. Su, S. Shi, H. Zhang, X. Yan, "Water in salt/ionic liquid" electrolyte for 2.8 V aqueous lithium-ion capacitor, Science Bulletin. 65(21) (2020) 1812–1822.

[50] Z. Chang, Y. Qiao, H. Deng, H. Yang, P. He, H. Zhou, A liquid electrolyte with de-solvated lithium ions for lithium-metal battery, Joule. 4(8) (2020) 1776–1789.

18 Non-Aqueous Electrolytes in Metal-Air Batteries

Pravin N. Didwal, An-Giang Nguyen, Satyanarayana Maddukuri, and Rakesh Verma

CONTENTS

18.1 Introduction 265
18.2 Requirements for Nonaqueous Electrolytes 268
18.3 Types of Nonaqueous Electrolytes 269
    18.3.1 Carbonate-Based Electrolytes 269
    18.3.2 Ether-Based Electrolytes 269
    18.3.3 DMSO-Based Electrolyte 270
    18.3.4 Nitriles-Based Electrolytes 271
    18.3.5 Amides-Based Electrolytes 271
18.4 Component in Non-Aqueous Electrolyte 271
    18.4.1 Effect of Solvent 272
    18.4.2 Effect of Lithium Salt 273
    18.4.3 Effect of Additives 274
18.5 Characteristics of Non-Aqueous Electrolyte 275
    18.5.1 Stability with Li-Metal 275
    18.5.2 Electrochemical and Chemical Stability 275
    18.5.3 Physical Stability 276
    18.5.4 Li-ion and Oxygen Reagent Transport Issues 276
    18.5.5 Charge-Transfer Issues 276
18.6 Conclusion and Outlook 277
References 277

18.1 INTRODUCTION

With an increased demand for portable electronics and electric vehicles, significant investment has been made in the research and development of energy storage and conversion technologies that have high power density, long lifetimes, exceptional energy efficiency, and sustainability [1–2]. Due to their ultrahigh theoretical energy density, comparable to gasoline, rechargeable Li-air batteries (LABs) have emerged as promising electrochemical energy storage devices [3]. Nevertheless, LABs are still in their early stages and encounter many challenges, including drastic capacity decay and poor rate performance [4–5]. The electrolyte in LABs is analogous to blood in humans, and it influences factors such as discharge-specific capacity, cycling stability, rate performance, and round-trip efficiency in determining battery performance [6–8]. As a result, appropriate electrolytes must be developed that work in tandem with the entire battery system.

Li-air batteries, as illustrated in Figure 18.1, are made up of air or O_2 cathodes, electrolytes, and Li metal anodes. To accommodate the insoluble discharge product Li_2O_2 [9], the cathode should be porous. To achieve a high capacity, it is assumed that O_2, the active material of the cathodes, is acquired from the open air [10]. Pure O_2 is extensively used in primarily laboratory-scale experiments in non-aqueous systems to prevent other components of air, particularly H_2O and CO_2, from

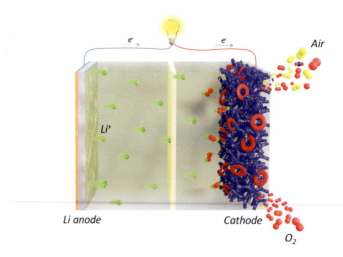

FIGURE 18.1 Li-O_2 battery operation mechanisms with nonaqueous electrolytes.

interrupting the preferred electrochemical behavior. These pure O_2 systems are also known as LABs. Furthermore, electrocatalysts are commonly used to aid the O_2 reduction reaction (ORR) and O_2 evolution reaction (OER) throughout the discharge/charge operations.

LABs electrolytes are classified into four kinds: nonaqueous liquid, aqueous, solid-state, and hybrid. Over the last two decades, non-aqueous electrolytes have received the most attention. Organic carbonates were initially investigated as nonaqueous electrolytes for LABs, but they were later found to be unsuitable due to nucleophilic attack by oxygen reduction species [11]. Ethers were given special consideration because they are more stable than alkyl carbonates. The electrochemical reaction that occurs in the Li anodes of LABs during discharge. According to the half-cell reaction, Li metal is oxidized to Li-ions, which liberate electrons [12]:

$$Li \rightleftharpoons Li^+(sol) + e^- \qquad [1]$$

Depending on the electrolyte, the chemistry at the cathode differs. Among these electrolytes, non-aqueous electrolytes have received the most attention in recent decades due to their ability to provide exceptionally high energy density with a straightforward battery structure. For nonaqueous liquid LABs, equation (2) provides the preferred electrochemical reaction with Li_2O_2 as the discharge product [13–14]:

$$2Li^+ + O_2 + 2e^- \rightleftharpoons Li_2O_2, \quad E^0 = 2.96 \text{ V vs. } Li/Li^+ \qquad [2]$$

The Li_2O_2 reaction product, as shown in Figure 18.2, is insoluble in the electrolyte and deposits on the solid surface of the cathode. According to Andrei et al. [15], the major reason for the relatively short life of current LABs is the formation of Li_2O_2, which fills in the channels and interrupts the flow of O_2 in the cathode. Furthermore, the relatively low value of the diffusion coefficient of O_2 in the electrolyte reduces the power density of these batteries significantly.

The basic steps, however, are difficult, and the discharge/charge mechanisms are still debated. Based on experiments and theoretical calculations, two primary discharge reaction mechanisms for ORR on the cathode surface of a nonaqueous liquid system have been proposed. In the first steps, the mechanism of the solution-mediated reaction is depicted in equations (3–5) as follows [13, 14, 16–17]:

$$O_2(sol) + e^- \rightarrow O_2^-(sol) \qquad [3]$$

Non-Aqueous Electrolytes in Metal-Air Batteries

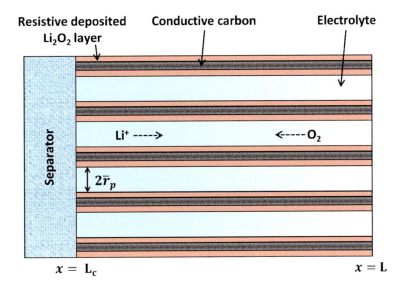

FIGURE 18.2 Oxygen diffusion and Li_2O_2 formation modeling (color online) in a porous carbon cathode.

$$2Li^+(sol) + 2O_2^-(sol) \rightarrow Li_2O_2 + O_2 \quad [4]$$

$$2Li^+(sol) + O_2^-(sol) + e^- \rightarrow Li_2O_2 \quad [5]$$

The preceding solution-mediated pathway, wherein a huge portion of Li_2O_2 increases as toroidal-shaped particles from solution, has a high specific capacity but sometimes low cyclability due to insulation from big Li_2O_2 particles. In addition, the surface-mediated mechanism occurs in the second step via equations (6–8) [18]:

$$Li^+ + O_2(sol) + e^- \rightarrow LiO_2^* \quad [6]$$

$$Li^+ + LiO_2^* + e^- \rightarrow Li_2O_2^* \quad [7]$$

$$2LiO_2^* \rightarrow Li_2O_2^* + O_2 \quad [8]$$

Where, (*) denotes a Li_2O_2 surface site.

The surface-mediated mechanism, which involves the formation of film-like Li_2O_2 on the cathode surface, has a low capacity due to surface clogging, but a good cycle performance when capacity is limited, which is ascribed to the few defect sites in the Li_2O_2 layer. Consequently, attempting to control the route and morphology of Li_2O_2 formation is critical for improving the cyclability of LABs. According to recent research, even slight tweaks in electrolytes, such as distinctions in lithium salts, solvents, and additives, can drastically change the discharge reaction path. The aforementioned classification and mechanisms demonstrate that electrolyte is important when considering the cycling stability of LABs. Furthermore, electrolytes must be optimized to address other concerns, such as electrolyte decomposition, poor kinetics of Li^+/O_2 dissolution and transportation, the negative impact of moisture and CO_2 in the air, and Li anode instability. As a result, we have concentrated on current research progress and prospects on non-aqueous electrolytes in LABs, which are studied in the sections that follow, including criteria for selecting electrolyte materials, new electrolyte formulations, efficient design, and use of electrolyte additives, and battery structure innovation. The purpose of this chapter is to provide information to aid in the choice of optimal nonaqueous electrolytes for LABs, as well as to highlight the existing challenges in finding ways to boost the electrochemical performance of LABs via electrolyte optimization.

18.2 REQUIREMENTS FOR NONAQUEOUS ELECTROLYTES

The non-aqueous electrolyte in lithium-air batteries (LABs) is an environment for ions transportation and generally consists of two main components: salt and organic solvent. In addition, LABs are a catalog of lithium-ion batteries (LIBs), therefore, the requirements for nonaqueous electrolytes should be covered all requirements for LIBs batteries as follow [13]:

1. High ionic conductivity and non/low electron conductivity
2. Highly chemically and electrochemically stable toward both anode and cathode
3. Safety (low volatility, non-flammability, thermal stability)
4. Environmentally friendly
5. Low cost

In addition, LABs are the half-open system with air-cathode and lithium metal anode to achieve high energy density. Therefore, some additional requirements should be satisfied [13, 14, 16]:

1. High O_2 diffusivity and solubility
2. Under O_2-rich environments, there is excellent chemical and electrochemical stability in the presence of Li_2O_2/LiO_2
3. Increasing efficiency with intermediates leads to the formation of toroidal-shaped Li_2O_2 through a solution-mediated mechanism
4. A powerful solvating effect that is not required but can aid in the decomposition of Li_2O_2

Furthermore, in LABs, the oxidation-reduction reaction is affected by the donor number (DN) or acceptor number (AN) of solvents. According to hard-soft acid-base theory, Li^+ as a hard acid has a strong affinity with a hard base (O_2^- and O_2^{2-}). In the electrolyte solution, the acidity of Li^+ can be adjusted by interaction with solvents to form solvate structures [16]. The high DN solvents can decrease the acidity of Li^+ and tend to strongly solvate Li^+, promote the formation of toroidal-shaped Li_2O_2, then increase discharge capacity, and decrease overpotential while low DN solvents limit the discharge reaction by promoting the reaction only on the surface, leading to premature cell death, as shown in Figure 18.3 [4, 17–18]. However, high DN solvents are generally unstable against Li anode

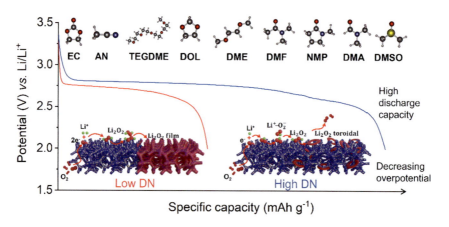

FIGURE 18.3 Schematic of discharge mechanisms in solvents with different DN.

TABLE 18.1
Physical Properties of the Various Organic Solvents Used in LABs [2, 19–21]

Solvents	Abbreviation	Viscosity [MPa/s] 25 °C	Melting point [°C]	Boiling point [°C]	AN (kcal/mol)	DN (kcal/mol)
Ethylene carbonate	EC	0.1825*	36.4	248	–	16.4
Propylene carbonate	PC	2.5130	−49.0	242	18.3	15.1
Dimethyl carbonate	DMC	0.5805	4.6	90	–	16.0
Dimethoxyethan	DME	0.4550	−58.0	82	10.2	23.9
Tetraethylene glycol dimethyl ether	TEGDME	3.3800	−30.0	272	11.7	16.6
1,3-dioxolane	DOL	0.5882	−95.0	75	–	21.2
Dimethyl sulfoxide	DMSO	1.9960	18.0	189	19.3	29.8
Acetonitrile	AN	0.3410	−48.8	82	18.9	14.1
Dimethylformamide	DMF	0.8020	−60.4	153	16.0	26.6
N,N-Dimethylacetamide	DMA	0.9270	−20.0	165	13.6	27.8
N-methyl-2-pyrrolidenone	NMP	1.6630	−24.4	202	13.3	27.3

* This test was performed at 40 °C

and easily attacked by oxygen species [4]. Therefore, optimizing the trade-off between the cycle life and capacity, and overcoming this challenge are the key parameters for LABs' electrolytes. The physical properties of the various organic solvents generally used in LABs are summarized in Table 18.1.

18.3 TYPES OF NONAQUEOUS ELECTROLYTES

18.3.1 Carbonate-Based Electrolytes

LABs used organic carbonate as solvents at the beginning, which was borrowed from LIBs because of the high boiling point and ion conductivity. In 1996, Abraham's group used a pouch cell with an electrolyte solution of 1M $LiPF_6$ in a solvent of (1:1:1) ethylene carbonate/dimethyl carbonate/diethyl carbonate. However, this battery is of the primary type, with a discharge time of 100 hours and a specific capacity of 91 mAh/g [21]. Over ten years later, several kinds of carbonate-based electrolytes were investigated, revealing their LABs instability [22–24]. The main reason is that nucleophilic strike at the carbonyl carbon and H_2 abstraction at O-alkyl carbon to promote Li_2CO_3 formation rather than Li_2O_2, which causes cell overpotential, capacity fading, and eventually cell failure [24, 25]. According to K. Amine's group, the ring-opening reaction of propylene carbonate in the presence of solvated species including O_2^-, LiO_2, LiO_2^-, and Li_2O_2 is spontaneous, facilitating the formation of Li_2CO_3 and lithium alkyl carbonate [26]. LABs with carbonate-based electrolytes always exhibit inadequate specific capacity (~3000 mAh/g), a maximum charge potential (≥ 4.4 V), and lesser energy efficiency (< 60%) [25]. As a result, organic carbonate electrolytes for LABs are no longer used.

18.3.2 Ether-Based Electrolytes

The dramatic instability of organic carbonates in the LABs led to the investigation of other more stable electrolytes. Ether-based electrolytes were found to be suitable candidates due to their high oxygen solubility and, more importantly, their stability against O_2^-. In 2006, J. Read first reported the use of a mixture of dimethoxyethane (monoglyme) and 1,3-dioxolane as compatible electrolytes for LABs. They contributed these properties to the high oxygen solubility and low viscosity, which results in improved oxygen transport in LABs [27]. In discharging, lithium peroxide is a reversible discharge product. However, when batteries charge at high potential, dimethoxyethane is oxidized in the presence of Li_2O_2 [28]. The properties of ether-based electrolytes depend on the chain length, the number of oxygens, and the steric effects of glyme that offer different reactivity toward the oxygen species formed in LABs. Since YK. Sun's report [9], tetraethylene glycol dimethyl ether (TEGME) has been a popular LABs electrolyte solvent. These compounds are also significantly more stable than carbonates because they are less affected by nucleophilic substitution by the superoxide anion radical and can withstand oxidation potentials of up to 4.5 V vs. Li/Li^+ [29]. However, due to the presence of electronegative ether groups, they are still not completely stable again O_2^- and decompose to formate, acetate, and polyester [30]. Additionally, dimethoxyethane has a low boiling point of 82 °C while TEGDME owns high viscosity. Therefore, they promote the formation of the thin layer of Li_2O_2 insulation on the surface of the electrode that limits the rates and cyclability of LABs [13]. To solve this issue, the concept of high-concentration electrolytes (HCEs) has been developed. For example, Y. Liu et al. early reported that 3M LiTFSI in TEGDME delivered the highest capacity of 13245 mAg_{carbon}^{-1}; B. Liu et al. stated that the performance of LABs greatly improved by employing 3M LiTFSI in DME electrolyte [2, 31]. The secret behinds the improvement of HCEs in LABs is no free solvent molecules because they tend to coordinate with Li^+ cations, the C-H bonds are more stable against the attacking of nucleophilic and then mitigate the electrolyte decomposition. In thermodynamic aspects, HCEs exhibited not only lower E_{HOMO}, which favours stability against electrochemical oxidation on cathode side, but also higher Gibbs activation energy of Li_2^+-DME_3 complexes, which can resist the attack of superoxide radical anions to improve of the electrolyte in comparison to DME molecules [2]. However, there are some disadvantages that hindered the practical application of HCEs such as high cost, high viscosity, poor wettability with electrodes, limited oxygen diffusion. The new electrolytes system is named localized high-concentration electrolytes (LHCEs); they are explored to suppress the aforementioned disadvantages. The concept of LHCEs is adding a diluent in HCEs system, then highly concentrated salt-solvent clusters are locally distributed in the diluent molecules to reduce the salt's concentration, viscosity, and improve the oxygen solubility. It is noted that diluent means nonsolvating "*inert*" solvents which do not dissolve Li salt. The F-containing ether solvents such as hydrofluoroethers (e.g. 1,1,2,2-tetrafluoroethyl 2,2,3,3-tetrafluoropropyl ether (TTE), 1H,1H,5H-octafluoropentyl 1,1,2,2-tetrafluoroethyl ether (OTS), etc.) were investigated as dilutents [1, 5]. The contribute to the great improvement of localized high-concentration electrolytes (LHCEs) in rechargeable Li-air batteries (LABs) is the low viscosity of this LHCEs leads to higher discharge capacity of LABs than high-concentration electrolytes. Second, LHCEs shown more stable with Li metal and against parasitic reactions because a part of solvent was replaced by the high stability diluent [1, 5, 6].

18.3.3 DMSO-Based Electrolyte

Dimethyl sulfoxide (DMSO) has favorable solvent properties, such as a high boiling point (189 °C), good oxygen diffusion, and a high DN (29.8) [32]. Several groups reported that DMSO exhibited good capability of LABs with Au electrodes that are not suitable for practical approaches [32, 10]. When using high pore volume carbon electrodes, DMSO was decomposed by nucleophilic attack of oxygen species to sulfone and dimethyl sulfone [33]. In addition, DMSO is unstable with bare lithium anodes; therefore, the additive is required when used in LABs (e.g. $LiNO_3$) but the performance is still unsatisfactory [7, 34]. Same as with ether-based electrolytes, the concept of HCEs was used for DMSO-based electrolyte systems (e.g. LiTFSI-3DMSO electrolyte (1:3 molar ratio)). The mechanism is also similar to the ether-based, as DMSO completely forms $TFSI^-$-Li^+-$(DMSO)_3$ complexes with the salt anion and there are no free DMSO molecules. To decrease the probability of nucleophilic attack by O_2^- species, no free DMSO molecules are preferred. As the result, HCEs exhibited excellent stability and reversibility even with a bare Li anode. The density functional theory calculation also revealed that complexes with the high Gibbs activation energies are more stable in superoxide radical anions medium than free DMSO molecules, which improves electrolyte stability [8]. However, as with ether-based electrolytes, LHCEs appeared more appealing, mitigating the drawbacks of HCEs. On the other hand, F-contained diluents such as TTE and OTS are expensive, which limited their practical application. Another strategy is using cosolvent by mixing a strong solvent (DMSO with high DN) with a weak solvent (TEGDME with low DN) to generate a local strong solvation effect electrolyte (LSSE). As was mentioned, high DN of DMSO solvents can significantly improve battery performances; however, they are also easily attacked by oxygen species [17, 18, 35]. In addition, as mentioned earlier, high DN (e.g. DMSO) are unstable to Li anode. Therefore, by adjusting the ratio of DMSO and TEGDME cosolvent, J. Lai et al. reported the LSSE not only combines both advantages of DMSO and TEGDME but also suppresses their disadvantages [4].

18.3.4 Nitriles-Based Electrolytes

Nitriles-based electrolytes also have some properties for LABs such as both high electrochemical stability and chemical stability against the nucleophilic attack and H abstraction [36]. P. Bruce's group claimed that when using an acetonitrile-based electrolyte, during discharging, LiO_2 on the surface was formed as an intermediate, which then disproportionated to Li_2O_2. During charging, Li_2O_2 was decomposed directly without passing through the LiO_2 intermediate [37]. Because of its low DN, control of the Li_2O_2 formation process is required to improve the capacity when using nitriles-based electrolytes for LABs. In addition, acetonitrile has a low boiling point of 82 °C; thus, the higher molecular nitriles or functionalized nitriles were studied by both computational and experimental [38]. Trimethylacetonitrile was screened as the most stable solvent in this group [38]. Unfortunately, it still reacts with lithium metal which limits further research about nitriles-based electrolytes for LABs [36].

18.3.5 Amides-Based Electrolytes

Amides such as *N,N*-dimethylacetamide, dimethylformamide, and *N*-methyl-2-pyrrolidone were next investigated as non-aqueous electrolyte solvents because they resist chemical degradation in the O_2 electrode. However, these solvents react vigorously with Li metal and do not form a stable solid-electrolyte interphase (SEI) on the Li anode. To suppress these issues, $LiNO_3$ salt was employed to facilitate the Li_2O-rich SEI layer on the surface of metallic lithium, thereby inhibiting the reaction between dimethylformamide and Li metal [39]:

$$2Li + LiNO_3 \rightarrow Li_2O + LiNO_2 \quad [9]$$

Furthermore, during cycling, the $LiNO_2$ in reaction (1) can be regenerated to $LiNO_3$ in the O_2-rich environment, allowing for long-term cycle stability [39]. However, this is amid further degradation to various species including dimethylamine, acetate, carbonate, and various N-O species in superoxide radical medium, which may not be suitable for oxygen reduction in LABs [40].

18.4 COMPONENT IN NON-AQUEOUS ELECTROLYTE

For prominent Li-air batteries, an optimal non-aqueous electrolyte should possess not only high chemical and electrochemical stability but also limited or no volatility to ensure high air (specifically O_2) solubility, long life, and be immobile to superoxide radicals. However, the electrolytes that have been reported could not fully meet these needs. An organic electrolyte is primarily contained in an organic solvent with a proper concentration of lithium salt. The interfacial structure of solid-gas-liquid (electrodes-O_2-electrolyte) is governed by the properties of formulated electrolytes, which in turn affect the performance of LABs. The effect is the outcome of the electrolyte's components: highly pure organic solvents, Li salts, and electrolyte additives, as well as their proper combination.

18.4.1 Effect of Solvent

The impact of oxygen radicals on non-aqueous solvent molecules is one of the big issues with LABs [40, 41]. Therefore, it is essential to find the solvent for a non-aqueous electrolyte that could be ultra-stable against O_2 radicals. A few main characteristics are expected of an electrolyte solvent for LABs for high performance, (Figure 18.4a) such as (i) a high dielectric constant that allows some Li salt to dissolve in; (ii) low viscosity, which allows for quick Li^+ ion transport in cells; (iii) high stability against all cell components, particularly O_2 radicals during cell operation; (iv) low vapor pressure, allowing for minimal loss when exposed to an O_2 flow; and (v) nontoxicity and cost-effectiveness. To enhance rapid Li^+/O_2 movement and facilitate electrochemical reactions between Li and O_2, solvent molecules coordinate with Li^+/O_2. Carbonates, ethers, amides, sulfones, and nitriles are among the aprotic solvents studied for use in LABs (Table 18.1). The main reason for capacity fading is electrolyte decomposition. Reactive O_2 species such as LiO_2, Li_2O_2, and notably O_2^- jeopardize solvent stability [3]. In general, there are five types of decomposition pathways for solvents in LABs: auto-oxidation, nucleophilic assault, acid/base reactions, Li reduction, and proton-mediated reactions (Figure 18.4c) [14].

Although, because of pretty good ionic conductivity and high boiling point, ethylene carbonate (EC) and propylene carbonate (PC) are often utilized as solvents, the oxygen radicals produced during the discharge process could damage carbonate solvents and boost the formation of lithium carbonate (Li_2CO_3) and alkyl carbonates ($RO-(C=O)-OLi$) product [42]. Furthermore, alkyl carbonate electrolyte decomposed into $C_3H_6(OCO_2Li)_2$, Li_2CO_3, HCO_2Li, CH_3CO_2Li, CO_2, and H_2O at the cathode during discharge and oxidation of Li_2CO_3, $C_3H_6(OCO_2Li)_2$, HCO_2Li, and CH_3CO_2Li during charging with H_2O, and CO_2 [41, 43]. After organic carbonates were shown to decompose rapidly, ether solvents were considered electrolyte solvents for LABs. Since then, ether solvents have been studied in the lab. The stability of ether solvents was found to be higher than that of PC. McCloskey discovered Li_2O_2 as the principal discharge product in Li-air cells based on dimethoxyethane (DME) [28]. However, as the research progressed, it was discovered that ether electrolyte solvents are unstable. The autoxidation of ethers solvent by O_2 has been proposed as the process for solvent deterioration in the presence of O_2. It has been proposed that substituting reactive hydrogens on solvent molecules can improve autoxidation resistance. Moreover, ester-based electrolytes are regularly hampered by extreme volatility or viscosity, which is detrimental to LAB's life and excellent rate performance [44]. It allows for the rapid evaporation of solvents in a short time, resulting in a fading in the capacity. Despite being more resilient to O_2 radicals than organic carbonates, DME's low vapor pressure will limit its use in LABs due to long-term performance concerns. The first time, Xu et al. investigated the performance of LABs using DMSO as an electrolyte solvent and carbon in Ni foam as a cathode. Li foils were pre-treated in 1 M $LiClO_4$ in PC to enable cycling because DMSO can react with Li metal [10]. Unfortunately, such an electrolyte works best with a gold electrode, and if carbon components are present, breakdown reactions can be dangerous. Other electrolytes, such as silane- and amide-based solvents with various substitutions, have been validated experimentally and conceptually as LABs replacements. Although in comparison to PC, Zhang and colleagues found that oligoether-functionalized silane solvent has a better ability, there was considerable electrolyte decomposition during charging [26].

Mixed solvents have the potential in LABs, although they haven't gained much recognition yet. Mixed solvents are made up of two or more solvents that can complement each other and have synergistic effects in some cases. Because of the high solubility and fast dissolution kinetics of O_2 in fluorinated solvent-containing electrolytes, a composition of 0.2M $LiSO_3CF_3$ in 1:3:1 PC/DME/methyl nonafluorobutyl ether has been reported to have a high specific capacity [45]. A cell containing 0.1M $LiClO_4$ in a 5:5 (DMSO+FE1) electrolyte has a high specific capacity and quick O_2

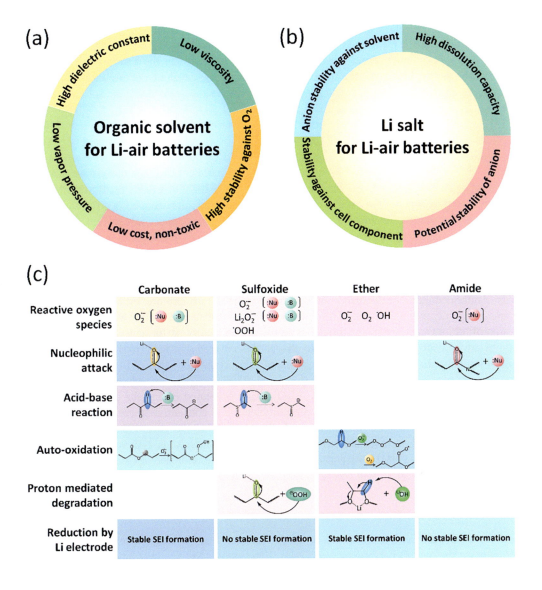

FIGURE 18.4 (a) A few main characteristics are expected of an electrolyte solvent for Li-air batteries. (b) The important characteristics are expected of a Li-salt for Li-air batteries. (c) Five solvents and related reactive O_2 species decomposition pathways.

transport [46]. Because of the higher O_2 solubility and diffusivity in the electrolyte, another moderately fluorinated compound, 3-[2-(perfluorohexyl) ethoxy]-1,2-epoxypropane, has been employed as a cosolvent in LABs with enhanced rate performance. A long-term stable LAB necessitates the discovery of a stable cycling and low vapor pressure solvent [47].

18.4.2 Effect of Lithium Salt

Before being employed as an essential component of electrolytes in LABs, Li salts must satisfy the following requirements (Figure 18.4b): (i) the Li salts must be soluble in solvents and attain a certain concentration to support rapid Li^+ ion transport; (ii) the anions must be stable at the requisite potentials; (iii) the anions must be solvent inert; and (iv) the anions must be stable against cell components such as separators, current collectors, and the package. Furthermore, the presence of Li salts greatly influences the ORR and OER mechanisms in electrolytes, resulting in reaction mechanisms that differ significantly from those in pure solvents [48]. As per the hard and soft (Lewis) acids and bases theory, Li^+ makes O_{2-} unsafe and encourages the development of non-soluble reaction products/intermediates, which could passivate the cathode and break down the solvent [49].

In Li-air batteries, Li salts such as $LiPF_6$, $LiClO_4$, $LiCF_3SO_3$, $LiN(SO_2CF_3)_2$, $LiN(SO_2C_2F_5)_2$, LiBr, LiI, and others have been usually utilized. Li salts have received less attention due to their scarcity, especially when contrasted with the interest in choosing solvents for Li-air battery electrolytes. Lithium salts have been demonstrated to react with Li_2O_2 during discharge in several recent investigations [50]. Veith and co-workers discovered contamination of Li_2O_2 on the surface of Li_2O_2 by decomposed halide species from the electrolyte anions in $LiClO_4$, $LiBF_4$, $LiPF_6$, and LiTFSI after examining the discharge product in tetraglyme with various salts using XPS [51]. In addition, LiF was found in discharged electrodes with LiTFSI and a wide range of organic solvents in a recent work by Xu et al., which was attributed to salt and/or binder (PTFE) decomposition [51].

The dissociation level of the salt in the electrolyte is just as essential as the impact of the solvent DN in the discharge process. As shown in Figure 18.5a, the first group consists of mainly dissociated (low DN) salts such as LiTFSI and lithium bis(fluorosulfonyl)imide, whereas the second group consists of intermediate DN salts such as lithium triflate (LiTf). Furthermore, the last group includes salts with a high degree of association (high DN), including $LiNO_3$ and CH_3COOLi [52]. The concentration of lithium salts in the electrolyte has an impact on battery performance. The molar ratios of lithium salt and solvent can be changed to change the solvated structure from solvent separated to contact ion pairs. In various forms of glymes, a high concentration (1M) of LiTFSI led to better cycling performance and slower degradation than low LiTFSI concentration solutions. Furthermore, the impacts of the Li^+ concentration on the Li_2O_2 shape generated during the discharge process have been studied (Figure 18.5b) [31]. An optimum electrolyte concentration (between 2 and 3M) enhanced the volume utilization ratio of the cathode attributed to the generation of porous three-dimensional (3D) Li_2O_2 [31]. In the absence of free DME solvent molecules, cells with a concentrated electrolyte (3M LiTFSI in DME) were also investigated [2]. The influence of Li-air battery cycling performance on concentrations will be useful for the exploitation of stable electrolyte systems.

18.4.3 Effect of Additives

Additives are substances that are found in small concentrations in electrolytes but can significantly increase the overall performance of LABs. Electrolyte additives may have a major impact on response processes, discharge capacity, and reversibility. Redox mediators (RM) have gained a

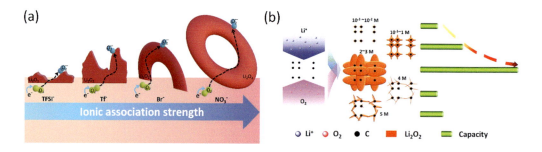

FIGURE 18.5 (a) Ionic association strength of various anions in Li salts, as well as discharge products of 1M Li salts in diglyme using sputtered gold cathodes. (b) Illustration of the effects of Li^+ ion concentrations on discharge capacities and morphologies. (Adapted with permission from Reference [31], Copyright 2014 Royal Society of Chemistry.)

lot of attention as the most popular electrolyte additive because they may efficiently minimize the charge or discharge polarization of LABs. As a result, a lot of effort has gone into researching redox mediator additives, which is seen to be the most promising idea for achieving both high capacity and excellent reversibility for LABs by lowering charge/discharge overpotentials.

To improve the solubility of O_2^{2-} in electrolytes, strong Lewis acids like $C_{18}F_{15}B$ (tris(pentafluorophenyl)borane) and boron esters were used [53]. Unfortunately, the majority of these additives have been evaluated in carbonate electrolytes, which were shown to interact with active O_2. ORR mediators (ORRMs), as well as OER mediators (OERMs), seem to be two types of RMs that are associated with ORR and OER reactions, respectively, based on the reaction mechanism. OERMs were shown to be effective in reducing charge overpotentials. They seem to be reversible redox couples that move between the cathode and the Li_2O_2 surface and can oxidize Li_2O_2 chemically rather than electrochemically. During the discharge process, ORRMs facilitate electron transport between the cathode and O_2 in the electrolyte. ORRMs must have a reduction potential that is somewhat lower than Li_2O_2's theoretical formation potential but greater than the actual discharge plateau. The Perfluorotriphenyl borane (TPFPB) $B(C_6F_5)_3$ has been added to carbonate-based electrolytes to improve the solubility of the discharge product Li_2O_2 but also allow for more oxygen reduction reactions at the produced active sites [54]. Perfluorotributylamine (FTBA) was added to propylene carbonate to improve its O_2 solubility. When compared to the electrolyte without FTBA, the discharge capacity may be substantially increased [54]. The approach of holding the discharge product Li_2O_2 in the electrolyte appears to be promising for increasing specific capacity and preventing channel blockage in the cathode, however, the system's stability must be carefully assessed.

18.5 CHARACTERISTICS OF NON-AQUEOUS ELECTROLYTE

Finding a non-aqueous electrolyte that can endure the long-term operation in a strongly oxidative environment while also allowing for the complete formation and decomposition of Li_2O_2 over several cycles has purposes outlined.

18.5.1 STABILITY WITH LI-METAL

Non-aqueous electrolytes usually react with Li metal, resulting in the formation of a layer on the surface. The idea is that a few electrolytes form a passivizing surface film of the lithium metal, preventing further reactions. When Li metal comes into contact with a polar aprotic solvent, an SEI layer is produced by the reductive decomposition reactions of organic solvents, which slows down Li-metal corrosion; some nonaqueous electrolytes allow Li metal cycling, but lithium deposition is frequently associated with dendrite formation and short circuit. The presence of oxygen(O), H_2O, CO_2, and nitrogen(N) at the electrolyte/lithium surface is to be expected in Li-air cells; these chemicals can enter the cell through the air cathode and cross across the electrolyte to the anode. This means that, in addition to all of the foregoing, the lithium metal could react with these impurities, as well as oxygen and water, in reactions with the electrolyte at the anode [55]. The sole extensive analysis of these phenomena is reported in the instance of oxygen reactions with ether- and glyme-based electrolytes at the Li anode; it is possible, however, that some other non-aqueous electrolytes are also susceptible to comparable oxygen interactions in the availability of Li.

18.5.2 ELECTROCHEMICAL AND CHEMICAL STABILITY

It was shown that non-aqueous electrolyte solvents go through decomposition processes involving, first and foremost, the intermediate of the cathode oxygen reduction reaction superoxide anion (O_2^-) which is a powerful nucleophile in polar solvents [11]. As a solid foundation, O_2 can also accelerate the auto-oxidation of (weakly) acidic solvents (Figure 18.6a), suggesting that aprotic solvents with lower acidity are more stable [56]. All of these processes are (expectedly) irreversible since the

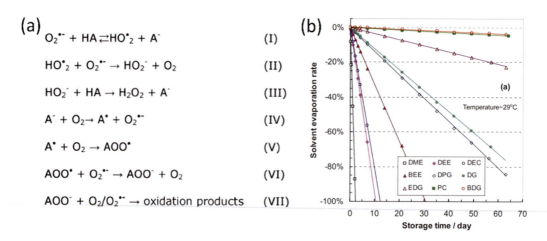

FIGURE 18.6 (a) The methodology of superoxide-induced autooxidation of weakly as well as moderately acidic C-H acid. (Adapted with permission from Reference [56], Copyright 2013 Elsevier.) (b) Solvent evaporation rate in pure solvents inside the dry box at 29°C in two months. (Adapted with permission [57]. Copyright 2009, The Electrochemical Society.)

electrolytes can react with oxygen and Li-oxides. All of the preceding reactions should be stable in a good Li-air cell electrolyte.

18.5.3 Physical Stability

In a Li-air battery, the cathode is exposed to the atmosphere, therefore the electrolyte's low evaporation rate is an important factor to consider when selecting a non-aqueous electrolyte. The majority of commonly used electrolytes have a high solvent vapor pressure and, as a result, a high solvent evaporation rate. The electrolyte losses due to evaporation in coin-type Li-air cells with numerous electrolytes are depicted in Figure 18.6b [57]. The results show that in a year, the cell with the least fugitive commonly used butyl diglyme (organic solvent) loses 22% of its electrolyte. The excess solvent can compensate for electrolyte loss; however, the price in terms of volume and mass may be prohibitively high.

18.5.4 Li-ion and Oxygen Reagent Transport Issues

Because most organic electrolytes quickly wet all cathode pores whereas flooding air pathways, only solubilized O_2 actively participated in the real ORR charge-transfer reaction, which occurs in a two-phase boundary reaction chamber [12]. Flooding is a negative phenomenon that occurs in the nonaqueous Li-air system and reduces oxygen ease at the cathode reaction centers, leading to lower current density. Finally, oxygen accessibility is determined by the diffusion and dissolution of the electrolyte within the cathode routes [15]. As a result, the potential of aprotic polar solvents to dissolve and transport oxygen is an important criterion. Given the preceding, the electrolyte of selection for Li-air cells must have excellent O_2-transport characteristics as well as excellent Li-ion conductivity.

18.5.5 Charge-Transfer Issues

Inside the porous structure of the air electrode, insoluble Li_2O formed. The oxides, on the other hand, clog the cathode pore orifices, preventing oxygen and Li-ion transit within the air cathode; however, oxides deposit on reaction centers, preventing ORR due to electron transfer inhibition

(Li_2O are insulators) [58]. Because these insoluble precipitates are involved in the Li-air cell charge reaction (OER), it can only occur at the one-dimensional interface between the conductive cathode material, the electrolyte, and Li_2O. As a result, the local charge current density is high, resulting in a high OER-overvoltage and charge resistance [59]. Electrolytes that enable (at least partially) Li-oxides dissolution may raise charge current rate (and consequently, charge overvoltage is reduced) by substituting one-dimensional OER-interface ([solid product]/[conductive cathode]/[electrolyte] interface) for 2D OER-interface ([dissolved product]/[conductive cathode] interface). Additionally, such electrolytes may reduce cathode pore system congestion as well as cathode/electrolyte interface deactivation.

18.6 CONCLUSION AND OUTLOOK

We summarized current research and understanding of nonaqueous electrolytes investigated for rechargeable Li-air batteries (LABs) in this chapter. The electrolyte is critical to the cyclability and rate capability of LABs. A slight change in the electrolyte can cause a considerable difference in the internal mechanism, affecting battery performance even more. The effects of various solvents, additives, and lithium salts on the cyclic stability performance of LABs in non-aqueous liquid systems have been studied. Individual components have little influence on improving LABs performance, and the majority of them are associated with side reactions and decomposition issues. As a result, the architecture of the LAB's electrolyte must take into account the electrolyte benefit of the entire system. For instance, low DN solvents have high stability but less discharge-specific capacity in LABs. A Li salt with a high degree of association, ORRMs, or a detectable amount of H_2O would be used in this case to encourage the creation of a large quantity of toroidal-shaped Li_2O_2, thus also raising the discharge-specific capacity of the LABs. Moreover, OER is very useful in decomposing huge Li_2O_2 donut-shaped particles at less charge potential, enhancing the energy efficiency and cyclability of the battery. In open systems, volatile electrolytes should not be used. Mixed solvents take the advantage of various solvents and can occasionally have synergistic effects. Given the abundance of aprotic solvents and RTILs, a plethora of combinations is possible. Furthermore, the concentrations of additives and Li salts related to the production of stable SEI films need to be optimized further. Continuous innovation should lead to the discovery of new solvents, Li salts, and additives.

REFERENCES

[1] W.J. Kwak, S. Chae, R.Z. Feng, P.Y. Gao, J. Read, M.H. Engelhard, L.R. Zhong, W. Xu, J.G. Zhang, Optimized electrolyte with high electrochemical stability and oxygen solubility for lithium-oxygen and lithium-air batteries, ACS Energy Lett. 5(7) (2020) 2182–2190.

[2] B. Liu, W. Xu, P.F. Yan, X.L. Sun, M.E. Bowden, J. Read, J.F. Qian, D.H. Mei, C.M. Wang, J.G. Zhang, Enhanced cycling stability of rechargeable Li-O-2 batteries using high-concentration electrolytes, Adv. Funct. Mater. 26(4) (2016) 605–613.

[3] N.G. Araez, P. Novák, Critical aspects in the development of lithium—air batteries, Solid State Electrochem. 17 (2013) 1793–1807.

[4] J.N. Lai, H.X. Liu, Y. Xing, L.Y. Zhao, Y.X. Shang, Y.X. Huang, N. Chen, L. Li, F. Wu, R.J. Chen, Local strong solvation electrolyte trade-off between capacity and cycle life of $Li-O_2$ batteries, Adv. Funct. Mater., 31(40) (2021) 2101831.

[5] Q. Zhao, Y.H. Zhang, G.R. Sun, L.N. Cong, L.Q. Sun, H.M. Xie, J. Liu, Binary mixtures of highly concentrated tetraglyme and hydrofluoroether as a stable and nonflammable electrolyte for $Li-O_2$ batteries, ACS Appl. Mater. Interfaces, 10(31) (2018) 26312–26319.

[6] W.J. Kwak, H.S. Lim, P.Y. Gao, R.Z. Feng, S.J. Chae, L.R. Zhong, J. Read, M.H. Engelhard, W. Xu, J.G. Zhang, Effects of fluorinated diluents in localized high-concentration electrolytes for lithium-oxygen batteries, Adv. Funct. Mater. 31(2) (2021) 2002927.

[7] M. Roberts, R. Younesi, W. Richardson, J. Liu, T. Gustafsson, J.F. Zhu, K. Edstrom, Increased cycling efficiency of lithium anodes in dimethyl sulfoxide electrolytes for use in $Li-O_2$ batteries, ECS Electrochem. Lett. 3(6) (2014) A62–A65.

[8] B. Liu, W. Xu, P.F. Yan, S.T. Kim, M.H. Engelhard, X.L. Sun, D.H. Mei, J. Cho, C.M. Wang, J.G. Zhang, Stabilization of Li metal anode in DMSO-based electrolytes via optimization of salt-solvent coordination for Li-O_2 batteries, Adv. Energy Mater. 7(14) (2017) 1602605.

[9] H.G. Jung, J. Hassoun, J.B. Park, Y.K. Sun, B. Scrosati, An improved high-performance lithium-air battery, Nat. Chem. 4(7) (2012) 579–585.

[10] Z.Q. Peng, S.A. Freunberger, Y.H. Chen, P.G. Bruce, A reversible and higher-rate Li-O-2 battery, Science. 337(6094) (2012) 563–566.

[11] M. Balaish, A. Kraytsbergb, Y.E. Eli, A critical review on lithium-air battery electrolytes, Phys.Chem. Chem.Phys. 16 (2014) 2801–2822.

[12] R. Padbury, X. Zhang, Lithium-oxygen batteries—limiting factors that affect performance, Journal of Power Sources. 196 (2011) 4436–4444.

[13] W.J. Kwak, D. Sharon, C. Xia, H. Kim, L.R. Johnson, P.G. Bruce, L.F. Nazar, Y.K. Sun, A.A. Frimer, M. Noked, S.A. Freunberger, Lithium–oxygen batteries and related systems: Potential, status, and future, Chem. Rev. 120 (2020) 6626–6683.

[14] J.N. Lai, Y. Xing, N. Chen, L. Li, F. Wu, R.J. Chen, Electrolytes for rechargeable lithium-air batteries, Angew. Chem. Int. Ed. 59(8) (2020) 2974–2997.

[15] P. Andrei, J.P. Zheng, M. Hendrickson, E.J. Plichta, Some possible approaches for improving the energy density of li-air batteries, Journal of the Electrochemical Society. 157(12) (2010) A1287–A1295.

[16] L. Wang, Y.T. Zhang, Z.J. Liu, L.M. Guo, Z.Q. Peng, Understanding oxygen electrochemistry in aprotic Li-O_2 batteries, Green Energy Environ. 2(3) (2017) 186–203.

[17] D. Aurbach, B.D. McCloskey, L.F. Nazar, P.G. Bruce, Advances in understanding mechanisms underpinning lithium-air batteries, Nat. Energy. 1(9) (2016) 16128.

[18] Y.L. Zhang, X.M. Zhang, J.W. Wang, W.C. McKee, Y. Xu, Z.Q. Peng, Potential-dependent generation of O_2^- and LiO_2 and their critical roles in O^{-2} reduction to Li_2O_2 in aprotic Li-O_2 batteries, J. Phys. Chem. C. 120(7) (2016) 3690–3698.

[19] Y. Li, X.G. Wang, S.M. Dong, X. Chen, G.L. Cui, Recent advances in non-aqueous electrolyte for rechargeable Li-O-2 batteries, Adv. Energy Mater. 6(18) (2016).

[20] B. Giner, H. Artigas, M. Haro, C. Lafuente, M.C. Lopez, Viscosities of binary mixtures of 1,3-dioxolane or 1,4-dioxane with isomeric chlorobutanes, J. Mol. Liq. 129(3) (2006) 176–180.

[21] A. Henni, P. Tontiwachwuthikul, A. Chakma, Densities, viscosities, and derived functions of binary mixtures: (Tetraethylene glycol dimethyl ether plus water) from 298.15 K to 343.15 K, J. Chem. Eng. Data., 49(6) (2004) 1778–1781.

[22] R.R. Arthur Dobley, K.M. Abraham, High capacity cathodes for lithium-air batteries, 1996.

[23] T. Ogasawara, A. Debart, M. Holzapfel, P. Novak, P.G. Bruce, Rechargeable Li_2O_2 electrode for lithium batteries, J. Am. Chem. Soc. 128(4) (2006) 1390–1393.

[24] J. Read, Characterization of the lithium/oxygen organic electrolyte battery, J. Electrochem. Soc. 149(9) (2002) A1190–A1195.

[25] S.A. Freunberger, Y. Chen, Z. Peng, J.M. Griffin, L.J. Hardwick, F. Barde, P. Novak, P.G. Bruce, Reactions in the rechargeable lithium-O_2 battery with alkyl carbonate electrolytes, J. Am. Chem. Soc. 133(20) (2011) 8040–8047.

[26] Z.C. Zhang, J. Lu, R.S. Assary, P. Du, H.H. Wang, Y.K. Sun, Y. Qin, K.C. Lau, J. Greeley, P.C. Redfern, H. Iddir, L.A. Curtiss, K. Amine, Increased stability toward oxygen reduction products for lithium-air batteries with oligoether-functionalized silane electrolytes, J. Phys. Chem. C. 115(51) (2011) 25535–25542.

[27] J. Read, Ether-based electrolytes for the lithium/oxygen organic electrolyte battery, J. Electrochem. Soc. 153(1) (2006) A96–A100.

[28] B.D. McCloskey, D.S. Bethune, R.M. Shelby, G. Girishkumar, A.C. Luntz, Solvents' critical role in nonaqueous lithium-oxygen battery electrochemistry, J. Phys. Chem. Lett. 2(10) (2011) 1161–1166.

[29] S.A. Freunberger, Y.H. Chen, N.E. Drewett, L.J. Hardwick, F. Barde, P.G. Bruce, The lithium-oxygen battery with ether-based electrolytes, Angew. Chem. Int. Ed. 50(37) (2011) 8609–8613.

[30] B.D. McCloskey, R. Scheffler, A. Speidel, D.S. Bethune, R.M. Shelby, A.C. Luntz, On the efficacy of electrocatalysis in nonaqueous Li-O_2 batteries, J. Am. Chem. Soc, 133(45) (2011) 18038–18041.

[31] Y. Liu, L. Suo, H. Lin, W. Yang, Y. Fang, X. Liu, D. Wang, Y.S. Hu, W. Han, L. Chen, Novel approach for a high-energy-density Li-air battery: Tri-dimensional growth of Li_2O_2 crystals tailored by electrolyte Li$^+$ ion concentrations, J. Mater. Chem. A. 2(24) (2014) 9020–9024.

[32] D. Xu, Z.L. Wang, J.J. Xu, L.L. Zhang, X.B. Zhang, Novel DMSO-based electrolyte for high performance rechargeable Li-O_2 batteries, ChemComm. 48(55) (2012) 6948–6950.

[33] D. Sharon, M. Afri, M. Noked, A. Garsuch, A.A. Frimer, D. Aurbach, Oxidation of dimethyl sulfoxide solutions by electrochemical reduction of oxygen, J. Phys. Chem. Lett. 4(18) (2013) 3115–3119.

[34] B. Sun, X.D. Huang, S.Q. Chen, J.Q. Zhang, G.X. Wang, An optimized $LiNO_3$/DMSO electrolyte for high-performance rechargeable $Li-O_2$ batteries, RSC Adv. 4(22) (2014) 11115–11120.

[35] W.J. Kwak, J.B. Park, H.G. Jung, Y.K. Sun, Controversial topics on lithium superoxide in $Li-O_2$ batteries, ACS Energy Lett. 2(12) (2017) 2756–2760.

[36] B.D. McCloskey, D.S. Bethune, R.M. Shelby, T. Mori, R. Scheffler, A. Speidel, M. Sherwood, A.C. Luntz, Limitations in rechargeability of $Li-O_2$ batteries and possible origins, J. Phys. Chem. Lett. 3(20) (2012) 3043–3047.

[37] Z.Q. Peng, S.A. Freunberger, L.J. Hardwick, Y.H. Chen, V. Giordani, F. Barde, P. Novak, D. Graham, J.M. Tarascon, P.G. Bruce, Oxygen reactions in a non-aqueous Li^+ electrolyte, Angew. Chem. Int. Ed. 50(28) (2011) 6351–6355.

[38] V.S. Bryantsev, J. Uddin, V. Giordani, W. Walker, D. Addison, G.V. Chase, The identification of stable solvents for nonaqueous rechargeable Li-air batteries, J. Electrochem. Soc. 160(1) (2013) A160–A171.

[39] J. Uddin, V.S. Bryantsev, V. Giordani, W. Walker, G.V. Chase, D. Addison, Lithium nitrate as regenerable SEI stabilizing agent for rechargeable Li/O_2 batteries, J. Phys. Chem. Lett. 4(21) (2013) 3760–3765.

[40] Y. Chen, S.A. Freunberger, Z. Peng, F. Barde, P.G. Bruce, $Li-O_2$ battery with a dimethylformamide electrolyte, J. Am. Chem. Soc. 134(18) (2012) 7952–7957.

[41] M.J. Trahan, Q. Jia, S. Mukerjee, E.J. Plichta, M.A. Hendrickson, K.M. Abrahama, Cobalt phthalocyanine catalyzed lithium-air batteries, Journal of the Electrochemical Society. 160(9) (2013) A1577–A1586.

[42] X. Liu, B. Cui, S. Liu, Y. Chen, Progress of non-aqueous electrolyte for li-air batteries, Journal of Materials Science and Chemical Engineering. 3 (2015) 1–8.

[43] W. Xu, K. Xu, V.V. Viswanathan, S.A. Towne, J.S. Hardy, J. Xiao, Z. Niea, D. Huc, D. Wanga, J.G. Zhang, Reaction mechanisms for the limited reversibility of li-O2 chemistry in organic carbonate electrolytes, Journal of Power Sources. 196 (2011) 9631–9639.

[44] V.S. Bryantsev, F. Faglioni, Predicting autoxidation stability of ether- and amide-based electrolyte solvents for li-air batteries, Journal of Physical Chemistry A. 116 (2012) 7128–7138.

[45] S.S. Zhang, J. Read, Partially fluorinated solvent as a co-solvent for the non-aqueous electrolyte of Li/air battery, J. Power Sources. 196 (2011) 2867–2870.

[46] H. Wan, Y. Mao, Z.X. Liu, Q.Y. Bai, Z. Peng, J.J. Bao, G. Wu, Y. Liu, D.Y. Wang, J.Y. Xie, Influence of enhanced O2 provision on the discharge performance of Li-air batteries by incorporating fluoroether, ChemSusChem. 10 (2017) 1385–1389.

[47] H. Wan, Q. Bai, Z. Peng, Y. Mao, Z. Liu, H. He, D. Wang, J. Xie, G. Wu, A high power Li-air battery enabled by a fluorocarbon additive, J. Mater. Chem. A 5 (2017) 24617–24620.

[48] F. De Giorgio, F. Soavi, M. Mastragostino, Effect of lithium ions on oxygen reduction in ionic liquid-based electrolytes, Electrochem. Commun. 13 (2011) 1090–1093.

[49] C.O. Laoire, S. Mukerjee, K.M. Abraham, E.J. Plichta, M.A. Hendrickson, Influence of nonaqueous solvents on the electrochemistry of oxygen in the rechargeable lithium-air battery, J. Phys. Chem. C 114 (2010) 9178–9186.

[50] W. Xu, J. Hu, M.H. Engelhard, S.A. Towne, J.S. Hardy, J. Xiao, J. Feng, M.Y. Huc, J. Zhang, F. Ding, M.E. Gross, J.G. Zhang, The stability of organic solvents and carbon electrode in no aqueous $Li-O_2$ batteries, Journal of Power Sources. 215 (2012) 240–247.

[51] G.M. Veith, J. Nanda, L.H. Delmau, N.J. Dudney, Influence of lithium salts on the discharge chemistry of Li-air cells, Journal of Physical Chemistry Letters. 3 (2012) 1242–1247.

[52] D. Sharon, D. Hirsberg, M. Salama, M. Afri, A.A. Frimer, M. Noked, W. Kwak, Y.K. Sun, D. Aurbach, Mechanistic role of Li^+ dissociation level in aprotic $Li–O_2$ battery, ACS Appl. Mater. Interfaces. 8 (2016) 5300–5307.

[53] D. Shanmukaraj, S. Grugeon, G. Gachot, S. Laruelle, D. Mathiron, J.M. Tarascon, M. Armand, Boron esters as tunable anion carriers for non-aqueous batteries electrochemistry, J. Am. Chem. Soc. 132 (2010) 3055–3062.

[54] B. Xie, H.S. Lee, H. Li, X.Q. Yang, J. McBreen, L.Q. Chen, New electrolytes using Li_2O or Li_2O_2 oxides and tris(pentafluorophenyl) borane as boron based anion receptor for lithium batteries, Electrochemistry Communications. 10 (2008) 1195–1197.

[55] A. Kraytsberg, Y. Ein-Eli, Review on Li-air batteries—Opportunities, limitations and perspective, J. Power Sources. 196 (2011) 886–893.

[56] V.S. Bryantsev, F. Faglioni, Predicting the stability of aprotic solvents in Li-air batteries: pKa calculations of aliphatic C—H acids in dimethyl sulfoxide, Chem. Phys. Lett., 558 (2013) 42–47.
[57] W. Xu, J. Xiao, J. Zhang, D. Wang, J.G. Zhang, Optimization of nonaqueous electrolytes for primary lithium/air batteries operated in ambient environment, J. Electrochem. Soc. 156 (2009) A773–A779.
[58] P. Albertus, G. Girishkumar, B. McCloskey, R.S. Sa´nchez-Carrera, B. Kozinsky, Identifying capacity limitations in the Li/oxygen battery using experiments and modeling, J. Christensen and A.C. Luntz, J. Electrochem. Soc. 158 (2011) A343–A351.
[59] A. Kraytsberg, Y. Ein-Eli, The impact of nano-scaled materials on advanced metal—air battery systems, Nano Energy. 2 (2013) 468–480.

19 Ionic Electrolytes

Vandana, Fabeena Jahan, Anjali Paravannoor, and Baiju Kizhakkekilikoodayil Vijayan

CONTENTS

19.1 Introduction 281
19.2 Oxygen Electrode Process..282
19.3 Ionic Electrolytes in Various Metal-Air Batteries 284
    19.3.1 Ionic Electrolyte in Zinc-Air Batteries 284
    19.3.2 Ionic Electrolytes in Lithium-Air Batteries 285
    19.3.3 Ionic Electrolytes in Aluminum-Air Batteries 286
19.4 Conclusion..287
References..288

19.1 INTRODUCTION

Metal-air batteries represent a well-established technology that offers several advantages like high energy densities, availability and low cost of electrode materials, and environmental concern. The active material for one of the half-cell redox reactions in these batteries is the air provided by the exterior of the system which makes the chief difference from other battery engineering types [1][2]. Despite these characteristics and the use of oxygen as the electrode reactant, metal-air batteries suffer from many complications that arise when the cell is recharged, which include poor efficacy and loss of electrolyte that results from the higher stability and side reactions of intermediates with components in the air. Irreversibility of electrochemical processes, along with the kinetic restrictions of cathodic reaction, makes these devices limited to primary batteries [3]. Thus, these limitations should be overcome for the commercialization of secondary batteries, producing a rechargeable metal-air battery that can be exploited for storing renewable energy efficiently. Choice of the right electrolyte could resolve most of these problems. However, the reactivity of metals made different explanations regarding the electrolyte that could be used for different metal-air batteries. As the chemistry of anodic and cathodic processes in aqueous and organic electrolytes are already known, the limitations of these electrolytes directed the metal-air battery research toward the formulation of IL electrolytes in the perspective that they are highly stable, non-volatile electrolytes and thus are interesting substitutes to traditional aqueous and organic electrolytes that could support the processes in rechargeable metal-air battery [4].

ILs are liquids comprising entirely of ions and are defined as molten salts that exist in the liquid phase with their melting points below 100°C. The difference in the structure of ions causes a decrease in the lattice energies of the crystal structure, which in turn results in weak interionic interaction that lowers the melting point. IL's properties can be finely controlled by altering the structures of bulky organic cations and organic/inorganic anions present in it and thus provide a wide range of combinations resulting in an unlimited range of IL designs that may be fine-tuned for the required application. As there is an increasing demand for sustainable and clean energy, the most substantial research areas for IL application are their energy storage application considering their unique properties [5].

In metal-air batteries, the use of conventional carbonate-based electrolytes causes several problems due to the electroactive material dissolution, side reactions, and solvent evaporation that dries out the cell. These problems have prompted the metal-air battery research to direct towards ILs as electrolyte material. Certain unique properties of ILs like nonvolatility, low vapor pressure, non-inflammability, electrochemical and chemical stability towards superoxide anion and metals, good thermal stability, hydrophobicity, wide liquid temperature ranges, etc., are desirable for safer electrolytes and have made them a substitute for conventional carbonate-based electrolytes. ILs in metal-air batteries possess the following benefits: increased safety (due to high electrochemical stability, absence of toxicity, usage in strenuous conditions, very low vapor pressure, minimizing the danger of overpressure-prompted exposures and release to the environment, non-flammability), high solubilizing power that allows improved solubility of active material and intermediates, tunability of system properties, and compatibility with other components of the system. Also, the absence of hydroxide ions prevents the formation of carbonate in presence of IL electrolyte [6][7][8].

However, ILs also have several disadvantages as they are highly viscous, have low metal ion/O_2 solubility resulting in lower diffusion, and low conductivity resulting in low energy density. These parameters rely on the structure of ions, and in general, they exhibit conductivity and viscosity within the ranges of 0.1–10 mS/cm and 10–500 mPa.s, respectively; but ILs with higher viscosity and lower conductivity than this range are also known (e.g. the viscosity and conductivity of *N*-octyl-*N,N,N*-tri-butylammonium tri-fluoromethanesulfonate, are 2030mPa.s and 0.017 mS/cm, respectively, at 25°C). Moreover, metal-air cells possess an open structure, which allows moisture from the air to enter the cell and affect the electrochemical behavior of the electrolyte, which would result in reduced viscosity and better conductivity in the case of ILs [9]. The O_2 solubility in ILs is of a similar magnitude as in organic solvents, while that of diffusivity of O_2 is found to be lower in magnitude. Despite existing merely as ions the intrinsic ionic conductivity of ILs is expected to be a desirable value but the presence of comparatively large constituent ions and the growth and aggregation of ion pairs in ILs will considerably hinder the mobility of the electrolyte components which in turn lowers the conductivity [10].

19.2 OXYGEN ELECTRODE PROCESS

The discovery of metal-air batteries was the result of the search for an alternative approach to the intercalation reaction mechanism in conventional batteries like primary Zn-MnO_2, Lithium-Ion batteries, secondary lead-acid batteries, etc. which limits the energy density. Metal-air batteries replace the intercalation cathode material with an electrode that could follow catalytically active oxygen reduction reaction (ORR) and oxygen evolution reaction (OER). In metal-air batteries, electricity is generated by a redox reaction between the metal anode and the oxygen cathodewhereas oxygen is supplied continuously and infinitely from outside of the system i.e., from the air, which in turn would remarkably enhance the energy density [1]. The ORR mechanism is extremely intricate and relies on many factors, such as the electrode material, electrolyte solvent, catalysts, and activity of the proton. Theearlier research on the metal-air systems were focusing on the ORR forming either water or hydrogen peroxide in aqueous electrolytes of different pH. The thermodynamics, potentials, and reaction mechanism were also thoroughly investigatedalong with electrocatalysis studies. In general, a simplified mechanism for ORR in acidic and basic aqueous solutions is proposed by Song et al. and Zhang et.al, which depends on the nature of the working electrode and catalysts that constitute three steps as (a) direct 4-electron reduction from O_2 to H_2O, (b) the peroxo route from O_2 to H_2O, generating hydrogen peroxide (H_2O_2) as an intermediate and iii)2-electron reduction pathway from O_2 to H_2O_2[8][11][12].

Even though the aqueous electrolytes are advantageous in many aspects, chemical reaction pathways involved in the whole process are too fast, making it hard to detect the intermediates. Thus, to evaluate the reaction mechanism in detail, non-aqueous electrolytes that support both ORR and OER should be studied, especially when reactive metals have to be used in a rechargeable metal-air battery.

The use of solvents that are aprotic and non-aqueous facilitates the understanding of the mechanism by stabilizing the superoxide anions ($O_2^{\bullet-}$) intermediate. Sawyer et al. studied the thermodynamic and kinetic aspects of O_2 and its electrogenerated species in non-aqueous aprotic solvents and compared that with aqueous aprotic solvents. In non-aqueous aprotic solvents, oxygen reduction follows a reversible 1-electron reduction pathway from O_2 to $O_2^{\bullet-}$ ion [13][14].

As the superoxide anion is nucleophilic, it is a highly reactive species, and thus acetonitrile, dimethylformamide, dimethylsulfoxide (DMSO), etc., have been selected for studying the $O_2/O_2^{\bullet-}$ redox couple, as superoxides get stabilized in these media. However, the high volatility, as well as the lower solubility of reaction products in organic electrolytes, have raised many safety concerns, which is why the mechanisms of ORR in IL electrolytes have begun to be explored during the last two decadesas an alternative to conventional aqueous and organic media. Moreover, the outstanding properties of ILs in terms of their physical properties, structure, and stability will affect the stability of the electrogenerated species of oxygen [15]. The ILs can be broadly classified into families like imidazolium, quaternary ammonium, and quaternary phosphonium.

In most of the early studies of ORR in aprotic ILs, it was seen that the reduction of dioxygen proceeds through a 1-electron quasi-reversible reaction to form superoxide. It is also found that the change in diffusivity of O_2 and superoxide is responsible for the decrease in coulombic ability as the protonation and reduction steps seem to be only sparingly reversible as shown in equation (7). Carter et al. reported the 1-electron reduction of oxygen in 1-ethyl-3-methylimidazolium chloride-aluminum chloride IL for the first time. The electrolyte preparation and cell assembly were carried out in a helium-filled dry box. Oxygen and argon were continuously introduced to the cell through a gas dispersion tube dipped in the electrolyte after passing via a drying column. Glassy-carbon and Pt discs polished with diamond paste were used as working electrodes, and a Pt wire was employed as the counter electrode. A vycor frit isolated the reference electrode from the working solution, and all electrochemical tests were carried out in a single and gas-tight chamber cell. The superoxide species generated was found to be unstable and an irreversible oxidation process was observed in cyclic voltammetry carried out at varying scan rates ranging from 5 to 200 mV/s. This irreversibility was attributed to the protic impurities generated when IL is saturated with O_2 [16]. The first evidence for the electrochemical formation of stable superoxide ions in ILs was reported by Al Nashef et al. in 1-n-butyl-3-methylimidazolium hexafluorophosphate, but the presence of impurities made it irreversible [17]. The relevance of exceptionally pure ILs and the probable side reactions between the electrogenerated species and impurities could be explained by the differences found in the electrochemical behavior of various imidazolium-based IL saturated with O_2. The reactive $O_2^{\bullet-}$ will be initially attacked by atoms with the highest positive charge. The physical and transport properties also have a significant impact on electrochemical behavior. When anhydrous IL or aprotic solvent are employed, the changes in the solvent polarity, electrolyte nature, or the presence of acidic additives are thought to be the reason for the change in the value of formal potential for a redox couple; these deviations are attributed to nonspecific values of solvation energies, pairing of ions, and equilibria related with protonation, respectively [18]. Several investigations, including those by Barnes et al. and Islam et al., have found evidence for a pairing of electrogenerated superoxide anion and the imidazolium, phosphonium, or ammonium cations of IL. They proposed that the quasi-reversibility of 1-electron ORR is due to ion-pairing interaction that stabilizes the produced superoxide radical, letting it be oxidized back instead of reacting with the adjacent molecules during the reverse potential scan. The reversibility of the $O_2/O_2^{\bullet-}$ redox pair is influenced by the chemical structure of the cations in the IL. Imidazolium-based ILs are vulnerable to degradation due to the attack of the superoxide anionowing to its nucleophilic nature with the acidic proton on the 2-position of the aromatic ring of imidazolium cation.However, thepyrrolidinium or quaternary ammonium-based ILs do not have positively charged carbon atoms susceptible to the superoxideattack. Thus, the superoxide anion was discovered to be more stable in the presence of aliphatic and alicyclic cations, such as quaternary ammonium or pyrrolidinium ions [19][20]. Katayama et al. used an Au and Pt macroelectrode disc to study 1-Butyl-1-ethylpyrrolidinium bis(trifluoromethanesulfonyl) imide and found a virtually reversible $O_2/O_2^{\bullet-}$ redox pair with an anodic to cathodic current density ratio of 0.97, indicating that $O_2^{\bullet-}$ is substantially stable in this IL. The counter electrode was made of Pt and was placed in a separate chamber with a glass filter. A silver wire dipped in Silver trifluoromethanesulfonate/1-ethyl-3-methylimidazolium bis(trifluoromethanesulfone) imide solution isolated from the bulk solution by the porous glass was used as the reference electrode. The coulombic attraction of the organic ions is likely to influence $O_2^{\bullet-}$ diffusion and the diffusion coefficient of $O_2^{\bullet-}$ is found to be very small about 8.6×10^{-7} cm^2/s than that of O_2 [21].

Pure ILs such as pyrrolidinium and piperidinium are less susceptible to O_2-attack due to the poor leaving groups linked to the N atom. It has been discovered that the working electrode's material and nature have an impact on the ORR mechanism. The measured current density and over-potentials of the reduction reaction, which are important indicators of the potential and kinetics performance of the same in an electrochemical cell, could be varied dramatically by modifying the substrate or catalyst materials utilized. These investigations also raise the query of how the superoxide ion, which appears to be reactive, can be stable enough during the cell operation.

The OER has been thoroughly investigated from the viewpoint of water splitting, which involves the oxidation of water to produce oxygen. The prohibitive over-potential required for effective water to dioxygen oxidation is the major difficulty, which is why the usage of catalysts has become a major focus of attention for researchers in this sector. However, little is known about how these reactions occur in ILs consisting of ORR products [22]. Izgorodin et al. in 2012 investigated the use of a hydrated protonic IL tetrabutylammonium bisulfate to improve water oxidation by utilizing MnO_x as a catalyst. The anode used was an Au electrode coated with MnO_x films, and the cathode was a Pt counter electrode. The electrode's working area was masked using Kapton tape, and standard calomel electrodes served as the reference electrode. All electrochemical measurements were carried out at room temperature of around 22 °C. The capacity of cation to bind with H_2O molecules being oxidized allowed a current of 1 mA/cm^2 to be generated at an overpotential of 150 mV when 1:1 ratio of IL/H_2O mix at a pH of 10 was used. This hydrated IL mixture outperformed the usual alkaline electrolyte by a significant margin. The potential of employing RTILs and diverse mixes to assist the OER in a rechargeable metal-air battery is highlighted in this study [23].

The reversibility of 4-e reduction of O_2 to H_2O in IL has yet to be demonstrated, and this is an area where the electrochemical community should concentrate its efforts. However, the O_2 4-electron reduction isn't required to make rechargeable secondary metal-air batteries commercially in practice. The ability to stabilize the peroxide product and use its shown reversibility in RTILs could be the most successful road ahead for the OER [24].

19.3 IONIC ELECTROLYTES IN VARIOUS METAL-AIR BATTERIES

There are various types of metal-air batteries. Among them, the most prevalent are zinc-air batteries, lithium-air batteries battery, and aluminum-air batteries. In the upcoming section, we will be discussing the typically used electrolytes in those metal-air batteries and the importance of ionic electrolytes in them.

19.3.1 Ionic Electrolyte in Zinc-Air Batteries

Zinc-air battery (ZAB) is a booming technology due to its safety, reasonable cost, comparatively better stability, and reversibility of the zinc material [25][26]. They possess the maximum theoretical specific energy density of 1350 Wh/kg within the non-air-cathode primary batteries, and their specific energy density is at its peak when compared to other metal-air battery systems [9]. Environmental friendliness, cheaper price, and high security are the main beneficial attributes owned by Zinc-air batteries. Oxygen constitutes the fundamental part of the cathode portion in a ZAB system, where its usage leads to the production and distribution of electrochemical energy. The basic operating strategy of ZABs is the exposure of the cathodic compartment of the cell to the atmosphere, which results in the diffusion of oxygen through the previous structure of the cathode which is in touch with the electrolyte, thereby forming hydroxyl ions by reductive reaction with the catalyst [27]. Dissimilar to other non-air batteries, the cathode doesn't undergo physical or chemical conversion when the oxygen undergoes catalytic reduction. ZAB possesses the topmost specific and volumetric energy-dense primary aqueous system since the cathodic part constitutes the least volume and mass of the total cell as a result of owing boundless reserve of O_2.

Aqueous alkaline solutions are the frequently used electrolytes in ZAB since the ORR and OER take place at their best under an alkaline atmosphere [28]. As a consequence of displaying the super most ionic conductivity and kinetic behavior, Potassium hydroxide is used predominantly instead of sodium hydroxide. However, it is a fact that the use of typical alkaline electrolytes has its demerits. Firstly, there will be a loss of water by evaporation since the cathode compartment is open to the atmosphere. Consequently, the electrolytes start to concentrate thereby decreasing the mobility of hydroxyl ions. The second demerit of using alkaline electrolytes is the formation of carbonate species. The carbon dioxide from the atmosphere enters the cell through an air cathode followed by the formation of carbonate crystals, hence depleting the hydroxide ion concentration which in turn makes the zinc anode inert. It also restricts the entry of air due to the precipitation of carbonate in the pores of the air cathode.

The morphological changes of zinc active material like the development of zinc dendrites and the shape change of zinc material (namely redistribution) are the key issues that curb the overall efficiency and lifetime of zinc batteries when a large number of charge-discharge cycles are applied [29][30]. The main causes for the morphology issues of zinc are: (a) better electrochemical kinetics in alkaline electrolytes and (b) production of zincate ions due to the greater solubility of zinc which in turn passes into the electrolyte causing the redistribution of zinc. Consequently, the zinc deposits get irregularly spread over the electrode surface or bring about dendrite formation, thereby cutting down the efficiency of the battery, or can even cause a short circuit within the batteries.

Hence a need comes to investigate suitable electrolytes in place of typically used aqueous alkaline electrolytes. Some of the currently used non-aqueous electrolytes are ILs, solid-state polymer electrolytes, and ceramic/glassy electrolytes [31][32].

RTILs, known as "Designer solvents", have emerged as a green chemistry solvent alternative for commonly used toxic organic solvents [33][34]. The emergence of ILs paved the way to eliminate the typically used hazardous chlorinated solvents. The advantageous attributes of ILs are they are less volatile, non-flammable, thermally stable, and possess a wide electrochemical window. These interesting properties make ILs suitable electrolytes in zinc-air batteries since it evades some of the drawbacks of ZAB [35]. ILs aid in reducing the morphological issues noted earlier for ZABs, attain a longer life since they prevent drying out of the electrolyte, eliminate the redeposition of zinc, and then eliminate carbonation. The wide electrochemical window of ZAB helps to eliminate hydrogen evolution. The ILs commonly used in ZAB consist of imidazolium and pyrrolidinium cations together with trifluoromethane-sulfonyl imide (TFSI-) and dicynamide anions.

19.3.2 Ionic Electrolytes in Lithium-Air Batteries

The idea of lithium-oxygen batteries emerged in the late 1970s for use in automobiles, but it became a booming innovation only in this decade. The open cell system of lithium-air batteries utilizes oxygen directly from the atmosphere, which in turn leads to a reduction in its weight. The lithium-air battery became an attractive idea due to the high theoretical energy density which is much better than typical lithium-ion batteries [36]. Compared with petrol, Lithium-air batteries havea higher specific energy density that makes them useful in electric vehicles. The lifespan of lithium-oxygen batteries is about 2000 charging cycles, but if it is properly used and maintained, they can last for more time.

Lithium-air battery exhibits an extraordinary theoretical energy density when compared to all the other metal-air batteries. The overall energy density of a cell is inversely proportional to the weight of the cell. The cathodic material of the Li-O$_2$ battery is oxygen which is directly taken from the atmosphere and thus aids in reducing the overall weight of the cell. Hence the theoretical energy density of Li-O$_2$ battery is exceptionally high of the order of 11586 Whkg^{-1} with the mass of lithium alone and 3505Wh/kgwhen the mass of lithium and Li$_2$O$_2$ is considered.

Carbonate-based electrolytes were the typically used electrolytes in the research studies of lithium-oxygen batteries. But the carbonates are prone to attack from nucleophiles like superoxide anions and other related species leading to the formation of bulk quantities of lithium carbonate (Li$_2$CO$_3$) and other unwanted organic side products. These issues that emerged because of the usage of carbonate electrolytes pave way for the investigation of suitable electrolytes which remain inert in the presence of intermediate products of oxygen. The current research studies of lithium-oxygen batteries focus on the usage of nonaqueous electrolyte solutions. The nonaqueous lithium-oxygen battery consists of a lithium anode, an organic aprotic electrolyte, and an air cathode. Solid electrolytes were also used as an electrolyte in some of the research works on lithium-oxygen batteries, but they didn't become a novel electrolyte due to their instability when in contact with lithium metal [37].

ILs are used as promising electrolytes in lithium-air batteries due to their novel properties [38] [39]. The hydrophobic ILs have negligible volatility which slows down the dehydration of electrolytes. ILs also eliminate the hydrolysis reaction of metal anode thereby increasing the overall efficiency of the lithium-air battery. These interesting characteristics make ILs a convenient electrolyte for Li-O$_2$ batteries [38]. Kuboki et al. were the first to use imidazolium-based IL electrolytes in lithium-air cells. Later, pyrrolidinium and ammonium-based electrolytes were used as suitable ionic electrolytes in nonaqueous Li-O$_2$ cells. Kuboki et al. studied the discharge characteristics of lithium-air cells in an IL of 1-ethyl-3-methyl imidazolium bis(trifluoromethylsulfonyl) imide and compared it with other ILs and is depicted in Figure 19.1. The battery exhibited a constant voltage at 0.01 mA/cm for approximately two months under ambient conditions. This positive outcome shows that ILs remain inert in the presence of superoxides [40].

Still, ILs have their drawbacks. The lower chemical stability and higher viscosity make the use of ILs controversial in many situations. Thus, studies have to be done for the proper customization of ILs that eliminates the issues mentioned formerly. More attempts are required to search for unique ionic electrolytes that possess higher stability and lower viscosity.

19.3.3 IONIC ELECTROLYTES IN ALUMINUM-AIR BATTERIES

The concept of aluminum-air batteries (Al-ABs) budded in the 1960s. These batteries possess a large number of advantageous attributes like high theoretical specific energy, high theoretical negative electrode potential, high volumetric capacity, ability to pass over three electrons for ORR at the

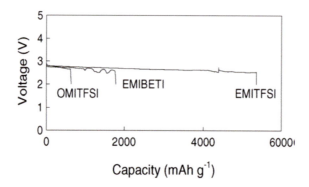

FIGURE 19.1 Discharge curves of the lithium-air cell using hydrophobic ILs. (Adapted with permission [40]. Copyright (2005), Elsevier.)

cathode, and higher security [41]. The Al-AB system consists of an aluminum anode (or Al alloy), an oxygen cathode(atmosphere), and a suitable electrolyte.

Al-ABs find wide varieties of applications from small-scale systems to highly sophisticated operations. They are used in electric vehicles, powering ships, low-power micro-batteries, and tiny electronic devices. The main issue that depletes the overall performance and lifetime of Al-AB is corrosion of the aluminum electrode. These issues related to corrosion prevent the commercialization of the battery [40].

Both aqueous and nonaqueous electrolytes are employed for Al-ABs in accordance with the conditions of operations. Three forms of aqueous electrolytes are proposed for Al-AB: alkaline, aqueous, and neutral medium. Aqueous alkaline solutions were the most prominently utilized electrolytes due to their highly favorable reaction kinetics. They also maintain the low overpotential condition which is necessary for an active ORR. Potassium hydroxide and sodium hydroxide are the most frequently used options. During the discharge process, the Al reacts with excess hydroxide concentration in the electrolyte to produce aluminate ions thereby transferring electrons to the air cathode. Consequently, an anodic electrode gets passivated followed by the masking of the active material lying beneath. The cathode pores get blocked due to the collection of $Al(OH)_3$ during the discharge process [42]. All these issues lead to the reduction of the overall efficiency of the battery.

Hydrochloric acid (HCl) and sulfuric acid (H_2SO_4) are the commonly used acidic electrolytes in Al-ABs. The acidic electrolytes aid in the reduction of the unwanted carbonation reaction at the oxygen electrode. The third major classification of aqueous electrolytes is the neutral salt electrolytes like NaCl solutions. However, the neutral salt solutions also failed in minimizing the corrosion rate of electrodes. To decrease the overall corrosion rate, different alloys of aluminum are used as an anode. Still, researchers were not successful in reducing the corrosion rate to the acceptance level. To enhance the efficiency and performance of a battery, the elimination of self-corrosion of anode and surface passivation seems mandatory.

Currently, nonaqueous electrolytes are used over aqueous electrolytes to deal with the issues related to the usage of aqueous electrolytes. The two main categories of nonaqueous electrolytes are ILs and polymer gels [43]. They were successful in dealing with the corrosion issues, dendrite formation, and volatility of aqueous electrolytes. RTILs possess a delocalized charge system which is responsible for their very low melting point.

Hydrogen evolution takes place from Al in the presence of room temperature ILs. Also, the electrodeposition of Al which is absent in typical aqueous electrolytes becomes possible in RTILs. But in alkaline electrolytes, the hydrogen evolution occurs before the deposition of aluminum. These are the main advantageous attributes that make IL beneficial over aqueous electrolytes. Some of the reported ionic electrolytes used in Al-ABs are 1-ethyl-3-methyl imidazolium chloride/Aluminum chloride ($EMImCl/AlCl_3$) and (1-ethyl-3-methylimidazolium oligo-fluoro hydrogenate) $EMIm(HF)_{2.3}$ which exhibited moderately high current densities as shown in Figure 19.2. These electrolytes were successful in reducing the common corrosion issues to an acceptable level. Chloroaluminate is the other reported IL in which the chloro-acidity is the determining factor of electrochemistry of the electrolyte. It is the composition of the electrolyte melt that determines its chloro-acidity [44]. The utilization of room temperature ILs as Al-AB electrolytes is gaining attraction currently due to their beneficial properties over aqueous electrolytes. Hence it is essential to search for novel ILs that eliminate corrosion and surface passivation, thereby paving the way to a new era of efficient Al-AB.

19.4 CONCLUSION

ILs could be able to match the demanding criteria for numerous energy storage applications due to their unique qualities like nonvolatility, excellent thermal stability, and ionic conductivity. Many of the previously recognized mechanisms of oxygen electrode processes were also relevant in RTIL-based electrolytes, allowing the development of new solutions for the irreversibility reported in

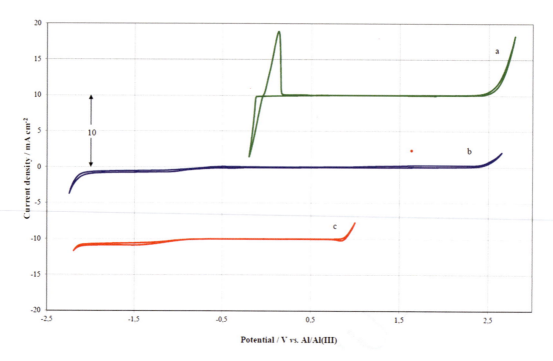

FIGURE 19.2 Cyclic voltammetry at 20 mV/s of $AlCl_3$/EMImCl melts: (a) acidic, (b) neutral, and (c) basic. (Adapted with permission [44]. Copyright (2014), Elsevier.)

ORR. For ILs, our understanding of crucial aspects including catalyst chemistry, surface structure, proton activity, and interfacial interactions is still in its early stage. Also, greater research into innovative ILs with higher stability and lower viscosity, as well as specialized applications for which these characteristics become unique, is required for further advancement. Thus, these areas should be focused on by electrochemists and battery engineers, to advance the understanding of physical and electrochemical processes in RTILs as electrolytes for various types of secondary metal-air batteries.

REFERENCES

[1] Y. Li, J. Lu, Metal-air batteries: Will they be the future electrochemical energy storage device of choice? ACS Energy Letters. 2(6) (2017) 1370–1377.
[2] B. Dunn, H. Kamath, J.M. Tarascon, Electrical energy storage for the grid: A battery of choices, Science. 334(6058) (2011) 928–935.
[3] L.J. Hardwick, C.P. De León, Rechargeable multi-valent metal-air batteries, Johnson Matthey Technology Review. 62(2) (2018) 134–149.
[4] H.F. Wang, Q. Xu, Materials design for rechargeable metal-air batteries, Matter. 1(3) (2019) 565–595.
[5] D.R. MacFarlane, N. Tachikawa, M. Forsyth, J.M. Pringle, P.C. Howlett, G.D. Elliott . . . C.A. Angell, Energy applications of ionic liquids, Energy & Environmental Science. 7(1) (2014) 232–250.
[6] A.R. Mainar, E. Iruin, L.C. Colmenares, A. Kvasha, I. de Meatza, M. Bengoechea . . . J.A. Blazquez, An overview of progress in electrolytes for secondary zinc-air batteries and other storage systems based on zinc, Journal of Energy Storage. 15(2018) 304–328.
[7] Y. Liu, Q. Sun, W. Li, K.R. Adair, J. Li, X. Sun, A comprehensive review on recent progress in aluminum-air batteries, Green Energy & Environment. 2(3) (2017) 246–277.
[8] M. Kar, T.J. Simons, M. Forsyth, D.R. MacFarlane, Ionic liquid electrolytes as a platform for rechargeable metal-air batteries: A perspective, Physical Chemistry Chemical Physics. 16(35) (2014) 18658–18674.

[9] M. Xu, D.G. Ivey, Z. Xie, W. Qu, E. Dy, The state of water in 1-butly-1-methyl-pyrrolidinium bis (trifluoromethanesulfonyl) imide and its effect on Zn/Zn (II) redox behavior, Electrochimica Acta. 97(2013) 289–295.

[10] J. Goldstein, I. Brown, B. Koretz, New developments in the Electric Fuel Ltd. zinc/air system, Journal of Power Sources. 80(1–2) (1999) 171–179.

[11] C. Song, J. Zhang, Electrocatalytic oxygen reduction reaction, in PEM fuel cell electrocatalysts and catalyst layers (pp. 89–134). Springer, London, 2008.

[12] D.T. Sawyer, A. Sobkowiak, J.L. Roberts, Electrochemistry for chemists. Wiley, 1995.

[13] C.P. Andrieux, P. Hapiot, J.M. Saveant, Mechanism of superoxide ion disproportionation in aprotic solvents, Journal of the American Chemical Society. 109(12) (1987) 3768–3775.

[14] D.T. Sawyer, G. Chiericato, C.T. Angelis, E.J. Nanni, T. Tsuchiya, Effects of media and electrode materials on the electrochemical reduction of dioxygen, Analytical Chemistry. 54(11) (1982) 1720–1724.

[15] D.T. Sawyer, J.S. Valentine, How super is superoxide?, Accounts of Chemical Research. 14(12) (1981) 393–400.

[16] M.T. Carter, C.L. Hussey, S.K. Strubinger, R.A. Osteryoung, Electrochemical reduction of dioxygen in room-temperature imidazolium chloride-aluminum chloride molten salts, Inorganic Chemistry. 30(5) (1991) 1149–1151.

[17] I.M. AlNashef, M.L. Leonard, M.C. Kittle, M.A. Matthews, J.W. Weidner, Electrochemical generation of superoxide in room-temperature ionic liquids, Electrochemical and Solid-State Letters. 4(11) (2001) D16.

[18] A. Rene, D. Hauchard, C. Lagrost, P. Hapiot, Superoxide protonation by weak acids in imidazolium based ionic liquids, The Journal of Physical Chemistry B. 113(9) (2009) 2826–2831.

[19] M.M. Islam, T. Ohsaka, Roles of ion pairing on electroreduction of dioxygen in imidazolium-cation-based room-temperature ionic liquid, The Journal of Physical Chemistry C. 112(4) (2008) 1269–1275.

[20] A.S. Barnes, E.I. Rogers, I. Streeter, L. Aldous, C. Hardacre, G.G. Wildgoose, R.G. Compton, Unusual voltammetry of the reduction of O2 in [C4dmim][N (Tf) 2] reveals a strong interaction of $O_2^{•-}$ with the [C4dmim]+ cation, The Journal of Physical Chemistry C. 112(35) (2008) 13709–13715.

[21] Y. Katayama, K. Sekiguchi, M. Yamagata, T. Miura, Electrochemical behavior of oxygen/superoxide ion couple in 1-butyl-1-methylpyrrolidinium bis (trifluoromethylsulfonyl) imide room-temperature molten salt, Journal of the Electrochemical Society. 152(8) (2005) E247.

[22] M. Hayyan, M.A. Hashim, I.M. AlNashef, Superoxide ion: Generation and chemical implications, Chemical Reviews. 116(5) (2016) 3029–3085.

[23] A. Izgorodin, E. Izgorodina, D.R. MacFarlane, Low overpotential water oxidation to hydrogen peroxide on a MnO x catalyst, Energy & Environmental Science. 5(11) (2012) 9496–9501.

[24] C. Pozo-Gonzalo, C. Virgilio, Y. Yan, P.C. Howlett, N. Byrne, D.R. MacFarlane, M. Forsyth, Enhanced performance of phosphonium based ionic liquids towards 4 electrons oxygen reduction reaction upon addition of a weak proton source, Electrochemistry Communications. 38(2014) 24–27.

[25] P. DerekGregory, Metal-air batteries (Vol. 6). Mills and Boon, 1972.

[26] D. Linden, Linden's handbook of batteries. McGraw-Hill, 2010.

[27] A.R. Mainar, L.C. Colmenares, H.J. Grande, J.A. Blázquez, Enhancing the cycle life of a zinc-air battery by means of electrolyte additives and zinc surface protection, Batteries. 4(3) (2018) 46.

[28] C. Wang, H. Zhang, S. Dong, Z. Hu, R. Hu, Z. Gu . . . L. Chen. High polymerization conversion and stable high-voltage chemistry underpinning an in situ formed solid electrolyte, Chemistry of Materials. 32(21) (2020) 9167–9175.

[29] H. Zhang, J. Zhang, J. Ma, G. Xu, T. Dong, G. Cui, Polymer electrolytes for high energy density ternary cathode material-based lithium batteries, Electrochemical Energy Reviews. 2(1) (2019) 128–148.

[30] M. Petkovic, K.R. Seddon, L.P.N. Rebelo, C.S. Pereira, Ionic liquids: A pathway to environmental acceptability, Chemical Society Reviews. 40(3) (2011) 1383–1403.

[31] K.R. Seddon, A. Stark, M.J. Torres, Influence of chloride, water, and organic solvents on the physical properties of ionic liquids, Pure and Applied Chemistry. 72(12) (2000) 2275–2287.

[32] G.G. Eshetu, D. Mecerreyes, M. Forsyth, H. Zhang, M. Armand, Polymeric ionic liquids for lithium-based rechargeable batteries, Molecular Systems Design & Engineering. 4(2) (2019) 294–309.

[33] D. Zhou, R. Liu, J. Zhang, X. Qi, Y.B. He, B. Li . . . F. Kang, In situ synthesis of hierarchical poly (ionic liquid)-based solid electrolytes for high-safety lithium-ion and sodium-ion batteries, Nano Energy. 33(2017) 45–54.

[34] A.F. De Anastro, N. Lago, C. Berlanga, M. Galcerán, M. Hilder, M. Forsyth, D. Mecerreyes, Poly (ionic liquid) iongel membranes for all solid-state rechargeable sodium battery, Journal of Membrane Science. 582(2019) 435–441.

[35] E.L. Littauer, K.C. Tsai, Anodic behavior of lithium in aqueous electrolytes: ii. Mechanical passivation, Journal of the Electrochemical Society. 123(7) (1976) 964.

[36] R. Murugan, V. Tangadurai, W. Weppner, Fast lithium ion conduction in garnettype Li7La3Zr2O12, Angew, Chem. Int. Ed. 46 (2007) 7778–7781.

[37] S. Randström, G.B. Appetecchi, C. Lagergren, A. Moreno, S. Passerini, The influence of air and its components on the cathodic stability of N-butyl-N-methylpyrrolidinium bis (trifluoromethanesulfonyl) imide, Electrochimica Acta. 53(4) (2007) 1837–1842.

[38] M. Piana, J. Wandt, S. Meini, I. Buchberger, N. Tsiouvaras, H.A. Gasteiger, Stability of a pyrrolidinium-based ionic liquid in Li-O2 cells, Journal of the Electrochemical Society. 161(14) (2014) A1992.

[39] G.A. Elia, J. Hassoun, W.J. Kwak, Y.K. Sun, B. Scrosati, F. Mueller . . . J. Reiter, An advanced lithium-air battery exploiting an ionic liquid-based electrolyte, Nano Letters. 14(11) (2014) 6572–6577.

[40] T. Kuboki, T. Okuyama, T. Ohsaki, N. Takami, Lithium-air batteries using hydrophobic room temperature ionic liquid electrolyte, Journal of Power Sources. 146(1–2) (2005) 766–769.

[41] D.R. Egan, C. Ponce de Leó n, R.J.K. Wood, R.L. Jones, K.R. Stokes, F.C. Walsh, Developments in electrode materials and electrolytes for aluminium air batteries, J Power Sources. 236(2013) 29.

[42] S. Zaromb, The use and behavior of aluminum anodes in alkaline primary batteries, Journal of the Electrochemical Society. 109(12) (1962) 1125.

[43] D. Gelman, B. Shvartsev, I. Wallwater, S. Kozokaro, V. Fidelsky, A. Sagy . . . Y. Ein-Eli, An aluminum—ionic liquid interface sustaining a durable Al-air battery, Journal of Power Sources. 364(2017) 110–120.

[44] R. Revel, T. Audichon, S. Gonzalez, Non-aqueous aluminium—air battery based on ionic liquid electrolyte, Journal of Power Sources. 272(2014) 415–421.

20 Hybrid-Electrolyte Metal-Air Batteries

Yifei Wang, Xinhai Xu, Mingming Zhang,
Meng Ni, and Dennis Y.C. Leung

CONTENTS

20.1 Introduction 291
20.2 Hybrid-Electrolyte Li-Air Battery 292
     20.2.1 Oxygen Catalyst 294
     20.2.2 Catholyte pH 294
     20.2.3 Separation Membrane 295
     20.2.4 Systematic Study 295
20.3 Hybrid-Electrolyte Na-Air Battery 296
     20.3.1 Oxygen Catalyst 296
     20.3.2 Electrolyte Material 296
20.4 Hybrid-Electrolyte Al-Air, Zn-Air, and Mg-Air Batteries 298
     20.4.1 Hybrid-Electrolyte Al-Air Battery 298
     20.4.2 Hybrid-Electrolyte Zn-Air Battery 300
     20.4.3 Hybrid-Electrolyte Mg-Air Battery 300
20.5 Conclusion and Future Outlook 301
References 302

20.1 INTRODUCTION

Metal-air batteries (MABs) can be classified by the activity of their metal anode. The Li-air and Na-air battery are vulnerable to ambient moisture because Li and Na can react violently with water. As for Al-air, Zn-air, and Mg-air batteries, Al, Zn, and Mg are more stable in the ambient environment. In this manner, water-free organic electrolytes are generally adopted for Li-air and Na-air batteries, while aqueous electrolytes are commonly used for Al-air, Zn-air, and Mg-air batteries. This will lead to different problems for the two types of MAB. For the organic Li-air and Na-air battery, the ionic conductivity is greatly sacrificed, leading to low current density operation. In addition, the reaction products are difficult to dissolve in the organic electrolyte, which will accumulate and clog the pores of the air-breathing cathode. As for the aqueous Al-air, Zn-air, and Mg-air batteries, the limited voltage output is the major problem, which is generally below 1.5 V at a practical current density. To tackle these issues, a single electrolyte may not be sufficient, and the concept of hybrid electrolyte is developed to couple the advantage of different electrolyte species, as shown in Figure 20.1.

    Considering the limitations of organic electrolytes, an organic-aqueous hybrid electrolyte has been proposed for Li-air and Na-air batteries, which uses the same organic electrolyte for the metal anode and an aqueous electrolyte for the air cathode. In the middle, a Li^+/Na^+ conducting membrane such as LISICON and NASICON can separate the two electrolytes from mixing while permitting the transportation of Li and Na ions, respectively. In this manner, the ionic conductivity can be partially improved, and the cathode clogging problem is also alleviated. On the other hand, an acid-alkaline hybrid electrolyte has been proposed for Al-air, Zn-air, and Mg-air batteries (acid-salt

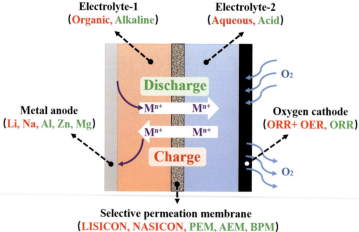

FIGURE 20.1 Schematic diagram of the hybrid-electrolyte MAB.

for Mg-air actually), which employs a low-pH acid electrolyte for the cathode side and high-pH alkaline electrolyte for the anode side. An anion exchange membrane (AEM) is frequently adopted for separation purposes in addition to the cation exchange membrane (CEM) and bipolar membrane (BPM). Such a dual-pH configuration can greatly expand the equilibrium potential difference of the battery (by about 0.83 V), so that a discharge voltage higher than 2 V is feasible. Nevertheless, there are also several unsolved problems associated with the hybrid electrolyte. First of all, the battery system is inevitably complicated, which may induce extra stability problems due to the mixing of different electrolytes (especially in portable applications). Secondly, the electrolyte pH is difficult to maintain, which will gradually deteriorate the discharge performance. For the Li-air and Na-air battery, the catholyte will encounter continuous alkalization, which is detrimental to the separation membrane material. As for the Al-air, Zn-air, and Mg-air batteries, the gradual neutralization of acid and alkaline is inevitable, which will lower the pH difference gradually.

Since its first appearance in 2010, the hybrid-electrolyte MAB is still in its infancy stage and needs more research efforts to improve its performance and durability [1]. In literature, there are already dozens of studies focusing on its material development, structure innovation, and mechanism investigation, which are worth a thorough review. In this chapter, the hybrid-electrolyte Li-air battery is introduced in Section 20.2, and the hybrid-electrolyte Na-air battery is summarized in Section 20.3. As for the hybrid-electrolyte Al-air, Zn-air, and Mg-air batteries, they are reviewed together in Section 20.4 due to the limited research on the latter two cases. Finally, conclusions are given and potential future research directions are also provided.

20.2 HYBRID-ELECTROLYTE LI-AIR BATTERY

Li-air battery, also known as Li-O_2 battery, was first proposed in the 1990s and is regarded as a strong competitor against Li-ion battery in terms of energy density. Due to the elimination of cathodic energy storage materials, the Li-air battery has an impressive theoretical energy density of 3500 Wh/kg, which is an order of magnitude higher than that of a Li-ion battery. On the other hand, the utilization of Li metal anode requires a water-free environment, which is accomplished by using an organic electrolyte. However, this also leads to several problems including limited ionic conductivity, insoluble Li_2O_2 product, and potential explosion/fire hazards. To tackle this, an organic-aqueous hybrid electrolyte configuration was proposed for a Li-air battery in 2010, with a Li-ion conducting membrane in between for separation purposes as shown in Figure 20.2a [2].

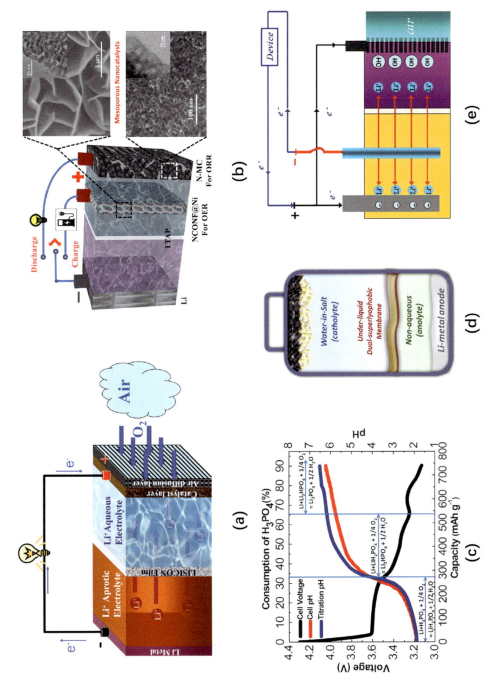

FIGURE 20.2 The organic-aqueous Li-air battery: (a) Basic battery structure. (Adapted with permission [5]. Copyright (2016), American Chemical Society.) (b) Research on its oxygen catalyst. (Adapted with permission [6]. Copyright (2014), The Royal Society of Chemistry.) (c) Research on its catholyte pH. Adapted with permission [7]. Copyright (2012), The Owner Societies. (d) Research on its membrane material. (Adapted with permission [8]. Copyright (2019), Elsevier Inc.) (e) Research on its structure innovation. (Adapted with permission [9]. Copyright (2011), The Royal Society of Chemistry.)

Afterward, more and more related works have been conducted on its oxygen catalyst, catholyte pH, membrane material, and systematic study. Moreover, a few modeling studies were proposed to study the detailed working mechanism [3, 4]. He et al. [5] published a review paper on this specific topic in 2016, while this section will provide a more updated review.

20.2.1 Oxygen Catalyst

The cathode side has the major responsibility for the high charge-discharge overpotential loss of Li-air batteries due to the sluggish oxygen reaction kinetics. To tackle this, some researchers developed efficient yet low-cost non-noble metal catalysts for the battery cathode, including both transition metal oxides and carbonaceous materials. He et al. [10] used titanium nitride (TiN) as the cathode catalyst, which exhibited considerable ORR activity in an acidic electrolyte of saturated lithium acetate in 90% acetic acid. The battery could be discharged at 0.5 mA/cm^2 for more than a week with a voltage plateau of 2.85 V. Wang et al. [11] employed the mixture of graphene oxide (GO) and carbon nanotube (CNT) as a bifunctional oxygen catalyst. The GO with enriched oxygen functional groups possessed high activity towards ORR, while the CNT provided an electronically conductive framework for efficient electron transport, leading to a low charge-discharge voltage gap of 0.17 V at 0.1 mA/cm^2. Yoo et al. [12] investigated the effect of different carbon materials for supporting the Fe phthalocyanine (FePc) catalyst, including graphene nanosheet, CNT, and acetylene black (AB), which all exhibited a 4-electron pathway for ORR. Among them, the FePc/CNT catalyst achieved the highest discharge voltage and good cycling stability. They also tried to use carbon material alone as cathode catalysts, such as the N-doped and P-doped graphene nanosheet [13]. It was found that the N-doped graphene nanosheet obtained higher discharge voltage at all current densities, but the P-doped graphene nanosheet exhibits better cycling stability. Nevertheless, the voltage gap of around 1 V at 0.5 mA/cm^2 was still not as effective as noble-metal catalysts such as Pt-IrO$_2$. Instead of using one bifunctional oxygen electrode, Li et al. [6] employed two independent cathodes, one of NiCo$_2$O$_4$ nanoflake grown on Ni foam for OER and the other of N-doped carbon on carbon paper for ORR as shown in Figure 20.2b. This novel cell could be cycled for 100 times at 0.5 mA/cm^2 (400 hours), and the voltage gap increased only 0.08 V, leading to a high efficiency retention rate of 97.8%. However, the initial voltage gap of 0.92 V was quite high, which was probably due to the complicated cathode compartment. Peng et al. [14] employed graphene as support for the Co$_3$O$_4$ catalyst and optimized their ratio. Among three catalyst samples, the 48.2 wt% Co$_3$O$_4$ catalyst demonstrated the least voltage gap increase during the 120 hrs cycling test, proving its best durability due to the well-balanced catalytic activity and charge transfer pathway.

20.2.2 Catholyte pH

The catholyte pH also has a strong influence on the battery performance. He et al. [15] studied the effect of LiOH concentration and found that both the OCV and discharge voltage decreased with the increase of this parameter. Also, the pH of the catholyte kept increasing due to the generation of hydroxyl ions on the cathode side. To alleviate this issue, 1M LiClO$_4$ was added to the catholyte as a buffer. By using acidic electrolyte instead of alkaline, the cathode potential can be further increased, leading to a higher battery voltage output. Li et al. [16] employed a phosphate buffer solution of 0.1M H$_3$PO$_4$ and 1M LiH$_2$PO$_4$ as catholyte, which achieved an OCV of 3.9 V and a peak power density of 11.9 mW/cm^2. The charge-discharge performance was also stable for 20 cycles at 0.5 mA/cm^2. They also tried 0.1M H$_3$PO$_4$ and 1M Li$_2$SO$_4$ as catholyte to avoid the influence of protons in LiH$_2$PO$_4$ [7], and a typical 3-plateau discharge curve was observed as shown in Figure 20.2c, which was related to the gradual loss of a proton from H$_3$PO$_4$. Good cycling stability was also observed at 0.5 mA/cm^2 for 20 cycles. To further improve the power output, they studied the effect of cathode catalyst, ambient temperature, and membrane conductivity on the hybrid-electrolyte Li-air battery with phosphate catholyte [17]. The peak power density was as high as 30 mW/cm^2 at 40 °C, with the classic Pt-IrO$_2$

as cathode catalyst and a LATP membrane of 2.5×10^{-4} S/cm as a separator. Nevertheless, due to the stability issue of LISICON membrane in a strongly acidic environment, only weak acid with mild pH can be adopted, which strongly limits the discharge capacity. To tackle this, Li et al. [18] also proposed the addition of imidazole into the acidic electrolyte, which could buffer the strong acid into a near neutral condition. With 6M imidazole additive, the 6M HCl could obtain a mild pH of 5, leading to a continuous cell discharge of 110 hrs at 0.5 mA/cm^2.

20.2.3 Separation Membrane

The separation membrane affects both the stability of the Li metal anode and the ohmic resistance of the whole battery, which should also be stable in contact with the aqueous electrolyte. In addition to the classic LISICON membrane such as LATP and LAGP, other novel membrane materials were also proposed. Wang et al. [19] developed a $Li_{6.5}La_3Zr_{1.5}Ta_{0.5}O_{12}$ membrane to separate two organic electrolytes, to alleviate the corrosion of Li anode. This solid electrolyte membrane can not only stop the crossover of oxygen to the anode but also reduce the potential difference between the two organic electrolytes and suppress their decomposition. As a result, three times more cycling durability than the conventional battery was demonstrated. Qiao et al. [8] designed a dual superlyophobic membrane composed of a polyacrylonitrile (PAN) nanofibrous membrane coated with 4-cyan-Ph-terminated thin film composite, which served as a good separator between the organic anolyte and a water-in-salt catholyte as shown in Figure 20.2d. Their battery exhibited a remarkably low voltage gap of 0.47 V at 1 mA/cm^2, and the cycling durability was also improved to 250 cycles. Deng et al. [20] developed a lithiated Nafion membrane to separate the organic anolyte and a dual mediator tetraglyme-based catholyte. A lithiated Si was also used as anode instead of Li metal. Their novel battery obtained a stable performance for 70 charge-discharge cycles at 1 A/g, with the charge voltage below 3 V and discharge voltage above 2 V. Bao et al. [21] improved chemical stability of the conventional LAGP membrane by sputtering an extra layer of Ti on its anode side. A liquid anode of lithium biphenyl dissolved in dimethoxyethane was also used to replace both the Li metal anode and the organic anolyte. It was confirmed that the Ti-coated LAGP achieved almost three times of cycling number than the untreated one, which is because the Ti layer could prevent the side reaction between LAGP and lithium biphenyl.

20.2.4 Systematic Study

The continuous generation of LiOH in the cathode compartment will increase the catholyte pH, which is detrimental to the LISICON membrane. He et al. [22] proposed a catholyte recycling strategy in which the cathode compartment was further divided by a cation exchange membrane. The inner catholyte in contact with the LISICON was kept static while the outer catholyte was recycled and treated to remove the excess LiOH so that the pH of the inner catholyte was kept constant. With this new structure, the battery was able to discharge at 0.5 mA/cm^2 for one week. Wang et al. [9] designed a hybrid lithium-air battery and lithium-carbon capacitor, which had an extra capacitor electrode in the organic electrolyte compartment as shown in Figure 20.2e. In this manner, a high peak power density of 120 mW/cm^2 could be delivered by the capacitor part while a high energy density could be achieved by the Li-air battery part. He et al. [23] studied the effect of catholyte alkalinity and ambient temperature on the performance of hybrid-electrolyte Li-air batteries. When the LiOH electrolyte concentration increased from 0.01 to 2M, the OCV decreased from 3.46 to 3.16 V, and the peak power density kept increasing due to the reduction of inner resistance. The battery performance could also be improved by the elevation of temperature. Yoo et al. [24] studied the influence of CO_2 in ambient air on battery discharge durability. It was found that the graphene nanosheet catalyst could achieve longer discharge than the commercial graphene, which was because the Li_2CO_3 was more likely to form on defects of the graphene nanosheet, leading to a homogenous distribution and better oxygen diffusion.

20.3 HYBRID-ELECTROLYTE NA-AIR BATTERY

In addition to Li, Na is also promising for developing MABs. The theoretical energy density of the Na-air battery is 1683 Wh/kg, which is lower than the Li-air battery due to the higher atomic weight of Na. However, Na is much more abundant in the earth's crust, which makes the Na-air battery a potentially more cost-efficient choice. Similar to a Li-air battery, a Na-air battery also requires an organic electrolyte to protect the metal anode from moisture in the ambient air, so that the electrolyte conductivity, product solubility, and safety issues are of great concern. With an organic-aqueous hybrid electrolyte configuration as shown in Figure 20.3a, these problems can also be alleviated, which have been investigated extensively in recent years since its first appearance in 2015 [25]. This section will summarize relevant research works on the organic-aqueous Na-air battery, especially on the oxygen catalyst and electrolyte material. For more details, readers are recommended to a recent review paper by Xu et al. [26]

20.3.1 Oxygen Catalyst

In literature, there are already extensive research works on the bifunctional oxygen catalyst which can be applied in both fuel cells and MABs. Therefore, only those catalysts employed in Na-air batteries are introduced in this section. Hashimoto et al. [30] prepared the nano-porous Au via annealing and chemical etching for the battery cathode, which achieved an OCV of 2.5 V and a peak power density of 12.4 mW/cm². However, the cycling test could be maintained for only 10 cycles due to inefficient electrodeposition of metallic Na during battery charging. Cheon et al. [31] embedded the graphitic nano-shell in mesoporous carbon via the pyrolysis method, which possessed multiple structural motifs for the oxygen reaction. A peak power density of 78.2 mW/g and a voltage gap of 115 mV were achieved, but the operating current density was not provided. On the other hand, the cycling time of 10 was also quite limited for practical applications. Khan et al. [27] grew a carambola-shaped VO_2 (Figure 20.3b) on reduced GO, which was further coated on carbon paper. When employed in the Na-air battery, such a binder-free cathode generated a peak power density of 104 mW/g, but the areal power density was only 0.26 mW/cm². Nevertheless, the battery was successfully operated for 50 cycles. Liang et al. [32] prepared graphene sheets via liquid exfoliation together with sonication treatment, which achieved a peak power density of 13.8 mW/cm² and a discharge voltage plateau of 2.48 V at 1 mA/cm². Such a good performance was attributed to numerous active edge sites in the graphene sheet. Khan et al. [33] synthesized 3D SnS_2 nano-petal catalysts via a facile solvothermal method, which possessed exposed active sites for the diffusion of air and electrolyte. A voltage gap of 0.52 V was achieved at 5 mA/g, and 40 charge-discharge cycles were demonstrated. Kim et al. [34] employed the single crystalline $Tl_2Rh_2O_7$ nanoparticle as an efficient oxygen catalyst, which exhibited high catalytic activity due to the oxidized Tl and Rh ions during ORR/OER. At a relatively low current density of 0.01 mA/cm², a voltage gap of 0.208 V and a lifetime of 50 cycles were achieved. They also synthesized a single crystalline $Bi_2Ru_2O_7$ catalyst, which showed similar battery performance with the $Tl_2Rh_2O_7$ [35]. Wu et al. [36] employed the metal-organic framework-derived N-doped CNT in the battery cathode, which maintained a discharge voltage plateau of 2.81 V and a cycling lifetime of 35 cycles at 0.1 mA/cm². This good performance was attributed to the increased catalytic active sites provided by the N dopant and confined Co in the N-CNT catalyst.

20.3.2 Electrolyte Material

For the catholyte, 0.1–1M NaOH was frequently employed, while other researchers have explored the feasibility of neutral and acidic solutions as catholyte. Abirami et al. [28] used seawater directly for the cathode side, together with hard carbon as anode and cobalt manganese oxide as a cathode catalyst. The voltage gap of 0.73 V was relatively high at 0.01 mA/cm², but a longer cycling lifetime

Hybrid-Electrolyte Metal-Air Batteries

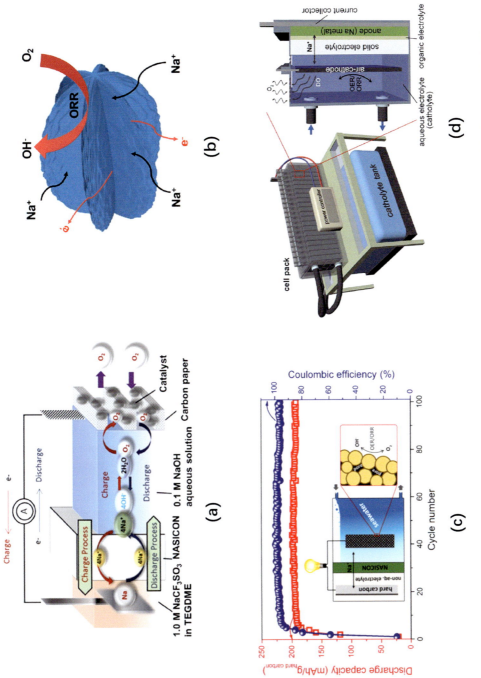

FIGURE 20.3 The organic-aqueous Na-air battery: (a) basic battery structure. (Adapted with permission [25]. Copyright (2015), Elsevier B.V.) (b) Research on its oxygen catalyst. (Adapted with permission [27]. Copyright (2017), The Royal Society of Chemistry.) (c) Research on its catholyte of neutral seawater. (Adapted with permission [28]. Copyright (2016), American Chemical Society.) (d) Research on its catholyte of circulating acid. (Adapted with permission [29]. Copyright (2017), The Royal Society of Chemistry.)

of 100 cycles was successfully demonstrated as shown in Figure 20.3c. Hwang et al. [29] circulated an acidic catholyte of 1M $NaNO_3$ and 0.1M citric acid to the cathode side as shown in Figure 20.3d. With the decreasing pH of catholyte, the battery OCV kept increasing, and the peak power density reached 5.4 mW/cm^2. The voltage gap at 0.1 mA/cm^2 was 0.4 V, and stable cycling for 20 times was also achieved. Kang et al. [37] also employed the acidic catholyte (0.1M H_3PO_4 + 0.1M Na_2SO_4), and a remarkable peak power density of 34.9 mW/cm^2 was reported. In addition, 30 stable charge-discharge cycles were also demonstrated at 0.13 mA/cm^2, with a small voltage gap of 0.3 V and energy efficiency up to 90%. Nevertheless, due to the continuous consumption of H^+ during battery discharge, the catholyte pH will be kept increasing, which is a common issue for acidic catholyte Na-air batteries. As for the anolyte, typically the $NaCF_3SO_3$ in TEGDME or $NaClO_4$ in EC/DMC were adopted. Kang et al. [38] employed a novel anolyte of $Na[FSA-C_2C_1im][FSA]$ ionic liquid, which achieved a high power density of 13.6 mW/cm^2. They also studied the effect of operating temperature, and the peak power density was further improved to 27.6 mW/cm^2 at 50 °C. Nevertheless, only 5 cycles were maintained at 1 mA/cm^2 at this high temperature.

20.4 HYBRID-ELECTROLYTE AL-AIR, ZN-AIR, AND MG-AIR BATTERIES

Unlike Li and Na which are very unstable in ambient air and aqueous electrolyte, other metals such as Al, Zn, and Mg can form a metal oxide protective layer on their surface, which can be used as anode in aqueous MABs with mild corrosion only. In this manner, the organic anolyte is less requisite, leading to improved ionic conductivity and cost-efficiency. On the other hand, the equilibrium potential of Al, Zn, and Mg are only −2.3, −1.2, and −2.7 V vs SHE, generating lower OCVs than the Li-air and Na-air batteries. To tackle this issue, a dual-pH hybrid electrolyte configuration was proposed in the literature, with the acid-alkaline electrolyte frequently adopted for Al-air and Zn-air batteries, and the acid-salt electrolyte for Mg-air battery. The following sections will introduce them separately.

20.4.1 Hybrid-Electrolyte Al-Air Battery

Al can be corroded by strong alkaline electrolytes, leading to reduced energy efficiency. To alleviate this issue, the organic-aqueous hybrid electrolyte was still effective, but the materials used in the literature are quite different from Li-air and Na-air batteries. As shown in Figure 20.4a, Wang et al. [39] employed KOH dissolved in methanol as an anolyte, which was separated from the 3M KOH catholyte by an AEM. The peak power density was only 33% of the all-aqueous Al-air battery, but the Al discharge capacity was improved by 30–50 times due to the suppression of Al corrosion in methanol. Chen et al. [40] also used methanol as the organic solvent for the anolyte, but the separation of two electrolytes was achieved by a microfluidic technology instead of any physical membrane as shown in Figure 20.4b, leading to lower battery cost. The microfluidic electrolytes had a very low Reynolds number, so they will not mix even in direct contact. A high Al discharge capacity of 2507 mAh/g was reported, which was 84.1% of the theoretical value. Teabnamang et al. [41] further induced a recycling process for the methanol-based anolyte, and the aqueous catholyte was gelled into polymer form. An OCV of 1.44 V and a peak power density of 7.5 mW/cm^2 were reported, and the Al discharge capacity could reach 2328 mAh/g. However, methanol solvent is volatile and toxic, which may not be a suitable choice in practical applications.

The corrosion of Al can be alleviated by using inhibition additives in the aqueous electrolyte so that the organic anolyte is not a requisite. Instead, a dual-pH all-aqueous Al-air battery can be developed to enhance the battery performance. Chen et al. [42] developed an acid-alkaline dual-electrolyte Al-air battery with microfluidic technology. The OCV was 2.2 V and the peak power density was 176 mW/cm^2, which were 37.5% and 104.6% higher than the single-alkaline battery. Feng et al. [43] investigated the feasibility of electrolyte recycling by

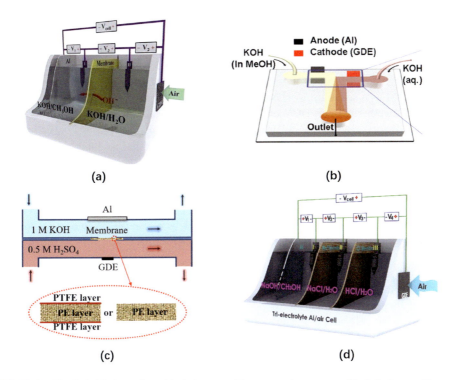

FIGURE 20.4 The hybrid-electrolyte Al-air battery: (a) an organic-aqueous Al-air battery with membrane separator. (Adapted with permission [39]. Copyright (2014), The Royal Society of Chemistry.) (b) An organic-aqueous Al-air battery without membrane separator. (Adapted with permission [40]. Copyright (2016), Elsevier B.V.) (c) An acid-alkaline Al-air battery with a composite membrane separator. Adapted with permission [43]. Copyright (2020), Elsevier B.V.) (d) An acid-alkaline Al-air battery with tri-electrolyte. (Adapted with permission [46]. Copyright (2020), Elsevier Ltd.)

adding an extra composite membrane between the microfluidic acid and alkaline electrolytes (Figure 20.4c). The membrane had a PTFE/PE/PTFE sandwich structure, which could further slow down the neutralization. In this manner, a stable OCV of 2.18 V and a peak power density of 300–500 mW/cm^2 were observed, which could be maintained for 10 times of electrolyte recycling. Wen et al. [44] studied two different membranes for separating the acid and alkaline. When BPM was used, the OCV was 2.53 V and the peak power density was 88 mW/cm^2. When AEM was used instead, the OCV increased a little to 2.56 V, but the peak power density increased evidently to 114.6 mW/cm^2. This was mainly due to the much lower resistance of AEM (4.76 Ω cm^2) than that of BPM (10.97 Ω cm^2). Furthermore, Wang et al. [45] developed an all-solid-state Al-air battery with an acid-alkaline electrolyte. This was accomplished by gelling the acid and alkaline solution into two polymer electrolytes, which can contact each other without violent neutralization. In this manner, a discharge voltage plateau of 2.4–1.8 V at 1 mA/cm^2 could be maintained for 4.7 hours.

To combine the advantage of both organic-aqueous and acid-alkaline hybrid electrolytes, Wang et al. [46] further proposed a tri-electrolyte aluminum-air battery as shown in Figure 20.4d. On the anode side, a NaOH/methanol solution was used as anolyte; on the cathode side, an HCl aqueous solution was used as catholyte; and in the middle, a NaCl aqueous solution was used as a buffer. A Na$^+$ exchange membrane and Cl$^-$ exchange membrane were also used as separators, respectively. Such a system obtained an OCV of 2.2 V and a peak power density of around 60 mW/cm^2, but the advantage on Al discharge capacity was not revealed.

20.4.2 Hybrid-Electrolyte Zn-Air Battery

Zn has been used as a battery anode for more than 100 years due to its fine balance between activity and stability. Therefore, the hybrid-electrolyte Zn-air battery study focuses on acid-alkaline rather than an organic-aqueous configuration. Cai et al. [47] developed an early dual-pH Zn-air battery using BPM as a separator as shown in Figure 20.5a. The peak power density was as high as 380 mW/cm² together with an OCV of 2.25 V when acid was used as catholyte (Figure 20.5b), and the energy density reached 1.52 kWh/kg. Zhao et al. [48] prepared an acid-alkaline hydrogel electrolyte for a flexible Zn-air battery as shown in Figure 20.5c. The Pluronic® F127 powders were added into solutions to form the acid and alkaline hydrogel, which were then combined and assembled in the battery. The OCV reached 2 V but the peak power density was only 4.2 mW/cm², which was probably due to the high ionic resistance of the hydrogel (Figure 20.5d). Wang et al. [45] also used acid and alkaline hydrogels to form the dual-pH electrolyte, with polyvinyl alcohol for the catholyte gel and sodium polyacrylate for the anolyte gel. A higher electrolyte concentration was also adopted, leading to an improved power output of 80 mW/cm².

20.4.3 Hybrid-Electrolyte Mg-Air Battery

Mg is stable in alkaline solution due to the formation of $Mg(OH)_2$ protective layer. Therefore, a salt electrolyte was commonly adopted instead, and the hybrid electrolyte should be the acid-salt combination to improve the OCV and power output. Relevant research is quite scarce at the moment. Leong et al. [49] first studied an acid-salt Mg-air battery using $NaNO_3$ as an anolyte and H_2SO_4

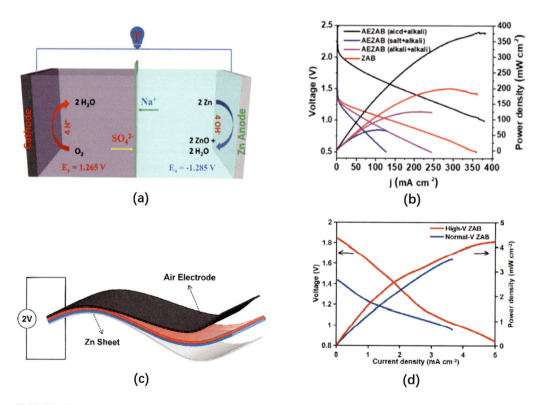

FIGURE 20.5 The acid-alkaline Zn-air battery: (a-b) an aqueous-electrolyte Zn-air battery with membrane separator and its performance with different types of catholyte. (Adapted with permission [47]. Copyright (2018), Wiley-VCH.) (c-d) A gel-electrolyte Zn-air battery without membrane separator and its performance compared with a single alkaline gel battery. (Adapted with permission [48]. Copyright (2021), Elsevier B.V.)

Hybrid-Electrolyte Metal-Air Batteries

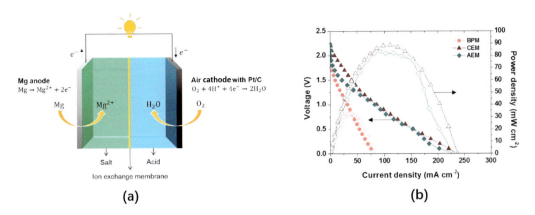

FIGURE 20.6 The acid-salt Mg-air battery: (a) cell structure and electrode reactions; (b) Comparison of different membrane separators. (Adapted with permission [49]. Copyright (2021), Elsevier.)

as catholyte (Figure 20.6a). They also investigated the effect of different separation membranes including AEM, CEM, and BPM. It was found that the acidic catholyte could not only increase the voltage and power output but also slow down the passivation of the Mg anode by neutralizing the generated OH- in the anolyte. With CEM as a separator, the battery achieved an OCV of 2.23 V, the highest peak power density of 89 mW/cm^2, and a coulombic efficiency of 84% at 50 mA/cm^2 (Figure 20.6b), proving the effectiveness of this hybrid system. Wang et al. [45] further solidify the acid and salt electrolytes to develop an all-solid-state dual-pH Mg-air battery. A lower power density of 25 mW/cm^2 was obtained due to the sacrificed ionic conductivity in exchange for higher safety in portable applications.

20.5 CONCLUSION AND FUTURE OUTLOOK

The hybrid-electrolyte MAB has already been studied for more than a decade with fruitful outcomes in the literature. Two types of them are frequently adopted, including the organic-aqueous electrolyte and the acid-alkaline electrolyte, while a membrane in between is generally needed for separation purposes. The former case is suitable for Li-air, Na-air, and sometimes Al-air batteries, whose anode was relatively unstable in an aqueous environment. The latter case utilizes the pH difference at different electrodes to expand the theoretical battery OCV so that a higher voltage and power output can be obtained. Till now, the organic-aqueous Li-air and Na-air batteries still suffer from limited current density and cycling lifetime, which should be significantly improved before any practical applications. As for the acid-alkaline Al-air, Zn-air battery, and acid-salt Mg-air battery, high power density is already achievable, and the key issue lies in the stability of the dual-pH environment, which requires an advanced membrane material to prevent the crossover of protons and hydroxyl ions effectively.

Despite a decade's R&D, the hybrid-electrolyte MAB technology is still in the infancy stage. In the future, more research efforts should be paid to its material improvement, structure innovation, and mechanism study. For the organic-aqueous Li-air and Na-air battery, a more conductive organic electrolyte and solid-state membrane should be developed to improve the battery ohmic resistance, while an effective control of the catholyte pH should be achieved to improve its high-voltage discharge ability. In addition, the anodic dendrite formation and cathodic overpotential loss are also well-known obstacles to its cycling durability and energy efficiency, which should be improved via electrode material innovation as well. As for the acid-alkaline Al-air, Zn-air, and Mg-air battery, the neutralization of electrolytes should be slowed down as long as possible to improve the discharge lifetime, which can be either achieved by advanced separation membrane or accomplished by an electrolyte recycling system. On the other hand, the regeneration of acid and alkaline via battery recharge may also be worth investigating, which is especially the case for secondary Zn-air batteries. In the future, more research breakthroughs can be anticipated in these two hybrid electrolyte configurations, while novel hybrid electrolyte designs may also worth a try, such as the dual-concentration and dual-temperature electrolytes.

REFERENCES

[1] H. Zhou, Y. Wang, H. Li, P. He, The development of a new type of rechargeable batteries based on hybrid electrolytes, ChemSusChem. 3 (2010) 1009–1019.
[2] Y. Wang, H. Zhou, A lithium-air battery with a potential to continuously reduce O2 from air for delivering energy, Journal of Power Sources. 195 (2010) 358–361.
[3] P. Andrei, J.P. Zheng, M. Hendrickson, E.J. Plichta, Modeling of Li-air batteries with dual electrolyte, Journal of The Electrochemical Society. 159 (2012) A770.
[4] T. Zhang, M. Yu, J. Li, Q. Li, X. Zhang, H. Sun, Effect of porosity gradient on mass transfer and discharge of hybrid electrolyte lithium-air batteries, Journal of Energy Storage. 46 (2022) 103808.
[5] P. He, T. Zhang, J. Jiang, H. Zhou, Lithium—air batteries with hybrid electrolytes, The Journal of Physical Chemistry Letters. 7 (2016) 1267–1280.
[6] L. Li, S.-H. Chai, S. Dai, A. Manthiram, Advanced hybrid Li-air batteries with high-performance mesoporous nanocatalysts, Energy & Environmental Science. 7 (2014) 2630–2636.
[7] L. Li, X. Zhao, Y. Fu, A. Manthiram, Polyprotic acid catholyte for high capacity dual-electrolyte Li-air batteries, Physical Chemistry Chemical Physics. 14 (2012) 12737–12740.
[8] Y. Qiao, Q. Wang, X. Mu, H. Deng, P. He, J. Yu, H. Zhou, Advanced hybrid electrolyte Li-O2 battery realized by dual superlyophobic membrane, Joule. 3 (2019) 2986–3001.
[9] Y. Wang, P. He, H. Zhou, A lithium-air capacitor-battery based on a hybrid electrolyte, Energy & Environmental Science. 4 (2011) 4994–4999.
[10] P. He, Y. Wang, H. Zhou, Titanium nitride catalyst cathode in a Li-air fuel cell with an acidic aqueous solution, Chemical Communications. 47 (2011) 10701–10703.
[11] S. Wang, S. Dong, J. Wang, L. Zhang, P. Han, C. Zhang, X. Wang, K. Zhang, Z. Lan, G. Cui, Oxygen-enriched carbon material for catalyzing oxygen reduction towards hybrid electrolyte Li-air battery, Journal of Materials Chemistry. 22 (2012) 21051–21056.
[12] E. Yoo, H. Zhou, Fe phthalocyanine supported by graphene nanosheet as catalyst in Li-air battery with the hybrid electrolyte, Journal of Power Sources. 244 (2013) 429–434.
[13] E. Yoo, H. Zhou, Hybrid electrolyte Li-air rechargeable batteries based on nitrogen- and phosphorus-doped graphene nanosheets, Rsc Advances. 4 (2014) 13119–13122.
[14] S.-H. Peng, T.-H. Chen, C.-H. Lee, H.-C. Lu, S.J. Lue, Optimal cobalt oxide (Co3O4): Graphene (GR) ratio in Co3O4/GR as air cathode catalyst for air-breathing hybrid electrolyte lithium-air battery, Journal of Power Sources. 471 (2020) 228373.
[15] H. He, W. Niu, N.M. Asl, J. Salim, R. Chen, Y. Kim, Effects of aqueous electrolytes on the voltage behaviors of rechargeable Li-air batteries, Electrochimica Acta. 67 (2012) 87–94.
[16] L. Li, X. Zhao, A. Manthiram, A dual-electrolyte rechargeable Li-air battery with phosphate buffer catholyte, Electrochemistry Communications. 14 (2012) 78–81.
[17] L. Li, A. Manthiram, Dual-electrolyte lithium—air batteries: Influence of catalyst, temperature, and solid-electrolyte conductivity on the efficiency and power density. Journal of Materials Chemistry A. 1 (2013) 5121–5127.
[18] L. Li, Y. Fu, A. Manthiram, Imidazole-buffered acidic catholytes for hybrid Li-air batteries with high practical energy density, Electrochemistry Communications. 47 (2014) 67–70.
[19] J. Wang, Y. Yin, T. Liu, X. Yang, Z. Chang, X. Zhang, Hybrid electrolyte with robust garnet-ceramic electrolyte for lithium anode protection in lithium-oxygen batteries, Nano Research. 11 (2018) 3434–3441.
[20] H. Deng, Y. Qiao, S. Wu, F. Qiu, N. Zhang, P. He, H. Zhou, Nonaqueous, metal-free, and hybrid electrolyte Li-Ion O2 battery with a single-ion-conducting separator, ACS Applied Materials & Interfaces. 11 (2019) 4908–4914.
[21] J. Bao, C. Li, F. Zhang, P. Wang, X. Zhang, P. He, H. Zhou, A liquid anode of lithium biphenyl for highly safe lithium-air battery with hybrid electrolyte, Batteries & Supercaps. 3 (2020) 708–712.

[22] P. He, Y. Wang, H. Zhou, A Li-air fuel cell with recycle aqueous electrolyte for improved stability, Electrochemistry Communications. 12 (2010) 1686–1689.

[23] P. He, Y. Wang, H. Zhou, The effect of alkalinity and temperature on the performance of lithium-air fuel cell with hybrid electrolytes, Journal of Power Sources. 196 (2011) 5611–5616.

[24] E. Yoo, H. Zhou, Influence of CO_2 on the stability of discharge performance for Li-air batteries with a hybrid electrolyte based on graphene nanosheets, RSC Advances. 4 (2014) 11798–11801.

[25] S.H. Sahgong, S. Senthilkumar, K. Kim, S.M. Hwang, Y. Kim, Rechargeable aqueous Na—air batteries: Highly improved voltage efficiency by use of catalysts, Electrochemistry Communications. 61 (2015) 53–56.

[26] X. Xu, K. San Hui, D.A. Dinh, K.N. Hui, H. Wang. Recent advances in hybrid sodium—air batteries, Materials Horizons. 6 (2019) 1306–1335.

[27] Z. Khan, B. Senthilkumar, S.O. Park, S. Park, J. Yang, J.H. Lee, H.-K. Song, Y. Kim, S.K. Kwak, H. Ko. Carambola-shaped VO_2 nanostructures: A binder-free air electrode for an aqueous Na-air battery, Journal of Materials Chemistry A. 5 (2017) 2037–2044.

[28] M. Abirami, S.M. Hwang, J. Yang, S.T. Senthilkumar, J. Kim, W.-S. Go, B. Senthilkumar, H.-K. Song, Y. Kim, A metal—organic framework derived porous cobalt manganese oxide bifunctional electrocatalyst for hybrid Na-Air/Seawater batteries, ACS Applied Materials & Interfaces. 8 (2016) 32778–32787.

[29] S.M. Hwang, W. Go, H. Yu, Y. Kim. Hybrid Na-air flow batteries using an acidic catholyte: Effect of the catholyte pH on the cell performance, Journal of Materials Chemistry A. 5 (2017) 11592–11600.

[30] T. Hashimoto, K. Hayashi, Aqueous and nonaqueous sodium-air cells with nanoporous gold cathode, Electrochimica Acta. 182 (2015) 809–814.

[31] J.Y. Cheon, K. Kim, Y.J. Sa, S.H. Sahgong, Y. Hong, J. Woo, S.D. Yim, H.Y. Jeong, Y. Kim, S.H. Joo, Graphitic nanoshell/mesoporous carbon nanohybrids as highly efficient and stable bifunctional oxygen electrocatalysts for rechargeable aqueous Na-air batteries, Advanced Energy Materials. 6 (2016) 1501794.

[32] F. Liang, T. Watanabe, K. Hayashi, Y. Yao, W. Ma, B. Yang, Y. Dai, Liquid exfoliation graphene sheets as catalysts for hybrid sodium-air cells, Materials Letters. 187 (2017) 32–35.

[33] Z. Khan, N. Parveen, S.A. Ansari, S. Senthilkumar, S. Park, Y. Kim, M.H. Cho, H. Ko, Three-dimensional SnS_2 nanopetals for hybrid sodium-air batteries, Electrochimica Acta. 257 (2017) 328–334.

[34] M. Kim, H. Ju, J. Kim, Single crystalline thallium rhodium oxide pyrochlore for highly improved round trip efficiency of hybrid Na-air batteries, Dalton Transactions. 47 (2018) 15217–15225.

[35] M. Kim, H. Ju, J. Kim, Single crystalline $Bi_2Ru_2O_7$ pyrochlore oxide nanoparticles as efficient bifunctional oxygen electrocatalyst for hybrid Na-air batteries, Chemical Engineering Journal. 358 (2019) 11–19.

[36] Y. Wu, X. Qiu, F. Liang, Q. Zhang, A. Koo, Y. Dai, Y. Lei, X. Sun, A metal-organic framework-derived bifunctional catalyst for hybrid sodium-air batteries, Applied Catalysis B: Environmental. 241 (2019) 407–414.

[37] Y. Kang, F. Su, Q. Zhang, F. Liang, K.R. Adair, K. Chen, D. Xue, K. Hayashi, S.C. Cao, H. Yadegari, Novel high-energy-density rechargeable hybrid sodium-air cell with acidic electrolyte, ACS Applied Materials & Interfaces. 10 (2018) 23748–23756.

[38] Y. Kang, F. Liang, K. Hayashi, Hybrid sodium-air cell with Na [FSA—C2C1im][FSA] ionic liquid electrolyte, Electrochimica Acta. 218 (2016) 119–124.

[39] L. Wang, F. Liu, W. Wang, G. Yang, D. Zheng, Z. Wu, M.K. Leung, A high-capacity dual-electrolyte aluminum/air electrochemical cell, Rsc Advances. 4 (2014) 30857–30863.

[40] B. Chen, D.Y. Leung, J. Xuan, H. Wang, A high specific capacity membraneless aluminum-air cell operated with an inorganic/organic hybrid electrolyte, Journal of Power Sources. 336 (2016) 19–26.

[41] P. Teabnamang, W. Kao-ian, M.T. Nguyen, T. Yonezawa, R. Cheacharoen, S. Kheawhom, High-capacity dual-electrolyte aluminum-air battery with circulating methanol anolyte, Energies. 13 (2020) 2275.

[42] B. Chen, D.Y. Leung, J. Xuan, H. Wang, A mixed-pH dual-electrolyte microfluidic aluminum—air cell with high performance, Applied Energy. 185 (2017) 1303–1308.

[43] S. Feng, G. Yang, D. Zheng, L. Wang, W. Wang, Z. Wu, F. Liu, A dual-electrolyte aluminum/air microfluidic cell with enhanced voltage, power density and electrolyte utilization via a novel composite membrane, Journal of Power Sources. 478 (2020) 228960.

[44] H. Wen, Z. Liu, J. Qiao, R. Chen, G. Qiao, J. Yang, Ultrahigh voltage and energy density aluminum-air battery based on aqueous alkaline-acid hybrid electrolyte, International Journal of Energy Research. 44 (2020) 10652–10661.

[45] Y. Wang, W. Pan, S. Luo, X. Zhao, H.Y. Kwok, X. Xu, D.Y. Leung. High-performance solid-state metal-air batteries with an innovative dual-gel electrolyte, International Journal of Hydrogen Energy. 47 (2022) 15024–15034.

[46] L. Wang, R. Cheng, C. Liu, M.C. Ma, W. Wang, G. Yang, M.K. Leung, F. Liu, S. Feng, Tri-electrolyte aluminum/air cell with high stability and voltage beyond 2.2 V, Materials Today Physics. (2020) 100242.

[47] P. Cai, Y. Li, J. Chen, J. Jia, G. Wang, Z. Wen. An asymmetric-electrolyte Zn-air battery with ultrahigh power density and energy density, ChemElectroChem. 5 (2018) 589–592.

[48] S. Zhao, T. Liu, Y. Dai, Y. Wang, Z. Guo, S. Zhai, J. Yu, C. Zhi, M. Ni, All-in-one and bipolar-membrane-free acid-alkaline hydrogel electrolytes for flexible high-voltage Zn-air batteries, Chemical Engineering Journal. 430 (2022) 132718.

[49] K.W. Leong, Y. Wang, W. Pan, S. Luo, X. Zhao, D.Y. Leung, Doubling the power output of a Mg-air battery with an acid-salt dual-electrolyte configuration, Journal of Power Sources. 506 (2021) 230144.

21 Polymer Electrolytes

Changlin Liu, Shasha Li, Abuliti Abudula, and Guoqing Guan

CONTENTS

- 21.1 Introduction 305
- 21.2 Ion Transfer Mechanism 307
- 21.3 Categories of SPEs 308
  - 21.3.1 Dual-Ion Conducting SPEs 309
  - 21.3.2 Single-Ion Conducting SPEs 312
- 21.4 Compatibility of SPEs 312
  - 21.4.1 Compatibility with Anode 312
  - 21.4.2 Compatibility with Cathode 313
- 21.5 Preparation Methods of SPEs 313
- 21.6 Conclusions 314
- Acknowledgments 314
- References 314

21.1 INTRODUCTION

Metal-air batteries are promising devices for energy storage because of their ultra-high theoretical energy density. Among all the popular metals (e.g., Li, Na, K, Mg, Al, and Zn), Li-air batteries possess high theoretical energy density (~3500 Wh/kg), which is comparable to gasoline [1]. However, severe capacity fading, poor rate performance, and safety problems still hinder the application of Li-air batteries [2]. The schematic of Li-air batteries is shown in Figure 21.1 [3], in which the cathode reaction is the reduction of oxygen obtained from the open air and the anode reaction is the oxidation of metal during the discharge process. The reactions on the two sides strictly demand high-performance electrolytes in many aspects.

Electrolytes of metal-air batteries can be mainly divided into four types, (I) aprotic electrolyte, (II) aqueous electrolyte, (III) solid-state electrolytes (SSEs), and (IV) mixed aqueous/aprotic electrolyte. In particular, Li-air batteries based on the mixed electrolyte have the architecture with an aqueous electrolyte immersing the cathode and an aprotic electrolyte immersing the anode. The aprotic one-based Li-air battery was firstly investigated by Abraham and Jiang [4], in which the Li metal contacts with the electrolyte directly to form a stable solid-electrolyte interface (SEI) layer. At the cathode side, insoluble Li_2O_2 or Li_2O is thought to be formed via oxygen reduction reaction (ORR) and the Li_2O_2 always undergoes the oxygen evolution reaction (OER) at high recharge voltages [5], and aggregation of insoluble matters could hinder the electrochemical reactions at the cathode. The Li-air batteries with an aqueous electrolyte developed by Polyplus [6] or the mixed electrolytes developed by Polyplus [6] and Wang and Zhou [7] have alleviated this problem to some extent because the discharge reaction product is soluble in water (H_2O) [8]. However, protecting the anode from reacting with H_2O is still a challenge. A suitable electrolyte must have sufficient ionic conductivity, low volatility, low flammability, high oxygen solubility, and diffusivity as well as high resistivity to oxygen reduction species and the superoxide anion radical [9].

DOI: 10.1201/9781003295761-21

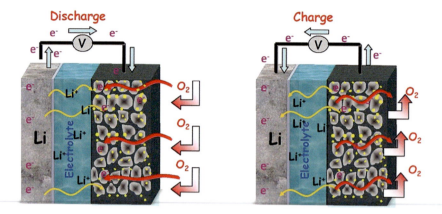

FIGURE 21.1 The schematic operation proposed for the rechargeable aprotic Li-air batteries. (Adapted with permission [3]. Copyright (2010), American Chemical Society.)

SPEs composed of polymers and Li salts are not prone to volatilization or leakage compared with liquid electrolytes. In addition, by protecting the Li metal from contacting with air using SPEs, reactions between lithium metal and moisture, carbon dioxide (CO_2), and reactive oxygen (O_2) species can also be suppressed. Thus, polymers such as poly (ethylene oxide) (PEO), polyacrylonitrile (PAN), poly(vinylidene fluoride) (PVDF), PVDF-HFP (HFP = hexafluoropropylene), poly(methyl methacrylate) (PMMA), poly(tetrafluoroethylene) (PTFE), poly(vinyl alcohol) (PVA), and polydimethylsiloxane (PDMS) are widely used in Li-air batteries. The polymer as an ion conductive matrix in SPEs has been investigated by P.V. Wright [10] in the early 1970s. But the technological interest in polymer electrolytes increased until M. Armand et al. [11] proposed them as a new class of electrolytes in solid-state batteries (SSBs). Although numerous polymers have been adopted for Li-air batteries, PEO hosting a Li salt, LiX, e.g., Lithium bis(trifluoromethanesulfonyl)imide (LiTFSI), Lithium bisfluorosulfonimide (LiFSI), Lithium trifluoromethanesulfonate ($LiCF_3SO_3$), Lithium perchlorate ($LiClO_4$), Lithium tetrafluoroborate ($LiBF_4$), and Lithium hexafluoroarsenate ($LiAsF_6$), is the most studied one. Shao-Horn et al. [12] investigated the stability of PEO and revealed that it is prone to auto-oxidation in O_2. Thus, the stability of PEO-based electrolytes in Li-air batteries should be further improved. In addition, Scrosati et al. [13] reported a PEO-$LiCF_3SO_3$-based Li-air battery, realizing a small ORR and OER overvoltage of 400 mV without using a catalyst. However, even though PEO has a superior ability to dissolve Li salts, the ionic conductivity of PEO-based SPEs is still less than 10^{-5} S/cm at room temperature (RT) due to the high crystallinity, which constraints the segmental motion of the polymer. Fortunately, the ionic conductivity of SPEs can be improved by doping inorganic fillers such as passive fillers (ZrO_2, Al_2O_3, SiO_2, TiO_2) and solid inorganic electrolytes (LLTO, LLZO, LATP, LLGP). Composite polymer electrolytes (CPEs) always display the advantages of the respective components and compensate for those drawbacks.

Nanometer-sized ceramic fillers are thought to not only act as the solid plasticizer, inhibiting the crystallization of PEO, but also promoting specific interactions between the surface groups within PEO and electrolyte ionic species [14]. It is reported that the conductivity of PEO-based CPEs with TiO_2 and Al_2O_3 passive fillers can be promoted by an order of magnitude to about 10^{-5} S cm^{-1} at 30 °C [15]. In addition, solid inorganic electrolyte fillers with inherent high ionic conductivity are considered to have greater potential in enhancing the ionic conductivity of CPEs. Li et al. [16] realized a high ionic conductivity of 7.2×10^{-5} S/cm at 30 °C by dispersing $Li_{6.25}Ga_{0.25}La_3Zr_2O_{12}$ (Ga-LLZO) nanoparticles in the PEO matrix. It is reported that using nanofibers and three-dimensional (3D) garnet network fillers can result in a higher performance of CPEs [17], [18]. What's more, blending, copolymerization, and crosslinking of polymers are all effective solutions to improve the ionic conductivity of SPEs or the performance of SSBs [19], [20], [21]. In addition, all the aforementioned methods for improving ionic conductivity are also effective for Li dendrite suppression due to the concomitant increase of SPE modulus, which is supported by the shear modulus theory proposed by Newman and Monroe [22].

Generally, SPEs are dual-ion conductors, and Li-ion and its counter anions can both move in the SPEs, causing a Li ionic transfer number (t_{Li+}) usually lower than 0.5 [23]. However, single Li-ion conductive SPEs as a strategy to improve the properties of electrolytes has attracted great attention for avoiding anion polarization deleterious effects [24].

This chapter discusses the research progress of SPEs for metal-air batteries in terms of classification of SPEs, ion transfer mechanism in SPEs, preparation method of SPEs, and compatibility of anode and cathode. Then, an outlook on the research and development of SPEs based on how to suppress Li dendrite growth, improve the ionic conductivity, and long-term stability will be given.

21.2 ION TRANSFER MECHANISM

For the SSEs, considerable ionic conductivity is a prerequisite for their application in battery devices. Generally, the SPEs are composed of inorganic salts and polymer matrix usually containing functional groups such as -O-, -S-, -N-, -P-, -C = O, and -NH$_2$. Those functional groups can promote the dissociation of salts into ions via electrostatic interaction. Figure 21.2 [25] illustrates the ion transfer mechanism in amorphous regions of the polymer matrix. The motion of ions in the polymer matrix can be described as follows: hopping between coordinating sites is one of the ways of ion transportation. In addition, ions that interact with polymer functional groups can also transport by coupling with the segments of polymer chains. Therefore, it is believed that the amorphous region in the polymer matrix mainly contributes to the transport of ions due to the mobility of polymer segments in amorphous regions. Thus, the ionic conductivity of SPEs is strongly dependent on the crystallinity of the polymer.

The ionic conduction properties of SPEs can be studied by electrochemical impedance spectroscopy (EIS) with a small potential (5~10 mV) to ensure a linear current-voltage relationship, and the ionic conductivity σ is calculated by equation (1):

$$\sigma = \frac{L}{RA} \tag{1}$$

where R (Ω) represents the obtained resistance measured by the EIS technique, L (cm) is the thickness of the electrolyte, and A (cm^2) stands for the electrolyte effective area.

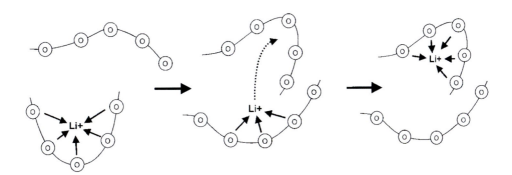

FIGURE 21.2 Schematic illustration of Li-ion-conduction mechanisms in the amorphous phase of SPE. (Adapted with permission [25]. Copyright (1998), John Wiley and Sons.)

Generally, with the decrease in glass transition temperature (T_g), the flexibility of polymer chains can be enhanced. Hence, the selection of polymer matrix is crucial in the development of SPEs. The complex structure-property correlation of polymer matrix determines the complex ion-conducting mechanism. The Vogel-Tammann-Fulcher (VTF) equation derived by quasi-thermodynamics with free volume and configurational entropy is suitable for SPEs [26]. The relationship between ionic conductivity and temperature is described by equation (2):

$$\sigma(T) = \sigma_0 T^{\frac{1}{2}} exp\left(-\frac{B}{T-T_0}\right) \qquad (2)$$

where B (EA/k) is the pseudo-activation energy, T_0 is the reference temperature (10~50 K below T_g), σ_0 is the pre-exponential factor.

However, the ion transfer mechanism in CPEs is more complex when compared with those in the SPEs composed of polymer matrix and salts. As aforementioned, the CPEs are divided into two categories based on the type of fillers. The ion-conducting mechanism of CPEs with passive fillers (SiO_2, Al_2O_3, and ZrO_2) is the same as that of SPEs because the filler can insult Li ions. In this type of CPEs, crystallinity, and T_g of polymer are decreased due to the modification of polymer structure. Moreover, the functional groups of passive fillers can enhance the dissociation of salts based on the Lewis acid-base theory and further increase the content of free ions. Although the ionic conductivity of SPEs can be improved greatly with the addition of passive fillers, the ion conduction of CPEs is still caused by the mobile segments of the polymer matrix. For another category of CPEs, active fillers (LLZO, LLTO, and LATP) are adopted. In this case, the ion transfer mechanism becomes extremely complex due to the combination of both organic and inorganic conductors. In addition to the same function as passive fillers, an additional Li-ion pathway is thought to be constructed to contribute to the improvement of ionic conductivity. As shown in Figure 21.3a [16], a surface layer (~3 nm) is observed by the transmission electron microscope (TEM), which can serve as the space charge region. Herein, the formation of the space charge region can be simulated with the phase-field method based on the Poisson-Cahn equations [27]. The schematic illustration of Ga-LLZO is shown in Figure 21.3b [16], in which the interface between the active particles and polymer matrix is the space charge region as the Li-ion pathway.

21.3 CATEGORIES OF SPEs

SPEs with alkali-metal salts dissociated in the polymer matrix were first discovered by P.V. Wright in 1973 [10]. Then, in 1979, Armand and co-workers [28] envisioned the potential of SPEs through the

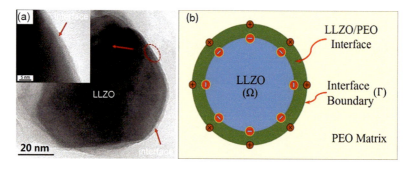

FIGURE 21.3 Space charge region at the Ga-LLZO/PEO interface. (a) TEM images of the Ga-LLZO/PEO interface. (b) Schematic illustration of Ga-LLZO nanoparticle in the PEO:Ga-LLZO composite. The domain of the Ga-LLZO nanoparticle Ω is surrounded by the Ga-LLZO/PEO interface Γ. (Adapted with permission [16]. Copyright (2019), American Chemical Society.)

investigation of PEO-based electrolytes. Since then, research on SPEs for Li batteries has improved tremendously. While Li-air batteries based on SPEs are first investigated in 1996 by Abraham and co-workers [4]. Since that time, SPEs are thought to be safe due to the absence of liquid solvents and flexible with various shapes, and their light weight can also improve the energy density of the battery. However, for the practical application of Li-air batteries in electrochemical devices, the SPEs must fulfill some other prerequisites [29]:

1) Ionic conductivity higher than 10^{-4} S/cm.
2) Chemical and electrochemical stability with the Li anode or cathode material during cycling.
3) Thermal and mechanical stability during charge and discharge processes.
4) Low cost, simple fabrication process, and environmentally friendly.

SPEs can be simply divided into dual-ion conducting electrolytes and single-ion conducting electrolytes according to the type of ions that transfer inside the polymer matrix. Different architectures of SPEs normally lead to differences in interfacial compatibility, SEI composition, Li dendrite suppression, and ionic conductivity. Therefore, an in-depth understanding of the composition and structure of SPEs is critical for metal-air battery development.

21.3.1 Dual-Ion Conducting SPEs

SPEs are composed of polymer matrix and salts. The salt dissociates into anions and cations under the interaction with polymer, and the ions can be conducted through the polymer chains under an electric field. The SPEs containing both the mobile cations and their counter anions are dual-ion conductors. To make the polymer function as a host in the SPEs, the polymer must possess polar groups with lone-pair electrons for bonding with the cation. Various polymers such as PEO, PAN, PVDF, PVDF-HFP, PMMA, and cellulose, as shown in Figure 21.4 [30] have been adopted as SPEs matrix.

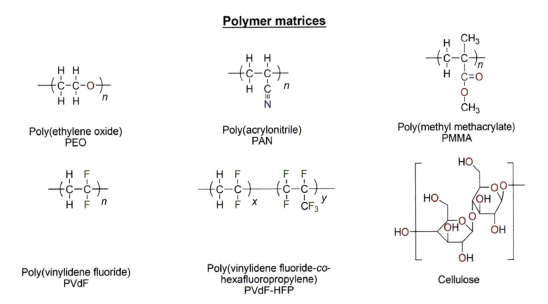

FIGURE 21.4 Chemical structures of commonly studied polymer matrices for solid composite electrolytes. (Adapted with permission [30]. Copyright (2017), IOP Publishing.)

Ideally, the SPEs should be thermodynamically stable against anode and cathode. As shown in Figure 21.5a, the potential range of electrodes is from μ_a to μ_c, and the energy gap (E_g) between the lowest unoccupied molecular orbital (LUMO) and the highest occupied molecular orbital (HOMO) is defined as the electrochemical window of SPEs. To avoid the electrochemical reactions between the SPE and electrodes, the potential range of electrodes should be within E_g. The electrochemical windows of several representative SPEs and inorganic electrolytes are summarized in Figure 21.5b [31].

PEO-based electrolytes have been reported to be stable even at 4.2 V [32], however, the upper limit of its electrochemical window may be overestimated due to the imperfect contact between electrolyte and cathode, and testing methods also make a great difference [33]. The electrochemical stability of other polymers has also been studied. Ebadi et al. [34] revealed that PAN is unstable in contact with Li metal due to the strong electron-withdrawing nitrile groups, and polymers with the strong polarity of the carbonyl C = O bonds are also unstable with Li metal. Additionally, different Li salts as the component of SPEs also make a great difference in interfacial stability [32]. As the popular salts, LiTFSI, LiFSI, LiCF$_3$SO$_3$, LiClO$_4$, and LiBF$_4$ have been studied extensively. In the PVDF-based electrolyte, it is found that the addition of LiFSI always shows better results than that of LiTFSI due to the formation of a stable LiF-Li$_x$SO$_y$ sulfur compounds-LiOH-Li$_2$CO$_3$-Li$_2$O mosaic interface [35]. Characteristics of typical polymers and properties of popular salts used in SPEs are summarized and compared in [36].

Normally, the motion of ions can only occur in the amorphous region of the polymer matrix due to a couple of ions with polymer chains. Nevertheless, it will result in low ionic conductivity of SPEs and further hinder the practical application of metal-air batteries. As a simple solution, the ionic conductivity of SPEs can be effectively improved by increasing the content of Li salts. However, in this case, it will result in a decrease in mechanical strength so that it cannot meet the requirements of Li dendrites suppression. Newman and Monroe considered the effects of electrolyte modules and surface tension on Li deposition kinetics and predicted that Li dendrite growth can be well suppressed when the SPEs with the high shear modulus (G´>7 GPa) are used [22]. Thus, the tradeoff

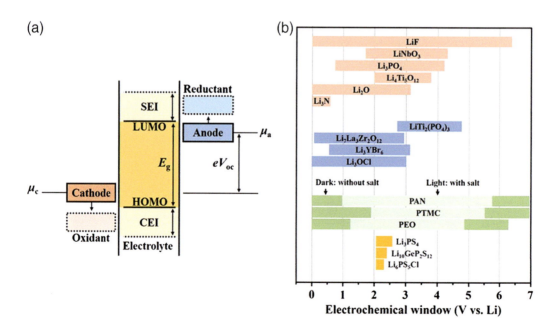

FIGURE 21.5 (a) Relation between energy gap (μ_a-μ_c) and the LUMO-HOMO window of an SSE. (b) Electrochemical windows of various electrolytes. (Adapted with permission [31]. Copyright The Authors, some rights reserved; exclusive licensee John Wiley and Sons.)

effect between mechanical strength and ionic conductivity is still the major contradiction during the development of SPEs.

CPEs have been extensively studied due to their ability to simultaneously improve the ionic conductivity and shear modulus of SPEs. As a representative strategy, inorganic nanoparticles which can disorder the crystallization of polymers are adopted as the filler to facilitate the dissociation of Li salts due to its plastification effect. In addition, it can also improve the mechanical property of SPEs. The inorganic fillers are usually divided into passive fillers (ZrO_2, Al_2O_3, SiO_2, TiO_2, CuO) and solid inorganic electrolytes (LLTO, LLZO, LATP, LLGP). Croce et al. [37] obtained ionic conductivity of 10^{-4} S/cm at 50 °C and 10^{-5} S/cm at 30 °C, respectively, with the doping of TiO_2 and Al_2O_3 fillers in PEO-$LiClO_4$ electrolyte. Cui's group [38] prepared SiO_2 nanoparticles with high monodispersity through an *in-situ* hydrolysis technology, by which the crystallinity of PEO can be decreased so that high ionic conductivity (10^{-4} S/cm to 10^{-5} S/cm at RT) was achieved. However, Tan et al. [39] compared the effect of SiO_2 and Al_2O_3 addition on PMMA-based CPEs, and found that the addition of Al_2O_3 enhanced the ionic conductivity of CPEs, but the addition of SiO_2 made no difference.

Jayathilaka et al. [14] further revealed that chemical groups on the Al_2O_3 nanoparticles play the main role in the enhancement of ionic conductivity of SPEs. Acidic groups always exhibit the highest enhancement followed by basic, neutral, and weakly acidic ones. In general, passive nanoparticle fillers enhance the ionic conductivity and mechanical properties of SPEs. Moreover, to improve the performance of SPEs further, solid inorganic electrolyte nanoparticles are more effective due to the intrinsic high ionic conductivity (higher than 10^{-4} S/cm at RT) apart from impeding crystallization and improving mechanical properties of SPEs. Among all the solid inorganic fillers, LLZO is favored for the chemical stability with Li anode and high oxidation potential [40]. Chen et al. [41] obtained the ionic conductivity of 1.9×10^{-5} S/cm at 30 °C when using PEO_{18}-LiTFSI-7.5%LLZO CPEs, which is much higher than that using PEO_{18}-LiTFSI (2.92×10^{-6} S/cm at 25 °C). Electrospinning $Li_7La_3Zr_2O_{12}$ (LLZO) fibers are also adopted to blend with PEO, and the hybrid electrolytes also exhibit enhanced Li-ion conductivity of 1.59×10^{-4} S/cm at ambient temperature [42]. In addition, continuous fast Li-ion conduction pathways can be formed by blending PEO with 3D garnet framework, by which the enhanced Li-ion conductivity (1.2×10^{-4} S/cm at 30 °C), excellent thermal, mechanical, and electrochemical stabilities are achieved [43]. To optimize the content of inorganic nanoparticles, Li et al. [16] simulated the transportation of Li-ion in the CPEs by the Monte Carlo simulation, and revealed the optimum content of LLZO for ionic conductivity enhancement.

In addition to the aforementioned solutions, blending, copolymerization, and crosslinking are all promising methods to improve the ionic conductivity as well as mechanical properties of SPEs. Polymer blending by combining two or more polymers is the simplest way to improve the ionic conductivity, mechanical properties, and electrochemical stability of SPEs due to the combination of advantages from different polymers. Wu et al. [19] blended PVDF-HFP with Poly(3-{2-[2-(2-hydroxyethoxy) ethoxy] ethoxy}methyl-3′-methyloxetane) (PHEMO) together as a new polymer matrix, and the ionic conductivity can reach up to 1.64×10^{-4} S/cm at RT. Besides, cross-linked SPEs are also effective to enhance ionic conductivity and suppress Li dendrite growth contributing to the amorphous phase increase and 3D network. Cross-linked polyethylene (PE)/PEO electrolytes with low-modulus ($G' \approx 1.0 \times 10^5$ Pa at 90 °C) can also exhibit excellent Li dendrite suppression ability with high ionic conductivity ($>1.0 \times 10^{-4}$ S/cm at 25 °C) [20]. What should be pointed out is that a high modulus is not necessary for suppressing Li dendrites. Perhaps the interfacial SEI stability caused by the confinement migration of polymer segments is the key to suppressing dendrites. Moreover, the copolymer contains at least two types of monomers that could provide ionic conduction and mechanical support functionalities separately. Excellent Li dendrite suppression ability is demonstrated by using polystyrene-block-polyethylene oxide (SEO) electrolytes [21], while the SPE prepared by mixing poly(ether block amide) (PEBA) 4011 and $LiClO_4$ has an ionic conductivity of 1.0×10^{-6} S/cm [44]. As discussed earlier, although there are many solutions to improve the ionic conductivity, mechanical properties, and electrochemical stability of SPEs, the migration of anions can result in severe concentration polarization and further cause cell impedance enlargement and even rapid Li dendrite growth. Thus, single Li-ion conductive SPEs as a strategy to avoid anion polarization deleterious effect has attracted great attention.

21.3.2 SINGLE-ION CONDUCTING SPEs

To avoid the shortages of dual-ion conducting SPEs, single-ion conducting SPEs have been developed by setting anion traps or connecting anions to a polymer backbone. In the single-ion conducting SPEs, cations are the only mobile species. However, the single-ion conductors with high t_{Li+} near 1.0 are simultaneously combined with low ionic conductivity. Thus, the battery can only work at a high temperature with the relatively high t_{Li+} as well as ionic conductivity. Notably, for suppressing the mobility of anionic species in SPEs, Armand et al. [45] proposed an ether-functionalized anion (EFA$^-$) and realized nano-sized self-agglomeration and trapping of anions. Through impeding the motion of anions in the SPEs, a high t_{Li+} (~0.43), as well as an ionic conductivity of 1.2×10^{-4} S/cm, are realized based on LiEFA/PEO electrolyte. Ma et al. [46] reported a PEO-based single-ion conducting electrolyte with a high t_{Li+} of 0.91 by dissolving poly[(4-styrenesulfonyl)(trifluoromethyl(S-trifluooromethylsulfony-limino)sulfonyl)imide] (PSsTFSI$^-$) in PEO matrix. In addition, the high ionic conductivity of 1.35×10^{-4} S/cm was also realized at 90 °C. To date, the development of single-ion conductors showing both high ionic conductivity and high t_{Li+} at ambient temperature remains a challenge. For achieving sufficient ionic conductivity, more cationic groups need to be attached to a polymer backbone. However, the bottleneck for metal-air battery development may be that further improvement of the ionic conductivity becomes more and more difficult while, how to improve the compatibility for an electrolyte-electrode interface is also full of challenges.

21.4 COMPATIBILITY OF SPEs

The ionic conductivity, mechanical properties, and electrochemical stability all can be improved greatly through various strategies; however, the interface issues are the greatest challenge for the application of metal-air batteries on electrochemical devices. Hereafter, the Li-air battery is taken as an example to discuss interface compatibility.

21.4.1 COMPATIBILITY WITH ANODE

Li metal with the lowest formal potential of −3.045 V vs. Standard hydrogen electrode (SHE) is very active to react with salt anions, liquid solvents, and impurities in the electrolyte to form a stable SEI layer, which can effectively hinder the continuation of the reaction. However, during the repeated charge-discharge process, the SEI layer will be damaged and rebuilt again and again, which could result in the continuous consumption of anions, and further lead to the decrease in ionic conductivity of SPEs. In addition, the increased roughness of the electrode surface inevitably causes the inhomogeneous distribution of electrical fields and Li plating. As such, Li dendrite will grow from these defects firstly and result in low coulombic efficiency, bad interface contact, short cycling life, and safety problems. In particular, as shown in Figure 21.6 [47], "dead Li" deposits at the anode surface could hinder the movement of Li ions.

Although SPEs can generally make intimate interfacial contact with Li anode, they still cannot inhibit Li dendrites. As aforementioned, high modulus contributes to the suppression of Li dendrites but leads to poor interface contacts with both anode and cathode, which reflects the trade-off between the interface contact and Li dendrite suppression. Therefore, to realize the application of Li-air batteries, there are at least two requirements: (1) the interface between anode and SPEs must be stable enough without destroying the SEI layer and keep intimate contact during the repeated

Polymer Electrolytes

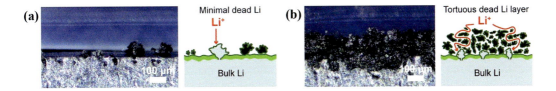

FIGURE 21.6 (a) Earlier cycles and (b) later cycles show that a more tortuous pathway is present after the accumulation of a thick dead Li layer on the electrode surface. (Adapted with permission [47]. Copyright (2012), Royal Society of Chemistry.)

charge-discharge process; and (2) the SPEs must possess high modulus to hinder the growth of Li dendrites at considerable current density.

Various strategies have been adopted to suppress the growth of Li dendrites, such as using inorganic fillers, blending different polymers, copolymerization, and crosslink. However, the Li dendrite problem has only been alleviated to a certain extent and is still not solved. Thus, the development of novel SPEs remains extremely urgent and important, especially new-type salts, additives, and combination of polymer groups. Moreover, it must be pointed out that the selection of solutions usually requires multiple considerations.

21.4.2 COMPATIBILITY WITH CATHODE

For the Li-air batteries, a suitable electrolyte that can withstand a highly oxidative environment, overcome the volume effects, and support the formation and decomposition of Li_2O_2 is of paramount importance. In the Li-air batteries, the nature of ORR and OER mechanisms at the cathode is well acknowledged as electrolytes play a key role in determining the nature of discharge products. The battery chemistry for the SPEs-based Li-air battery is still not clear now, but almost the same as the aprotic electrolyte. In addition, the cathode discharge reaction can be expressed as:

$$2Li + O_2 \rightarrow Li_2O_2$$

However, Li_2O_2 is generated and aggregated at the cathode during the discharge process, which impedes the transportation of oxygen as well as Li ions and inhibits ORR due to the inhibition of electron transfer. Therefore, ORR and OER can only take place at the 1D interface between the SPE and cathode material. In addition, the capacity of the Li-air battery is also greatly limited because of the inhibition of reaction by Li_2O_2. In addition to the aforementioned aspects, the SPEs must be stable with air and can hinder the diffusion of oxygen, nitrogen, carbon dioxide, and water to the anode. In a summary, there are still many obstacles on the cathode side for the applications of Li-air batteries.

21.5 PREPARATION METHODS OF SPEs

Generally, the properties of SPEs are closely related to the preparation method and fabrication process. Solution casting, electrospinning, and phase inversion are several popular methods for SPE preparation. Among all the mentioned methods, the solution cast is the most conventional and easy way. At first, the solution containing both polymer and salts is cast onto a suitable substrate. Then, the solvent is evaporated at RT or elevated temperature in an oven. Sometimes, the casting procedure is also conducted in a glovebox. After drying, the as-prepared SPE can be used directly. The electrospinning method can be used to produce films with a fiber morphology, and the thermal stability of electrospinning films is much better than that prepared by the solution-cast method.

However, many parameters relating to the preparation such as solution viscosity, surface tension, and polymer molecular weight need to be optimized in the electrospinning process. In addition, the setup conditions (e.g., flow rate, applied voltage) of electrospinning are also very complex.

The phase inversion technique is suitable to prepare SPEs with porous structures. Solangi et al. [48] prepared the PVDF-HFP-based SPEs by using this method. Herein, firstly, the solution containing PVDF-HFP, CeO_2, and N-Methylpyrrolidone (NMP) solvent was stirred for 24 h and then coated onto the glass plate. Hereafter, it was immersed in water for solvent extraction and then soaked in $LiClO_4$ solution for 6 h. Finally, the as-prepared SPE possessed a conductivity of 2.5×10^{-5} S/cm at 25 °C while, dry-mixing is a new way to prepare SPEs without using solvent, which is usually combined with hot-pressing. The PEO-based SPEs have been prepared with this method and ionic conductivity of 10^{-5} S/cm is achieved at RT [49]. As stated earlier, the solution casting should be more convenient than the electrospinning, dry-mixing, and phase inversion techniques for SPE preparation.

21.6 CONCLUSIONS

In this chapter, the development of various SPEs with suitable ionic conductivity for metal-air batteries has been introduced, and the tradeoff effect between conductivity and mechanical properties of SPEs is discussed. Blending, copolymerization, and crosslinking of polymers as effective methods have been adopted to enhance the mechanical properties of SPEs. Inorganic nanoparticles are also doped into a polymer matrix to prepare CPEs, which can improve both the ionic conductivity and mechanical properties greatly. However, it is still full of challenges for the development of ambient temperature Li-air batteries due to the intrinsic low mobility of polymer chains determined by the ion transport mechanism inside the polymer matrix. The conduction of ions within polymers generally occurs in two ways: (1) hopping between coordinating sites and (2) transporting by coupling with the segments of polymer chains.

Moreover, the compatibility of the electrolyte-electrode interface is still considered the bottleneck for metal-air battery development. The theory of suppressing Li dendrite growth by increasing the modulus has been proposed, by which the CPEs with enhanced mechanical strength should be a promising solution for dendrite suppression, while the adoption of Li salts is also crucial for dendrite suppression through changing SEI composition. As such, it is expected to develop the ambient temperature Li-air batteries with both the adjustment of Li salts and the improvement of the tradeoff effect.

ACKNOWLEDGMENTS

This work is supported by Hirosaki University, Japan. C. Liu greatly acknowledges China Scholarship Council (CSC), China.

REFERENCES

[1] J. Lu, L. Li, J.-B. Park, Y.-K. Sun, F. Wu, K. Amine, Aprotic and aqueous $Li-O_2$ batteries, Chemical Reviews. 114(11) (2014) 5611–5640.

[2] J. Cabana, L. Monconduit, D. Larcher, M.R. Palacín, Beyond intercalation-based Li-ion batteries: The state of the art and challenges of electrode materials reacting through conversion reactions, Advanced Materials. 22(35) (2010) E170–E192.

[3] G. Girishkumar, B. McCloskey, A.C. Luntz, S. Swanson, W. Wilcke, Lithium-air battery: Promise and challenges, The Journal of Physical Chemistry Letters. 1(14) (2010) 2193–2203.

[4] K.M. Abraham, Z. Jiang, A polymer electrolyte-based rechargeable lithium/oxygen battery, Journal of the Electrochemical Society. 143(1) (1996) 1–5.

[5] T. Ogasawara, A. Débart, M. Holzapfel, P. Novák, P.G. Bruce, Rechargeable Li_2O_2 electrode for lithium batteries, Journal of the American Chemical Society. 128(4) (2006) 1390–1393.

[6] S.J. Visco, B.D. Katz, Y.S. Nimon, L.C. De Jonghe, Protected active metal electrode and battery cell structures with nonaqueous interlayer architecture. Patent No. 7282295. U.S. Patent and Trademark Office, Washington, DC, 2007.

[7] Y. Wang, H. Zhou, A lithium-air battery with a potential to continuously reduce O_2 from air for delivering energy, Journal of Power Sources. 195(1) (2010) 358–361.

[8] I. Kowalczk, J. Read, M. Salomon, Li-air batteries: A classic example of limitations owing to solubilities, Pure and Applied Chemistry. 79(5) (2007) 851–860.

[9] S.A. Freunberger, Y. Chen, Z. Peng, J.M. Griffin, L.J. Hardwick, F. Bardé, P. Novák, P.G. Bruce, Reactions in the rechargeable lithium-O_2 battery with alkyl carbonate electrolytes, Journal of the American Chemical Society. 133(20) (2011) 8040–8047.

[10] D.E. Fenton, Complexes of alkali metal ions with poly (ethylene oxide), Polymer. 14 (1973) 589.

[11] M.B. Armand, Polymer electrolytes, Annual Review of Materials Science. 16(1) (1986) 245–261.

[12] J.R. Harding, C.V. Amanchukwu, P.T. Hammond, S.-H. Yang, Instability of poly (ethylene oxide) upon oxidation in lithium-air batteries, The Journal of Physical Chemistry C. 119(13) (2015) 6947–6955.

[13] J. Hassoun, F. Croce, M. Armand, B. Scrosati, Investigation of the O_2 electrochemistry in a polymer electrolyte solid-state cell, Angewandte Chemie. 123(13) (2011) 3055–3058.

[14] P.A.R.D. Jayathilaka, M.A.K.L. Dissanayake, I. Albinsson, B.E. Mellander, Effect of nano-porous Al_2O_3 on thermal, dielectric and transport properties of the $(PEO)_9$ LiTFSI polymer electrolyte system, Electrochimica Acta. 47(20) (2002) 3257–3268.

[15] Z.H. Li, H.P. Zhang, P. Zhang, Y.P. Wu, X.D. Zhou, Macroporous nanocomposite polymer electrolyte for lithium-ion batteries, Journal of Power Sources. 184(2) (2008) 562–565.

[16] Z. Li, H.-M. Huang, J.-K. Zhu, J.-F. Wu, H. Yang, L. Wei, X. Guo, Ionic conduction in composite polymer electrolytes: case of PEO: Ga-LLZO composites, ACS Applied Materials & Interfaces. 11(1) (2018) 784–791.

[17] F. Chen, W. Zha, D. Yang, S. Cao, Q. Shen, L. Zhang, D.R. Sadoway, All-solid-state lithium battery fitted with polymer electrolyte enhanced by solid plasticizer and conductive ceramic filler, Journal of the Electrochemical Society. 165(14) (2018) A3558–A3565.

[18] S. Song, X. Qin, Y. Ruan, W. Li, Y. Xu, D. Zhang, J. Thokchom, Enhanced performance of solid-state lithium-air batteries with continuous 3D garnet network added composite polymer electrolyte, Journal of Power Sources. 461 (2020) 228146.

[19] F. Wu, T. Feng, Y. Bai, C. Wu, L. Ye, Z. Feng, Preparation and characterization of solid polymer electrolytes based on PHEMO and PVDF-HFP, Solid State Ionics. 180(9–10) (2009) 677–680.

[20] R. Khurana, J.L. Schaefer, L.A. Archer, G.W. Coates, Suppression of lithium dendrite growth using cross-linked polyethylene/poly (ethylene oxide) electrolytes: A new approach for practical lithium-metal polymer batteries, Journal of the American Chemical Society. 136(20) (2014) 7395–7402.

[21] G.M. Stone, S.A. Mullin, A.A. Teran, D.T. Hallinan, A.M. Minor, A. Hexemer, N.P. Balsara, Resolution of the modulus versus adhesion dilemma in solid polymer electrolytes for rechargeable lithium metal batteries, Journal of the Electrochemical Society. 159(3) (2011) A222–A227.

[22] C. Monroe, J. Newman, The impact of elastic deformation on deposition kinetics at lithium/polymer interfaces, Journal of the Electrochemical Society. 152(2) (2005) A396–A404.

[23] A. Ghosh, C. Wang, P. Kofinas, Block copolymer solid battery electrolyte with high Li-ion transference number, Journal of the Electrochemical Society. 157(7) (2010) A846–A849.

[24] R. Bouchet, S. Maria, R. Meziane, A. Aboulaich, L. Lienafa, J.P. Bonnet, T.N.T. Phan, D. Bertin, D. Gigmes, D. Devaux, R. Denoyel, M. Armand, Single-ion BAB triblock copolymers as highly efficient electrolytes for lithium-metal batteries, Nature Materials. 12(5) (2013) 452–457.

[25] W.H. Meyer, Polymer electrolytes for lithium-ion batteries, Advanced Materials. 10(6) (1998) 439–448.

[26] M.H. Cohen, D. Turnbull, Molecular transport in liquids and glasses, The Journal of Chemical Physics. 31(5) (1959) 1164–1169.

[27] D.S. Mebane, R.A. De Souza, A generalised space-charge theory for extended defects in oxygen-ion conducting electrolytes: From dilute to concentrated solid solutions, Energy & Environmental Science. 8(10) (2015) 2935–2940.

[28] M.B. Armand, J.M. Chabagno, M.J. Duclot, Poly-ethers as solid electrolytes in fast ion transport in solid electrodes and electrolytes, P. Vashishta, J. N. Mundy, and G. K. Shenoy, eds. Elsevier, Amsterdam, North Holland, 1979.

[29] D.Y. Voropaeva, S.A. Novikova, A.B. Yaroslavtsev, Polymer electrolytes for metal-ion batteries, Russian Chemical Reviews. 89(10) (2020) 1132–1155.

[30] X. Judez, H. Zhang, C. Li, G.G. Eshetu, J.A. González-Marcos, M. Armand, L.M. Rodriguez-Martinez, Solid electrolytes for safe and high energy density lithium-sulfur batteries: Promises and challenges, Journal of the Electrochemical Society. 165(1) (2017) A6008–A6016.

[31] X. Chen, J. Xie, X. Zhao, T. Zhu, Electrochemical compatibility of solid-state electrolytes with cathodes and anodes for all-solid-state lithium batteries: A review, Advanced Energy and Sustainability Research. 2(5) (2021) 2000101.

[32] C.F.N. Marchiori, R.P. Carvalho, M. Ebadi, D. Brandell, C.M. Araujo, Understanding the electrochemical stability window of polymer electrolytes in solid-state batteries from atomic-scale modeling: the role of Li-ion salts, Chemistry of Materials. 32(17) (2020) 7237–7246.

[33] J. Kasnatscheew, B. Streipert, S. Röser, R. Wagner, I.C. Laskovic, M. Winter, Determining oxidative stability of battery electrolytes: Validity of common electrochemical stability window (ESW) data and alternative strategies, Physical Chemistry Chemical Physics. 19(24) (2017) 16078–16086.

[34] M. Ebadi, C. Marchiori, J. Mindemark, D. Brandell, C.M. Araujo, Assessing structure and stability of polymer/lithium-metal interfaces from first-principles calculations, Journal of Materials Chemistry A. 7(14) (2019) 8394–8404.

[35] X. Zhang, S. Wang, C. Xue, C. Xin, Y. Lin, Y. Shen, L. Li, C.-W. Nan, Self-suppression of lithium dendrite in all-solid-state lithium metal batteries with poly (vinylidene difluoride)-based solid electrolytes, Advanced Materials. 31(11) (2019) 1806082.

[36] L.P. Teo, M.H. Buraidah, A.K. Arof, Development on solid polymer electrolytes for electrochemical devices, Molecules. 26(21) (2021) 6499.

[37] F. Croce, G.B. Appetecchi, L. Persi, B. Scrosati, Nanocomposite polymer electrolytes for lithium batteries, Nature. 394(6692) (1998) 456–458.

[38] D. Lin, W. Liu, Y. Liu, H.R. Lee, P.C. Hsu, K. Liu, Y. Cui, High ionic conductivity of composite solid polymer electrolyte via in situ synthesis of monodispersed SiO_2 nanospheres in poly (ethylene oxide), Nano Letters. 16(1) (2016) 459–465.

[39] S. Chen, K. Wen, J. Fan, Y. Bando, D. Golberg, Progress and future prospects of high-voltage and high-safety electrolytes in advanced lithium batteries: From liquid to solid electrolytes, Journal of Materials Chemistry A. 6(25) (2018) 11631–11663.

[40] J.H. Choi, C.H. Lee, J.H. Yu, C.H. Doh, S.M. Lee, Enhancement of ionic conductivity of composite membranes for all-solid-state lithium rechargeable batteries incorporating tetragonal $Li_7La_3Zr_2O_{12}$ into a polyethylene oxide matrix, Journal of Power Sources, 274 (2015) 458–463.

[41] W. Zha, F. Chen, D. Yang, Q. Shen, L. Zhang, High-performance $Li_{6.4}La_3Zr_{1.4}Ta_{0.6}O_{12}$/poly(ethylene oxide)/succinonitrile composite electrolyte for solid-state lithium batteries, Journal of Power Sources. 397 (2018) 87–94.

[42] M. Jing, H. Yang, C. Han, F. Chen, L. Zhang, X. Hu, F. Tu, X. Shen, Synergistic enhancement effects of LLZO fibers and interfacial modification for polymer solid electrolyte on the ambient-temperature electrochemical performances of solid-state battery, Journal of the Electrochemical Society. 166(13) (2019) A3019–A3027.

[43] Z. Li, W.-X. Sha, X. Guo, Three-dimensional garnet framework-reinforced solid composite electrolytes with high lithium-ion conductivity and excellent stability, ACS Applied Materials & Interfaces. 11(30) (2019) 26920–26927.

[44] R.A. Zoppi, C.M.N.P. Fonseca, Marco-A. De Paoli, S.P. Nunes, Solid electrolytes based on poly (amide 6-b-ethylene oxide), Solid State Ionics. 91(1–2) (1996) 123–130.

[45] H. Zhang, F. Chen, O. Lakuntza, U. Oteo, L. Qiao, M. Martinez-Ibañez, H. Zhu, J. Carrasco, M. Forsyth, M. Armand, Suppressed mobility of negative charges in polymer electrolytes with an ether-functionalized anion, Angewandte Chemie. 131(35) (2019) 12198–12203.

[46] Q. Ma, H. Zhang, C. Zhou, L. Zheng, P. Cheng, J. Nie, W. Feng, Y.S. Hu, H. Li, X. Huang, L. Chen, M. Armand, Z. Zhou, Single lithium-ion conducting polymer electrolytes based on a super-delocalized polyanion, Angewandte Chemie International Edition. 55(7) (2016) 2521–2525.

[47] K.-H. Chen, K.N. Wood, E. Kazyak, W.S. LePage, A.L. Davis, A.J. Sanchez, N.P. Dasgupta, Dead lithium: Mass transport effects on voltage, capacity, and failure of lithium metal anodes, Journal of Materials Chemistry A. 5(23) (2017) 11671–11681.

[48] M.Y. Solangi, U. Aftab, M. Ishaque, A. Bhutto, A. Nafady, Z.H. Ibupoto, Polyvinyl fibers as outperform candidature in the solid polymer electrolytes, Journal of Industrial Textiles. 0(0) (2020) 1–13.

[49] T. Eriksson, A. Mace, Y. Manabe, M.Y. Fujita, Y. Inokuma, D. Brandell, J. Mindemark, Polyketones as host materials for solid polymer electrolytes, Journal of the Electrochemical Society. 167(7) (2020) 070537.

22 Hydrogel Electrolytes

Siyuan Zhao, Tong Liu, and Meng Ni

CONTENTS

- 22.1 Introduction 317
- 22.2 Fundamentals of Hydrogel Electrolytes 318
  - 22.2.1 Hydrogels: Polymer Chains and Water 318
  - 22.2.2 History of the Hydrogel Electrolytes for MABs 319
  - 22.2.3 Commonly Used Hydrogel Electrolytes for MABs 320
- 22.3 Multi-Functional Hydrogel Electrolytes 322
  - 22.3.1 Anti-Freezing Hydrogel Electrolytes 322
  - 22.3.2 Anti-Dehydration Hydrogel Electrolytes 323
  - 22.3.3 Anti-CO_2 Hydrogel Electrolytes 324
  - 22.3.4 Robust Hydrogel Electrolytes 326
  - 22.3.5 Self-Healing Hydrogel Electrolytes 326
- 22.4 Recent Advances in Hydrogel-Based MABs 329
  - 22.4.1 Dual-Electrolytes for High-Voltage and High-Energy-Density MABs 329
  - 22.4.2 Eco-Friendly Hydrogel Electrolytes for MABs 330
- 22.5 Summary and Future Perspective 330
- References 332

22.1 INTRODUCTION

Flexible electronics with deformable nature have received increasing attention [1, 2]. Compared to conventional rigid electronics, they are more adaptable to various environments and thus applicable to broader working conditions [3, 4]. Over the last two decades, the world has witnessed the boom of flexible electronics in healthcare, sensing, soft robotics, and wearable devices. Accordingly, to support the rapid development of flexible electronics, integrated power sources have also experienced a transformative shift from hardness to flexibility [5, 6]. Nonaqueous flexible lithium-ion batteries (LIBs) are popular energy storage devices due to their satisfactory cycle life and energy efficiency [7, 8]. However, they are always denounced for safety issues caused by the instability of internal electrodes and flammability of organic electrolytes, which may become more severe during the daily repeated battery bending, stretching, and twisting [9]. Besides, LIBs are very costly because of the limited lithium reserve. Therefore, further exploration for alternative aqueous flexible batteries with high safety, low cost, and long lifetime is critical and urgent.

Recently, aqueous flexible metal-air batteries (MABs) with the aforementioned advantages have attracted great interest from researchers [10, 11]. For the anode side, ductile metals with vast reserves and high specific capacity, usually aluminum (Al, 2980 mAh/g), magnesium (Mg, 2200 mAh/g), and zinc (Zn, 820 mAh/g), are used as fuel [12]. Due to the unique semi-open system of MABs, infinite air from the ambient can be inhaled into the flexible catalyst-loaded air cathode for oxygen reduction reaction (ORR) and oxygen evolution reaction (OER) for rechargeable Zn-air batteries (ZABs). In the middle, the flexible polymer electrolyte not only acts as the separator but also simultaneously serves as the messenger to bridge the ion transport between the metal anode and air cathode [13]. By the solid-state and nonflammable nature of the electrolyte, the fabricated MABs will not undergo leakage or explosion.

During the last decade, extensive research has concentrated on catalyst and flexible electrode design in MABs, while the development of flexible electrolytes is relatively limited [14]. However, from the perspective of ion conducting, flexible electrolytes are more significant because they correlate to the ionic transport efficiency when the electrochemical reaction takes place [15]. Among various flexible electrolytes, hydrogel electrolytes with multiple advantages are very promising for aqueous flexible MABs. First, the water-saturated hydrogel network can maintain the aqueous property and satisfactory ionic conductivity as liquid-state electrolytes. Second, since hydrogel electrolytes are soft, they can be easily applied to flexible MABs. Third, most hydrogel materials are inexpensive and easy to obtain. Fourth, hydrogel materials can be modified to achieve multiple functions to satisfy various battery application scenarios. Therefore, it is highly significant to develop hydrogel electrolytes to facilitate the development of aqueous flexible MABs.

This chapter provides our readers with comprehensive knowledge about hydrogel electrolytes for aqueous flexible MABs. In the fundamentals of hydrogel electrolytes section, the definition and development history of hydrogel electrolytes are given, and the current widely used hydrogels are introduced. In the multi-functional hydrogel electrolytes section, strategies for realizing hydrogels with various functions will be provided. The reason for preparing this section is that the flexible MABs may work under different practical situations like subzero environments, hot temperatures, extreme deformations, etc. In the recent advances in hydrogel-based MABs section, we selected some recent noteworthy work on hydrogel-based MABs that may represent future research directions. Finally, a summary with a future perspective is also provided. We hope this chapter can attract more researchers to engage in and contribute to the hydrogel-based MABs research field.

22.2 FUNDAMENTALS OF HYDROGEL ELECTROLYTES

22.2.1 Hydrogels: Polymer Chains and Water

Hydrogels are the most reported and used polymer electrolyte for aqueous flexible MABs [16]. The crosslinked polymer chains with abundant hydrophilic functional groups weave as 3D hydrogel networks to provide sufficient interspace for water and lock them in tightly [17]. For the hydrogels used as electrolytes in flexible MABs, some significant properties should be considered. First, a higher ionic conductivity determines better battery electrochemical performances because ion transportation is faster. Second, a better water absorption/retention capability can endow the battery longer operation duration. Moreover, tough and elastic hydrogels can endure the battery's daily deformations.

Two main hydrogel components, the polymer chains, and water determine the properties of hydrogels [18]. Various polymer chains have various functional groups and deliver various properties. For example, the polarity of the carboxyl group (-COOH) is stronger than the hydroxyl group (-OH), leading to a stronger interaction with water molecules. Therefore, the sodium polyacrylate (PANa) hydrogel with carboxyl groups can absorb more water than other hydrogels with hydroxyl groups [19]. The functional groups on the polymer chains can also be modified to achieve anticipated functions. In addition, the crosslinking degree will affect the mechanical property and water absorption of hydrogels. For instance, under a high-level crosslinking degree, the polymer chains will crowd in the hydrogel substrate, which may occupy the space for water. Moreover, high-density crosslinking points will prevent the free stretching of polymer chains and then lower the hydrogel stretchability. On the other hand, insufficient crosslinking may cause chain dissolution in the water, leading to unsuccessful hydrogel synthesis. Therefore, when synthesizing hydrogels for flexible MABs, the crosslinking degree should be carefully controlled and optimized.

Water is also essential, and that is why it is called "hydro-gel." When salts are added to water, the hydrogel becomes a hydrogel electrolyte. By virtue of the 3D hydrogel networks, the fluidity of water, and various salt ions, the ionic conductivities of some hydrogel electrolytes are comparable to liquid electrolytes. Moreover, water is intrinsically safe compared to flammable organic

Hydrogel Electrolytes

electrolytes. Therefore, soft hydrogel electrolytes are perfect for flexible MABs. Moreover, certain ions may bring new functions to hydrogels. For example, when a ferric ion is introduced, the hydrogel will gain self-healing properties because the ferric ion can physically crosslink the polymer chains of the cut hydrogel parts [20]. Apart from ions, some inorganic nanoparticles like graphene oxide (GO) or aluminum oxide (Al_2O_3) can also disperse uniformly in hydrogels to become hydrogel composites, which will gain enhanced mechanical properties and water retention capability [21]. Finally, the water in the hydrogel can also be partially replaced by other solvents, such as polyols, which can lower the freezing point of hydrogel electrolytes and stabilize the metal anode when applied in flexible MABs [22].

22.2.2 History of the Hydrogel Electrolytes for MABs

Before 2012, research on flexible MABs was very limited. Only a few works ever reported flexible ZABs with poor electrochemical performances [23–25]. From 2013 to 2016, flexible MABs started arousing interest from researchers. Not only multiple hydrogel electrolytes, including polyacrylic acid (PAA) hydrogel, polyvinyl alcohol (PVA) hydrogel, copolymer PVA/polyethylene oxide (PEO) hydrogel, and gelatin are developed to use in flexible MABs, but also several MAB prototypes,

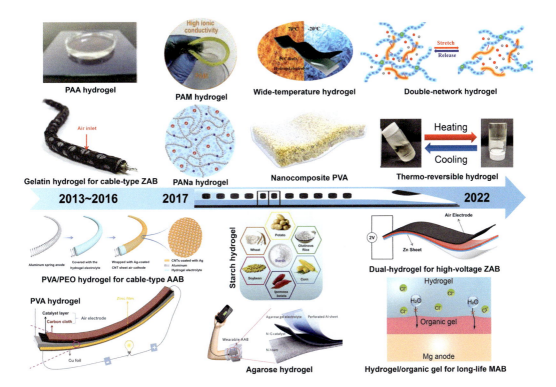

FIGURE 22.1 The development history of hydrogel electrolytes and their application in MABs. (Adapted with permission [26]. Copyright 2014, Elsevier. Adapted with permission [27]. Copyright 2015, Wiley. Adapted with permission [28]. Copyright 2015, Wiley. Adapted with permission [29]. Copyright 2015, Wiley. Adapted with permission [30]. Copyright 2020, Elsevier. Adapted with permission [31]. Copyright 2018, Wiley. Adapted with permission [32]. Copyright 2020, American Chemical Society. Adapted with permission [33]. Copyright 2019, Elsevier. Adapted with permission [34]. Copyright 2019, Wiley. Adapted with permission [35]. Copyright 2021, American Chemical Society. Adapted with permission [36]. Copyright 2022, Elsevier. Adapted with permission [37]. Copyright 2021, Elsevier. Adapted with permission [38]. Copyright 2022, Elsevier. Adapted with permission [39]. Copyright 2021, Wiley.)

such as cable-type and sandwich-type, are proposed. At this stage, the electrochemical properties of hydrogels and battery performances were improved but not comparable to the liquid electrolyte and liquid MABs. In the last five years, the development of hydrogel electrolytes and flexible MABs has been astonishing. More hydrogels with higher ionic conductivity and better water adsorption/retention capability like PANa and polyacrylamide (PAM) hydrogels have been developed and used in flexible MABs for enhanced electrochemical performances and longer operation duration. Moreover, to create a sustainable world, researchers also focused on biodegradable hydrogels such as starch, agarose, alginate, etc. These hydrogels are acquired from nature and can be reused and finallyback to nature, whose sustainable essence is human consensus. The final goal for flexible MABs is commercialization. Researchers were trying to make flexible MABs work under a wide temperature range and able to tolerate daily bending, stretching, and twisting. Therefore, corresponding multi-functional hydrogel electrolytes (i.e., anti-freezing hydrogels, anti-dehydration hydrogels, and tough hydrogels) have been designed by solvent and polymer chain modification. Apart from single hydrogels, dual hydrogels in flexible MABs also received attention because they can provide MABs with higher voltage and energy density.

22.2.3 Commonly Used Hydrogel Electrolytes for MABs

The commonly used hydrogel electrolytes for aqueous flexible MABs are summarized in Table 22.1. Among these hydrogels, PVA, PAA, PAM, and PANa are representative homopolymer hydrogels which distinguish from each other by their different hydrophilic functional groups. The synthesis approach of PVA hydrogels can be physical crosslinking (hydrogel bonding) or chemical crosslinking (chemical crosslinking agent), while the PAA, PAM, and PANa are mainly synthesized by chemical crosslinking. These homopolymer hydrogels can combine to synthesize as copolymer hydrogels and gain better performances. Besides, they can also synthesize with other biopolymer hydrogels like sodium alginate and cellulose to form double-network hydrogels with toughness and high stretchability. There are also block polymer hydrogels poly(ethylene oxide)-poly(propylene

TABLE 22.1
The Commonly Used Hydrogels for Aqueous Flexible MABs

Name	Molecular structure	Functional groups	Features	Crosslinking
Polyvinyl alcohol (PVA)	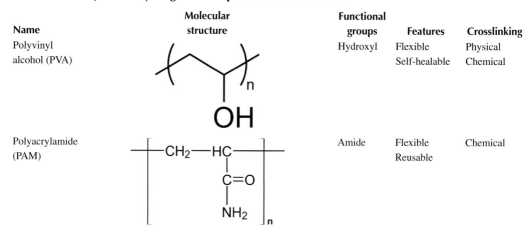	Hydroxyl	Flexible Self-healable	Physical Chemical
Polyacrylamide (PAM)		Amide	Flexible Reusable	Chemical

Name	Molecular structure	Functional groups	Features	Crosslinking
Polyacrylic acid (PAA)		Carboxyl	Flexible Self-healable	Physical (Self-healing) Chemical
Sodium polyacrylate (PANa)		Sodium carboxylate	Flexible Self-healable	Physical (Self-healing) Chemical
Gelatin		Peptides	Flexible Salt-unstable Biodegradable	Physical
Agarose		Hydroxyl	Flexible Biodegradable	Physical
Sodium alginate		Sodium carboxylate Hydroxyl	Flexible Biodegradable	Physical
Cellulose		Hydroxyl	Flexible Reusable Biodegradable	Physical Chemical
Poly(ethylene oxide)-poly(propylene oxide)-poly(ethylene oxide) (PEO-PPO-PEO)		Hydroxyl	Thermo-responsive	Micellization

oxide)-poly(ethylene oxide) (PEO-PPO-PEO), which is treated as a smart hydrogel because of its unique thermo-responsible property. The listed hydrogels are all basic hydrogels, where further modifications like polymer chain grafting and solvent replacement can be applied.

22.3 MULTI-FUNCTIONAL HYDROGEL ELECTROLYTES

Towards practical application, flexible MABs may work under various conditions like subzero temperatures, high temperatures, CO_2 environments, and extreme deformations. These may bring problems to the hydrogel electrolytes. Therefore, it is important to develop multi-functional hydrogel electrolytes to overcome these issues.

22.3.1 Anti-Freezing Hydrogel Electrolytes

The water-saturated property of hydrogel electrolytes brings about good ionic conductivity. However, it also becomes a problem in the subzero environment because water may freeze, and MABs will be out of operation. For tackling the freezing issue, the key lies in decreasing the water activity, that is, replacing the hydrogen bonds between water-water with stronger hydrogen bonds of water-X (X can be the following subjects) [40], and three main approaches have been developed as follows:

(1) Establishing stronger polymer chain-water hydrogen bonds. The polarization degrees of polymer chains with different functional groups are varied. From the density functional theory (DFT) calculation results in Figure 22.2a [41], the interaction energy (E_{int}) between the terminal hydroxyl group of polyvinyl alcohol (PVA) and nearby water molecules is −6.11 Kcal/mol, very close to the E_{int} between two water molecules of −5.75 Kcal/mol. By contrast, the E_{int} value increases to −12.92 Kcal/mol in polyacrylic acid (PAA) with a more polar carboxyl group. Moreover, the carboxyl group in PAA can be further alkalified (A-PAA) to acquire a stronger interaction with water molecules (−16.96 Kcal/mol). As a result, the freezing point of A-PAA with 10 wt.% KOH reaches −25 °C, while PVA with the same KOH concentration is only −13 °C. Although a decrease in ionic conductivity at −20 °C is inevitable, it still retains as high as 199 mS/cm, which is close to that at room temperature and sufficient for stable battery operation. Therefore, selecting hydrogels with highly polar polymer chains is crucial for anti-freezing and high-performance flexible MABs.

(2) Employing high concentration salt solutions. This way, hydrogen bonds between free water molecules are reduced by hydrated ions, and ice crystallization can be suppressed [42, 43]. Generally, solutions with a higher salt concentration may have a lower freezing point. As shown in Figure 22.2b, the anti-freezing ability of the hydrogel enhances remarkably as the KOH concentration increase from 0 M to 6 M. At −20 °C and −50 °C, the 6M KOH PANa hydrogel can be stretched to 1400% and 900%, respectively, while the pure PANa hydrogel freeze. For ZABs and Al-air batteries (AABs) commonly using 6 M and 4 M KOH electrolytes, the high concentration electrolyte itself can help resist the low temperature of −20 °C. Therefore, when employing hydrogel saturated with high-concentration KOH electrolytes, flexible ZABs and AABs can deliver satisfactory electrochemical performances at a low temperature. For example, at 0 °C and −20 °C, flexible ZABs employing 6 M KOH PANa hydrogel electrolyte behave well under various current densities, only slightly degrading compared to the room temperature (25 °C) [41]. Moreover, the battery also exhibits good cycle stability over a wide temperature range. For flexible AABs, it is regretful that there is no report focusing on their low-temperature operations. Nevertheless, a study on the liquid-state AAB with 25 wt.% KOH suggests that the battery discharge duration can be unexpectedly elongated at −15 °C since the severe H_2 evolution parasitic reaction on the Al surface is significantly suppressed at that

low temperature [44]. Mg-air batteries (MgABs) are not lucky as those two because they usually employ low concentration salt solutions (e.g., 3.5 wt.% NaCl) due to the high activity of Mg metal. Therefore, the first way of anti-freezing is to raise the NaCl concentration. Another possible approach is to substitute the NaCl with a more hydrated salt like LiCl and $MgCl_2$, which bonds more intensely with water molecules to lower the freezing point [45]. Nevertheless, the effect of these two methods on flexible MgABs' low-temperature operation needs further evaluation considering the absent research. Considering the effectiveness of this "high-concentration" strategy in flexible ZABs, it is anticipated the flexible AABs and MgABs with high concentration hydrogel electrolytes can show comparable performance.

(3) Constructing stronger organic solvent-water hydrogen bonds. Water mixing with organic solvents (e.g., dimethylsulfoxide (DMSO), methanol, ethylene glycol, glycerol, sorbitol, etc.) have been widely used as cryoprotectants. The intense interaction between organic solvents and water molecules significantly diminishes the water-water hydrogen bonds, thus suppressing the ice crystallization (Figure 22.2c). By partially or fully substituting the water with organic solvents during hydrogel synthesis, it is also available to achieve anti-freezing hydrogels [46]. Since organic solvents can also lower the water activity, the corrosion of the active metal's surface can be alleviated, especially with Al and Mg. For example, the corrosion current density of Al in a methanol-KOH solution is significantly lower than in an aqueous KOH solution [47]. The mitigated parasitic reaction in organic solvents can further contribute to a higher battery discharge capacity, which is about 2.7 times that of the aqueous counterpart [48]. Yet, the organic solvents may slightly decrease the ionic conductivity of the hydrogel electrolyte at room temperature, leading to a decline in discharge voltage. Compared to metal Al, the Zn corrosion in aqueous 6 M KOH can be ignored, and the high alkaline concentration can endow the hydrogel with remarkable anti-freezing properties. Therefore, the addition of organic solvents is more suitable for long-life flexible AABs and may be redundant for flexible ZABs.

22.3.2 Anti-Dehydration Hydrogel Electrolytes

The water within hydrogels may readily evaporate to the ambient because of the semi-open structure of MABs. Unfortunately, with the decline in water content, the ionic conductivity of the hydrogel electrolyte will decrease [49]. Salt will precipitate and then block the air inlet of the air cathode, leading to battery performance degradation. Moreover, hydrogel electrolytes may shrink and detach from the electrode, resulting in direct battery failure [49]. The water evaporation in hydrogels is more severe at high temperatures, which may affect the use of aqueous flexible MABs. Therefore, it is crucial to develop anti-dehydration hydrogel electrolytes.

The key to anti-dehydration hydrogel is to lower the activity of water molecules and lock them in the hydrogel matrix. Therefore, stronger interaction between hydrophilic functional groups/hydrated ions and water molecules is preferred, similar to strategies for anti-freezing hydrogels. Since most organic solvents used in flexible MABs are volatile, they may not be suitable for anti-dehydration hydrogels.

Recently, Zhong's research group proposed to add inorganic nanoparticles in the hydrogel electrolyte to enhance the water retention ability [33, 50]. Normally, these nanoparticles can lower the crystallization degree of the hydrogel, meaning the originally twisted polymer chains can stretch out and expose more hydrophilic functional groups to interact with water molecules. Besides, the surface of some nanoparticles like GO is also hydrophilic. Therefore, the water retention ability can be improved. For instance, Figure 22.2d shows the digital photo and microstructure schematic of PVA/PAA-GO (PVAA-GO) nanocomposite hydrogel. The uniform brown color demonstrates the uniform dispersion of GO nanoparticles. The PVAA-GO hydrogel can retain 95% electrolyte content after exposure to the ambient for 12 hours, higher than the PVAA and PVA hydrogels

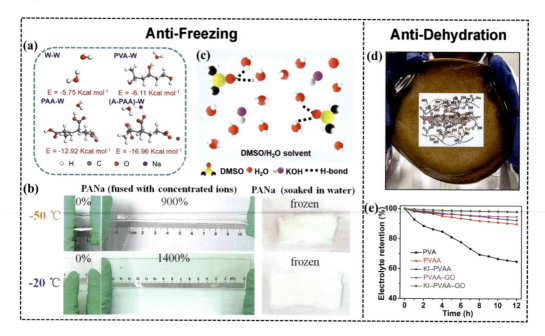

FIGURE 22.2 (a) Interaction energy between water-water and water-polymer chains. (Adapted with permission [41]. Copyright 2020, Wiley.) (b) Anti-freezing capability of PANa hydrogels with and without 6 M KOH concentrations at −20 °C and −50 °C. (Adapted with permission [43]. Copyright 2019, American Chemical Society.) (c) Schematic of the organic solvent-water binary solution. (Adapted with permission [46]. Copyright 2022, Wiley.) (d) Digital photo and structure schematic of the PVAA-GO hydrogel. (e) Electrolyte retention of various hydrogel electrolytes. (Adapted with permission [50]. Copyright 2020, Wiley.)

(Figure 22.2e). It is worth mentioning that besides water retention ability, the ionic conductivity and mechanical strength of the PVAA-GO hydrogel are also enhanced from PVAA hydrogel. This phenomenon can also be explained by the untwisted polymer chains. Consequently, the fabricated flexible ZAB can stably cycle for over 200 hours without performance degradation. Apart from GO, other inorganic nanoparticles, such as reduced GO (rGO), carbon nanotube (CNT), silica (SiO_2), aluminum oxide (Al_2O_3), etc., can also deliver similar functions in hydrogel electrolytes [51]. However, the weight percent of the adding nanoparticles should be optimized because too few or too many nanoparticles may harm hydrogel electrolytes. The reason is that most polymer chains are still twisted under a low weight percentage, while nanoparticles will aggregate and destroy the hydrogel structure under a high weight percentage.

22.3.3 ANTI-CO_2 HYDROGEL ELECTROLYTES

The unique semi-open structure of MABs not only allows O_2 but also CO_2 from the air inhaling in. The high concentration alkaline electrolyte in ZABs and AABs can readily react with CO_2, resulting in declined ionic conductivity and degraded battery performance [52]. Although this issue has existed since the alkaline electrolyte was applied, research on the anti-CO_2 alkaline MABs is still limited, especially for flexible MABs. Here, three effective anti-CO_2 strategies for hydrogel electrolytes from the physical and chemical perspectives are introduced.

CO_2 in the air may easily diffuse into the liquid electrolyte with low viscosity. However, the diffusion coefficient of CO_2 decreases when increasing the solution viscosity [53]. Therefore, Bae and co-workers proposed to employ a highly viscous hydrogel Pluronic F127 to physically resist the CO_2 in the environment [35]. The viscosity of the Pluronic F127 hydrogel is about 10^7 times of

Hydrogel Electrolytes

the aqueous solution, thus the anti-CO_2 capabilities of these two are also significantly varied. The result showed that the resistance of the liquid alkaline electrolyte rises obviously after 24 hours of CO_2 exposure, indicating the alkaline electrolyte is severely contaminated by CO_2 and becomes carbonate. On the other hand, the resistance of the viscous Pluronic F127 hydrogel slightly increased, demonstrating a high-level anti-CO_2 capability. By virtue of the good anti-CO_2 property, the flexible ZAB with Pluronic F127 hydrogel delivered an elongated discharge lifetime of 133 hours, about 2.5 times the conventional liquid-state ZAB.

In addition to physical resistance, it has recently been shown that CO_2 can be successfully repelled chemically by pre-fixing CO_2 on the PVA chains [54]. As shown in Figure 22.3a, PVA chains can be ionized by CO_2 in the presence of tetramethylguanidine (TMG). This synthesis step allows the formation of a new functional group -OCO^{2-} which interacts closely with KOH and avoids the reaction of KOH with ambient CO_2 to produce carbonates. The content of TMG also affects the hydrogel structure. From Figure 22.3b, a higher ratio of TMG can form a more porous structure, showing higher ionic conductivity, stronger mechanical property, and better water retention capability. More

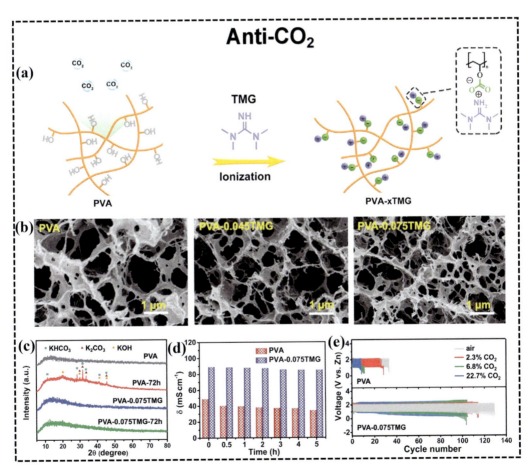

FIGURE 22.3 (a) Schematic of the synthesis steps of the PVA-TMG hydrogel. (b) Scanning electron microscope (SEM) images of PVA hydrogels with different amounts of TMG. (c) XRD patterns of the PVA and PVA-TMG before and after 72-hour exposure in the ambient. (d) Ionic conductivity change of the PVA and PVA-TMG in CO_2 atmosphere for various exposure times. (e) Cycle life of the flexible ZABs with PVA and PVA-TMG hydrogel electrolytes under different CO_2 concentrations. (Adapted with permission [54]. Copyright 2021, Wiley.)

importantly, after exposing untreated PVA to air for 72 hours, the presence of potassium carbonates is detected by X-ray Diffraction (XRD), demonstrating CO_2 corrosion. However, under the same conditions, no potassium carbonates formed on the pretreated PVA-0.075TMG hydrogel, confirming its resistance to CO_2 (Figure 22.3c). In addition, the ionic conductivity of the PVA hydrogel electrolyte gradually declines in a pure CO_2 atmosphere, while the PVA-0.075TMG hydrogel electrolyte remains almost unchanged (Figure 22.3d). As a result, by virtue of the good anti-CO_2 property, the cycle life of the flexible ZAB with PVA-0.075TMG hydrogel electrolytes can reach over 100 cycles at different CO_2 concentrations, significantly superior to the unmodified PVA of less than 40 cycles (Figure 22.3e).

Moreover, flexible MABs employing neutral hydrogel electrolytes can intrinsically show anti-CO_2 capability [55]. For example, ZABs can use NH_4Cl and $ZnCl_2$ as salts, AABs and MgABs can use NaCl as salt. Yet, the discharge voltages of neutral flexible MABs are about 0.3~0.4 V lower than those with alkaline electrolytes. Moreover, the rate performances and peak power densities of neutral flexible MABs are also not comparable to those with alkaline electrolytes.

22.3.4 Robust Hydrogel Electrolytes

Compared to the metal anode and air cathode, the hydrogel electrolyte is the most vulnerable part of MABs. During daily use, the battery may undergo various deformations, including bending, stretching, twisting, and crumpling, which may damage the fragile hydrogel electrolyte. Therefore, robust hydrogel electrolytes are urgently needed.

It is reported that adding polyol solvents (e.g., ethylene glycol, glycerol, and sorbitol) can significantly improve the mechanical property of hydrogels. This is because the polyols can enhance the interactions among polymer chains by plenty of hydrogen bonds, and the hydrogel becomes stronger [56]. However, the mixture of polyol and water is highly volatile, which will shorten the working life of flexible MABs. Hence, alternative strategies should be further explored.

Recently, researchers developed double-network hydrogels with enhanced mechanical properties [57]. Generally, the first network is formed by highly covalent cross-linking long chains, making it tough and strong. Meanwhile, the second network is loosely and physically crosslinked by short chains, endowing its softness and elasticity. Zhi and co-workers have designed a PANa-cellulose double-network hydrogel electrolyte for flexible ZABs [34]. The short cellulose chains intertwine with the long PANa chains and interact with them by hydrogen bonds, facilitating a uniform and robust double hydrogel network (Figure 22.4a). Therefore, the maximum strain and stress of the PANa-cellulose hydrogel can be extended several times to the pure PANa hydrogel (Figure 22.4b). Moreover, the PANa-cellulose hydrogel exhibits a good alkaline-tolerance property. After soaking in 6 M KOH solution, the stress-strain profile remains nearly unchanged compared to the unsoaked PANa-cellulose hydrogel (Figure 22.4b). As shown in Figure 22.4c, the PANa-cellulose hydrogel can be freely folded, rolled, twisted, and crumpled without damage. Furthermore, there is also no distinct degradation in the discharge and charge performances of the fabricated flexible ZABs after continuous deformations (Figure 22.4d). Apart from PANa-cellulose hydrogels, other double-network hydrogels are also reported, such as PANa-starch hydrogels, PAM-cellulose hydrogels, PAM-alginate hydrogels, PVA-bacterial cellulose hydrogels, etc. [18]. These double-network hydrogels show enhanced mechanical properties than the single-network hydrogels and can be further used in other aqueous flexible MABs besides ZABs.

22.3.5 Self-Healing Hydrogel Electrolytes

When suffering from more extreme deformations like cutting, the hydrogel will tear apart, leading to direct battery failure. Therefore, researchers have developed self-healing hydrogels to solve this problem. Normally, hydrogels are synthesized by chemically crosslinking with covalent bonds, which cannot automatically crosslink again once cut. For this reason, reversible physical bonds

Hydrogel Electrolytes

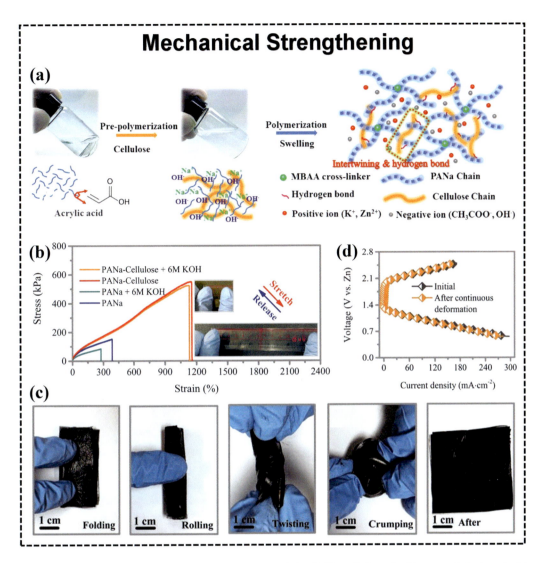

FIGURE 22.4 (a) Schematic of the synthesis steps of the PANa-cellulose double-network hydrogel. (b) Stress-strain profiles of the PANa, PANa+ 6M KOH, PANa-cellulose, PANa-cellulose+ 6M KOH hydrogels. (c) Various deformations of the PANa-cellulose hydrogel electrolyte. (d) Charge and discharge polarization profiles of the PANa-cellulose hydrogel electrolyte before and after continuous deformations. (Adapted with permission [34]. Copyright 2021, Wiley.)

like hydrogen and ionic bonds are employed. Huang and co-workers introduced Fe^{3+} ions in PANa hydrogel network to endow hydrogel with self-healing ability (Figure 22.5a) [58]. As shown in Figure 22.5b, the healed PANa-Fe^{3+} hydrogel showed a satisfactory mechanical property with a maximum 1000% strain and 200 kPa stress. For comparison, the pure PANa hydrogel didn't exhibit self-healing ability. As shown in Figure 22.5c, the hydrogel cut in two parts can be healed to become integrated again and light up a bulb. Moreover, even after several times cutting, the PANa-Fe^{3+} hydrogel can maintain its self-healing property.

Apart from the physical crosslinking strategy, the thermo-reversible hydrogel with sol-gel transition property can also be employed as self-healing hydrogels for MABs. For example, Pluronic F127 hydrogel is a solid-state gel at room temperature; however, it transforms to a liquid-state at subzero temperature (Figure 22.5d) [35]. Therefore, when the hydrogel is damaged, a simple cooling-recovery process can help the impaired hydrogel back to its initially intact state. The degraded battery performance by deformations can also get healed [59]. Moreover, the contact between hydrogel electrolytes and electrodes is not as good as between liquid electrolytes and electrodes. However, by virtue of the liquid state at subzero temperatures, the Pluronic F127 can build a conformal electrolyte-electrode interface, improving the battery performance (Figure 22.5e and f).

Although self-healing hydrogels are widely investigated, their applications in MABs are hardly reported [60]. The current issue for self-healing hydrogel is that the self-healing time is too long because it takes time to reconstruct hydrogel networks, which will affect the practical application in MABs. Besides, the high alkaline concentration may also limit the self-healing effect. Future flexible MAB researchers may focus on these research directions to develop alkaline-tolerant hydrogel electrolytes with rapid self-healing speed.

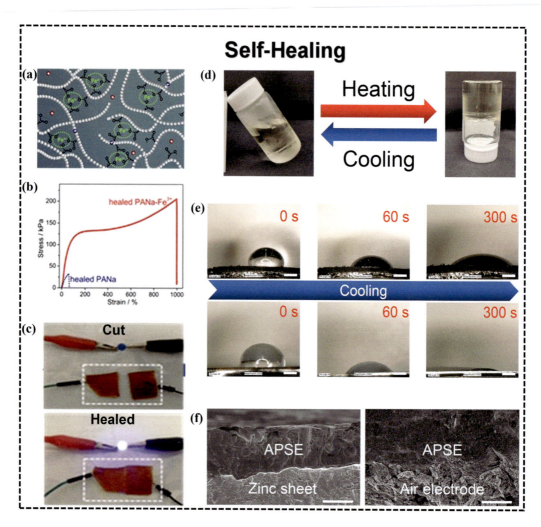

FIGURE 22.5 (a) Schematic of the PANa hydrogel network with Fe^{3+} ions. (b) Stress-strain profile comparison of the healed PANa and PANa-Fe^{3+} hydrogel electrolytes. (c) Optic photos of the cut and self-healed PANa-Fe^{3+} hydrogel electrolytes. (Adapted with permission [58]. Copyright 2018, Wiley.) (d) Sol-gel transition of Pluronic F127 hydrogel electrolyte. (e) Wetting process of the Pluronic F127 hydrogel electrolyte on the air electrode and zinc electrode, in a −5 °C environment. Scale bar: 100 μm. (f) SEM images of the conformal contact between Pluronic F127 hydrogel electrolyte and zinc anode/air cathode. Scale bar: 20 μm. (Adapted with permission [35]. Copyright 2021, American Chemical Society.)

Hydrogel Electrolytes

22.4 RECENT ADVANCES IN HYDROGEL-BASED MABs

22.4.1 DUAL-ELECTROLYTES FOR HIGH-VOLTAGE AND HIGH-ENERGY-DENSITY MABs

The low discharge voltage of the aqueous MABs has always been an issue prohibiting their wide application. Recently, researchers have proposed to decouple the electrolyte into an acid catholyte and an alkaline anolyte by using a bipolar membrane [61]. With this asymmetric structure, the cathodic potential can be significantly increased, and therefore the overall battery voltage is higher. For flexible MABs, it is reasonable to employ acid and alkaline hydrogel electrolytes. For example, Ni and co-workers proposed to use acid and alkaline Pluronic F127 hydrogels that can be decoupled but integrated to form an all-in-one acid-alkaline hydrogel electrolyte (Figure 22.6a) [38]. Besides, the high viscosity of Pluronic F127 hydrogel can effectively alleviate the mixture of acid and alkaline without the use of a bipolar membrane. Compared to the conventional alkaline normal-voltage ZAB, the asymmetric high-voltage ZAB exhibits a distinct increase in open-circuit voltage (OCV), which has reached over 2 V (Figure 22.6b). Besides, the discharge voltage of the high-voltage ZAB

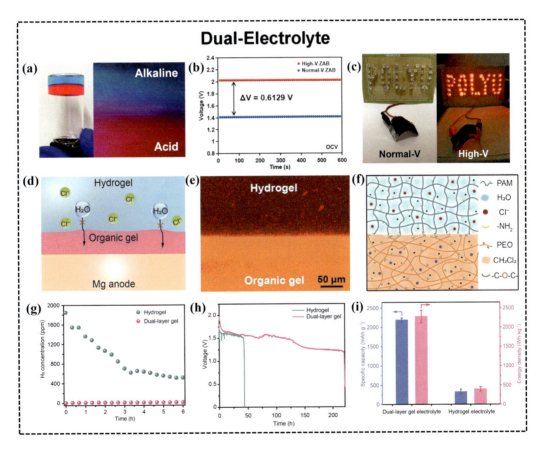

FIGURE 22.6 (a) Optic photos of the acid-alkaline hydrogel electrolyte. (b) Voltage comparison of the ZAB with acid-alkaline and alkaline hydrogel electrolytes. (c) Optic photo of the high-voltage ZAB lighting up a series of LED bulbs. (Adapted with permission [38]. Copyright 2022, Elsevier.) (d) Schematic of the mechanism of the dual-layer hydrogel in prohibiting Mg and water contact. (e) Microscopic image of the hydrogel-organic gel interface. (f) Schematic of the PAM hydrogel and PEO organic gel network. (g) Hydrogen concentration of the MgAB after discharged for 6 h using dual-layer gel electrolyte and hydrogel electrolytes. (h) Discharge profiles and (i) specific capacities and energy densities of the MgABs with dual-layer and hydrogel electrolytes. (Adapted with permission [39]. Copyright 2021, Wiley.)

is also 0.6532 V higher than the normal-voltage ZAB. As shown in Figure 22.6c, one high-voltage ZAB can light up a series of 1.5 V LED bulbs, while the normal-voltage ZAB cannot.

For AABs and MgABs with an active metal anode, the alkaline part can be substituted by a neutral or organic gel to obtain a higher energy density and longer battery discharge duration [39, 62]. For example, Peng and co-workers developed a hydrogel-organic gel dual-layer electrolyte for flexible MgABs. Since the Mg anode is very active, the use of an organic gel layer can effectively block the contact between the Mg anode and water molecules (Figure 22.6d). The integrated interface between the hydrogel and organic gel lowers the interfacial resistance because of the intimate interaction between PAM and PEO chains (Figures 22.6e and f). By virtue of the organic gel, the side reaction on the Mg anode surface, the hydrogen evolution reaction, can be greatly suppressed during battery discharge (Figure 22.6g). This will also lead to an improved battery voltage and lifetime. As shown in Figure 22.6h, the OCV of the MgAB with a dual-layer electrolyte is nearly 2 V, higher than the single hydrogel-based MgAB. Moreover, the discharge duration of the dual-layer MgAB is 220 hours, much higher than the single hydrogel-based MgAB. Finally, the specific capacity and energy density of the dual-layer MgAB are over 2000 mAh/g and 2000 Wh/kg, while that of the single hydrogel-based MgAB are less than 500 mAh/g and 500 Wh/kg (Figure 22.6i).

22.4.2 Eco-Friendly Hydrogel Electrolytes for MABs

The large-scale use of electronics will consume large amounts of batteries. To build a sustainable world, it is highly significant to recycle used batteries. For hydrogel electrolytes in MABs, biomaterials or biodegradable hydrogels are the first choices because they come from nature and can be recycled or reused. For example, Wang and co-workers proposed to use starch as hydrogel electrolytes for flexible ZABs [36]. Starch comes from a wide range of natural sources, such as corn, potato, wheat, etc. (Figure 22.7a). The cost of starch is as low as 0.005 RMB/g, which is very attractive for large-scale applications (Figure 22.7b). Moreover, the starch hydrogel electrolyte exhibits a high ionic conductivity of 111.5 mS/cm, and the fabricated flexible ZAB delivers a good discharge rate performance, comparable to the mainstream hydrogel electrolyte like PVA and PAA hydrogel electrolytes (Figure 22.7c). Most importantly, the used starch hydrogel can be degraded to CO_2 and H_2O by microorganisms or processed to powder for reuse. Apart from starch hydrogel, a PAM-alginate double-network hydrogel also shows similar regeneration properties [63]. When the ground PAM-alginate powders are dissolved in water, adding HCl can help the regeneration of a tough hydrogel (Figure 22.7d). Sun et al. also reported an agarose hydrogel for flexible AABs [37]. The synthesis process of the agarose hydrogel is very simple after the dissolved agarose powders in boiled water cool down to room temperature; the agarose chains can automatically crosslink into hydrogels. The obtained agarose hydrogel owns a satisfactory mechanical property that it can be freely bent and withstand a 200 g weight (Figure 22.7e). Moreover, as shown in Figure 22.7f, the $Fe(NO_3)_3$ and $CuSO_4$ solutions with red and blue colors diffuse quickly in the U-tube filled with agarose hydrogel. This phenomenon demonstrates the ionic migration within agarose hydrogel is very rapid, consistent with the high ionic conductivity of 318 mS/cm. In the future, these high-performance, eco-friendly, and low-cost bio hydrogel electrolytes may be widely used in flexible MABs.

22.5 SUMMARY AND FUTURE PERSPECTIVE

In this chapter, the fundamentals and development history of hydrogel electrolytes are systematically illustrated. In addition, their applications in MABs as multi-functional hydrogel electrolytes are discussed. Moreover, recent advances in hydrogel-based MABs are introduced. We have witnessed massive progress in hydrogel-based MABs from poor electrochemical performance to applicable-level performance. However, there is still space for improvement.

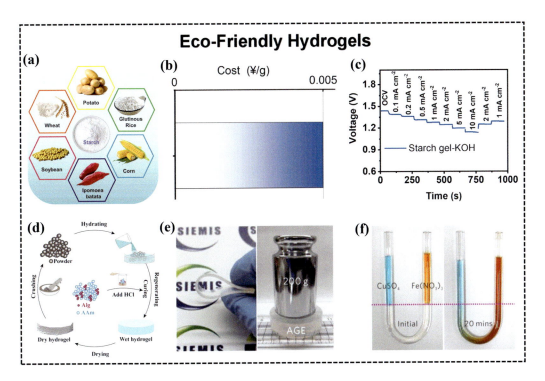

FIGURE 22.7 (a) Various sources of starch. (b) The cost of starch. (c) Discharge rate performance of the flexible ZAB with starch hydrogel with 6 M KOH. (Adapted with permission [36]. Copyright 2022, Elsevier.) (d) The recycling process of PAM-alginate double-network hydrogel. (Adapted with permission [63]. Copyright 2019, Wiley.) (e) Optic photos of the elastic and robust agarose hydrogel. (f) The mass transport and exchange of $Fe(NO_3)_3$ and $CuSO_4$ solutions in a U-tube filled with agarose hydrogel. (Adapted with permission [37]. Copyright 2021, Elsevier.)

(1) The interaction between the metal anode/air cathode and hydrogel electrolytes is unclear. For example, the functional groups on hydrogel chains may affect metal passivation/dendrite during discharge/charge. In addition, compared to the solid-liquid-gas (air cathode, liquid electrolyte, air) interface of liquid MABs, reactions on the solid-solid-gas (air cathode, hydrogel electrolyte, air) interface of flexible MABs are more complicated. Advanced characterization technologies are needed to help deeply understand the reaction mechanism on the interface, which can facilitate electrodes and hydrogel electrolytes design and further improve flexible MAB performance.

(2) All-round hydrogel electrolytes are needed. The current hydrogel electrolyte design only focuses on a single application scenario, such as subzero environments. However, other functions like anti-CO_2 and self-healing ability also should be taken into consideration simultaneously. To extend the working scenarios of flexible MABs, researchers should try to achieve all-around hydrogel electrolytes with good electrochemical properties, mechanical properties, environmental adaptability, etc.

(3) Future electronics may have various shapes. Therefore, apart from the current sandwich-type and cable-type MABs, other types of MABs designs are also needed to fit electronics. 3D printable hydrogels may become a possible way because their shape is patternable, albeit their application in MABs is still absent. Combined with 3D printed zinc anode and air cathode, it is believed the shape adaptable MABs can be achieved.

REFERENCES

[1] J.Y. Oh, Z. Bao, Second skin enabled by advanced electronics, Advanced Science. 6(11) (2019) 1900186.
[2] H. Li, Y. Ma, Y. Huang, Material innovation and mechanics design for substrates and encapsulation of flexible electronics: A review, Materials Horizons. (2021).
[3] P. Wang, M. Hu, H. Wang, Z. Chen, Y. Feng, J. Wang, W. Ling, Y. Huang, The evolution of flexible electronics: From nature, beyond nature, and to nature, Advanced Science. 7(20) (2020) 2001116.
[4] Z. Zhou, H. Zhang, J. Liu, W. Huang, Flexible electronics from intrinsically soft materials, Giant. 6 (2021) 100051.
[5] X. Fan, B. Liu, J. Ding, Y. Deng, X. Han, W. Hu, C. Zhong, Flexible and wearable power sources for next-generation wearable electronics, Batteries & Supercaps. 3(12) (2020) 1262–1274.
[6] X. Gong, Q. Yang, C. Zhi, P.S. Lee, Stretchable energy storage devices: From materials and structural design to device assembly, Advanced Energy Materials. (2020) 2003308.
[7] G. Qian, X. Liao, Y. Zhu, F. Pan, X. Chen, Y. Yang, Designing flexible lithium-ion batteries by structural engineering, ACS Energy Letters. 4(3) (2019) 690–701.
[8] Y. Zhao, J. Guo, Development of flexible Li-ion batteries for flexible electronics, InfoMat 2(5) (2020) 866–878.
[9] W. Shen, K. Li, Y. Lv, T. Xu, D. Wei, Z. Liu, Highly-safe and ultra-stable all-flexible gel polymer lithium ion batteries aiming for scalable applications, Advanced Energy Materials. 10(21) (2020) 1904281.
[10] P. Tan, B. Chen, H. Xu, H. Zhang, W. Cai, M. Ni, M. Liu, Z. Shao, Flexible Zn- and Li-air batteries: Recent advances, challenges, and future perspectives, Energy Environ. Sci. 10(10) (2017) 2056–2080.
[11] Y. Zhang, Y.-P. Deng, J. Wang, Y. Jiang, G. Cui, L. Shui, A. Yu, X. Wang, Z. Chen, Recent progress on flexible Zn-air batteries, Energy Storage Materials. 35 (2021) 538–549.
[12] S. Wu, S. Hu, Q. Zhang, D. Sun, P. Wu, Y. Tang, H. Wang, Hybrid high-concentration electrolyte significantly strengthens the practicability of alkaline aluminum-air battery, Energy Storage Materials. 31 (2020) 310–317.
[13] Y. Guo, J. Bae, F. Zhao, G. Yu, Functional hydrogels for next-generation batteries and supercapacitors, Trends in Chemistry. 1(3) (2019) 335–348.
[14] T. Liu, S. Zhao, Y. Wang, J. Yu, Y. Dai, J. Wang, X. Sun, K. Liu, M. Ni, In situ anchoring Co-N-C nanoparticles on Co4N nanosheets toward ultrastable flexible self-supported bifunctional oxygen electrocatalyst enables recyclable Zn-air batteries over 10 000 cycles and fast charging, Small. 18(7) (2022) 2105887.
[15] S. Hosseini, S.M. Soltani, Y.-Y. Li, Current status and technical challenges of electrolytes in zinc-air batteries: An in-depth review, Chem. Eng. J. 408 (2021) 127241.
[16] P. Zhang, K. Wang, P. Pei, M. Zuo, M. Wei, X. Liu, Y. Xiao, J. Xiong, Selection of hydrogel electrolytes for flexible zinc-air batteries, Materials Today Chemistry. 21 (2021) 100538.
[17] Y. Guo, J. Bae, Z. Fang, P. Li, F. Zhao, G. Yu, Hydrogels and hydrogel-derived materials for energy and water sustainability, Chem. Rev. 120(15) (2020) 7642–7707.
[18] S. Zhao, Y. Zuo, T. Liu, S. Zhai, Y. Dai, Z. Guo, Y. Wang, Q. He, L. Xia, C. Zhi, Multi-functional hydrogels for flexible zinc-based batteries working under extreme conditions, Advanced Energy Materials. (2021) 2101749.
[19] Y. Huang, Z. Li, Z. Pei, Z. Liu, H. Li, M. Zhu, J. Fan, Q. Dai, M. Zhang, L. Dai, C. Zhi, Solid-state rechargeable Zn//NiCo and Zn-air batteries with ultralong lifetime and high capacity: The role of a sodium polyacrylate hydrogel electrolyte, Advanced Energy Materials. 8(31) (2018) 1802288.
[20] Z. Wei, J. He, T. Liang, H. Oh, J. Athas, Z. Tong, C. Wang, Z. Nie, Autonomous self-healing of poly(acrylic acid) hydrogels induced by the migration of ferric ions, Polymer Chemistry. 4(17) (2013) 4601.
[21] Z. Song, J. Ding, B. Liu, X. Liu, X. Han, Y. Deng, W. Hu, C. Zhong, A rechargeable Zn-air battery with high energy efficiency and long life enabled by a highly water-retentive gel electrolyte with reaction modifier, Adv. Mater. 32(22) (2020) e1908127.
[22] F. Chen, D. Zhou, J. Wang, T. Li, X. Zhou, T. Gan, S. Handschuh-Wang, X. Zhou, Rational fabrication of anti-freezing, non-drying tough organohydrogels by one-pot solvent displacement, Angew. Chem. 130(22) (2018) 6678–6681.
[23] G. Wu, S. Lin, C. Yang, Alkaline Zn-air and Al-air cells based on novel solid PVA/PAA polymer electrolyte membranes, J. Membr. Sci. 280(1–2) (2006) 802–808.
[24] C.C. Yang, S. Lin, Preparation of alkaline PVA-based polymer electrolytes for Ni-MH and Zn-air batteries, J. Appl. Electrochem. 33(9) (2003) 777–784.

[25] C.-C. Yang, S.-J. Lin, Alkaline composite PEO-PVA-glass-fibre-mat polymer electrolyte for Zn-air battery, J. Power Sources. 112(2) (2002) 497–503.

[26] Z. Zhang, C. Zuo, Z. Liu, Y. Yu, Y. Zuo, Y. Song, All-solid-state Al-air batteries with polymer alkaline gel electrolyte, J. Power Sources. 251 (2014) 470–475.

[27] J. Park, M. Park, G. Nam, J.S. Lee, J. Cho, All-solid-state cable-type flexible zinc-air battery, Adv. Mater. 27(8) (2015) 1396–1401.

[28] J. Fu, D.U. Lee, F.M. Hassan, L. Yang, Z. Bai, M.G. Park, Z. Chen, Flexible high-energy polymer-electrolyte-based rechargeable zinc-air batteries, Adv. Mater. 27(37) (2015) 5617–5622.

[29] Y. Xu, Y. Zhao, J. Ren, Y. Zhang, H. Peng, An all-solid-state fiber-shaped aluminum-air battery with flexibility, stretchability, and high electrochemical performance, Angew. Chem. 128(28) (2016) 8111–8114.

[30] H. Miao, B. Chen, S. Li, X. Wu, Q. Wang, C. Zhang, Z. Sun, H. Li, All-solid-state flexible zinc-air battery with polyacrylamide alkaline gel electrolyte, J. Power Sources. 450 (2020) 227653.

[31] Y. Huang, Z. Li, Z. Pei, Z. Liu, H. Li, M. Zhu, J. Fan, Q. Dai, M. Zhang, L. Dai, Solid-state rechargeable Zn//NiCo and Zn-air batteries with ultralong lifetime and high capacity: The role of a sodium polyacrylate hydrogel electrolyte, Advanced Energy Materials. 8(31) (2018) 1802288.

[32] R. Chen, X. Xu, S. Peng, J. Chen, D. Yu, C. Xiao, Y. Li, Y. Chen, X. Hu, M. Liu, H. Yang, I. Wyman, X. Wu, A flexible and safe aqueous zinc-air battery with a wide operating temperature range from −20 to 70 °C, ACS Sustainable Chemistry & Engineering. 8(31) (2020) 11501–11511.

[33] X. Fan, J. Liu, Z. Song, X. Han, Y. Deng, C. Zhong, W. Hu, Porous nanocomposite gel polymer electrolyte with high ionic conductivity and superior electrolyte retention capability for long-cycle-life flexible zinc-air batteries, Nano Energy. 56 (2019) 454–462.

[34] L. Ma, S. Chen, D. Wang, Q. Yang, F. Mo, G. Liang, N. Li, H. Zhang, J.A. Zapien, C. Zhi, Super-stretchable zinc-air batteries based on an alkaline-tolerant dual-network hydrogel electrolyte, Advanced Energy Materials. 9(12) (2019) 1803046.

[35] S. Zhao, D. Xia, M. Li, D. Cheng, K. Wang, Y.S. Meng, Z. Chen, J. Bae, Self-healing and anti-CO2 hydrogels for flexible solid-state zinc-air batteries, ACS Applied Materials & Interfaces. (2021).

[36] Y. Zuo, K. Wang, M. Wei, S. Zhao, P. Zhang, P. Pei, Starch gel for flexible rechargeable zinc-air batteries, Cell Reports Physical Science. 3(1) (2022) 100687.

[37] P. Sun, J. Chen, Y. Huang, J.-H. Tian, S. Li, G. Wang, Q. Zhang, Z. Tian, L. Zhang, High-Strength agarose gel electrolyte enables long-endurance wearable Al-air batteries with greatly suppressed self-corrosion, Energy Storage Materials. 34 (2021) 427–435.

[38] S. Zhao, T. Liu, Y. Dai, Y. Wang, Z. Guo, S. Zhai, J. Yu, C. Zhi, M. Ni, All-in-one and bipolar-membrane-free acid-alkaline hydrogel electrolytes for flexible high-voltage Zn-air batteries, Chem. Eng. J. 430 (2022) 132718.

[39] L. Li, H. Chen, E. He, L. Wang, T. Ye, J. Lu, Y. Jiao, J. Wang, R. Gao, H. Peng, High-energy-density magnesium-air battery based on dual-layer gel electrolyte, Angew. Chem. Int. Ed. (2021).

[40] Y. Zhao, Z. Chen, F. Mo, D. Wang, Y. Guo, Z. Liu, X. Li, Q. Li, G. Liang, C. Zhi, Aqueous rechargeable metal-ion batteries working at subzero temperatures, Advanced Science 8(1) (2020) 2002590.

[41] Z. Pei, Z. Yuan, C. Wang, S. Zhao, J. Fei, L. Wei, J. Chen, C. Wang, R. Qi, Z. Liu, A flexible rechargeable zinc-air battery with excellent low-temperature adaptability, Angew. Chem. 132(12) (2020) 4823–4829.

[42] M. Zhu, X. Wang, H. Tang, J. Wang, Q. Hao, L. Liu, Y. Li, K. Zhang, O.G. Schmidt, Antifreezing hydrogel with high zinc reversibility for flexible and durable aqueous batteries by cooperative hydrated cations, Adv. Funct. Mater. 30(6) (2019) 1907218.

[43] H. Wang, J. Liu, J. Wang, M. Hu, Y. Feng, P. Wang, Y. Wang, N. Nie, J. Zhang, H. Chen, Q. Yuan, J. Wu, Y. Huang, Concentrated hydrogel electrolyte-enabled aqueous rechargeable NiCo//Zn battery working from −20 to 50 degrees C, ACS Applied Materials & Interfaces. 11(1) (2019) 49–55.

[44] Y. Zuo, Y. Yu, C. Zuo, C. Ning, H. Liu, Z. Gu, Q. Cao, C. Shen, Low-temperature performance of Al-air batteries, Energies. 12(4) (2019) 612.

[45] Y. Bai, B. Chen, F. Xiang, J. Zhou, H. Wang, Z. Suo, Transparent hydrogel with enhanced water retention capacity by introducing highly hydratable salt, Appl. Phys. Lett. 105(15) (2014) 151903.

[46] D. Jiang, H. Wang, S. Wu, X. Sun, J. Li, Flexible zinc-air battery with high energy efficiency and freezing tolerance enabled by DMSO-based organohydrogel electrolyte, Small Methods. 6(1) (2022) 2101043.

[47] J.-B. Wang, J.-M. Wang, H.-B. Shao, J.-Q. Zhang, C.-N. Cao, The corrosion and electrochemical behaviour of pure aluminium in alkaline methanol solutions, J. Appl. Electrochem. 37(6) (2007) 753–758.

[48] Y. Wang, W. Pan, K.W. Leong, S. Luo, X. Zhao, D.Y. Leung, Solid-state Al-air battery with an ethanol gel electrolyte, Green Energy & Environment. (2021).

[49] S. Zhao, K. Wang, S. Tang, X. Liu, K. Peng, Y. Xiao, Y. Chen, A new solid-state zinc-air battery for fast charge, Energy Technology. 8(5) (2020) 1901229.

[50] Z. Song, J. Ding, B. Liu, X. Liu, X. Han, Y. Deng, W. Hu, C. Zhong, A rechargeable Zn—air battery with high energy efficiency and long life enabled by a highly water-retentive gel electrolyte with reaction modifier, Adv. Mater. 32(22) (2020) 1908127.

[51] A. Alam, Y. Zhang, H.-C. Kuan, S.-H. Lee, J. Ma, Polymer composite hydrogels containing carbon nanomaterials—Morphology and mechanical and functional performance, Prog. Polym. Sci. 77 (2018) 1–18.

[52] N. Xu, Y. Zhang, M. Wang, X. Fan, T. Zhang, L. Peng, X.-D. Zhou, J. Qiao, High-performing rechargeable/flexible zinc-air batteries by coordinated hierarchical Bi-metallic electrocatalyst and heterostructure anion exchange membrane, Nano Energy. 65 (2019) 104021.

[53] K. Tan, R. Thorpe, Gas diffusion into viscous and non-Newtonian liquids, Chem. Eng. Sci. 47(13–14) (1992) 3565–3572.

[54] Y. Zhou, J. Pan, X. Ou, Q. Liu, Y. Hu, W. Li, R. Wu, J. Wen, F. Yan, CO_2 ionized poly (vinyl alcohol) electrolyte for CO_2-tolerant Zn-air batteries, Advanced Energy Materials 11(38) (2021) 2102047.

[55] Y. Li, X. Fan, X. Liu, S. Qu, J. Liu, J. Ding, X. Han, Y. Deng, W. Hu, C. Zhong, Long-battery-life flexible zinc—air battery with near-neutral polymer electrolyte and nanoporous integrated air electrode, Journal of Materials Chemistry A. 7(44) (2019) 25449–25457.

[56] J.W. Zhang, D.D. Dong, X.Y. Guan, E.M. Zhang, Y.M. Chen, K. Yang, Y.X. Zhang, M.M.B. Khan, Y. Arfat, Y. Aziz, Physical organohydrogels with extreme strength and temperature tolerance, Frontiers in Chemistry. 8 (2020) 102.

[57] J.P. Gong, Y. Katsuyama, T. Kurokawa, Y. Osada, Double-network hydrogels with extremely high mechanical strength, Adv. Mater. 15(14) (2003) 1155–1158.

[58] Y. Huang, J. Liu, J. Wang, M. Hu, F. Mo, G. Liang, C. Zhi, An intrinsically self-healing NiCo‖Zn rechargeable battery with a self-healable ferric-ion-crosslinking sodium polyacrylate hydrogel electrolyte, Angew. Chem. Int. Ed. Engl. 57(31) (2018) 9810–9813.

[59] Y. Zuo, K. Wang, S. Zhao, M. Wei, X. Liu, P. Zhang, Y. Xiao, J. Xiong, A high areal capacity solid-state zinc-air battery via interface optimization of electrode and electrolyte, Chem. Eng. J. 430 (2022) 132996.

[60] D.L. Taylor, M. in het Panhuis, Self-healing hydrogels, Adv. Mater. 28(41) (2016) 9060–9093.

[61] P. Cai, Y. Li, J. Chen, J. Jia, G. Wang, Z. Wen, An asymmetric-electrolyte Zn—air battery with ultrahigh power density and energy density, ChemElectroChem 5(4) (2018) 589–592.

[62] K.W. Leong, Y. Wang, W. Pan, S. Luo, X. Zhao, D.Y. Leung, Doubling the power output of a Mg-air battery with an acid-salt dual-electrolyte configuration, J. Power Sources. 506 (2021) 230144.

[63] G. Qu, Y. Li, Y. Yu, Y. Huang, W. Zhang, H. Zhang, Z. Liu, T. Kong, Spontaneously regenerative tough hydrogels, Angew. Chem. Int. Ed. 58(32) (2019) 10951–10955.

23 Wearable Metal-Air Batteries

Arpana Agrawal and Chaudhery Mustansar Hussain

CONTENTS

23.1 Introduction 335
23.2 Flexible Components of Wearable MABs 336
23.3 Structure Configuration of Wearable MABs 341
23.4 Conclusion 344
References 344

23.1 INTRODUCTION

Advancement in flexible, wearable electronic devices has sparked widespread interest in science and technology. Several inspiring ideas for flexible devices, for example, wearable sensors, smart clothes, smart watches, band-aids, and various healthcare devices, have been envisioned and depict potential application recommendations for the future [1–3]. In contrast to traditional electronic devices, the new generation of electronic devices are compact, versatile, stretchable, flexible, portable, and even wearable. Conventional batteries, because of being rigid, are commonly unsuitable for powering such devices, and hence it is imperative to develop more advanced power systems to facilitate their progress.

Metal-air batteries (MAB) have recently emerged as new and promising candidates for fabricating new generation wearable devices owing to their ability to provide long-lasting power supply because of high specific capacities, high energy density, cost-effectiveness, high durability, availability of raw materials, and environmental friendliness, flexibility, and excellent better adaptableness to uneven human body surfaces [4, 5]. A conventional MAB is composed of a metal anode, an air cathode, and an electrolyte where the metal (M) at the anode gets oxidized to form metal ions (M^+) which flow through the electrolyte to the cathode, where they react with O_2 to produce metal oxides (MO_{2x}). Mathematically,

$$M_{(s)} \leftrightarrow M^+ + e^- \text{ (Anode reaction)} \tag{1}$$

$$M^+ + xO_2 + e^- \leftrightarrow MO_{2x} \text{ (Cathode reaction)} \tag{2}$$

$$M + xO_2 \leftrightarrow MO_{2x} \text{ (Overall cell reaction)} \tag{3}$$

In contrast to this, in wearable MAB, all of these components are flexible. It should be noted that the air electrode is critical to the battery's functioning and not only affects the electrocatalysis properties, but also the mechanical flexibility necessary for flexible batteries. The basic working principle of MAB involves the dissolution and deposition of cations on the negative (metal) electrode, as well as the oxygen reduction reaction (ORR) and oxygen evolution reaction (OER) on the air (or positive) electrode. Depending upon the anode material, MAB can be a lithium-air battery (LAB), sodium-air battery, zinc-air battery (ZAB), aluminum-air battery, or magnesium-air battery with certain advantages as well as limitations. However, LAB and ZAB are more popular.

It is worth mentioning here that traditional MAB is bulky and inflexible, with no effectively flexible components or adjustable cell configuration. They usually utilize liquid electrolytes and rigid electrodes that are unsuitable for flexible applications. Furthermore, bringing flexible MAB into

FIGURE 23.1 Requirements (flexible materials and structural configurations) and advantages of a wearable MAB.

practical applications will be difficult due to the structural design and competent battery configuration. Until now, significant attempts have been devoted to bringing the flexibility of MAB by utilizing flexible battery components, mainly the electrolyte and electrodes, and modifying the structural configurations, so that these MABs can be effectively exploited for fabricating wearable electronic devices. These flexible battery components include flexible electrolytes e.g. gel/polymer [6–12] and flexible electrode materials such as carbon-based materials, conductive polymers, sheets/foils of metals or their oxides [13–15] while the structural configurations are generally the planar sandwich-type structure [16] or a fiber-type structure [17]. Figure 23.1 illustrates the basic requirements to fabricate a wearable flexible MAB including flexible materials and structural configurations and the various advantages offered by a wearable MAB.

Accordingly, the present chapter describes the factors including flexible components and structural configurations (planar sandwich structure and fiber structure) that are crucial for engineering flexible and wearable MAB, with particular focus on LAB and ZAB. Such MAB can withstand various deforming conditions such as twisting, folding, rolling, bending, and stretching, making them extremely flexible and wearable.

23.2 FLEXIBLE COMPONENTS OF WEARABLE MABs

Advanced wearables are usually applied straightforwardly on bowed and soft surfaces of the human body and can endure several deformable conditions. As a result, the advancement in novel electrode and electrolyte materials that are flexible and stretchable with extraordinary properties is critical. The electrolyte, as a fundamental component of MABs, is critical in flowing metal ions as well as oxygen species during battery operations. Regrettably, frequent mechanical deformations of the flexible battery during practical usage lead to leakage of the traditional liquid electrolyte. Furthermore, open structured MABs are extensively used. In these scenarios, leakage issues result in both battery failure and pollution of the environment, as well as a safety risk for humans. Considering such concerns, it is imperative to prepare flexible electrolytes simultaneously functioning as a separator and an ion conductor, avoiding leakage issues and easing the manufacturing process. The use of a gel or polymer approach in the construction of flexible electrolytes for MABs is an effective method that can be obtained by combining the solvent, metal salt ions, and precursor; after polymerization and solvent evaporation, the ions will be encapsulated in the cross-linked polymer, yielding a freestanding polymer electrolyte [18–20]. Such gel/polymer electrolytes include poly(acrylic acid) (PAA)-based aqueous electrolyte [6], poly(ethylene oxide) (PEO)-based aqueous and polymer electrolyte [7, 10], and poly(vinyl alcohol) (PVA)-based aqueous electrolyte [8, 9], poly(vinylidene fluoride) (PVDF), poly(vinylidene fluoride-co-hexafluoropropene) (PVDF-HFP) based polymer electrolyte [11], ionic liquid electrolyte [12], etc. These polymer electrolytes have exceptional flexibility and mechanical characteristics under multiple deformations suggesting their potentiality in flexible MABs.

Apart from the flexible electrolyte, flexible electrodes are equally important for the fabrication of wearable MABs. Because thin metal foil/sheets have some flexibility, they can be employed as an anode in flexible MABs. However, the quest for a flexible air cathode has become much more vital; because MABs rely on the ORR and OER, the cathode must have efficient air permeability. Numerous flexible substrates, regrettably, have low porosity or poor electrochemical catalytic activities, lowering the specific capacity. The commonly used cathode is typically a carbon black powder coated metallic current collector where the powder may fall off upon turning it on, resulting in battery malfunction. To meet the demands imposed by the production of flexible MABs, flexible materials with high electrical conductivity are being sought by researchers with fascinating catalytic capability. Consequently, carbon-based materials (carbon nanofibers (CNFs), carbon cloth (CC)/carbon textiles (CTs), carbon nanotubes (CNTs), graphene-based materials, etc.) and conductive polymers have been extensively employed as flexible electrode substrates for wearable MABs. Carbon-based materials are appealing due to their low cost, reasonably high electrical conductivity, and mechanical stability and remain stable under various deforming conditions. CC/CT is also more appealing for flexible cathodes than stainless steel mesh due to its greater flexibility and porosity. Polymer-based materials are naturally flexible and can withstand bending, folding, twisting, and other deformations.

Li et al. [21] have established an integrated structured flexible LAB practicable under unfavorable conditions by combining a three-dimensional open-structured Co_3O_4@MnO_2 cathode with a composite lithium anode wrapped in a gel electrolyte. Here, the anode is fabricated via a straightforward, inexpensive, and efficient rolling process which considerably reduces lithium electrode fatigue fracture and is enclosed by gel electrolyte. This serves as a protecting layer and facilitates the reduction of electrode/electrolyte interfacial resistance as well as avoids rusting of the anode. As a result, the battery demonstrates stable performance even after 100 cycles under ambient air conditions and can withstand harsh circumstances including bending, twisting, and cutting, suggesting their utility for wearable device applications. Lei et al. [22] have also discussed the fabrication of flexible LAB under ambient air conditions utilizing gel electrolytes. Dai et al. [23] have provided a comprehensive review on graphene-based electrodes for various flexible battery applications. Interestingly, wearable LABs utilizing bamboo slips have also been reported, endowing a very high energy density [24]. Liu et al. [25] have demonstrated the construction of flexible LABs employing the cathode containing arrays of TiO_2 nanowire deposited on CTs and designated as TiO_2 NAs/CT. Figure 23.2(a) depicts the manufacturing procedure of the cathode and Figures 23.2(b-d) show the scanning electron microscopy (SEM) images of the cathode which reveals that TiO2 NAs with diameters of about 50 nm were grown vertically on the CTs. Excellent electrochemical properties

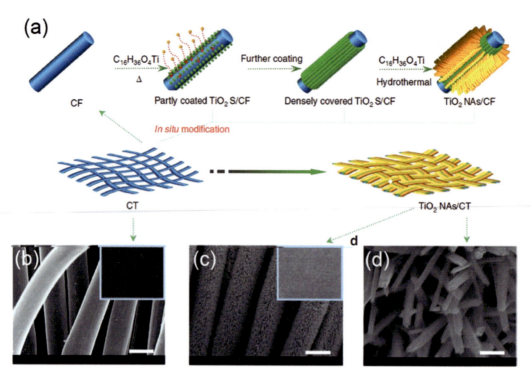

FIGURE 23.2 (a) Schematic depiction for the fabrication procedure of TiO$_2$ NAs/CT. SEM image with photograph (inset) of pristine-CT (b) and TiO$_2$ NAs/CT cathode (c). (d) Enlarged image of TiO$_2$ NAs/CT cathode. (Adapted with permission [25]. Copyright (2015), Copyright The Authors, some rights reserved. Distributed under a Creative Commons Attribution License 4.0 (CC BY).)

can be acquired for this LAB upon bending as well as twisting with outstanding recoverability, improved cycle life, and lower life cycle cost.

Jing et al. [26] demonstrated the performance of a flexible solid-state ZAB where nanoscale hair-like catalysts were straightforwardly and perpendicularly deposited on a flexible stainless steel mesh to create a rechargeable air electrode. The chemical vapor deposition technique was employed to develop nitrogen-incorporated multiwalled CNT, followed by electrodeposition and calcination of iron catalyst, and was then assembled with Co$_3$O$_4$ nanopetals. Even at a torsion angle of 360, this air electrode demonstrated superior flexibility. The fabricated flexible ZAB is highly stable and exhibits high energy density (~847.6 Wh/kg), allowing it to be used in smart and wearable devices. Huang et al. [27] have presented a novel flexible ZAB that makes use of flexible cotton textile waste (CTW) substrate metalized with nickel for the preparation of Zn electroplated and catalyst (NiFe hydroxide) electroplated electrodes as depicted in Figure 23.3(a). Both the electrodeposited electrodes are highly flexible with excellent sticking to CTW under several deformations and robust sonication. This would be advantageous in the development of a ZAB with high flexibility and mechanical characteristics. The cross-sectional SEM scan of the constructed CTW-based ZAB is revealed in Figure 23.3(b) depicting the stacked structure with an anode, cathode, and hydrogel electrolyte layer. Galvanostatic charging/discharging cycles of the fabricated ZAB based on CTW (at 1 mA/cm^2; discharge cycle period: 20 min; charging time: 20 min) is shown in Figure 23.3(c) which illustrates better stability of the assembled ZAB with no considerable potential change throughout the evaluation tenure, indicating charge and discharge process steadiness of the fabricated ZAB. Discharge and charge potentials of 0.94 V and 1.99 V are observed at a current density of 1 mA/cm^2, respectively, with a potential gap of about 1.05 V, comparable to that of

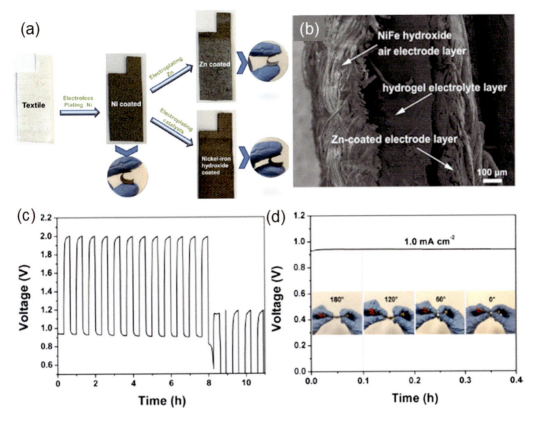

FIGURE 23.3 (a) Fabrication processes for the electroplated Zn-coated electrode and electroplated NiFe hydroxide electrode. (b) Cross-sectional SEM image of the constructed ZAB. (c) Galvanostatic charge and discharge cycles of the fabricated ZAB at a current density of 1 mA cm^{-2}. (d) Galvanostatic discharge curve of cotton textile-based ZAB at several bending deformation angles. (Adapted with permission [27]. Copyright (2019), Copyright The Authors, some rights reserved. Distributed under a Creative Commons Attribution License 4.0 (CC BY).)

commercially available Pt/C- and Ir/C-mixed catalyst-based ZAB, indicating the excellent catalytic activity of NiFe hydroxide. Furthermore, the failure time of ZAB during the cycling test can be seen, and is approximately 8 h, indicating its durability. Figure 23.3(d) depicts the discharge stability of assembled CTW-based ZAB at 0°, 60°, 120°, and 180° bending angles with no considerable loss in discharge behavior, signifying extraordinary flexibility with excellent discharge stability of the fabricated ZAB.

Meng et al. [28] have demonstrated the fabrication of ZAB (capacity: 774 mAh/g at 10 mA/cm^2) using flexible freestanding electrode comprising Co_4N, carbon fiber, and CC (designated as Co_4N/CNW/CC) with good ORR and OER actions indicating high and long-term catalytic performance. Co_4N/CNW/CC was created by pyrolyzing ZIF-67 on a polypyrrole nanofiber developed on CC, with the liberation of nitrogenous gases promoting the conversion of ZIF-67 into Co_4N. Resulted electrode exhibited great flexibility. Reduced overpotential with steady current density retention was observed in OER and ORR activities. The design of knittable ZAB using atomic layer Co_3O_4 nanosheets has also been discussed [29]. Liu et al. [30] have demonstrated the fabrication of flexible ZAB using bifunctional catalytic electrodes made up of nonporous carbon fiber films. A significantly stretchable ZAB array for wearable electronics possessing manipulative output voltage and current has also been illustrated by Qu et al. [31]. They have constructed a layer-by-layer structure of

ZAB array consisting of Co_3O_4 nanosheet showing unwavering electrochemical response in maximum strain conditions. This fabricated ZAB can get discharged under bending and stretching circumstances and could produce voltage ranging from 1 to 4 V after rearranging the electrode array. Because of the aforementioned properties, this ZAB has been employed to glow a green light band consisting of 60 light emitting diodes (LEDs).

Jin et al. [32] developed a new plasma-based approach for fabricating extremely bendable cathode prepared from carbon layer doped with nitrogen and encapsulated Co and FeCo nanoparticles hybrid nanowire arrays (designated as Co-FeCo/N-G) exhibiting excellent ORR/OER activity, anodic Zn foil, and hydrogel electrolyte. The open circuit voltage of the Co-FeCo/N-G cathode-based ZAB was found to be 1.419 V, as shown in Figure 23.4(a). Figure 23.4(b) shows the voltage versus current density and power density versus current density patterns for ZABs prepared using different cathode materials (Co-FeCo/N-G, Co/N-G, Co-FeCo/G) under atmospheric conditions. It is interesting to note that Co/FeCo/N-G cathode-based ZAB has a significantly greater peak power density (82 mW/cm²) than that of the Co/N-G-based ZAB (56 mW/cm²) and Co/FeCo/G ZAB (25 mW/cm²). They have also examined the practical utility of the fabricated Co-FeCo/N-G based ZAB under several bending conditions and found that it can be folded to differing degrees, including 180° while maintaining nearly unchanged discharging/charging efficiency (Figure 23.4(c)). The 180°-folded ZAB utilizing Co-FeCo/N-G cathode then restores its initial degree (0°), with no effect on battery performance. Moreover, the Co-FeCo/N-G ZAB can energize LEDs of several colors in both the unfolded (Figure 23.4(d)) and folded (Figure 23.4(e)) conditions. The desirable flexible character of Co-FeCo/N-G cathode ZAB is mainly ascribed to the strong sticking of catalysts and substrates in the cathode, preventing the detaching of active materials from CC under extreme deforming circumstances. Consequently, the flexible Co-FeCo/N-G based ZAB has a huge upside for use in wearable/flexible electronics.

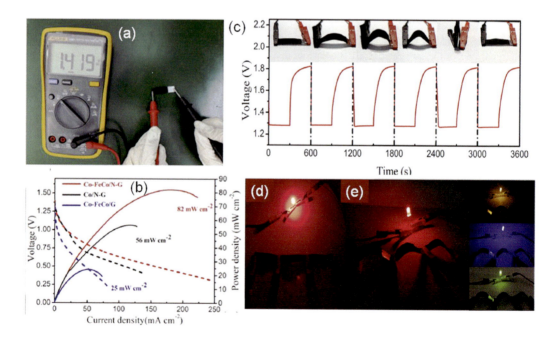

FIGURE 23.4 (a) Flexible ZAB open circuit voltage employing Co-FeCo/N-G cathode. (b) Discharge polarization and power density curves for the Co/N-G, Co/N-G, and Co/FeCo/G based ZAB. (c) Cyclic stability of Co-FeCo/N-G ZAB at 1 mA/cm² in different bending states. (d) LED powered in series using Co-FeCo/N-G ZABs. (e) Various LEDs powered by Co-FeCo/N-G ZABs under various binding conditions. (Adapted with permission [32], Copyright (2019), Elsevier.)

23.3 STRUCTURE CONFIGURATION OF WEARABLE MABs

It is worth mentioning here that the electrochemical performance under several deformable conditions such as rolling, bending, twisting, folding, and stretching, is one of the decisive factors for the flexibility of MAB and hence its wearability. Depending upon the structure configuration, wearable and flexible MABs can be classified mainly into two categories: (i) planar sandwich structure and (ii) fiber-shaped structure. The former is fabricated by sandwiching the flexible electrolyte between the electrodes and facilitates the bending of the flexible wearable MAB in on-dimension. In contrast to this, a fiber-structured MAB is simply a coaxial structure consisting of a flexible air cathode, polymer electrolyte, spiral/straight metal wire as anode, and a packaging insulator.

Fu et al. [33] have first proposed the flexible planar sandwich structured ZAB by sandwiching a PVA gel polymer electrolyte between the Zn film anode and $LaNiO_3$/NCNT loaded CC cathode. The fabricated flexible ZAB can withstand various deformable conditions (rolling, bending, and twisting) and shows steady performance for over 120 cycles at charging-discharging rates of 50 A/kg and excellent energy density (581 Wh/kg) at a current density of ~ 25 A/kg. Fan et al. [34] have fabricated a bendable sandwich type ZAB using a novel porous-structured PVA-based nanocomposite gel polymer electrolyte (GPE) with the optimum addition of silica (SiO_2). The porous PVA-based nanocomposite GPE was formed by gelling a mixture of PVA, PEG, and SiO_2 powder (in specific weight ratio), accompanied by solubilizing the pore-forming agent (i.e., PEG) in absolute ethanol. The obtained polymer membrane was then dipped in a KOH solution. By sandwiching the manufactured GPE between a zinc foil and a catalyst-loaded air electrode, the layered structured ZAB was constructed. Figure 23.5(a) depicts the schematic representation of the porous PVA-based nanocomposite GPE's porous ZAB structure and also the inner structure of this nanocomposite GPE. Because of the excellent electronic, thermal, and mechanical characteristics of the PVA-based nanocomposite GPE, the constructed flexible ZAB demonstrated outstanding cycle stability, discharge performance, and power density. Two as-assembled ZABs were then encapsulated in series by aluminum plastic films containing breathable holes. Two ZAB sets have an open circuit potential of 2.54 V, which is considerably large (Figure 23.5(b)) and can be used to power various electronic devices. By integrating the constructed ZABs in series, a bracelet-shaped ZAB pair can

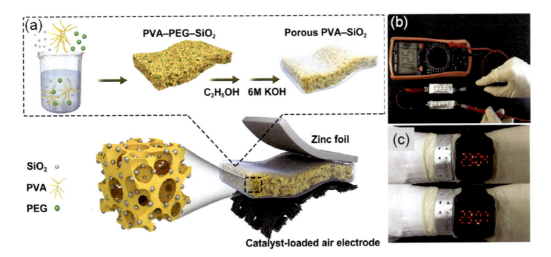

FIGURE 23.5 (a) Schematic representation of the porous PVA-based nanocomposite GPE's porous ZAB and synthesis process, as well as its inner structure. (b) An open circuit potential illustration using two ZABs connected in series. (c) An LED watch powered by a fabricated bracelet-type ZAB. (Adapted with permission [34], Copyright (2019), Elsevier.)

light up an LED watch (Figure 23.5(c)), demonstrating the excellent mechanical properties and flexibility of the produced porous PVA SiO_2 GPEs and the potential application of the manufactured ZAB for wearable electronics.

Sandwich structured flexible ZAB fabricated using Zn anode foil, gel polymer electrolyte, and nonporous CNF layer as air cathode has also been reported with stable charge and discharge voltages of 1.7 V and 1.0 V even under repeated bending conditions at 2 mA/cm² charge density, respectively [31]. A wearable band-aid serving as a flexible substrate and air-permeable consisting of miniaturized ZAB was proposed by Fu et al. [35]. The fabricated device is highly flexible and rechargeable and composed of a cellulose electrolyte membrane which is sandwiched between Co_3O_4 loaded CC cathode and Zn anode consisting of Zn powder, CNF, carbon black, and a polymer binder. This band-aid when wrapped around the finger can lighten up the LED under bending and exhibits a high specific capacity. Zhang et al. [16] have established a flexible ZAB made of CoNi alloy/NCNSAs/CC-800 in a three-dimensional hierarchical structure. Zn-air now has better mechanical cyclability, a high energy density of 98.8 mW/cm² with a high capacity of 879 mAh/g. Two ZABs connected in series can power 5l-red LED to light up "HBU" indicator, inferring that the ZAB was a reasonable alternative for new generation energy storage applications.

In addition to ZAB, sandwich architecture was also being utilized to fabricate flexible LAB. Liu et al. [25] manufactured a flexible rechargeable LAB comprising undoped TiO_2 NAs/CT and

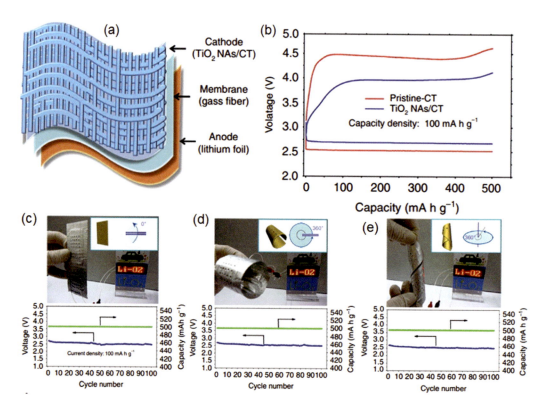

FIGURE 23.6 (a) Representation of the cell assembly made up of cathode, anode, and glass fiber. (b) First discharge-charge curves for Li-O_2 cells consisting of pristine-CT and TiO_2 NAs/CT cathodes at a capacity density and specific capacity of 100 mA/g and 500 mAh/g, respectively. (c–e) Bending and twisting behavior of the flexible Li-O_2 cells with TiO_2 NAs/CT cathode along with the respective dependence of terminal discharge voltage on cycle number. (Adapted with permission [25]. Copyright (2015) The Authors, some rights reserved; exclusive licensee Nature Publishing. Distributed under a Creative Commons Attribution License 4.0 (CC BY).)

lithium-foil as cathode and anode, respectively, and a glass-fiber membrane. Figure 23.6(a) illustrates the overall assembly mode. The first discharge-charge curve of ZAB with undoped-CT and TiO_2 NAs/CT cathodes at 100 mA/g has been shown in Figure 23.6(b) with specific capacity maintained constant at 500 mAh/g. To prove its promising implications in flexible electronics, a commercial red LED display screen was powered by the fabricated LAB under several bending and twisting circumstances as presented in Figures 23.6(c-e).

A $Li-O_2$ sandwich structured battery fabricated by layering scalable paper ink cathode acting as a conductor as well as a flexible substrate, glass fiber membrane, and lithium foil anode was also reported [36]. No notable effect was observed in the charging-discharging profiles even under 1000 time folding. The construction of a highly flexible wearable, foldable, and stretchable LAB with improved electrochemical performances which remains stable under repeatedly stretching, bending, and twisting or strain conditions has also been demonstrated [37]. This LAB consists of a polymer electrolyte, lithium anode, and a rippled air cathode which is made up of aligned multiwalled CNT sheets. Liu et al. [38] have proposed an extremely thin, lightweight, and wearable LAB with a novel segmented structure. Even after several repetitive deformations, better and stable battery operations with high specific capacity/energy density are obtained.

It is noteworthy to mention that sandwich structure allows bending in one dimension and hence lacks other deformation conditions such as twisting [39]. To overcome this drawback, fiber structures are introduced. Park et al. [40] proposed the first fiber-structured flexible ZAB consisting of a spiral anode made up by wrapping a Zn foil on a rod and kept inside a cellophane template containing a mixture of gelatin and KOH electrolyte solution. The obtained spiral anode was then wound with a Fe/N/C loaded air cathode. This fiber-based ZAB was found to exhibit a discharge voltage plateau of 0.92 V at a current density and discharge duration of 0.1 mA/cm^2 and 9 h, respectively. A three-dimensional flexible rechargeable ZAB using C_2N aerogels was proposed by Shinde et al. [41]. Xu et al. [42] demonstrated the production of a flexible and wearable fiber-shaped ZAB which is highly stretchable using cross-stacked porous CNT sheets that simultaneously serve as gas diffusion layer, catalyst layer, and current collector. The ZAB consists of PVA/PEO/KOH gel polymer electrolyte cross-linked in situ and enveloped on the zinc spring anode, then coated with RuO_2 and multiwalled CNT sheets to produce the air cathode. The resulting fiber-shaped ZAB might be encased in a punched tube to prevent excessive water evaporation in the electrolyte. It exhibited discharging and charging voltages of 1.0 V and 1.9 V for over 30 cycles at a current density of 1A/g. When the battery was bent at various angles, the discharge potential remained steady. Aside from flexibility, stretchability was first attained for the MAB, which could be stretched by 10% without evident structural harm or performance deterioration. Zhang et al. [43] created a coaxial fiber-shaped LAB with three layers: an exterior aligned multiwalled CNT sheet air electrode, a solid-state polymer electrolyte in the center, and an inside lithium wire anode. To maintain the lithium anode insulating from the air and avoid corrosion, an in-situ UV irradiation cross-linking of the electrolyte on the lithium wire anode was used. It could operate indefinitely in the air with a high specific capacity of 12470 mAh/g with a current density of 1400 mA/g. It also has a great degree of flexibility with superior performance which is evident from its twisting behavior without damaging the structure.

Despite such advancements in the fabrication of wearable flexible MAB, there remain a few challenges which include, (i) the creation of an interphase layer of solid electrolyte because of the reaction between anode material and electrolyte leading to an irretrievable loss in battery's efficiency [44]; (ii) dendrite growth on the anode electrode causing an internal short circuit in MAB and hence deterioration of the battery performance [45]; (iii) desired electrolyte with all favorable properties (high stability and oxygen solubility, low volatility, environmental friendliness, and broad electrochemical window); and (iv) stability of materials where the cathodic reaction occurs in MAB. Generally, the carbon-based materials serving as cathodes for such MAB applications show instability for charging and discharging voltages above 3.5 V which results in side reactions that diminish the cyclability of MABs and hinder battery performance [46].

23.4 CONCLUSION

In conclusion, factors including flexible components and structural configurations that are essential for the fabrication of wearable MABs, particularly LABs and ZABs, are discussed in detail. Flexible electrolytes either gel or polymer and flexible materials for manufacturing flexible electrodes including mainly carbon-based materials (graphene-based materials, CNTs, etc.), polymers mainly conducting, metal/metal oxide sheets/foils are widely employed for fabricating wearable MAB. Apart from these, structural configurations specifically the planar sandwich structure and fiber-type structure adopted have been demonstrated. Such flexible components and structural configurations facilitate the twisting, folding, rolling, bending, and stretching of wearable MABs. However, more efforts are still required to circumvent the raised challenges and hence realize the commercialization of wearable batteries which are critical prerequisites for developing wearable electronic devices/sensors with outstanding performance.

REFERENCES

[1] J. Zhang, J. Fu, X. Song, G. Jiang, H. Zarrin, P. Xu, K. Li, A. Yu, Z. Chen, Laminated cross-linked nanocellulose/graphene oxide electrolyte for flexible rechargeable zinc-air batteries, Adv. Energy Mater. 6 (2016) 1600476.
[2] H. Nishide, K. Oyaizu, Toward flexible batteries, Science. 319 (2008) 737–738.
[3] W. Liu, K. Feng, Y. Zhang, T. Yu, L. Han, G. Lui, M. Li, G. Chiu, P. Fung, A. Yu, Hair-based flexible knittable supercapacitor with wide operating voltage and ultra-high rate capability, Nano Energy. 34 (2017) 491–499.
[4] T. Peng, B. Chen, H. Xu, H. Zhang, W. Cai, M. Ni, M. Liu, Z. Shao, Flexible Zn- and Li-air batteries: Recent advances, challenges, and future perspectives, Energy Environ Sci. 10 (2017) 2056–2080.
[5] D. Ji, L. Fan, L. Li, S. Peng, D. Yu, J. Song, S. Ramakrishna, S. Guo, Atomically transition metals on self-supported porous carbon flake arrays as binder-free air cathode for wearable zinc-air batteries, Adv. Mater. 31 (2019) 1808267.
[6] Z. Zhang, C. Zuo, Z. Liu, Y. Yu, Y. Zuo, Y. Song, All-solid-state Al-air batteries with polymer alkaline gel electrolyte, J. Power Sources. 251 (2014) 470–475.
[7] N. Vassal, E. Salmon, J.F. Fauvarque, Electrochemical properties of an alkaline solid polymer electrolyte based on P (ECH-co-EO), Electrochim. Acta. 45 (2000) 1527–1532.
[8] C.C. Yang, S.J. Lin, Preparation of alkaline PVA-based polymer electrolytes for Ni-MH and Zn-air batteries, J. Appl. Electrochem. 33 (2003) 777–784.
[9] G.M. Wu, S.J. Lin, C.C. Yang, Preparation and characterization of PVA/PAA membranes for solid polymer electrolytes, J. Membr. Sci. 275 (2006) 127–133.
[10] M. Balaish, E. Peled, D. Golodnitsky, Y. Ein-Eli, Liquid-free lithium-oxygen batteries, Angew. Chem. 127 (2015) 446–450.
[11] H. Zhang, X. Ma, C. Lin, B. Zhu, Gel polymer electrolyte-based on PVDF/fluorinated amphiphilic copolymer blends for high performance lithium-ion batteries, RSC Adv. 4 (2014) 33713–33719.
[12] T. Kuboki, T. Okuyama, T. Ohsaki, N. Takami, Lithium-air batteries using hydrophobic room temperature ionic liquid electrolyte, J. Power Sources. 146 (2015) 766–769.
[13] Z. Wu, Y. Wang, X. Liu, C. Lv, Y. Li, D. Wei, Z. Liu, Carbon-nanomaterial-based flexible batteries for wearable electronics, Adv. Mater. 31 (2019) 1800716.
[14] S. Ozcan, M. Tokur, T. Cetinkaya, A. Guler, M. Uysal, M.O. Guler, H. Akbulut, Free standing flexible graphene oxide+ α-MnO_2 composite cathodes for Li-Air batteries, Solid State Ion. 286 (2016) 34–39.
[15] G. Fu, Y. Tang, J.M. Lee, Recent advances in carbon-based bifunctional oxygen electrocatalysts for Zn-air batteries, ChemElectroChem. 5 (2018) 1424–1434.
[16] W. Zhang, Z. Li, J. Chen, X. Wang, X. Li, K. Yang, L. Li, Three-dimensional CoNi alloy nanoparticle and carbon nanotube decorated N doped carbon nanosheet arrays for use as bifunctional electrocatalysts in wearable and flexible Zn-air batteries, Nanotechnology. 31 (2020) 185703.
[17] M. Liao, L. Ye, Y. Zhang, T. Chen, H. Peng, The recent advance in fiber-shaped energy storage devices, Adv. Electron. Mater. 5 (2019) 1800456.
[18] Y. Lin, J. Li, Y. Lai, C. Yuan, Y. Cheng, J. Liu, A wider temperature range polymer electrolyte for all-solid-state lithium ion batteries, RSC Adv. 3 (2013) 10722–10730.

[19] W.H. Meyer, Polymer electrolytes for lithium-ion batteries, Adv. Mater. 10 (1998) 439–448.
[20] N.A. Choudhury, S. Sampath, A.K. Shukla, Hydrogel-polymer electrolytes for electrochemical capacitors: An overview, Energy Environ Sci. 2 (2009) 55–67.
[21] J. Li, Z. Wang, L. Yang, Y. Liu, Y. Xing, S. Zhang, H. Xu, A flexible li-air battery workable under harsh conditions based on an integrated structure: A composite lithium anode encased in a gel electrolyte, ACS Appl. Mater. Interfaces. 13 (2021) 18627–18637.
[22] X. Lei, X. Liu, W. Ma, Z. Cao, Y. Wang, Y. Ding, Flexible lithium-air battery in ambient air with an in situ formed gel electrolyte, Angew. Chem. Int. Ed. 57 (2018) 16131–16135.
[23] C. Dai, G. Sun, L. Hu, Y. Xiao, Z. Zhang, L. Qu, Recent progress in graphene-based electrodes for flexible batteries, InfoMat. 2 (2020) 509–526.
[24] Q.C. Liu, T. Liu, D.P. Liu, Z.J. Li, X.B. Zhang, Y. Zhang, A flexible and wearable lithium—oxygen battery with record energy density achieved by the interlaced architecture inspired by bamboo slips, Adv. Mater. 28 (2016) 8413–8418.
[25] Q.C. Liu, J.J. Xu, D. Xu, X.B. Zhang, Flexible lithium-oxygen battery based on a recoverable cathode, Nat. Commun. 6 (2015) 1–8.
[26] F. Jing, F.M. Hassan, J. Li, D.U. Lee, A.R. Ghannoum, G. Lui, M.A. Hoque, Z. Chen, Flexible rechargeable zinc-air batteries through morphological emulation of human hair array, Adv. Mater. 28 (2016) 6421–6428.
[27] X. Huang, J. Liu, J. Ding, Y. Deng, W. Hu, C. Zhong, Toward flexible and wearable Zn-air batteries from cotton textile waste. ACS Omega, 4 (2019) 19341–19349.
[28] F. Meng, H. Zhong, D. Bao, J. Yan, X. Zhang, In situ coupling of strung Co_4N and intertwined N-C fibers toward free-standing bifunctional cathode for robust, efficient, and flexible Zn—air batteries, J. Am. Chem. Soc. 138 (2016) 10226–10231.
[29] X. Chen, C. Zhong, B. Liu, Z. Liu, X. Bi, N. Zhao, X. Han, Y. Deng, J. Lu, W. Hu, Atomic layer Co_3O_4 nanosheets: the key to knittable Zn–air batteries, Small. 14 (2018) 1702987.
[30] Q. Liu, Y. Wang, L. Dai, J. Yao, Scalable fabrication of nanoporous carbon fiber films as bifunctional catalytic electrodes for flexible Zn-air batteries, Adv. Mater. 28 (2016) 3000–3006.
[31] S. Qu, Z. Song, J. Liu, Y. Li, Y. Kou, C. Ma, X. Han, Y. Deng, N. Zhao, W. Hu, C. Zhong, Electrochemical approach to prepare integrated air electrodes for highly stretchable zinc-air battery array with tunable output voltage and current for wearable electronics, Nano Energy. 39 (2017) 101–110.
[32] Q. Jin, B. Ren, J. Chen, H. Cui, C. Wang, A facile method to conduct 3D self-supporting Co-FeCo/N-doped graphene like carbon bifunctional electrocatalysts for flexible solid-state zinc air battery, Appl. Catal. B. 256 (2019) 117887
[33] J. Fu, D.U. Lee, F.M. Hassan, L. Yang, Z. Bai, M.G. Park, Z. Chen, Flexible high-energy polymer-electrolyte-based rechargeable zinc-air batteries. Adv. Mater. 27 (2015) 5617–5622.
[34] X. Fan, J. Liu, Z. Song, X. Han, Y. Deng, C. Zhong, W. Hu, Porous nanocomposite gel polymer electrolyte with high ionic conductivity and superior electrolyte retention capability for long-cycle-life flexible zinc-air batteries, Nano Energy. 56 (2019) 454–462.
[35] J. Fu, J. Zhang, X. Song, H. Zarrin, X. Tian, J. Qiao, L. Rasen, K. Li, Z. Chen, A flexible solid-state electrolyte for wide-scale integration of rechargeable zinc—air batteries. Energy Environ Sci. 9 (2016) 663–670.
[36] Q.C. Liu, L. Li, J.J. Xu, Z.W. Chang, D. Xu, Y.B. Yin, X.Y. Yang, T. Liu, Y.S. Jiang, J.M. Yan, X.B. Zhang, Flexible and foldable $Li-O_2$ battery based on paper-ink cathode. Adv. Mater. 27 (2015) 8095–8101.
[37] L. Wang, Y. Zhang, J. Pan, H. Peng, Stretchable lithium-air batteries for wearable electronics. J. Mater. Chem. A. 4 (2016) 13419–13424.
[38] T. Liu, J.J. Xu, Q.C. Liu, Z.W. Chang, Y.B. Yin, X.Y. Yang, X.B. Zhang, Ultrathin, lightweight, and wearable $Li-O_2$ battery with high robustness and gravimetric/volumetric energy density, Small. 13 (2017) 1602952.
[39] W. Weng, P. Chen, S. He, X. Sun, H. Peng, Smart electronic textiles, Angew. Chem. Int. Ed. 55 (2016) 6140–6169.
[40] J. Park, M. Park, G. Nam, J.S. Lee, J. Cho, All-solid-state cable-type flexible zinc-air battery, Adv. Mater. 27 (2015) 1396–1401.
[41] S.S. Shinde, C.H. Lee, J.-Y. Yu, D.-H. Kim, S.U. Lee, J.-H. Lee, Hierarchically designed 3D holey C_2N aerogels as bifunctional oxygen electrodes for flexible and rechargeable Zn-air batteries, ACS Nano. 12 (2018) 596–608.

[42] Y. Xu, Y. Zhang, Z. Guo, J. Ren, Y. Wang, H. Peng, Flexible, stretchable, and rechargeable fiber-shaped zinc-air battery based on cross-stacked carbon nanotube sheets, Angew. Chem. 127 (2015) 15610–15614.

[43] Y. Zhang, L. Wang, Z. Guo, Y. Xu, Y. Wang, H. Peng, High-performance lithium-air battery with a coaxial-fiber architecture, Angew. Chem. Int. Ed. 55 (2016) 4487–4491.

[44] E. Mengeritsky, P. Dan, I. Weissman, A. Zaban, D. Aurbach, Safety and performance of tadiran TLR-7103 rechargeable batteries, J. Electrochem. Soc. 143 (1996) 2110.

[45] X. Zhang, X.-G. Wang, Z. Xie, Z. Zhou, Recent progress in rechargeable alkali metal-air batteries, Green Energy and Environment, 1 (2016) 4.

[46] M.M.O. Thotiyl, S.A. Freunberger, Z. Peng, P.G. Bruce, The carbon electrode in nonaqueous $Li-O_2$ cells, J. Am. Chem. Soc. 135 (2013) 494.

24 Flexible Metal-Air Batteries

Runwei Mo

CONTENTS

- 24.1 Introduction 347
- 24.2 Working Principle 348
- 24.3 Evaluation of FMABs..349
- 24.4 Battery Configuration..349
  - 24.4.1 Cable Type 349
  - 24.4.2 Sandwich Type..350
  - 24.4.3 Other Types...350
- 24.5 Recent Advancements in Flexible Metal-Air Batteries 351
  - 24.5.1 Flexible Electrodes..351
    - 24.5.1.1 Flexible Metal Anodes...351
    - 24.5.1.2 Carbon-Based Cathodes 351
    - 24.5.1.3 Non-Carbon-Based Cathodes 352
  - 24.5.2 Electrolytes 353
    - 24.5.2.1 Aqueous Gel Polymer Electrolytes 353
    - 24.5.2.2 Organic Gel Polymer Electrolytes...354
- 24.6 Advantages and Challenges of Flexible Metal-Air Batteries 355
  - 24.6.1 Advantages of FMABs..355
  - 24.6.2 Challenges of FMABs...355
- 24.7 Conclusions 356
- References..357

24.1 INTRODUCTION

In recent years, with the rapid development of wearable smart devices and electronics, flexible devices have shown increasing market demand, such as Google Glass, wearable Apple Watch, and Huawei Mate X flexible smartphones [1]. In the near future, it is foreseeable that flexible devices will greatly change human life. Especially in sports and medical industries they will be widely used, which puts forward higher requirements for the performance of flexible devices. Unlike traditional energy storage devices, the ideal flexible device needs to be able to operate continuously during bending, stretching, and twisting. It is worth noting that the safety performance of flexible devices under stretching, bending, and twisting conditions is crucial for practical applications.

To date, many flexible devices have been extensively studied, especially as flexible supercapacitors, solar cells, and batteries have been commercialized in the market [2–4]. From a research perspective, research on flexible devices began about 100 years ago, focusing on polymer and alkaline batteries. Since then, a lot of research work has been done on the structural design and working principle of flexible devices [5]. Among them, flexible lithium-ion batteries based on the principle of lithium-ion de/intercalation have been rapidly developed [6]. However, their application in high-capacity energy devices is greatly restricted due to their limited energy density. For the aforementioned reasons, flexible metal-air batteries (FMABs) have attracted the extensive attention of researchers due to their high energy density. The battery structure mainly includes metal as the anode and air as the cathode, which significantly reduces weight compared to other batteries [7]. Due to the different metal species, FMABs can be classified into lithium-air, zinc-air, potassium-air, aluminum-air, sodium-air, and so on [7, 8]. Among them, flexible lithium-air batteries exhibit higher theoretical energy density, which can be as high as 3500 Wh/kg [8]. Compared with other flexible batteries, flexible lithium-air batteries exhibit the advantages of being thinner and lighter under the same energy density requirement, which is very important for the practical application of flexible energy storage devices.

In addition, the design of the device architecture of FMABs is also very effective for increasing the overall energy density of the device. Researchers so far have done a lot of reported work on different types of device architecture, including cable, sandwich, and some other types [9–11]. However, there are still some urgent difficulties that need to be solved, which severely restrict its commercial application, mainly including the insulating properties of discharge products, the poor contact between the electrolyte and electrode interface, the decomposition of electrolytes, low flexibility of battery materials, and the dendrite growth of metal anodes [12]. In this chapter, we review the working principle, evaluation, battery configuration, and recent advancements in FMABs. Afterward, we discuss the advantages and challenges of FMABs for commercial applications.

24.2 WORKING PRINCIPLE

In recent years, with the advancement of science and technology, FMABs have developed rapidly. Different from traditional flexible batteries, the components of flexible metal-air batteries are mainly divided into air cathode, metal anode, and electrolyte. It should be noted here that metal-air batteries have different working principles according to different electrolytes. When the electrolyte is an aqueous liquid, the working principle of the FMABs during the discharge process is as follows (M = Mg/Zn/Fe/Al) by Eq. (1,2):

$$\text{Air cathode: } O_2 + 2H_2O + 4e^- \rightarrow 4OH^- \tag{1}$$

$$\text{Metal anode: } M + nOH^- \rightarrow M(OH)_n + ne^- \tag{2}$$

The oxygen in the environment is reduced to OH^- in the catalyst layer of the cathode, and at the same time, the aqueous metal ions released from the metal anode react with the OH^- in the electrolyte to form $M(OH)_n$ [13].

When the electrolyte is an organic electrolyte, the working principle of the FMABs is as follows (M = Na/Li/K) by Eq. (3,4):

$$\text{Air cathode: } O_2 + xM^{n+} + xe^- \leftrightarrow M_xO_2 \; (x = 1 \text{ or } 2) \tag{3}$$

$$\text{Metal anode: } M \leftrightarrow M^{n+} + ne^- \tag{4}$$

Reversible stripping/plating reactions occur on metal anodes during discharge/charge. At the same time, the oxidation-reduction reaction or oxygen evolution reaction, which is the reversible formation and decomposition of peroxides or peroxides, occurs on the air cathode [14]. Based on the working principle of FMABs, the air cathode and metal anode need to react with ambient air/O_2 during the reaction process, so the design of battery configuration has an important impact on improving its electrochemical performance.

Compared with other traditional flexible energy storage systems, the FMABs based on the aforementioned working principle have a higher theoretical energy density. Therefore, FMABs are promising energy storage device candidates for implantation in wearable flexible electronics. However, the development of flexible metal-air batteries is still in its infancy at the current research stage. It is worth noting here that the evaluation criteria, battery configuration, flexible electrodes, and electrolytes all have a great impact on the electrochemical performance and mechanical stability of FMABs.

24.3 EVALUATION OF FMABs

Flexible materials are of great interest in applications in wearable devices, smart sensors, and portable medical devices. It is worth noting that the unification of test standards is very critical to the accurate evaluation of the comprehensive performance of FMABs. Among them, bending is one of the most basic operating conditions when evaluating the performance of FMABs. In certain evaluation systems, FMABs tend to be bent around a cylinder of a given diameter to obtain a bend radius (r) of curvature. It is generally believed that the smaller the r-value, the better the flexibility of the FLAB [15]. It is worth noting here that for some specific applications, larger battery thicknesses (h) generally exhibit smaller bend radii, which requires consideration of the effect of actual battery thickness [16]. Stretching is another fundamental operating condition when evaluating the performance of FMABs. The commonly used method is by fixing the two ends of the FMABs on two moving fixtures controlled by linear actuators, respectively. Through this method, the stretching properties of FMABs can be accurately evaluated. Compared to these evaluation methods, the twisting test on FMABs is quite rough. So far, the commonly used method is only to deform it on a three-dimensional scale. Therefore, it is necessary to establish a more systematic test method to evaluate the twisting properties of FMABs.

24.4 BATTERY CONFIGURATION

To meet the practical application of flexible electronics, it is an important indicator for FMABs to still exhibit excellent electrochemical performance under repeated external forces. Therefore, the design of the battery configuration plays an important role in improving the comprehensive performance of the FMABs. In this section, several battery configurations of FMABs are detailed and the effects of different configurations on their performance are also discussed.

24.4.1 Cable Type

In recent years, this cable type has been one of the common configurations of FMABs. Its structure mainly includes metal electrodes, electrolyte membranes, and catalyst-loaded air electrodes. The configuration is made by first including a metal electrode in the center of the structure, which is surrounded by an electrolyte membrane, such as a gel polymer. Then, the catalyst-loaded air electrode was fixed on the electrolyte by winding. To improve the surface contact between the electrolyte and the electrodes, heat shrinkable rubber cables are often used when packaging the electrode assembly. It is worth noting that air needs to participate in the reaction between the positive and negative electrodes, so punching is required to increase air diffusion.

The cable type has attracted research interest in FMABs in recent years due to its excellent flexibility. For example, Zhang et al. innovatively designed a cable-type flexible zinc-air battery and tested its electrochemical performance [17]. The battery could maintain a discharge voltage plateau of about 1.23 V after up to 15 hours of standby time under a current density of 0.5 mA/cm². More importantly, this type of battery exhibited good stability, especially after 2000 bending/stretching cycles under different bending and twisting conditions, which still provide stable electrochemical performance. Furthermore, the application of the cable-type battery configuration to the flexible lithium-air battery also exhibits excellent electrochemical performance. For example, Peng et al. successfully introduced this cable-type battery configuration into a flexible lithium-air battery [9]. The battery exhibits a high specific capacity of 13055 mAh/g and long cycle life of more than 100 charge-discharge tests. It is worth noting that the discharge curve remains unchanged with increasing bending angle. In addition to the aforementioned twisting, bending, and long-term fatigue tests, the evaluation of flexibility can also be achieved through a unique design to achieve tensile testing. For example, Peng et al. studied the effect of the tensile state on the electrochemical performance of FMABs. The results show that this cable-type battery configuration can provide stable discharge voltage even when its elongation reaches 10% [18].

24.4.2 Sandwich Type

Sandwich type is another battery configuration of the FMABs. Specifically, to ensure good electrical contact, it is necessary to first connect the metal electrodes to metal foils as current collectors/substrates during the fabrication of this battery configuration. Subsequently, an electrolyte membrane was inserted between the air electrode and the metal electrode. It is worth noting here the connection to the porous current collector at the air electrode, which will facilitate electron transfer and gas diffusion. This sandwich type has certain similarities to the configuration structure of conventional batteries, which has been extensively studied in FMABs recently. For example, Fu et al. successfully introduced a sandwich type into a flexible solid-state zinc-air battery and systematically investigated its electrochemical properties [19]. The results show that its energy density and power density are as high as 847.6 Wh/kg and 160.7 mW/cm^2, respectively. The battery also exhibits excellent cycling stability, which can be cycled at a current density of 25 mA/cm^2 for over 500 hours without significant voltage loss.

In addition to testing the electrochemical properties under conventional conditions, this flexibility is also a very important measure for FMABs. Typically, flexibility testing requires performance testing under bending, twisting, and tensile conditions. Specifically, this bendability was measured by measuring the discharge/charge performance under bending conditions [20]. In a more systematic test, the electrochemical performance (e.g., rate capability, polarization, cycling stability) was performed under different measured bending angles, and stability was shown by comparison with the initial angle under planar conditions [20]. Likewise, twistability is a test of electrochemical properties under different twist angles [21]. Furthermore, the tensile properties are rarely tested in sandwich-type battery configurations, but in practice, the battery components also stretch to varying degrees during bending and twisting, which can have an impact on the electrochemical performance. It is worth noting here that in addition to the short-term deformation mentioned earlier, long-term fatigue testing is also an important indicator for FMABs. For example, Zhang et al. performed long-term fatigue tests on FMABs. The results showed that FMABs could maintain the first discharge-charge curves, rate capability, capacity, and cycle number after 1000 folding cycles [22].

24.4.3 Other Types

In addition to the aforementioned cable and sandwich types, to improve the mechanical and electrochemical properties of FMABs, researchers have recently carried out a large number of innovative designs for battery configuration. The researchers innovatively designed foldable FMABs based on a paper-ink air electrode. To make the battery foldable and bendable, an inexpensive flexible paper was chosen as the substrate for the air electrode, separator, and outer packaging in this battery configuration. Equally important, the paper-wrapped metal electrode can be shared by two air electrodes, which could significantly enhance the energy density of FMABs and the utilization of metal. Recently, Zhang et al. developed a wearable and flexible lithium-air battery inspired by ancient Chinese bamboo slips [23]. It is worth noting that the researchers have innovatively woven the air electrode and the lithium electrode together, which is very similar to the "leather rope" and "bamboo sheet" in bamboo slips. Compared with other reported FMABs, this battery configuration has many advantages: (i) the use of air diffusion layers and packaging materials is avoided in this battery configuration, which effectively increases the energy density of the battery; and (ii) this battery configuration allows gas to easily enter the reaction site through both sides, which can significantly improve the kinetics of the battery. The battery exhibited excellent electrochemical performance and flexibility, especially good cycling stability under various twisting, bending, and folding conditions. Equally important, this battery configuration has excellent gas permeability, which can allow gas to enter the air electrode through both sides of the woven structure.

Furthermore, Liu et al. constructed a flexible lithium-air battery with a segmented structure [24]. The battery consists of an array of small-scale lithium electrode disks and an array of carbon electrode disks, which are interconnected by copper wires and carbon ropes, respectively. This battery configuration has many advantages compared to other battery configurations for the fabrication of FMABs: (i) in this battery configuration, the materials used are all very light and thin, which greatly improves the mass-energy density and volumetric energy density of the battery; and (ii) the battery configuration is decomposed into many tiny cells, which relieves the stress of the battery and greatly improves the mechanical stability. To elucidate the effect of repeated external forces, the electrochemical performance of the battery was evaluated in its folded state. This innovative battery configuration enables the battery to exhibit excellent electrochemical performance under bending, twisting, and tensile conditions, especially the volumetric and weight energy densities of the battery are still as high as 274.06 Wh/L and 294.68 Wh/kg after 10000 cycles of folding and stretching, respectively.

Although great progress has been made in battery configuration and evaluation methods of FMABs in recent years, there are still some problems that need to be improved. In particular, there is a need to develop uniform testing standards, which will facilitate the commercialization of FMABs. Therefore, a series of innovative research work on the battery configuration and evaluation methods of FMABs is still needed in the future.

24.5 RECENT ADVANCEMENTS IN FLEXIBLE METAL-AIR BATTERIES

24.5.1 Flexible Electrodes

The development of flexible electrodes is an important part of the fabrication of FMABs. The electrodes of FMABs are mainly divided into anode and cathode. The anode is usually a flexible metal strip, metal foil, metal rod, or metal powder coating on a flexible metal substrate, which can be directly used in FMABs. Therefore, here we focus on the introduction of the cathode. It is well known that the components of a cathode include the active material and the current collector. Among them, the fabrication of flexible current collectors is crucial for realizing high-performance cathodes. It is worth noting here that factors such as electrical conductivity, air permeability, and flexibility need to be comprehensively considered in the design of the current collector.

24.5.1.1 Flexible Metal Anodes

In the early stages of research, metals such as lithium, sodium, potassium, zinc, and aluminum, which are well known to be deformable, have been directly used as flexible anodes for FMABs. With the deepening of research, researchers found that pure metal anodes can only achieve limited flexibility, which is far from meeting the needs of FMABs. In addition, the charge storage principle of anodes in FMABs is mainly based on the reversible metal stripping/plating process, which directly leads to serious safety problems caused by corrosion and dendrite growth [25]. To solve these problems, recent research work mainly focuses on the configuration design of alloy anodes and the preparation of flexible composite anodes. Recently, Fu et al. designed a free-standing zinc composite electrode, which was mainly composed of zinc powder, carbon black, and carbon nanofiber [26]. This free-standing composite anode has a three-dimensional interconnected network structure, which enables twisting, free rolling, and folding without any mechanical damage. More importantly, the presence of carbon helps to improve the electrical conductivity of the composite anode. Therefore, the composite anode exhibited excellent electrochemical performance and good mechanical stability.

24.5.1.2 Carbon-Based Cathodes

It is well known that the fabrication of flexible air cathodes is usually formed by uniformly loading active materials such as carbonaceous materials, noble metals, and metal oxides on flexible current collectors. A variety of carbon materials have been extensively studied due to their outstanding electrical conductivity, relative stability, well-defined catalytic properties, and porous structure.

Compared with other carbon materials, graphene and carbon nanotubes have attracted widespread interest due to their high tensile strength and Young's modulus. More importantly, both graphene and carbon nanotubes can be used as units to build a stable 3D network structure, which is beneficial to the structural design of flexible electrodes. In recent years, the structural design of flexible electrodes is often metal or metal oxide/nitride as a catalyst supported on a carbon substrate, which can greatly improve the catalytic performance of the cathode [27]. In addition, to further enhance the conductivity and catalytic performance of the cathode, active materials can also be grown in-situ on the flexible substrate by chemical or electrochemical methods to form a freestanding structure of the cathode. This freestanding cathode generally exhibits excellent electrochemical performance for the following main reasons: (i) this unique structure avoids the use of polymer binders, which could enhance the cycling stability of the cathode; (ii) this freestanding electrode is easy to carry out the design of the porous micro-nano structure, which can facilitate mass transfer; and (iii) this electrode structure promotes effective contact between the active material and current collector, which is beneficial to electron transport and kinetics improvement. Recently, Meng et al. successfully fabricated a flexible air cathode based on Co_4N and carbon fiber network by electrodeposition method [17]. In this electrode structure, Co_4N as the active material enhances the catalytic performance of the air cathode, and the network structure of carbon fibers as the substrate provides a three-dimensional continuous conductive network. The air cathode exhibits excellent flexibility, which provides the possibility for the assembly of FMABs.

Although carbon-based flexible air cathodes have made great progress in recent years, the problem of low utilization of active materials in air cathodes still exists. Therefore, how to further improve the utilization rate of the air cathode is still an urgent problem to be solved. To solve this problem, the structural design of free-standing air cathodes has been carried out in the field of FMABs. Both the current collector and the active material participate in the electrochemical reaction in the free-standing air electrode. In particular, free-standing air cathodes for FMABs need to satisfy the following properties, such as high electrical conductivity, good gas permeability, excellent flexibility, and outstanding catalytic performance. To achieve the aforementioned goals, Yin et al. developed a strategy to fabricate lightweight, free-standing, micro- and nano-porous-structured activated carbon fiber cathodes by electrospinning technology [11]. This electrode structure is formed by interweaving to form a porous framework, which can be directly applied to FMABs as an air cathode. The as-prepared air cathode structure has abundant surface openings and interconnected pore structures, which are beneficial to gas diffusion and electron transport. The electrode structure also exhibits excellent flexibility, which facilitates the assembly and testing of FMABs.

Chemical vapor deposition is an efficient synthesis technique of carbon materials, which has attracted extensive research [28]. Recently, Peng et al. successfully fabricated a free-standing carbon nanotube-based flexible air cathode by chemical vapor deposition [9]. The porous structure of the air cathode is composed of neatly arranged carbon nanotubes, which is conducive to the rapid transfer of electrons and the rapid diffusion of oxygen. Assembled into FMABs, the reaction efficiency was also greatly improved. More importantly, in addition to excellent flexibility, the electrode structure also has a certain stretchability, which lays the foundation for the fabrication of flexible and stretchable metal-air batteries.

24.5.1.3 Non-Carbon-Based Cathodes

As mentioned earlier, carbon-based materials have attracted extensive interest in the field of FMABs due to their excellent properties, but still suffer from poor stability, especially in non-aqueous FMABs. Under the condition of charging voltage higher than 3.5 V, the carbon-based material is prone to decomposition phenomenon and promotes the decomposition of electrolyte, which greatly reduces its cycle life [29]. Therefore, it is very important to find suitable non-carbon-based materials. In recent years, non-carbon-based materials have been extensively studied in the field of FMABs, among which the most common materials are metal oxides. Liu et al. developed a method to successfully fabricate recoverable and flexible air cathodes using a hydrothermal method

[30]. This flexible air cathode was fabricated by uniformly growing TiO_2 nanowire arrays (TiO_2 NAs) on flexible carbon fiber fabrics (CTs). It is worth noting here that the overall flexibility properties are not affected by the uniform growth of TiO_2 NAs on CTs. Compared with the previously mentioned carbon-based flexible cathodes, TiO_2 NAs as the active material in this work can effectively avoid the decomposition phenomenon caused by carbon materials, thus exhibiting excellent electrochemical performance.

Furthermore, Liu et al. successfully fabricated a flexible cathode air based on Co_3O_4 nanosheets and Ru nanoparticle composites (Co_3O_4 NSs-Ru) grown on flexible CTs [31]. The active material in this flexible air cathode is Co_3O_4 NSs-Ru, so this is classified as a non-carbon-based air cathode. In this work, the electrochemical properties of air cathodes can be further improved by compounding Co_3O_4 NSs with Ru. At the same time, the electrode structure also exhibited excellent flexibility. Compared with carbon-based air cathodes, the degradation problem of air cathodes is effectively alleviated. However, compared with carbon-based materials, non-carbon-based materials as active materials have higher weight, which will reduce the energy density of FMABs. Therefore, how to balance the cathode stability and energy density is an important issue to be solved in the future. This requires an in-depth study of the real reaction mechanism of non-carbon-based materials. In addition, non-carbon-based materials as active materials have been extensively studied in non-aqueous FMABs. However, the feasibility of non-carbon-based materials to operate in aqueous FMABs needs to be further explored.

24.5.2 Electrolytes

It is well known that the role of the electrolyte in FMABs is the key mediator for the transport of oxygen and metal ions during the process of charging and discharging. The use of liquid electrolytes in FMABs tends to leak easily when repeatedly deformed, leading to battery failure and even safety issues in severe cases. Therefore, non-liquid electrolytes have attracted widespread interest in recent years. In addition, electrolyte systems suitable for FMABs need to meet several basic requirements, such as high oxygen solubility and diffusivity, high ionic conductivity, outstanding electrochemical stability, excellent flexibility, and good coordination with other components of the battery [32]. Based on these requirements mentioned earlier, quasi-solid polymer-based gel electrolytes are considered an ideal electrolyte system for FMABs. First, the electrolyte has quasi-solid-state properties, which will be beneficial to suppress dendrite growth and avoid internal short circuits. Second, the electrolyte can also alleviate the leakage phenomenon of the liquid electrolyte, thereby achieving electrochemical stability under repeated flexibility tests. Third, the electrolyte can be made hydrophobic by structural design, which prevents the highly reactive metals from corroding carbon dioxide or moisture in the ambient air. These advantages make quasi-solid polymer-based gel electrolytes improve the electrochemical stability of FMABs under repeated flexibility tests.

24.5.2.1 Aqueous Gel Polymer Electrolytes

Recently, researchers have reported a variety of hydrogels with excellent flexibility, good kinetics, and outstanding mechanical stability, which can be highly compatible with flexible metal batteries [33]. Aqueous gel polymer electrolytes (GPEs) have been successfully used in flexible aluminum-air and zinc-air batteries. There are many synthetic methods for aqueous GPE. One of the most common methods is by first swelling the polymer body with a suitable aqueous solution and then adding a concentrated aqueous alkaline solution to the aforementioned reactants. It is worth noting here that the aqueous solution can form a stable gel with other components in the battery by homogeneous mixing. Subsequently, the as-prepared GPE is processed by an appropriate evaporation process, and the next step can be directly coated on the flexible electrode, and finally used to fabricate a flexible metal-air battery. In the preparation of aqueous GPE, KOH is the first choice among alkaline salts due to its desirable ionic conductivity, relatively low viscosity, good solubility of carbonate by-products, and large oxygen diffusion coefficient. Recently, researchers successfully prepared GPE by a simple method [34]. The specific procedure is to dissolve the PVA into the aqueous KOH solution with vigorous stirring. Then, the prepared GPE was coated on the electrodes multiple times by brush coating, and finally, a flexible zinc-air battery was successfully fabricated by the assembly. The battery exhibited excellent electrochemical stability, which can maintain stable electrochemical performance even at different bending angles from 0 to 90 degrees. Subsequently, the researchers designed an improved new GPE [18]. The specific strategy is to add poly(ethylene oxide) (PEO) to the KOH-PVA-based GPE. The novel GPE exhibits high ionic conductivity (0.3 S/m) and excellent mechanical properties (stretchable up to 300%). More importantly, the flexible Zn-air battery assembled from this electrolyte exhibits excellent flexibility, especially since the discharge voltage remains unchanged after 100 bending and stretching cycles.

In the preceding research work, the preparation of FMABs is usually divided into two steps. First, GPEs need to be synthesized, and then they are combined with flexible electrodes. Due to the high viscosity of GPEs, FMABs prepared by this two-step method often suffer from relatively poor interfacial contact between electrodes and GPEs, which easily leads to high resistance of FMABs. More importantly, when the flexible battery encounters severe deformation, it often leads to the problem of GPEs detachment in the FMABs. Based on the aforementioned problems, the one-step in-situ preparation in FMABs can effectively improve the interface contact between electrodes and GPEs. For example, Liu et al. innovatively proposed a one-step method for the synthesis of GPE based on in-situ polymerization of polyacrylic acid during the fabrication of flexible batteries [35]. The specific steps are to first spread the liquid polymer precursor solution on the electrode in seamless contact and then carry out the polymerization under the condition of room temperature. The prepared GPE exhibited excellent flexibility, which could be as high as 205.6% in tensile elongation and more than 99.5% in recovery. Inspired by previous research work, Park et al. also successfully developed a one-step strategy for in-situ formations of GPE using natural gels as gelling agents during the fabrication of flexible zinc-air batteries [36]. Excitingly, the solution and charge transfer resistance of this GPE can be compared with liquid electrolytes, especially its ionic conductivity can reach 3.1×10^{-3} S/cm, which can greatly promote the kinetics of flexible zinc-air batteries.

24.5.2.2 Organic Gel Polymer Electrolytes

In addition to the aforementioned aqueous GPE systems, organic GPE systems are also commonly used in flexible metal-air batteries. Among them, organic GPE systems have been widely studied in flexible lithium-air battery systems. To meet the requirements in the fabrication of flexible lithium-air batteries, the organic GPE needs to have the characteristics of high ionic conductivity, the high solubility of oxygen and lithium salts, and good stability with highly active lithium metal. It is worth noting here that metallic lithium is flammable, which makes it susceptible to corrosion by carbon dioxide or moisture in the ambient air [37]. Therefore, the synthesis of organic GPEs is very important for flexible Li-air batteries. On the one hand, it is necessary to avoid the problems of electrolyte leakage and short circuit during the process of repeated flexibility tests, and on the other hand, it is necessary to improve the protection and moisture resistance to lithium metal.

For example, researchers innovatively designed an organic GPE for flexible Li-air batteries, which was mainly composed of 1 M LiTF in TEGDME and based on PVDF-HFP/ETPTA framework composition [38]. The organic GPE shows good hydrophobic properties, which can greatly reduce the corrosion effect of moisture on Li metal electrodes. The organic GPE has good hydrophobic properties, which can greatly reduce the corrosion effect of moisture on Li metal electrodes. It is worth noting that the flexible lithium-air battery can still exhibit stable electrochemical performance after being immersed in water for 5 hours.

Similar to the aqueous GPE mentioned in the previous section, in addition to the aforementioned two-step fabrication process, researchers have recently developed a one-step in-situ self-gelling GPE that can be used in flexible lithium-air batteries [39]. The preparation process of this GPE is that a liquid TEGDME electrolyte containing 1% ethylenediamine (EDA) is gelled to form a protective layer through the gradual reaction of LiEDA and TEGDME on the Li metal surface. This one-step preparation process can greatly improve the interfacial contact between the electrolyte and lithium metal, which is beneficial to prevent the corrosion of the battery by carbon dioxide and moisture. To further verify the superiority of the organic GPE, the researchers tested the cycle stability of the battery in a certain humidity environment. Surprisingly, the battery can work for more than 50 days in ambient air with a humidity of 45–65%.

As mentioned earlier, organic GPE has made great progress in the field of FMABs, but there are still some problems that need to be solved urgently in the face of the complex use conditions of FMABs. For instance, organic solvents in GPE are prone to evaporation problems under high-temperature conditions, which can lead to unstable electrochemical properties and leakage of flammable gases [40]. Thus, further expanding the operating temperature range of organic GPEs is critical to improving the practicality of flexible metal-air batteries. Based on this, room temperature ionic liquids have attracted extensive attention as organic GPEs in recent years due to their wide electrochemical window, negligible vapor pressure, high thermal stability, and low flammability. Recently, researchers have successfully fabricated a new ionic liquid-based GPE system for flexible lithium-air batteries. The battery can achieve stable electrochemical performance under a temperature of 140 °C [41]. The results show that all components in the GPE exhibit excellent thermal stability under high-temperature test conditions. It is worth noting that when the temperature is increased from 25 to 100 °C, the ionic conductivity increases from 10^{-4} to 10^{-3} S/cm. The assembled batteries showed outstanding rate performance and high cycling stability. In addition, although the research of organic GPEs in flexible batteries has made great progress, future research directions still need to focus on designing novel GPE systems that can operate under wider temperature conditions.

24.6 ADVANTAGES AND CHALLENGES OF FLEXIBLE METAL-AIR BATTERIES

24.6.1 Advantages of FMABs

With the rapid development of modern society, people pay more and more attention to the development of lightweight, small, and portable flexible electronic devices. Therefore, it is crucial to design lightweight, small, and portable flexible energy storage devices. FMABs are considered to be one of the most promising approaches in flexible energy storage systems due to their high mass-specific energy and volume-specific energy. More importantly, flexible metal-air batteries can have excellent flexibility through rational design of electrode structure and electrolyte components, which has a strong role in promoting the development of flexible energy storage devices [42]. In addition, the raw materials used in the fabrication of flexible metal-air batteries are cheap to produce and easy to recycle, and most of the metals used are low-polluting or non-polluting.

24.6.2 Challenges of FMABs

With the continuous development of energy storage technology, there will be a more or less potential impact on the environment in the process of production and use, which also includes FMABs [43]. As mentioned earlier, the research on FMABs has made great progress, but there are still some challenges in the process of production and practical application. For example, the self-corrosion problem of metal electrodes still exists during the long-cycle charge-discharge process of FMABs [44, 45]. In particular, FMABs react with air during repeated bending, which can accelerate the corrosion of metal electrodes [46]. In addition to the hydrogen evolution reaction, self-discharge of the FMABs also occurs under alkaline conditions. Specifically, the reaction rate is greatly reduced due to the formation of a metal passivation layer in the air and the ability of the metal electrode to lose electrons is also reduced, which leads to a drop in the voltage of the battery.

As mentioned earlier, alkaline electrolytes are commonly used in aqueous FMABs. Alkaline electrolytes are volatile, which makes them susceptible to the working environment. When the ambient temperature is too high, it is easy to cause the volatilization of the electrolyte, which will seriously affect the electrochemical performance of the battery. In addition, the battery is prone to irreversible charging when carbon dioxide in the ambient air reacts with the electrolyte to form carbonate, which significantly reduces the cycling stability of the battery.

In addition to these aforementioned challenges, the clogging of air electrodes in FMABs is another key challenge. It is well known that the design of air electrodes in FMABs often adopts a sparse and porous structure, which can increase the contact area with air. However, after repeated charge-discharge cycles, a series of side reactions will occur at the air electrode, resulting in the generation of insoluble substances. With the continuous occurrence of side reactions, these insoluble substances will accumulate at the air electrode, which will block the original pore structure of the air electrode [47, 48]. This problem will seriously affect the electrochemical performance of the FMABs, and even lead to battery failure.

Furthermore, safety and cost issues related to the integration of organic electrolyte systems are also a big challenge for FMABs in practical commercial applications. For example, with the rapid development of flexible electronic devices, the global demand for FMABs will also rise substantially in the future. This demand can be expected to increase dramatically over the next decade as many countries move from traditional societies to smart societies. Therefore, the continuous development of high-performance flexible energy storage technology while maintaining environmental sustainability is of great significance for the sustainable development of the future world. When the FMABs are scrapped, the reasonable disposal of the battery as waste is also very important. The main reason for the analysis is that there are some harmful chemicals in the battery, so they cannot be handled simply. It should be noted here that the incineration of waste batteries also requires great care. For the preceding reasons, the end-of-life management of these non-functioning batteries is also a major challenge.

At present, researchers are still carrying out scientific research and application development to improve the electrochemical performance and mechanical stability of FMABs. However, from a practical application point of view, achieving safe and environmentally sustainable materials that simultaneously possess excellent electrochemical performance as well as strong structural stability and durability remains a key challenge in developing safe, cost-effective, and commercially viable flexible energy storage systems.

24.7 CONCLUSIONS

In summary, the rapid development of material science and nanotechnology in recent years has greatly promoted the development of FMABs. Compared with conventional secondary batteries, FMABs have a higher energy density. However, it is worth noting here that the air electrode belongs to an open system, which leads to problems that are prone to occur and urgently need to be solved, such as poor cyclability, low safety, and high cost. These problems will largely limit the further development and commercial application of FMABs. To solve the aforementioned problems, it is also crucial to deeply understand the reaction mechanism of FMABs during long-cycle charge-discharge processes. This requires experimental studies combined with theoretical models to gain a deep understanding of the electrochemical mechanism of FMABs. In addition to the need for scientific research on FMABs, technical research is also very important for the development of FMABs. For example, the use of different battery assembly methods will also greatly affect the electrochemical properties and mechanical stability of FMABs. So far, there are still some reported FMABs that have shown excellent comprehensive properties, even though industrial production still needs continuous efforts.

In addition, establishing effective evaluation criteria for electrochemical properties is very important for the development of flexible batteries. To date, the energy density of FMABs reported in many studies is usually calculated from the mass of the air electrode rather than the total mass of the battery device. This often results in a higher energy density calculated based on the mass of the air electrode, but for practical applications, the calculated energy density based on the total mass of the device is more important. After all, compared with other batteries, the energy density of the device is the most important indicator for FMABs to become a strong competitor of commercially viable batteries for future flexible electronics.

Although the preceding problems still need to be solved urgently, there is no doubt that FMABs has broad application prospects in smart bracelets, medical monitoring, smart cards, electronic textiles, flexible sensors, etc. It is also worth noting here that the effective combination of flexible energy harvesting devices with FMABs will greatly broaden the application of flexible electronics for future life.

REFERENCES

[1] S. Niu, N. Matsuhisa, L. Beker, J. Li, S. Wang, J. Wang, Y. Jiang, X. Yan, Y. Yun, W. Burnett, A.S.Y. Poon, J.B.H. Tok, X. Chen, Z. Bao, A wireless body area sensor network based on stretchable passive tags, Nat. Electron. 2 (2019) 361–368.

[2] Z. Gao, W. Yang, J. Wang, N. Song, X. Li, Flexible all-solid-state hierarchical $NiCo_2O_4$/porous graphene paper asymmetric supercapacitors with an exceptional combination of electrochemical properties, Nano Energy. 13 (2015) 306–317.

[3] Z. Yang, H. Sun, T. Chen, L. Qiu, Y. Luo, H. Peng, Photovoltaic wire derived from a graphene composite fiber achieving an 8.45% energy conversion efficiency, Angew. Chem., Int. Ed. 52 (2013) 7545–7548.

[4] X. Xiao, T. Li, P. Yang, Y. Gao, H. Jin, W. Ni, W. Zhan, X. Zhang, Y. Cao, J. Zhong, L. Gong, W.C. Yen, W. Mai, J. Chen, K. Huo, Y.L. Chueh, Z.L. Wang, J. Zhou, Fiber-based all-solid-state flexible supercapacitors for self-powered systems, ACS Nano. 6 (2012) 9200–9206.

[5] Y. Zhao, J. Guo, Development of flexible Li-ion batteries for flexible electronics, InfoMat. 2 (2020) 866–878.

[6] G. Qian, X. Liao, Y. Zhu, F. Pan, X. Chen, Y. Yang, Designing flexible lithium-ion batteries by structural engineering, ACS Energy Lett. 4 (2019) 690–701.

[7] H.F. Wang, Q. Xu, Materials design for rechargeable metal—air batteries, Matter. 1 (2019) 565–595.

[8] W.J. Kwak, D. Sharon, C. Xia, H. Kim, L.R. Johnson, P.G. Bruce, L.F. Nazar, Y.K. Sun, A.A. Frimer, M. Noked, S.A. Freunberger, D. Aurbach, Lithium–oxygen batteries and related systems: Potential, status, and future, Chem. Rev. 120 (2020) 6626–6683.

[9] Y. Zhang, L. Wang, Z. Guo, Y. Xu, Y. Wang, H. Peng, High-performance lithium——air battery with a coaxial-fiber architecture, Angew. Chem., Int. Ed. 55 (2016) 4487–4491.

[10] T. Cetinkaya, S. Ozcan, M. Uysal, M.O. Guler, H. Akbulut, Free-standing flexible graphene oxide paper electrode for rechargeable $Li—O_2$ batteries, J. Power Sources. 267 (2014) 140–147.

[11] Y.B. Yin, J.J. Xu, Q.C. Liu, X.B. Zhang, Macroporous interconnected hollow carbon nanofibers inspired by golden-toad eggs toward a binder-free, high-rate, and flexible electrode, Adv. Mater. 28 (2016) 7494–500.

[12] Y. Qiao, Q. Wang, X. Mu, H. Deng, P. He, J. Yu, H. Zhou, Advanced hybrid electrolyte $Li–O_2$ battery realized by dual superlyophobic membrane, Joule. 3 (2019) 2986–3001.

[13] Q. Liu, Z. Pan, E. Wang, L. An, G. Sun, Aqueous metal—air batteries: Fundamentals and applications, Energy storage mater. 27 (2020) 478–505.

[14] H. Song, H. Deng, C. Li, N. Feng, P. He, H. Zhou, Advances in lithium-containing anodes of aprotic $Li-O_2$ batteries: Challenges and strategies for improvements, Small Method. 1 (2017), 1700135.

[15] G. Zhou, F. Li, H.M. Cheng, Progress in flexible lithium batteries and future prospects, Energy Environ. Sci. 7 (2014) 1307–1338.

[16] J. Chang, Q. Huang, Z. Zheng, A figure of merit for flexible batteries, Joule. 4 (2020) 1346–1349.

[17] F. Meng, H. Zhong, D. Bao, J. Yan, X. Zhang, In situ coupling of strung Co_4N and intertwined N-C fibers toward free-standing bifunctional cathode for robust, efficient, and flexible Zn-air batteries, J. Am. Chem. Soc. 138 (2016) 10226–10231.

[18] Y. Xu, Y. Zhang, Z. Guo, J. Ren, Y. Wang, H. Peng, Flexible, stretchable, and rechargeable fiber-shaped zinc-air battery based on cross-stacked carbon nanotube sheets, Angew. Chem., Int. Ed. 54 (2015) 15390–15394.

[19] J. Fu, F.M. Hassan, J. Li, D.U. Lee, A.R. Ghannoum, G. Lui, M.A. Hoque, Z. Chen, Flexible rechargeable zinc-air batteries through morphological emulation of human hair array, Adv. Mater. 28 (2016) 6421–6428.

[20] Q.C. Liu, J.J. Xu, D. Xu, X.B. Zhang, Flexible lithium-oxygen battery based on a recoverable cathode, Nat. Commun. 6 (2015) 7892.

[21] L. Wang, Y. Zhang, J. Pan, H. Peng, Stretchable lithium-air batteries for wearable electronics, J. Mater. Chem. A. 4 (2016) 13419–13424.

[22] Q.C. Liu, L. Li, J.J. Xu, Z.W. Chang, D. Xu, Y.B. Yin, X.Y. Yang, T. Liu, Y.S. Jiang, J.M. Yan, X.B. Zhang, Flexible and foldable $Li-O_2$ battery based on paper-ink cathode, Adv. Mater. 27 (2015) 8095–8101.

[23] Q.C. Liu, T. Liu, D.P. Liu, Z.J. Li, X.B. Zhang, Y. Zhang, A flexible and wearable lithium-oxygen battery with record energy density achieved by the interlaced architecture inspired by bamboo slips, Adv. Mater. 28 (2016) 8413–8418.

[24] T. Liu, J.J. Xu, Q.C. Liu, Z.W. Chang, Y.B. Yin, X.Y. Yang, X.B. Zhang, A general strategy assisted with dual reductants and dual protecting agents for preparing Pt-based alloys with high-index facets and excellent electrocatalytic performance, Small. 13 (2017) 1602952.

[25] L. Ye, M. Liao, H. Sun, Y. Yang, C. Tang, Y. Zhao, L. Wang, Y. Xu, L. Zhang, B. Wang, F. Xu, X. Sun, Y. Zhang, H. Dai, P. Bruce, H. Peng, A sodiophilic interphase-mediated, dendrite-free anode with ultrahigh specific capacity for sodium-metal batteries, Angew. Chem. Int. Ed. 58 (2019) 2437–2442.

[26] J. Fu, J. Zhang, X. Song, H. Zarrin, X. Tian, J. Qiao, L. Rasen, K. Li, Z. Chen, A flexible solid-state electrolyte for wide-scale integration of rechargeable zinc-air batteries, Energy Environ. Sci. 9 (2016) 663–670.

[27] A.C. Luntz, B.D. McCloskey, Nonaqueous Li-air batteries: A status report, Chem. Rev. 114 (2014) 11721–11750.

[28] R.W. Mo, D. Rooney, K.N. Sun, H.Y. Yang, 3D nitrogen-doped graphene foam with encapsulated germanium/nitrogen-doped graphene yolk-shell nanoarchitecture for high-performance flexible Li-ion battery, Nat. Commun. 8 (2017) 13949.

[29] M.M.O. Thotiyl, S.A. Freunberger, Z. Peng, P.G. Bruce, The carbon electrode in nonaqueous $Li-O_2$ cells, J. Am. Chem. Soc. 135 (2013) 494–500.

[30] M.M.O. Thotiyl, S.A. Freunberger, Z. Peng, Y. Chen, Z. Liu, P.G. Bruce, A stable cathode for the aprotic $Li-O_2$ battery, Nat. Mater. 12 (2013) 1050–1056.

[31] Q.C. Liu, J.J. Xu, Z.W. Chang, D. Xu, Y.B. Yin, X.Y. Yang, T. Liu, Y.S. Jiang, J.M. Yan, X.B. Zhang, Growth of Ru-modified Co_3O_4 nanosheets on carbon textiles toward flexible and efficient cathodes for flexible $Li-O_2$ batteries, Part. Part. Syst. Charact. 33 (2016) 500.

[32] X. Cheng, J. Pan, Y. Zhao, M. Liao, H. Peng, Gel polymer electrolytes for electrochemical energy storage, Adv. Energy Mater. 8 (2018) 1702184.

[33] G. Liang, F. Mo, Q. Yang, Z. Huang, X. Li, D. Wang, Z. Liu, H. Li, Q. Zhang, C. Zhi, Commencing an acidic battery based on a copper anode with ultrafast proton-regulated kinetics and superior dendrite-free property, Adv. Mater. 31 (2019) 1905873.

[34] B. Lv, S. Zeng, W. Yang, J. Qiao, C. Zhang, C. Zhu, M. Chen, J. Di, Q. Li, In-situ embedding zeolitic imidazolate framework derived Co-N-C bifunctional catalysts in carbon nanotube networks for flexible Zn-air batteries, J. Energy Chem. 38 (2019) 170–176.

[35] G. Liu, J. Kim, M. Wang, J. Woo, L. Wang, D. Zou, J. Lee, Soft, highly elastic, and discharge-current-controllable eutectic gallium-indium liquid metal-air battery operated at room temperature, Adv. Energy Mater. 8 (2018) 1703652.

[36] J. Park, M. Park, G. Nam, J. Lee, J. Cho, All-solid-state cable-type flexible zinc-air battery, Adv. Mater. 27 (2015) 1396–1401.

[37] J. Shui, J. Okasinski, P. Kenesei, H. Dobbs, D. Zhao, J. Almer, D. Liu, Reversibility of anodic lithium in non-aqueous $Li-O_2$ batteries, Nat. Commun. 4 (2013) 2255.

[38] T. Liu, Q. Liu, J. Xu, X. Zhang, Cable-type water-survivable flexible $Li-O_2$ battery, Small. 12 (2016) 3101–3105.

[39] X. Lei, X. Liu, W. Ma, Z. Cao, Y. Wang, Y. Ding, Flexible lithium-air battery in ambient air with an insitu formed gel electrolyte, Angew. Chem. Int. Ed. 57 (2018) 16131–16135.

[40] J. Lai, Y. Xing, N. Chen, L. Li, F. Wu, R. Chen, Electrolytes for rechargeable lithium-air batteries, Angew. Chem. Int. Ed. 59 (2020) 2974–2997.

[41] J. Pan, H. Li, H. Sun, Y. Zhang, L. Wang, M. Liao, X. Sun, H. Peng, A lithium-air battery stably working at high temperature with high rate performance, Small. 14 (2018) 1703454.

[42] Y. Li, C. Zhong, J. Liu, X. Zeng, S. Qu, X. Han, Y. Deng, W. Hu, J. Lu, Atomically thin mesoporous Co_3O_4 layers strongly coupled with N-rGO nanosheets as high-performance bifunctional catalysts for 1D knittable zinc-air batteries, Adv. Mater. 30 (2018) 1703657.

[43] A.R. Dehghani-Sanij, E. Tharumalingam, M.B. Dusseault, R. Fraser, Study of energy storage systems and environmental challenges of batteries, Renew Sustain Energy Rev. 104 (2019) 192–208.

[44] C. Zhu, H. Yang, A. Wu, D. Zhang, L. Gao, T. Lin, Modified alkaline electrolyte with 8-hydroxyquinoline and ZnO complex additives to improve Al-air battery, J Power Sources. 432 (2019) 55–64.

[45] J. Ma, G. Wang, Y. Li, W. Li, F. Ren, Influence of sodium silicate/sodium alginate additives on discharge performance of Mg—air battery based on AZ61 alloy, J Mater Eng Perform. 27 (2018) 2247–2254.

[46] M.A. Rahman, X. Wang, C. Wen, A review of high energy density lithium-air battery technology, J Appl Electrochem. 44 (2014) 5–22.

[47] K.F. Blurton, A.F. Sammells, Metal/air batteries: their status and potential—a review, J Power Sources. 4 (1979) 263–279.

[48] M. Xu, D. Ivey, Z. Xie, W. Qu, Rechargeable Zn-air batteries: progress in electrolyte development and cell configuration advancement, J Power Sources. 283 (2015) 358–371.

25 Challenges in Metal-Air Batteries

Alexander Kube and Dennis Kopljar

CONTENTS

25.1 Introduction 361
25.2 Constitution of a Metal-Air Battery 361
25.3 Challenges in Metal-Air Battery 361
    25.3.1 Cathode 362
    25.3.2 Electrolyte 367
    25.3.3 Anode 371
25.4 Conclusion 372
References 373

25.1 INTRODUCTION

Along with an increasing interest in energy storage and battery technology, the topic of metal-air batteries has not fallen short of attention in the last few years. This is because depending on the specific chemistry, they might offer intriguing characteristics, in particular with regard to the combination of high energy density and absence of the critical raw materials found in state-of-the-art lithium-ion batteries. In light of the increasing diversification of cell chemistries that metal-air batteries have experienced and that nowadays goes far beyond the historically dominating lithium- and zinc-air chemistries, material abundance and sustainability have been invoked as additional and, probably, most important selling points for metal-air technologies. However, to manifest these promises, numerous technology-overarching and chemistry-specific challenges need to be overcome. In this chapter, an attempt is made to summarize and categorize some of these challenges according to the individual components of the battery cell—the metal, the air electrode as well as the electrolyte.

25.2 CONSTITUTION OF A METAL-AIR BATTERY

A metal-air battery (MAB) consists of a porous gas diffusion electrode, an electrolyte with a separator, and an anode. Generally, MABs can be divided by their interaction between the anode and the electrolyte. Figure 25.1 shows a classification according to the anode reaction, as it can either form protecting layers on the anode material like oxide for Zn-air batteries, building insulation layers on the cathode materials like e.g., Li_2O_2 for Li-air batteries, or solely dissolution into the bulk electrolyte (not shown separately). Examples for the last setup are a $Zn-O_2$ battery with an acidic electrolyte or a $Li-O_2$ battery with LiOH on the cathode side as an electrolyte.

25.3 CHALLENGES IN METAL-AIR BATTERY

All of these anode reaction pathways pose their challenges. For example, the formation of ZnO_2 is reversible but limits the current density with increasing thickness due to mass transport limitation of OH^- through the porous ZnO_2. Furthermore, special care has to be taken for host structures

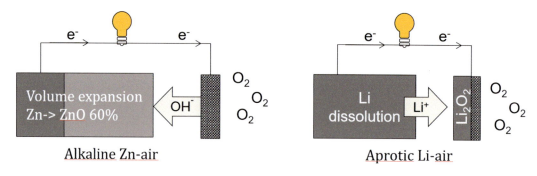

FIGURE 25.1 Differentiation of metal-air batteries by anode reaction.

sustaining the anode structure with channels for electrolyte and high surface area. Otherwise, the anode material cannot be used to its full extent, reducing the capacity of the battery. Ion dissolution into the electrolyte limits the capacity due to the solubility limit. This reduces the energy density of the battery. The precipitation of anode material onto the cathode material—as is the case in e.g. Li-O_2—leads to the deactivation of active centers and clogging of pores.

A second possible classification of metal-air batteries is the type of electrolyte, which is associated with specific and pronounced advantages and disadvantages. The different types are shown in Figure 25.2 together with two cell architecture possibilities. Aqueous electrolytes are considered environmentally friendly, cost-efficient, and safe, as they are not flammable, however, exhibit a limited stability window and react heavily with alkali metals. Non-aqueous electrolytes, on the other hand, allow for a wider choice of anode materials and show higher electrochemical and thermal stability windows. By using all solid-state electrolytes challenges due to crossover from cathode to anode and unwanted substances brought into the system by the gas supply can be mitigated, reducing the risk of unwanted side reactions and degradations like the attack of protective layers of the anode or hydrogen evolution. This comes with the costs of higher contact resistances and lower ionic conductivities of such electrolytes.

A third possibility would be a grouping based on the cathode layout, as shown in Figure 25.2 as well. The high overpotentials of metal-air batteries inflict stability issues on all materials the battery is built off, especially cathode materials which are oxidized and reduced during operation. By using two separate electrodes, one for oxygen reduction reaction and one for oxygen evolution reaction, this can be avoided and allows for cheaper less sophisticated materials. This comes with the trait of bigger cells and therefore reduced energy densities, as an additional electrode, as well as an additional gas supply, has to be included.

In this chapter, separation will be done on the battery components and the challenges these components are facing.

25.3.1 Cathode

The cathode is built of a porous material allowing electrolytes to enter from one and gas from the other side. The oxygen dissolves into the electrolyte and diffuses to the active center of the catalyst.

As shown in Figure 25.3, three regions can be identified: in orange the jR_s region where ionic transport is hindered, in green the region with the highest current density and the yellow area with lower current density due to diffusion limitation of the oxygen in the electrolyte. The region where gas, electrolyte, and catalyst are present is also called the triple phase boundary. Here, dissolved oxygen is transformed into oxygen anions by taking up electrons from the cathode. This process is very sluggish and involves 4 electrons, resulting in high overpotentials. The kinetics can be improved with catalysts, but even by using catalysts, high overpotentials are commonly observed.

Challenges in Metal-Air Batteries 363

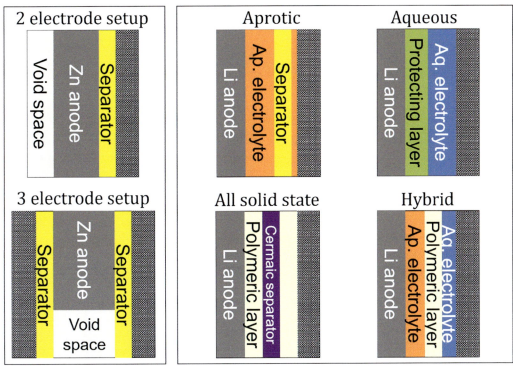

FIGURE 25.2 Different types of metal-air batteries by electrode configuration (left side) and type of electrolyte (right side).

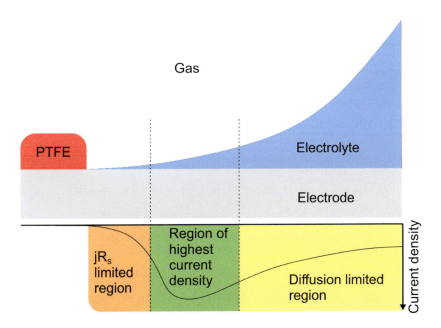

FIGURE 25.3 Current distribution near the triple phase region. (Adapted with permission [1]. Copyright (2022) The Authors, some rights reserved; exclusive licensee Wiley. Distributed under a Creative Commons Attribution License 4.0 (CC BY).)

This poses challenges to the stability of the used materials, as carbon corrosion is reported to start at potentials above 1.4 V vs. RHE [2]. This oxidation of carbon poses a major bottleneck for high cycle life as the porous network loses its backbone. The corrosion follows the reaction pathway [3]:

$$C \rightarrow C^+ + e^- \tag{1}$$

$$C^+ + H_2O \rightarrow CO + 2H^+ + e^- \tag{2}$$

$$2CO + H_2O \rightarrow CO + CO_2(g) + 2H^+ + 2e^- \tag{3}$$

Two possible paths can be taken to mitigate carbon corrosion. One approach is to use non-carbon backbones for GDEs like metals (Ni, Ag), transmission metal oxides (Ti_2O_3, Ti_4O_7), or nitrides (TiN) [3]. To overcome the carbon corrosion itself the amount of active sites to be attacked during carbon corrosion have to be reduced. Carbon blacks like Vulcan XC72 do show faster degradation due to the higher number of sites to be attacked than graphite which shows already good stability [4]. By using advanced carbons, the durability can be enhanced even further. Additionally, by doping these carbon supports catalytic activity for OER and ORR can be added.

As mentioned earlier, catalysts are essential to reduce overpotentials and therefore enhance cycle life and efficiency. Catalyst should meet the following requirements:

- **Activity:** The macroscopically observable activity of catalysts is determined by their intrinsic activity and number of accessible active sites. While the first can be tuned e.g. by the surface morphology or addition of co-catalysts, the latter is influenced by wetting characteristics of the electrode structure, accessibility to the reactant, and dispersion.
- **Stability:** The chemical and morphological/structural integrity of the catalyst is essential for high cycle life as it directly influences the pro-longed activity and therefore the energy efficiency of the battery. At high overpotentials, catalysts can undergo changes in oxidation states in each cycle which can result in catalyst leaching, agglomeration, and/or morphology changes on the surface.
- **Cost/environmental friendliness:** The use of alkaline media offers the possibility to use non-noble metal catalysts like spinels with the general formula AB_2X_4 like Co_3O_4 and $NiCo_2O_4$ [5], perovskites with general formula ABO_3 like $LaNiO_3$ and $La_{1-x}Ca_xCoO_3$, or even metal-organic frameworks (MOFs).

As there are two different ORR mechanisms, the stability of the catalyst and the whole GDE structure depends on the prevailing mechanism. The first possible ORR mechanism is the so-called four-electron pathway. For this, two oxygen atoms are coordinated onto the active center of the catalyst and directly reduced:

$$O_2(g) + H_2O(l) + 4e^- \rightleftharpoons 4OH^-(aq), \text{ with } 0.4V \text{ vs. SHE} \tag{4}$$

The alternative ORR mechanism is called the two-electron pathway where only one oxygen atom is coordinated onto the catalyst forming a peroxide, which is then further reduced according to [6]

$$O_2(g) + H_2O(l) + 2e^- \rightleftharpoons HO_2^-(aq) + OH^-(aq), \text{ with } -0.07V \text{ vs. SHE} \tag{5}$$

$$HO_2^-(aq) + H_2O(l) + 2e^- \rightleftharpoons 3OH^-(aq), \text{ with } 0.87V \text{ vs. SHE} \tag{6}$$

The four-electron pathway is preferred as the direct reduction is faster. One additional drawback of the two-electron pathway is the formation of HO_2^- species which is highly reactive and can lead to corrosion, reducing the stability of GDE materials. By using organic electrolytes, the reaction pathway changes to the formation of metal peroxides and superoxides which are precipitated onto the cathode surface. The amount of peroxide and superoxide formation depends on the type of anion. While K-O_2 systems form almost 99% superoxide, in Li-O_2 systems 59–93% peroxide is formed [7]. In general, reversibility becomes better if superoxide is formed. The reaction pathway in organic solvents for such systems is exemplarily shown for Li-O_2 batteries by:

$$O_2 + Li^+ + e^- \rightarrow LiO_2 \tag{7}$$

Further reduction can follow two pathways. Either disproportionation:

$$2LiO_2 \rightarrow Li_2O_2 + O_2 \tag{8}$$

Or reduction by receiving another electron:

$$LiO_2 + Li^+ + e^- \rightarrow Li_2O_2 \tag{9}$$

The formed lithium peroxide is insoluble and depending on the operation conditions and the aprotic electrolyte composition precipitates either on the surface as an insulating film, or in toroidal shape on the cathode surface. This limits the capacity as active centers are blocked and the conversion of Li_2O_2 to Li^+ and O_2 is hindered in the following charge step. One reason for irreversibility is the formation of oxygen intermediates during cyclings like oxygen singlets ($^1O_2$), which can react via nucleophilic substitution with electrolyte or cathode. Singlet oxygen is the lowest excited state of the diatomic oxygen molecule and is known to be very reactive towards organic compounds and therefore is a primary source for redox mediator decomposition [8, 9]. Oxygen singlet is mainly formed by disproportionation of lithium superoxide; another source of singlet oxygen is the electrochemical decomposition of Li_2CO_3 above 3.82 V vs. Li. Singlet oxygen was also observed in other metal-air batteries, but as most investigations were performed at Li-O_2 batteries explanations are given for this type of battery. Depending on the Lewis acidity of the electrolyte and the type of alkali metal the amount of oxygen singlets will vary. While strong Lewis acids generally produce higher amounts of $_1O^2$, weak Lewis acids undergo different pathways during disproportionation, bypassing most unfavorable reaction steps and hence strongly facilitating $_1O^2$ formation [7]. These pathways are active both during charge and discharge.

Both aqueous and organic electrolytes share the same characteristics if applied in metal-air batteries. They will penetrate the porous system until non-wetting structures/phases inside the gas diffusion electrode (GDE) hinder further penetration. Commonly GDEs are built from a conductive media (e.g. Carbon, or Ni), a non-wetting binder (e.g. PTFE), and a catalyst (e.g. metal oxides like CO_3O_4; or MnO_2). All of these components have their wetting characteristics influencing the penetration. To achieve stable cell performance, one needs to control the penetration behavior and distribution of electrolytes inside its porous structure. The penetration of electrolyte into the porous system is determined by the surface energy of the liquid and solid interface (e.g. roughness), the microstructural characteristics of the system as well as the environmental (gas/liquid pressure, temperature) and cycling conditions. If a liquid is in the wetting regime, imbibed liquid in the porous structure will spread, and if the wetting conditions are unfavorable eventually blocks gas diffusion completely to the active sites [1]. GDEs are a combination of material properties and pore structures with different wetting characteristics. Capillary forces that initiate the wetting process are a strong function of pore size; this leads to a mixed wettability where smaller pores can be flooded more easily than larger pores resulting in heterogeneous saturation throughout the porous structure. If cells get larger the impact of hydrostatic pressure on the saturation distribution increases. As a result, the local current density is reduced in the bottom part of the cell which is flooded more strongly whereas in higher regions of the cell a more optimal balance of saturation level can be achieved. This heterogeneous current and potential distribution are also reflected on the anode side, as can

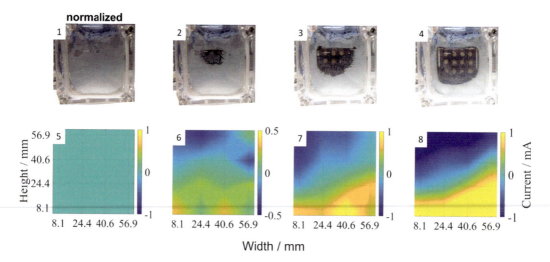

FIGURE 25.4 Upper part (1–4) shows pictures of the Zn-anode during discharge, and the lower pictures (5–8) the change in the current distribution of the silver cathode. (Adapted with permission [10]. Copyright (2021) The Authors, some rights reserved; exclusive licensee IOP Publishing Limited. Distributed under a Creative Commons Attribution License 4.0 (CC BY).)

be evaluated with a segmented cell setup. As shown in Figure 25.4, the anode dissolves first in the upper cell regions. By further discharge, the cathode current density follows the remaining metal on the anode side. This heterogeneous behavior results in heterogeneous degradation of both the anode and cathode.

As mentioned, wetting is determined by the surface energy of the liquid and solid interface. Organic electrolytes and ionic liquids show lower surface energies than aqueous electrolytes resulting in unstable triple-phase boundaries and therefore unstable performance in common cell setups. For Li-air batteries, Xia et al. showed improved reaction kinetics for a partially wetted electrode, which could be attributed to a better oxygen diffusion throughout the electrode thickness [11]. The oxygen diffusion length plays a key role in the achievable specific capacity [12], showing that the control of the wetting inside the porous system is mandatory both for high electrochemical performance as well as for high specific capacities. This correlation between oxygen diffusion length and capacity is believed to be due to the precipitation of Li_2O_2 at regions with high oxygen saturation, e.g., near the gaseous supply chamber, resulting in a blockage of oxygen diffusion paths into the porous network [1, 13].

To avoid heterogeneous wetting and allow for stable triple-phase boundaries a gradient in pore size is beneficial as known from fuel cells. Their double skeleton electrodes are used with larger pores on the gas side and smaller pores on the electrolyte side. As smaller pores have higher capillary pressures (p_1) as larger pores (p_2), they are filled by electrolytes first. By setting the gas pressure between p_1 and p_2 the electrolyte saturation can be adjusted to be only in the region with smaller pores. If larger pores are additionally filled with a nonwetting binder, pressure p_2 can be reduced even further, enhancing this effect [14].

At open circuit potential (OCV) a stable saturation point is reached determined by the aforementioned interaction between liquid electrolyte and GDE characteristics. Things change if a potential is applied. Depending on the potential applied the contact angle changes, as described by the Young-Lippmann equation

$$cos(\Theta) = cos(\Theta_0) + \eta \tag{10}$$

$$\eta = \frac{C_i\left(U_{pcz} - U\right)}{2\gamma_{lg}} \quad (11)$$

Where Θ_0 describes the contact angle of an uncharged surface and η the electrowetting number with the capacitance per unit area C_i and U_{pcz} describing the voltage at the point zero charges (pcz). As the electro wetting depends directly on the capacitance and the capacitance scales inversely with the dielectric layer, not all materials react the same to changes in potential [15]. As GDEs are made of conducting and non-conducting materials like PTFE, only conducting materials build double layers where electro wetting can inflict changes in wetting characteristics. This makes a comprehensive understanding of wetting behavior during cycling very complex. Still, several studies have tried to shed light on this using a variety of methods supplemented by modeling of the underlying phenomena [1, 15–19]. A deeper understanding of these wetting characteristics by direct interaction of liquid and solid materials or changes due to applied potentials has to be accomplished to increase the cycle life of GDEs.

Crossover of materials between the anode and the cathode might also alter the electrode behavior. One example is the potential intercalation of dissolved ions into the electrocatalyst structure, e.g. Zn into MgO_2 catalyst, reducing their performance. It could also precipitate onto the cathode surface altering the wetting and electrochemical behavior [20].

25.3.2 Electrolyte

The role of the electrolyte inside the system is to enable ionic transport and can be categorized into aqueous, non-aqueous, hybrid, and solid-state. One main challenge for all electrolytes used is the wide potential window of metal-air batteries. The stability of liquid and solid electrolytes is determined by the chemical interface. At equilibrium the electrode voltage (e.g. V vs. Li) depends on the chemical potential (μ_{Li}), the Faraday constant, and the charge q:

$$V = \frac{\mu_{Li}}{qF} \quad (12)$$

In contact with another material like electrolyte, the chemical potential changes as a double layer capacitance are formed between both layers in contact. At the anode mobile charge, carriers (e.g. Li^+) are driven into the electrolyte, while at the cathode mobile charge carriers are driven from the electrolyte into the cathode. As energy levels of both may differ drastically, this can lead to severe concentration gradients forming Li-rich interfaces at the anode and Li-poor interfaces at the cathode. This poses a challenge for electrolyte stability, and the choice of materials used has to be done carefully. The electrochemical stability of electrolytes can be determined by the energy difference between the highest occupied molecular orbital (HOMO) and the lowest unoccupied molecular orbital (LUMO). In contact with the anode, μ_{Li} should be greater than the LUMO, while at the cathode μ_{Li}<HOMO should apply [21].

The type of electrolyte is determined by the used anode material, e.g. for zinc alkaline electrolytes like KOH is commonly used. Aqueous alkaline electrolytes offer the possibility to use non-precious catalyst materials and zinc-air batteries favor ZnO formation. This is an essential characteristic of a high capacity as otherwise the capacity is limited by the solubility limit of the ions in the electrolyte. Another positive effect of using alkaline electrolytes is the lower amount of evolved hydrogen during operation, compared to acidic electrolytes. Besides these positive effects, zinc is highly soluble in alkaline media, forming $Zn(OH)_4^2$ which is strongly associated with dendrite formation and zinc shape change [22]. Dendrite growth can in a worst-case scenario lead to short circuits of the battery. Another drawback of alkaline electrolytes is that they react with CO_2

from the air, or decomposition of cathode materials, forming carbonates. These carbonates can precipitate onto the cathode materials, isolating active centers or even block pores completely resulting in blocked mass transport. Additionally, if the carbonate concentration in the electrolyte bulk is high enough the extraordinary conductivity and ionic transport mechanism (composed of structural and vehicular diffusion) are reduced as well. As shown by equations 13–15 hydroxide ions are consumed during reaction with CO_2 forming dibasic carbonic acid.

$$CO_2(g) \rightleftharpoons CO_2(aq) \tag{13}$$

$$CO_2(aq) + H_2O(l) \rightleftharpoons H_2CO_3(aq) \tag{14}$$

$$H_2CO_3(aq) + OH^-(aq) \rightleftharpoons HCO_3^- + H_2O(l) \tag{15}$$

Acidic electrolytes are not prone to such problems of carbonate formation as they exist within the protonated state which limits their tendency of precipitation. As mentioned earlier, only alkaline electrolytes favor ZnO precipitation. For acidic electrolytes, Zn stays in solution as Zn^{2+} and therefore the capacity is limited by the solubility of these ions. As commercial batteries are built without an excess of electrolyte this poses an issue. Minimizing the impact of the formation of carbonates near neutral electrolytes is of great interest. Ammonium chloride (NH_4Cl) has been investigated by many groups, as this kind of electrolyte also precipitates chloride-containing salts ($Zn(NH_3)_2Cl_2$ and $Zn_5(OH)_8Cl_2$). The drawback of ammonium chloride-based electrolytes is that precipitation of chloride-containing salts consumes the electrolyte and therefore pH might change. Additionally, the wide potential window of metal-air batteries poses stability issues onto the cathode materials, as they are attacked by chlorides, as well as onto the electrolyte resulting in chloride gas evolution. Alternative near neutral electrolytes containing organic salts like potassium citrate and glycine were investigated as they gel the electrolyte and reduce the pH to around 9 [23]. This electrolyte shows the necessity to keep all aspects of the battery in mind when designing electrolytes. Both organic additives show good control of pH during cycling. Unfortunately, citric acid is decomposed during charging by a Kolbe electrolysis in which adsorbed intermediate products block the surface.

$$2RCOO^- \rightarrow 2RCOO \rightarrow +2e^- \tag{16}$$

$$2RCOO \rightarrow 2R + 2CO_2 \tag{17}$$

$$R + R \rightarrow R-R \tag{18}$$

The blockage of the surface is due to the citric acid which is normally used as a ligand for nanoparticles and strongly adsorbs on metals. This blockage of active sites reduces the achievable electrochemical performance. Radicals formed during decomposition attack the cathode materials enhancing the aging in such systems [24]. Besides carbonate formation in alkaline electrolytes and precipitating onto the cathode material surface, crossover from the anode materials might precipitate as well onto the cathode. Besides the aqueous electrolytes which are non-flammable, less toxic, cheap, and have high ionic conductivity there is also the group of non-aqueous electrolytes. Compared to aqueous electrolytes they show fewer issues with evaporation but might take up ambient moisture. Non-aqueous electrolytes do show higher stability windows, as aqueous electrolytes are only stable in a small potential window (the open circuit potential of e.g. Zn-air batteries is 1.65 V, and the stability window of aqueous electrolytes is 1.23 V) [22, 25]. This instability causes some hydrogen evolution during charge and changes in the concentration of the electrolyte, leading to locally highly corrosive environments. Non-aqueous electrolytes additionally show elimination of anode corrosion reactions, but also higher wetting of cathode materials and lower conductivities and therefore lower achievable current densities. Another challenge in reducing current densities is the lower mobility and/or solubility of dissolved oxygen in some organic electrolytes (Table 25.1).

TABLE 25.1

Solubility and Diffusion Constants for Some Organic Electrolytes and KOH for Two Molarities

Electrolyte	Solubility (mM)	Diff. Constant (10^{-6} cm^{2}s^{-1})	Temperature (°C)	Concentration (mol/l)
KOH [26]	0.8	14	25	1
KOH [26]	0.1	8.2	25	4
[Emim][NTF] [27]	3.9	7.3	25	–
[BMIN][PF$_6$] [27]	3	2.5	25	–
BMIM][BF$_4$] [27]	1.1	12	25	–

The reduction in current density is because the exchange current density depends on the surface concentration of reactants as shown by equation 19.

$$j_0 = Fk_0 \left(C_{oxy}^{1-\beta} C_{red}^{\beta} \right) \quad (19)$$

Where C_{oxy} describes the concentration of the oxidized species and C_{red} the concentration of the reduced species, F the Faraday constant, k_0 the reaction rate and β a symmetry factor [28]. The exchange current density directly influences the current as shown by the Butler-Vollmer equation.

$$j = j_0 \left(e^{\frac{(1-\alpha)zF\eta}{RT}} + e^{\frac{\alpha zF\eta}{RT}} \right) \quad (20)$$

With the current density j, the activation overpotential η, the temperature T, the universal gas constant R, the Faraday constant, and the number of electrons involved in the reaction z. While the lower gas diffusion and solubility of oxygen in the electrolyte reduce the surface coverage and therefore j_0, the higher wetting characteristics of non-aqueous electrolytes lead to flooding of the porous structure and therefore to lower amounts of reduced actives sites where reactions can take place during ORR. One possibility to reduce the impact of flooding could be higher gas pressures, pushing back electrolytes into the battery. The amount of water in the feed gas controls if the aqueous electrolyte either loses or gains water, resulting in an either dried or flooded system. Water uptake also dilutes aqueous electrolytes. If water is gained in non-aqueous electrolytes, the characteristics and reaction mechanisms are altered [29]. To accommodate for unwanted changes in the amount of water in electrolytes, polymers can be added to gel the electrolyte. For this, PEO was used by many researchers at the beginning for aqueous electrolytes. Due to the low water absorption capacity and interfacial properties of PEO, other polymers like PAA are more suitable. Such gelled electrolytes also reduce unwanted crossover of species between both electrodes like Zn(OH)$_4^{2-}$ for Zn-air batteries, or water, nitrogen, and oxygen from the cathode to the anode for Li-air batteries. Reducing or completely stopping such crossovers would allow for higher cycle life as these pose severe degradation to the system.

The inhibition of crossover would also reduce the degradation of redox mediators in e.g. Li-air batteries, but as shown in literature is not an easy task [30, 31]. As explained earlier, Li-air batteries are mostly operated with organic electrolytes, leading to the formation of Li$_2$O$_2$ in the cathode which passivates the active surface. Decomposition of such Li$_2$O$_2$ back to Li requires high overpotentials threatening the stability of electrolytes and cathode materials. Redox mediators (RM) are molecules solved in the electrolyte with redox potential slightly above the potential needed for Li$_2$O$_2$ oxidation. RMs are oxidized on open sites of the cathode surface via electron transfer. This oxidized RM can then diffuse to Li$_2$O$_2$ and oxidize these and by this all Li$_2$O$_2$ can be fully oxidized back to molecular oxygen, bypassing the insolating nature of Li_2O_2 [31]. While RMs propose a good solution for challenges during charging, improving the stability of other battery materials, they are not stable and get deactivated, e.g. by shuttle reactions with the Li anode, or are consumed by side reactions. RMs can be classified as organic RMs and halide-based RM. While organic electrochemical RM like e.g. tetramethyl-piperidinyloxyl (TEMPO), tetrathiafulvalene (TTF), and dimethylphenazine (DMPZ) suffer from intrinsic instability due to the wide potential window in Li-air batteries (2.3–4 V vs. Li), Li halides like LiBr show corrosive behavior [31].

The evaporation of liquid organic and aqueous electrolytes during long-term operation increases the interfacial resistance. This can be overcome by using all solid-state electrolytes. This class of electrolytes has its challenges. The ionic conductivity is in the region of 10^{-6} to 10^{-4} S/cm which is lower than liquid electrolyte (10^{-2} S/cm) [21]. This low ionic conductivity limits the current densities and therefore the power output of the battery. As solid-state electrolytes are incorporated into the porous structure of the cathode as well, they must be stable against ambient air. This excludes high ionic conductive all solid-state electrolytes like sulfidic electrolytes. Additionally, all-solid-state (ASS) electrolytes are designed to transport ions between the electrodes but not to transport oxygen. This poses a challenge for GDE design, as pores are also needed for gas transport. Some developments using porous ASS electrolytes to enable gas transport were already made to mitigate this problem. Although gas transport limitations can be reduced with such an approach, high interface resistance in the electrode/electrolyte contact limits the performance. Another challenge arises from the fact that e.g. redox mediators cannot be used in such systems. Discharge products precipitated onto the cathode active centers insulate these parts. Therefore, precipitation occurs at the same active spots every cycle, if charging was successful, producing mechanical stress onto the ASS electrolyte as the discharge products build thin films pushing the ASS electrolyte aside.

A similar challenge arises at the anode side where volume changes can be observed. These can be either due to intercalation of e.g. Li in silicon or graphite or from energetically favored spots where Li deposition starts. In Figure 25.5 three cases are demonstrated. The first case describes the influence of heterogeneous current distribution e.g. due to heterogeneities in the conduction of the ASS. This leads to regions with higher metal dissolution than other regions. If the battery is charged again, metal is plated at regions that are energetically favored. These are close to the ASS electrolyte and further plating of anode material would lead to voids inside the anode, which then would need additional space for anode material during charging, resulting again in mechanical stress for the ASS electrolyte [32]. The second case demonstrates the influence of defects in either the anode

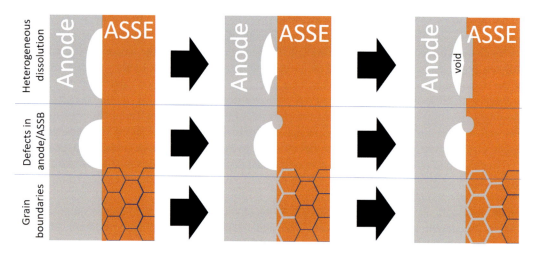

FIGURE 25.5 Schematic representation of void and dendritic growth in the anode during the charge of the battery.

material or the ASS electrolyte. These defects can be the start e.g. Li nucleation. The third case is grain boundaries inside the ASS electrolyte. These can as well be the source for nucleation and the end of Li dendrite formation. As a result of these effects, it was found that unless perfect ASS electrolytes and defect-free interfaces are used, ASS electrolytes are prone to dendrite growth and can even show more severe dendrite growth than liquid electrolytes [21].

25.3.3 ANODE

The most common anode materials are Li, K, Na, Zn, Fe, or Mg, whereas most research is done for Li, followed by Zn. The metal dissolves as ions during discharge into the electrolyte and is precipitated back onto the anode, or current collector during charge. The metallic anodes face many challenges such as the formation of passivation layers, hydrogen formation, and corrosion of metallic anodes in the presence of water if an aqueous electrolyte is used. The stability and reactivity of the anode materials determine the appropriate electrolyte used, e.g. Li reacts heavily with water and therefore must either be shielded by a protective layer or organic electrolytes must be used. Hydrogen evolution is a challenge for almost all MABs as they have a standard voltage below that of hydrogen evolution, favoring hydrogen evolution and reducing the coulombic efficiency of the MAB.

Equation 22 shows the reaction path of the corrosion reaction on the anode side where M describes the metal, while equation 23 describes the direct oxidation reaction.

$$2M + 2H_2O \rightleftharpoons 2M(OH) + H_2 + ne^- \tag{22}$$

$$yM + xH_2O \rightleftharpoons yMO_x + xH_2 + ne^- \tag{23}$$

Both reaction paths lead to loss of electrolytes which is detrimental to cycle life. Additionally, the reaction of the metal anode with water leads to self-discharge of the battery reducing both the capacity as well as the coulombic efficiency. In 2013, Siahrostami showed that hydrogen evolution takes place at kink sites of the metal anode. At these kink sites, atoms are not bound as strongly as in the bulk [33]. By adding dopants to the zinc the hydrogen evolution reaction (HER) can be reduced or completely suppressed. If such doped zinc anodes are used in rechargeable metal-air batteries, they have to be selected carefully as the anode is dissolved and precipitated again. By this, doping atoms would be inside the zinc bulk diminishing their positive effect. Typically, Indium and Bismuth are used as dopants for Zn-air batteries as they favor kink and edge sites where HER would take place otherwise. DRT studies showed that both dopants can rearrange on the surface and therefore stay on the surface at kink and edge sites allowing for higher cycle life than without these dopants [34]. The effect of dopants is also essential if crossover from cathode materials to the anode takes place. One possibility for this would be highly mobile silver ions. If these adsorb onto the zinc surface, they enhance the HER but do not rearrange and therefore are only effective during the cycle they are adsorbed [34]. In zinc-air batteries, HER evolution decreases on ZnO films and therefore reduces the self-discharge rate during ongoing discharge. Dopants with similar effects for other metal-air battery systems are also known, e.g. Copper decreases the anodic potential on Al electrodes.

Besides the integration of dopants into the metal bulk during charging, the rearrangement of anode material itself has to be well controlled as a type of electrolyte as well as current density influences reprecipitation and therefore leads to changes of the surface morphology like the evolution of dendrites. As described earlier there are two kinds of metal-air batteries. The ones building a passivation layer on the anode side by e.g. formation of ZnO, or the ones diffusing into the bulk electrolyte or to the cathode. The ZnO in Zn-air batteries is formed at places where the solubility limit of $Zn(OH)_4^{2-}$ is reached. As these ions diffuse the place of precipitation might differ from the place it went into the solution and even in the bulk electrolyte, ZnO could form. Both effects lead to the accumulation of anode material at the bottom of the cell, reducing the 3D morphology of such anodes. The accumulated zinc shows less to no porosity and a high thickness of the ZnO layer. Although ZnO in Zn-air batteries stays doped, appearing blue and not white, and therefore still has acceptable electric conductivity, gravimetrical forces keep it at the cell bottom. These effects lead to reduced capacity, as the active surface of the anode is reduced. Another effect of ZnO formation is the hindrance to diffusion of hydroxide ions through the porous ZnO layer to the zinc particle leading to reduced current densities. This applies to type I zinc oxide which is formed by the dissolution-precipitation model. Especially for higher current densities, zinc oxide type II is formed by nucleation and growth model. While type I zinc oxide is reversible, type II is very dense and cannot be dissolved again and hence is irreversible [22].

Continuous stripping and plating of metal onto the anode can lead to morphology changes due to electrical, chemical, and gravitational forces. As metal anodes are not smooth down to the atomistic level nucleation centers lead to the formation of dendrites and heterogeneous plating which then is energetically favored for further metal deposition. The growth of such dendrites is more severe for higher current densities because of the limited diffusion of metals on the anode surface. As dendrites grow with each cycle they pose threads to batteries: one is puncturing of the separator leading to short circuits; the other is that dendrites could break leading to lower capacities. Dendrites have lower coordination of atoms compared to bulk material and a more defective crystal lattice plane [35]. These defects in the crystal structure can also promote side reactions e.g. HER in aqueous electrolytes.

To overcome challenges arising from heterogeneous plating the battery has to be considered complete. The current density of the cathode must be as homogeneous as possible, allowing for homogeneous plating at the anode. Electrolytes, may it liquid or solid, must cover the anode material completely without defects. Additionally, they should block contaminants from e.g. ambient air diffusing to the anode. Furthermore, they should aid direct plating e.g. Zn-air batteries to avoid morphology changes reducing the accessible area e.g. Li-air batteries allow for homogeneous Li$^+$ conducting. By this, a homogeneous defect-free plating and therefore high cycle life can be achieved.

25.4 CONCLUSION

In this contribution, some of the general and technology-specific challenges have been summarized and classified according to overarching topics that need further development efforts which are partially independent of the specific technology as well as chemistry-specific. Complicating the general assessment is the fact that the state of development for metal-air batteries is very broad ranging from high-TRL technologies like zinc-air close to commercialization to lithium-air which although it has seen tremendous research efforts in the past is still plagued by intrinsic limitations associated with its sluggish kinetics at the cathode and severe degradation on the anode all down to new and emerging chemistries which are still at the proof-of-principle level. Accordingly, the challenges also range from questions related to the fundamental understanding of mechanistic and material-level aspects to such that become relevant when upscaling the technology, for example, the capability to produce components at a reasonable scale and cost or the development of suitable cell configurations and GDE architectures. The latter points lend themselves to be regarded as partially chemistry-independent development steps from which the various technologies can benefit at a certain development stage. In addition, when looking for reoccurring themes that show distinct potential for cross-fertilization, one can find it in the development of the bifunctional air electrode including suitable and stable electrocatalysts and electrode architectures as well as getting a better understanding of rate- and stability-limiting issues. The same is true for enabling stable cycling of the metal anodes for which interface engineering by electrolyte additives or ex situ protection layers are overarching development tasks. For these things to be true one needs to step from sole empirical testing to more knowledge-based approaches that make the use of highly advanced operando methods and accurate modeling of underlying phenomena indispensable.

Finally, as in any low- to mid-TRL technology, the definition of a development roadmap needs to include an early and realistic assessment of what the technologies are capable of and what they are not, to manage expectations and embrace good research practice: an issue that cannot be overemphasized enough in a field particularly susceptible to unrealistic claims and overstating potential performance metrics. Standardized reporting methodologies—as increasingly offered by journals—can help to better benchmark and compare the different technologies and research results against each other and therefore understand and prioritize the specific challenges. Besides the quantitative performance metrics, that also needs to take into account qualitative aspects such as sustainability, intrinsic safety, toxicity, and cost. Although difficult to assess for a low-TRL technology it is important to be evaluated at an initial stage at least to a certain degree to judge if a technology path is worth further elaborating on, on the contrary, if reallocating resources to more promising technologies would be more expedient. After all, the development goal that defines all other challenges is to leverage the significant potential of metal-air batteries as an important piece of the future sustainable and climate-neutral energy system.

REFERENCES

[1] A. Kube, F. Bienen, N. Wagner, K.A. Friedrich, Wetting behavior of aprotic Li-air battery electrolytes, Advanced Materials Interfaces. 9 (2022) 2101569.

[2] S. Möller, S. Barwe, J. Masa, D. Wintrich, S. Seisel, H. Baltruschat, W. Schuhmann, Online monitoring of electrochemical carbon corrosion in alkaline electrolytes by differential electrochemical mass spectrometry, Angewandte Chemie International Edition. 59 (2020) 1585–1589.

[3] S. Zhang, M. Chen, X. Zhao, J. Cai, W. Yan, J.C. Yen, S. Chen, Y. Yu, J. Zhang, Advanced noncarbon materials as catalyst supports and non-noble electrocatalysts for fuel cells and metal-air batteries, Electrochemical Energy Reviews. 4 (2021) 336–381.

[4] A. Pandy, Z. Yang, M. Gummalla, V.V. Atrazhev, N.Y. Kuzminyh, V.I. Sultanov, S. Burlatsky, A carbon corrosion model to evaluate the effect of steady state and transient operation of a polymer electrolyte membrane fuel cell, Journal of The Electrochemical Society. 160 (2013) F972–F979.

[5] L. Jörissen, Bifunctional oxygen/air electrodes, Journal of Power Sources. 155 (2006) 23–32.

[6] F. Cheng, J. Chen, Metal-air batteries: From oxygen reduction electrochemistry to cathode catalysts, Chemical Society Reviews. 41 (2012) 2172–2192.

[7] W.J. Kwak, Rosy, D. Sharon, C. Xia, H. Kim, L.R. Johnson, P.G. Bruce, L.F. Nazar, Y.K. Sun, A.A. Frimer, M. Noked, S.A. Freunberger, D. Aurbach, Lithium-oxygen batteries and related systems: Potential, status, and future, Chem Rev. 120 (2020) 6626–6683.

[8] J. Al-Nu'airat, I. Oluwoye, N. Zeinali, M. Altarawneh, B.Z. Dlugogorski, Review of chemical reactivity of singlet oxygen with organic fuels and contaminants, The Chemical Record. 21 (2021) 315–342.

[9] W.-J. Kwak, H. Kim, Y.K. Petit, C. Leypold, T.T. Nguyen, N. Mahne, P. Redfern, L.A. Curtiss, H.-G. Jung, S.M. Borisov, S.A. Freunberger, Y.-K. Sun, Deactivation of redox mediators in lithium-oxygen batteries by singlet oxygen, Nature Communications. 10 (2019) 1380.

[10] A. Kube, J. Meyer, D. Kopljar, N. Wagner, K.A. Friedrich, A segmented cell measuring technique for current distribution measurements in batteries, exemplified by the operando investigation of a Zn-air battery, Journal of The Electrochemical Society. 168 (2021) 120530.

[11] C. Xia, C.L. Bender, B. Bergner, K. Peppler, J. Janek, An electrolyte partially-wetted cathode improving oxygen diffusion in cathodes of non-aqueous Li-air batteries, Electrochemistry Communications. 26 (2013) 93–96.

[12] P. Andrei, J.P. Zheng, M. Hendrickson, E.J. Plichta, Some possible approaches for improving the energy density of Li-air batteries, Journal of the Electrochemical Society. 157 (2010) A1287.

[13] J. Read, Characterization of the lithium/oxygen organic electrolyte battery, Journal of the Electrochemical Society, 149 (2002) A1190.

[14] K.J. Euler, Brennstoffzellen—moderne elektrochemische Stromquellen, Chemie in unserer Zeit. 1 (1967) 84–93.

[15] F. Bienen, M.C. Paulisch, T. Mager, J. Osiewacz, M. Nazari, M. Osenberg, B. Ellendorff, T. Turek, U. Nieken, I. Manke, K.A. Friedrich, Investigating the electrowetting of silver-based gas-diffusion electrodes during oxygen reduction reaction with electrochemical and optical methods, Electrochemical Science Advances. n/a (2022) e2100158.

[16] P. Sarkezi-Selsky, H. Schmies, A. Kube, A. Latz, T. Jahnke, Lattice Boltzmann simulation of liquid water transport in gas diffusion layers of proton exchange membrane fuel cells: Parametric studies on capillary hysteresis, Journal of Power Sources. 535 (2022) 231381.

[17] F. Mugele, J. Buehrle, Equilibrium drop surface profiles in electric fields, Journal of Physics: Condensed Matter. 19 (2007) 375112.

[18] N. Wagner, M. Schulze, E. Gülzow, Long term investigations of silver cathodes for alkaline fuel cells, Journal of Power Sources. 127 (2004) 264–272.

[19] M.C. Paulisch, M. Gebhard, D. Franzen, A. Hilger, M. Osenberg, N. Kardjilov, B. Ellendorff, T. Turek, C. Roth, I. Manke, Operando laboratory X-Ray imaging of silver-based gas diffusion electrodes during oxygen reduction reaction in highly alkaline media, Materials. 12 (2019) 2686.

[20] M. Prabu, P. Ramakrishnan, H. Nara, T. Momma, T. Osaka, S. Shanmugam, Zinc-air battery: Understanding the structure and morphology changes of graphene-supported $CoMn_2O_4$ bifunctional catalysts under practical rechargeable conditions, ACS Applied Materials & Interfaces. 6 (2014) 16545–16555.

[21] S. Lou, F. Zhang, C. Fu, M. Chen, Y. Ma, G. Yin, J. Wang, Interface issues and challenges in all-solid-state batteries: Lithium, sodium, and beyond, Advanced Materials. 33 (2021) 2000721.

[22] A.R. Mainar, E. Iruin, L.C. Colmenares, A. Kvasha, I. de Meatza, M. Bengoechea, O. Leonet, I. Boyano, Z. Zhang, J.A. Blazquez, An overview of progress in electrolytes for secondary zinc-air batteries and other storage systems based on zinc, Journal of Energy Storage. 15 (2018) 304–328.

[23] S. Clark, A.R. Mainar, E. Iruin, L.C. Colmenares, J.A. Blázquez, J.R. Tolchard, Z. Jusys, B. Horstmann, Designing aqueous organic electrolytes for zinc-air batteries: Method, simulation, and validation, Advanced Energy Materials. 10 (2020) 1903470.

[24] A. Kube, N. Wagner, K.A. Friedrich, Influence of organic additives for zinc-air batteries on cathode stability and performance, Journal of the Electrochemical Society. 168 (2021) 050531.

[25] Y. Yokoyama, T. Fukutsuka, K. Miyazaki, T. Abe, Origin of the electrochemical stability of aqueous concentrated electrolyte solutions, Journal of the Electrochemical Society. 165 (2018) A3299–A3303.

[26] R.E. Davis, G.L. Horvath, C.W. Tobias, The solubility and diffusion coefficient of oxygen in potassium hydroxide solutions, Electrochimica Acta. 12 (1967) 287–297.

[27] A. Khan, C.A. Gunawan, C. Zhao, Oxygen reduction reaction in ionic liquids: Fundamentals and applications in energy and sensors, ACS Sustainable Chemistry & Engineering. 5 (2017) 3698–3715.

[28] A.H.C.H. Hamann, W. Vielstich, Electrochemistry, Wiley-VCH, 2007.

[29] Z. Liu, S.Z.E. Abedin, F. Endres, Electrodeposition of zinc films from ionic liquids and ionic liquid/water mixtures, Electrochimica Acta. 89 (2013) 635–643.

[30] B.G. Kim, J.-S. Kim, J. Min, Y.-H. Lee, J.H. Choi, M.C. Jang, S.A. Freunberger, J.W. Choi, A moisture- and oxygen-impermeable separator for aprotic $Li-O_2$ batteries, Advanced Functional Materials. 26 (2016) 1747–1756.

[31] W.-J. Kwak, H. Kim, H.-G. Jung, D. Aurbach, Y.-K. Sun, Review: A comparative evaluation of redox mediators for $Li-O_2$ batteries: A critical review, Journal of the Electrochemical Society. 165 (2018) A2274–A2293.

[32] S. Luo, Z. Wang, X. Li, X. Liu, H. Wang, W. Ma, L. Zhang, L. Zhu, X. Zhang, Growth of lithium-indium dendrites in all-solid-state lithium-based batteries with sulfide electrolytes, Nature Communications. 12 (2021) 6968.

[33] S. Siahrostami, V. Tripković, K.T. Lundgaard, K.E. Jensen, H.A. Hansen, J.S. Hummelshøj, J.S.G. Mýrdal, T. Vegge, J.K. Nørskov, J. Rossmeisl, First principles investigation of zinc-anode dissolution in zinc—air batteries, Physical Chemistry Chemical Physics. 15 (2013) 6416–6421.

[34] S. Lysgaard, M.K. Christensen, H.A. Hansen, J.M. García Lastra, P. Norby, T. Vegge, Combined DFT and differential electrochemical mass spectrometry investigation of the effect of dopants in secondary zinc-air batteries, ChemSusChem. 11 (2018) 1933–1941.

[35] W. Du, E.H. Ang, Y. Yang, Y. Zhang, M. Ye, C.C. Li, Challenges in the material and structural design of zinc anode towards high-performance aqueous zinc-ion batteries, Energy & Environmental Science. 13 (2020) 3330–3360.

Index

A

active Sites 19, 32, 33, 46, 53, 54, 57, 58, 92, 93, 101, 123, 125, 131, 144, 152, 160, 161, 165, 167, 168, 170–74, 185, 186, 189, 196, 197, 199–201, 216, 218, 222, 233, 238, 239, 241, 275, 296, 364, 365, 368, 369
aluminum-air Battery 78–87, 114, 182, 195, 284, 299, 335
anode 1–5, 8, 15, 16, 19, 21–25, 29–32, 34, 36, 37, 39, 40, 66–74, 78–83, 85, 87, 90–102, 105, 106, 111, 114, 115, 124, 129, 136, 166, 181–85, 188, 210–12, 215, 216, 222, 227–31, 244, 250, 253–55, 257, 258, 260, 262, 263, 267, 268, 271, 275, 282, 284–87, 291, 292, 295, 296, 298–301, 305, 307, 309–13, 317, 319, 326, 328, 330, 331, 335, 337, 338, 341–43, 347, 348, 351, 362, 365–72

B

boron nitride 172, 174

C

carbides 8, 46, 129, 170, 236, 241, 244
carbon nanofibers 73, 138, 139, 142, 143, 153, 157, 203, 232, 235
carbon nanotubes 10, 21, 73, 97, 111, 112, 115, 138, 141, 142, 153, 157, 159, 167, 172, 184, 185, 198, 232, 337, 352
cathode 1–11, 16–21, 23, 24, 29–32, 34, 36–40, 45, 51, 53, 61, 66, 67, 69, 70, 73, 74, 78, 79, 82, 83, 85–87, 90–92, 95–99, 101, 102, 105–7, 110, 111, 114, 115, 123, 124, 126, 127, 129, 130, 135–39, 141–43, 146, 151, 153, 155–60, 165, 181, 183–85, 187–91, 203–5, 210–14, 216, 222, 228–33, 235–39, 241, 243, 244, 250, 253–55, 258, 265–68, 270, 272, 274–77, 282, 284–87, 291, 292, 294–96, 298, 299, 305, 307, 309, 310, 312, 313, 317, 323, 326, 328, 331, 335, 337, 338, 340–43, 347, 348, 351–53, 362, 365–72
chalcogenides 46, 126, 188, 213, 237, 238
chemical vapor ddeposition 156, 172, 184, 185, 232, 236, 338, 352
CNF 123, 153, 155, 157, 158, 203, 204, 232, 233, 342
CNT 10, 97, 123, 124, 153, 154, 156, 157, 159, 162, 172, 173, 189, 232, 233, 238, 294, 296, 324, 338, 343
COFs 173
donductive polymers 20, 109, 336, 337
covalent organic frameworks 173, 174
CVD 172, 236
cyclic stability 97, 98, 159, 185, 189, 239, 243, 253, 277, 340

D

density functional theory 46, 124, 199, 200, 203, 239, 271
DFT 46–48, 51–54, 56, 59, 60, 127–29, 198, 201, 219, 239, 262, 322
doping 1, 20, 58, 97, 136, 153, 157, 167, 168, 170–72, 174, 185, 186, 196, 198–202, 204, 211, 213, 214, 218, 221, 222, 232, 306, 311, 364, 371

E

electrocatalyst 17, 19, 20, 26, 34, 37–39, 42, 46, 54, 73, 79, 86, 95, 97, 106, 110, 123, 124, 131, 136–38, 143, 144, 151–53, 157, 159, 160, 165–67, 171, 172, 174, 181, 184–86, 188, 189, 191, 195, 213, 216, 217, 219, 222, 231, 257, 367
electrolytes 1, 4–7, 10, 11, 16, 17, 19, 21–26, 29–34, 36–41, 51, 65, 66, 72, 74, 81–85, 90, 92, 97, 98, 111, 112, 114, 115, 117, 124, 136, 141, 142, 156, 165, 180–84, 188, 189, 210–12, 214–16, 222, 228–30, 250–55, 257–62, 266–77, 282–88, 291, 292, 295, 298, 299, 301, 302, 306, 307, 309–11, 313, 318–20, 322–26, 328–31, 335–37, 344, 348, 353, 354, 356, 362, 364–72
energy density 1, 2, 11, 15, 16, 26, 29, 30, 36, 37, 40, 41, 65, 78, 79, 89, 92–94, 96, 99, 110, 111, 123, 129, 130, 136, 141, 151, 179, 181, 182, 184, 191, 195, 200, 204, 210, 213, 227–30, 244, 249–51, 253–55, 260–62, 265, 266, 268, 282, 285, 286, 292, 295, 296, 300, 305, 309, 320, 330, 335, 337, 338, 341, 343, 347, 348, 350, 351, 353, 356, 357, 361, 362
energy storage systems 30, 65, 135, 179, 181, 204, 249, 251, 348
ESSs 135

F

flexible 25, 30, 41, 106, 107, 109–12, 114, 116, 126, 142, 143, 160, 183, 185, 187–89, 203, 204, 244, 251, 258, 260, 262, 263, 300, 309, 317–26, 328–31, 335–44, 347–57

G

gel electrolyte 37, 73, 83–85, 111, 115, 116
GO 233, 294, 296, 319, 323, 324
graphene 20, 21, 23, 54, 56, 57, 59, 72, 73, 86, 90, 94, 97, 102, 115, 119, 124, 127, 129, 133, 137, 138, 141, 143, 145, 159, 162, 166–68, 170, 172, 184, 185, 189, 190, 198–201, 213, 218, 223, 231–35, 238–41, 254, 294–96, 319, 352
grpahene oxide 23, 141, 213, 233, 234, 294

H

HER 22, 26, 36, 74, 93, 95, 98, 101, 105, 115, 127, 133, 138, 145, 216, 219, 221, 254, 255, 371, 372
hybrid-electrolyte 292, 294–96, 298–301
hydrogels 112, 300, 317–20, 322–31, 353
hydrogen evolution reaction 22, 36, 93, 101, 105, 216, 330, 355

I

ionic-electronic 243

375

L

LABs 228–33, 235–39, 241, 243, 244, 265–68, 270–75, 277, 337, 344
Li-air batteries 16, 22, 23, 25, 29, 37–40, 93, 95, 96, 99, 111, 112, 153, 155–60, 181, 182, 191, 209, 210, 213, 222, 250, 251, 265, 270–77, 292–96, 298, 301, 305, 306, 309, 312–14, 354, 361, 366, 369, 370, 372

M

MABs 15–17, 19–21, 23–26, 29–32, 34, 36–41, 45, 46, 51, 60, 89–95, 97, 99, 100, 102, 105–7, 109, 113, 117, 121–23, 131, 165, 166, 168, 171, 172, 174, 195–201, 203–5, 209–17, 222, 251–53, 257, 291, 296, 298, 317–20, 322–24, 326–31, 335–37, 341, 343, 344, 361, 371
membrane 5, 32, 36, 39, 83, 112, 115, 117, 126, 161, 203, 235, 244, 292–95, 298–301, 329, 341–43, 349, 350
metal-air batteries 2, 3, 6, 16–18, 25, 26, 29, 31, 37, 39, 40, 45, 77, 78, 90–92, 95, 105–7, 109–12, 114, 117, 127, 136, 143, 145, 146, 151, 153, 157, 159, 160, 162, 163, 167, 174, 175, 179–84, 188, 189, 191, 195, 204, 209, 210, 212, 235, 243, 249–51, 253–55, 257, 258, 262, 281, 282, 284, 285, 291, 305, 307, 309, 310, 312, 314, 317, 348, 351–55, 361–63, 365, 367, 368, 371–73
metal-organic framework 59, 99, 157, 160, 171, 174, 184, 188–90, 216, 241, 296, 364
MOF 157, 160, 189, 190, 241, 244
multiwalled carbon nanotube 97, 137, 159, 338, 343
MWCNT 137, 153, 154, 159, 232
MXenes 171

N

nitrides 8, 129, 133, 170, 171, 184, 196, 236, 241, 243, 244, 364
noble metals 20, 33, 46, 95, 102, 122, 123, 127, 135–37, 141–46, 166, 170, 184, 195, 196, 200, 203, 204, 213, 216, 217, 234, 351

O

OER 3, 16–20, 26, 31, 34–37, 40, 45–52, 54–60, 92, 95, 106, 110, 121–27, 129, 131, 133, 135, 136, 138, 139, 141, 144–46, 151, 152, 157, 159, 160, 165–68, 170–74, 176, 184–86, 189–91, 195–205, 211–14, 216–22, 227–29, 231, 233, 234, 236–41, 243, 244, 250, 257, 260, 266, 274, 275, 277, 282, 284, 285, 294, 296, 305, 306, 313, 317, 335, 337, 339, 340, 364
orr 3, 6, 10, 16–20, 26, 31–37, 40, 45–60, 73, 82, 83, 85, 86, 91, 92, 95, 97, 101, 102, 106, 110, 111, 121–25, 127, 129, 131, 133, 135, 136, 138, 141, 144–46, 151, 152, 157, 159, 160, 165–68, 170–74, 184–86, 189–91, 195–205, 209, 211–14, 216–22, 227–29, 231–34, 236–41, 243, 244, 250, 253, 257, 266, 274–76, 282–88, 294, 296, 305, 306, 313, 317, 335, 337, 339, 340, 364, 369
overpotential 8, 10, 19, 38, 39, 49, 50, 52, 54, 56–60, 65, 91, 95, 98, 101, 125, 127, 129, 137, 138, 142, 144, 145, 156, 157, 159, 165, 166, 168, 170, 171, 173, 174, 186, 188, 197, 200, 201, 211–13, 216, 221, 222, 228–31, 234–36, 238, 239, 241, 244, 250, 253, 268, 269, 284, 287, 294, 301, 339, 369
oxygen evolution reaction 3, 16, 31, 45, 106, 121, 135, 136, 165, 184, 195, 211, 223, 227, 250, 260, 266, 282, 305, 317, 335, 348
oxygen reduction reaction 3, 16, 31, 45, 73, 82, 83, 87, 91, 92, 106, 121, 135, 136, 151, 165, 185, 211, 227, 254, 275, 282, 305, 317, 335, 362

P

perovskites 20, 59, 123, 125, 214, 216, 218, 222, 240, 364
phosphides 129
power density 30, 31, 80, 83, 84, 86, 91, 109, 114–17, 122–24, 127, 129, 130, 142, 145, 168, 181, 183–86, 188–91, 200–202, 214, 241, 257, 258, 265, 266, 294–96, 298–301, 332, 340, 341, 350

R

RGO 23, 90, 94, 98, 102, 127, 232–34, 236, 239, 324

S

separator 31, 39, 81, 83, 84, 96–98, 180–82, 211, 215, 227, 228, 231, 244, 257, 258, 295, 299–301, 317, 337, 350, 361, 372
solid electrolyte 1, 8, 23, 39, 84, 98, 99, 137, 184, 239, 252, 253, 295, 343
solvent effect 56

T

TMDs 170
TMNs 129
TMO 123
TMPs 129
transition metal dichalcogenides 170
transition metal nitrides 129, 171, 243, 244
transition metal oxides 20, 33, 46, 50, 73, 99, 102, 170, 186, 216, 237, 238, 294
transition metal phosphides 129

W

wearable 30, 93, 105, 107–10, 115, 117, 204, 260, 317, 335–44, 347–50

Z

Zn-air battery 15, 17, 21, 24, 29, 30, 34–36, 40, 53, 99, 106, 110, 112, 113, 116, 121, 157–59, 165, 168, 180, 186, 189, 191, 209, 210, 213, 218, 222, 250, 253–55, 258, 263, 291, 292, 298, 300–302, 317, 354, 361, 368, 369, 371, 372